Preface

The primary objective in revising the first edition of this volume was to bring it up to date. The organization is essentially the same, and it covers, as did the first edition, all phases of the structure, properties, origin, and occurrence of the clay minerals. Further, it supplies necessary information for the identification of the clay-mineral components of argillaceous materials.

Since the publication of the first edition, the great expansion of clay-mineral investigations that began in the late 1920s and 1930s, with the development of new research tools, has continued. A vast literature has accumulated and has provided significant advances in our knowledge of all phases of clay mineralogy. Especially important contributions have been made regarding the atomic structure of these minerals and the relation of the atomic structure to the properties of the clay minerals themselves. The reaction of the clay minerals and organic compounds has been studied especially, and significant advances have been made in understanding the solid-state reactions that take place when the clay minerals are heated to elevated temperatures. The clay-mineral compositions of a vast number of Recent and ancient sediments, soils, and other argillaceous materials have been published so that the origin, occurrence, and distribution of the clay minerals in all types of clay materials are now better understood. Some conclusions listed as tentative and preliminary in the first edition are now more completely acceptable.

The continuing and expanding interest in clay-mineral investigations is reflected in the formation of clay-mineral societies and organizations in various countries (United States, Great Britain, France, Japan, Germany, Australia, etc.) and by the large attendance at their regular meetings. Also, an International Clay Minerals Association has been formed, which now holds international conferences attended by large numbers of people representing various disciplines from many countries.

Of particular significance, attesting to the growth of interest in clay-mineral studies, are the expanded research activities in laboratories in industries devoted to the production and preparation of clays for use in the paper, rubber, petroleum, and other industries, and in the laboratories associated with companies using clays for these and other purposes.

As in the first edition, this volume does not consider the larger properties of clay materials, such as plasticity, refractoriness, sensitivity, etc. These properties and the fundamental factors controlling them have been assembled in a companion volume, "Applied Clay Mineralogy," published in 1963.

The literature in the field of clay mineralogy is now so vast and is published in such a large variety of journals that it is extremely difficult to follow. The author can only hope that nothing very significant has been overlooked in the preparation of this revised volume.

As in the first edition, an attempt has been made to include substantial numbers of references on each subject so that persons interested in a particular aspect of clay mineralogy can pursue the subject further by means of these references.

The author wishes to express again his great indebtedness to Dr. W. F. Bradley, formerly of the Illinois State Geological Survey and now of the University of Texas. Doctor Bradley and the author have been associated in clay-mineral investigations for over 30 years and in that time have had many fruitful discussions. Doctor Bradley has read parts of the revised manuscript and has offered many important suggestions. Any omissions or errors, however, are the sole responsibility of the author.

The author also wishes to express his great appreciation to Dr. Kenneth M. Towe of the United States National Museum, who has supplied the electron micrographs included in the revision. Not only is Dr. Towe a superb technician with the electron microscope, but he is also a thoroughly capable mineralogist and geologist interested in clay mineralogy.

Finally, it is a pleasure to acknowledge the great help of my secretary, Mrs. Shirley Ate, in the preparation of the revised manuscript and seeing it through the press. Without her assistance, willingly and competently given, the revision would probably never have been completed.

Ralph E. Grim

Contents

One
Introduction

Clay is used as a rock term and also as a particle-size term in the mechanical analysis of sedimentary rocks, soils, etc. As a rock term it is difficult to define precisely, because of the wide variety of materials that have been called clays. In general the term clay implies a natural, earthy, fine-grained material which develops plasticity when mixed with a limited amount of water. By plasticity is meant the property of the moistened material to be deformed under the application of pressure, with the deformed shape being retained when the deforming pressure is removed. Chemical analyses of clays show them to be composed essentially of silica, alumina, and water, frequently with appreciable quantities of iron, alkalies, and alkaline earths.

The difficulty is that some material called clay does not meet *all* the above specifications. Thus, so-called *flint clay* has substantially no plasticity when mixed with water. It does, however, have the other attributes of clay.

The term clay has no genetic significance. It is used for material that is the product of weathering, has formed by hydrothermal action, or has been deposited as a sediment.

As a particle-size term, the clay fraction is that size fraction composed of the smallest particles. The maximum size of particles in the clay size

$< 2.0 \mu$

grade is defined differently in different disciplines. In geology the tendency has been to follow the Wentworth[1] scale and to define the clay grade as material finer than about 4 microns (μ). In soil investigations, the tendency is to use 2 μ as the upper limit of the clay size grade. Although there is no sharp universal boundary between the particle size of the clay minerals and nonclay minerals in argillaceous materials, a large number of analyses have shown that there is a general tendency for the clay minerals to be concentrated in a size less than about 2 μ, or that naturally occurring larger clay-mineral particles break down easily to this size when the clay is slaked in water. Also such analyses have shown that the nonclay minerals usually are not present in particles much smaller than about 1 to 2 μ. A separation at 2 μ is frequently about the optimum size for the best split of the clay-mineral and nonclay-mineral components of natural materials. There is, therefore, a fundamental reason for placing the upper limit of the clay size grade at 2 μ.

Clays contain varying percentages of clay-grade material and, therefore, varying relative amounts of nonclay-mineral and clay-mineral components. The writer knows of no clay which does not contain some nonclay-mineral material coarser than the clay grade, although the amount in some hydrothermal clays is extremely small (less than 5 percent). In many materials called clays the clay-grade and clay-mineral components make up considerably less than half the total rock. In such materials the nonclay is frequently not much coarser than the maximum for the clay grade, and the clay-mineral fraction may be particularly potent in causing plasticity. In general, fine-grained materials have been called *clay* so long as they had distinct plasticity and insufficient amounts of coarser material to warrant the appellations *silt* or *sand*. If particle-size analyses are made, the term clay would be reserved for a material in which the clay grade dominates. However, names have been and are applied most frequently solely on the basis of the appearance and bulk properties (e.g., plasticity) of the sample.

Shale is a fine-grained, earthy, sedimentary rock with a distinct laminated, or layered, character. The layering may be due to a general parallel arrangement of flake-shaped or elongate particles or to an alternation of beds of somewhat different composition. The lamination is parallel to the bedding and has not been developed by postdepositional metamorphic action. The requirements of composition are substantially the same for a shale as for a clay. Occasionally, however, natural materials are called shale with little regard to composition. Thus, thinly layered rocks composed essentially of quartz and/or carbonate with little clay-mineral component have been called shale. Sometimes, although by no means always, shales are more indurated

---

[1] Wentworth, C. K., A Scale of Grade and Class Terms for Clastic Sediments, *J. Geol.*, **30:** 377–392 (1922).

and are harder than clays. The term shale is sometimes used by engineers for any hard, indurated, argillaceous rock regardless of any lamination.

Argillite is a fine-grained argillaceous material that is massive and somewhat indurated and hard. It differs from shale in being massive rather than laminated and from clay by being harder.

The term *soil* is likely to have a considerably different meaning when used by a geologist, by an agronomist, and by civil engineer. Soil to a geologist is the weathered regolith at the earth's surface that supports vegetation. It is thought of generally as being loose, argillaceous, and with some organic content. To the agronomist it is the loose regolith at the earth's surface. It need not be weathered nor contain any vegetation; it may be gravel, for example. Also according to agronomists, a soil is likely to be composed of a series of horizons and have properties quite independent of the underlying parent bedrock. The civil engineer tends to divide the material at the earth's crust into two categories: (1) rock and (2) soils. Rock is defined as something that is hard and consolidated. Soil, according to Terzaghi and Peck,[2] "is a natural aggregate of mineral grains that can be separated by such gentle means as agitation in water." Substantially any loose material at the earth's crust, regardless of particle-size distribution, composition, or organic content, is soil to the engineer. It may or may not be weathered. Similarly to the engineer soil can extend to any depth below the surface so long as the material is not indurated substantially. To the engineer shale is similar to soil except that the term is applied to material that is slightly harder and is definitely argillaceous. The term *clay* is primarily a particle-size term to the engineer.

The author has found it convenient to use the expression *clay material* for any fine-grained, natural, earthy, argillaceous material. Clay material includes clays, shales, and argillites of the geologist. It would also include soils, if such materials were argillaceous and had appreciable contents of clay-size-grade material.

No attempt will be made herein to consider the definitions of relatively minor types of argillaceous materials with somewhat specific properties, such as loam, gumbo, etc. Description of such materials can be obtained from standard textbooks on soils and sedimentary rocks.

⌣ *Factors controlling the properties of clay materials*

The factors which control the properties of clay materials or the attributes which must be known to characterize completely a clay material may be classified as follows:

[2] Terzaghi, K., and R. Peck, "Soil Mechanics in Engineering Practice," Wiley, New York (1948).

Clay-mineral composition. This refers to the identity and relative abundance of _all_ the clay-mineral components. Since certain clay minerals which may be present in very small amounts may exert a tremendous influence on the attributes of a clay material, it is not adequate to determine only the major clay-mineral components. Thus a small amount (5%±) of smectite* in a clay is likely to provide a material very different from another clay with the same composition in all ways except for the absence of smectite. In order to make complete clay-mineral determinations, it is frequently necessary to fractionate the clay grade to concentrate minor constituents so that adequate analytical data can be obtained. Fortunately such a concentration can often be attained, because the various clay minerals frequently occur in particles of different sizes or break down easily in water to particles of different size. Also the clay minerals must be determined in their natural state. For example, care must be taken that the analysis will reveal the natural hydration state of the minerals and their ion-exchange composition. Clays composed of halloysite have very different physical properties, depending on whether the mineral is in the $4H_2O$ form, the $2H_2O$ form, or an intermediate state. Smectite clays have very different properties when Na^+ is the exchange cation and when Ca^{++} is the cation.

In clay materials containing a considerable amount of nonclay-mineral material, it is frequently necessary to remove the nonclay-mineral material before the clay minerals can be identified completely.[3] Frequently this involves merely a particle-size separation. Sometimes, as in the presence of pigmentary iron oxide or hydroxides, extremely fine carbonate, and pigmentary organic material, other methods must be attempted. Considerable caution is necessary to avoid significant change in the clay-mineral components in such separations. For example, the use of acids to remove the iron or carbonate, even if very dilute, may dissolve certain of the clay minerals if they are present (see Chap. 12). In the case of iron oxide or hydroxide, biological methods[4] of removal appear to be quite satisfactory. Strong oxidizing agents to eliminate the organic material are likely to alter the clay minerals significantly.

Some clay materials contain poorly organized or essentially amorphous material that is particularly vulnerable to change or removal by chemical treatment processes. Methods to "clean up" a sample to provide better analytical data may well eliminate such material which has a significant influence on properties.

---

* In the first edition montmorillonite was used as the name for this group (see page 33).

[3] Grim, R. E., P. F. Kerr, and R. H. Bray, Application of Clay Mineral Technique to Illinois Clay and Shale, *Bull. Geol. Soc. Am.,* **46:** 1909–1926 (1935).

[4] Allison, L. E., and G. D. Scarseth, A Biological Reduction Method for Removing Free Iron Oxides from Soils and Colloidal Clays, *J. Am. Soc. Agron.,* **34:** 616–623 (1942).

The perfection of crystallinity of the clay minerals is variable in clay materials. Thus, kaolinite may be well organized or poorly organized in a given sample; illite may be degraded or well crystallized. It is known that certain properties such as plasticity vary with the crystalline character of the component clay minerals.

Mixed-layer components (see Chap. 4) are now known to be common components of clay materials. The character of the mixed-layer components may be a diagnostic feature of a particular horizon in a sequence of argillaceous rocks.

The morphology of the clay-mineral components may be a distinctive property of a particular clay. For example, in some kaolins the flakes of kaolinite are large and thin, and in others they are thicker with less areal extent; in some bentonites the smectite particles are large and very thin whereas in other bentonites the particles are more granular.

The chemical composition of the clay-mineral components may be important. Thus, smectite for certain uses, e.g., catalyst manufacture, must have a very low iron content, and the abundance of trace amounts of boron, gallium, and rubidium may indicate the environmental conditions under which argillaceous material has accumulated.

To characterize a clay material completely and to provide adequate data to understand physical properties, therefore, clay-mineral analyses must be thorough and complete. The procedure must not alter the clay minerals. Frequently several lines of approach are necessary, e.g., diffraction, electron microscopy, and chemical methods. At times, even with a combination of methods, it is difficult to obtain complete data. For example, it is often difficult to establish the presence of allophane, and one can say only that the solubility data, diffraction data, etc., suggest the presence of very poorly organized material. Clay-mineral analyses of samples composed entirely of kaolinite are simple by any procedure. On the other hand, the analysis of a sample of mud from the Mississippi River Delta, composed of a complex mixture of poorly crystalline clay minerals and mixed-layer structures, is very difficult.

Nonclay-mineral composition. This refers to the identity of the nonclay minerals, their relative abundance, and the particle-size distribution of the individual species. Calcite, dolomite, large flakes of mica, pyrite, feldspar, gibbsite, and other minerals are very abundant in some clay materials.

Obviously, it is impossible or unjustifiably time-consuming to get all the data concerning the nonclay minerals in most investigations of clay materials. The lengths to which one can and must go depend largely on the problem at hand and the purpose of the investigation. It is frequently adequate to determine the identity of only the more abundant nonclay minerals and their sorting and particle-size distribution in a general way. Thus the "heavy minerals" may be of no significance in relation to the physical properties of a clay but may be of important diagnostic value in determining whether or not a clay has formed by the alteration of volcanic ash. As another example, the study of a soil from the point of view of soil mechanics demands that sorting within the silt size range be studied in considerable detail, since the presence of some silt materials may yield a material of unique physical properties of great importance to the construction engineer. The analysis of a clay material must be tailor-made to the material being studied and to the purpose of the investigation and must provide comparable results from one sample to another. One cannot blindly use a set analytical procedure for all materials and all problems and still get adequate data without a tremendous waste of time and effort.

The nonclay minerals in clay materials tend generally to be concentrated in particles coarser than about 2μ. There are, however, materials in which they are much finer-grained. In some Wyoming bentonites, for example,[5,6] a considerable amount of cristobalite is present in particles considerably less than $1\ \mu$ in diameter intimately mixed with the clay mineral smectite. Many clay materials contain extremely fine iron oxide or hydroxide, which acts as a pigment. $Fe_2O_3\ \ Al_2O_3$

The identification of the coarser nonclay minerals can be made with a petrographic microscope. The determination of those occurring in extremely fine particles requires X-ray techniques. Neither of these methods permits very precise quantitative determinations. In the case of extremely fine silica, the maximum accuracy by X-ray diffraction is about $\pm 2\%$ if the quantity is small and somewhat less if the quantity is large ($4\pm\%$). There is a distinct tendency to overestimate the abundance of quartz and carbonates from diffraction data. There have been numerous instances in the author's laboratory in which the intensity of diffraction lines for quartz suggested quantities of the mineral considerably larger than could possibly be present on the basis of complete chemical data. Numerous attempts[7,8] have been made to determine the amount of nonclay minerals chemically, e.g., the amount of free and combined silica in a clay material. Such methods are based on a difference in solubility of the constituents. Unfortunately variations of solubility with particle size cause the results to have questionable

[5] Gruner, J. W., Abundance and Significance of Cristobalite in Bentonites and Fuller's Earths, *Econ. Geol.,* **35:** 867–875 (1940).

[6] Roth, R. S., The Structure of Montmorillonite in Relation to the Occurrence and Properties of Certain Bentonites, Ph.D. thesis, University of Illinois (1951).

[7] Trostel, L. J., and D. J. Wynne, Determination of Quartz (Free Silica) in Refractory Clays, *J. Am. Ceram. Soc.,* **23:** 18–22 (1940).

[8] Sauzeat, H., Can Free Quartz Be Determined in a Rock, *Rev. Ind. Minerale,* **529:** 114–117 (1948).

value. The absence of accurate quantitative methods for determining the nonclay-mineral components of clay materials frequently makes it impossible to obtain exact data on the chemical composition of the clay minerals themselves in such materials.[9]

Organic material. This refers to the kind and amount of organic material contained in clay material. In general the organic material occurs in clay materials in several ways: It may be present as discrete particles of wood, leaf matter, spores, etc.; it may be present as organic molecules adsorbed on the surface of the clay-mineral particles; or it may be intercalated between the silicate layers (see Chap. 10). The discrete particles may be present in any size from large chunks easily visible to the naked eye to particles of colloidal size which act as a pigment in the clay-mineral materials.

The total amount of organic material can be determined simply by readily available standard analytical procedures. Values may be obtained from the difference between total loss on ignition and determination of loss of water, sulfur, and other inorganic volatiles. Such values are not precise but are usually adequate. Differential thermal analyses provide a crude determination of amount of organic material. Fine pigmentary organic material gives a dark-gray or black color to a clay material, but there is no direct relationship between the color and organic content. A very small amount of organic material may have a very large pigmenting effect.

Determination of the kind of organic material is a more difficult problem.[10] Sometimes, if the discrete particles are relatively large, they can be identified visually or microscopically. By means of X-ray-diffraction techniques, the presence of adsorbed intercalated organic molecules may usually be determined. The presence of organic molecules is indicated by the change in the c-axis dimension following treatment to remove the organic component. At the present stage of our knowledge, it is frequently impossible to go further and identify the organic components present in small amounts and of extremely fine size, either discrete or adsorbed. However, the large amount of fundamental work applying infrared-absorption and diffraction procedures in recent years has provided much information on the kind of organic molecules that may be adsorbed, the way in which they are tied to the clay-mineral surfaces, and the manner in which they are oriented on the clay-mineral surfaces (see Chap. 10).

The study of the organic content of clay materials is a problem worthy of intensive research for a variety of reasons. For example, the organic con-

[9] Kelley, W. P., Calculating Formulas for Fine Grained Minerals on the Basis of Chemical Analysis, *Am. Mineralogist,* **30:** 1–26 (1945).

[10] Francis, M., Sur la matière organique dans les argiles, *Verre Silicates Ind.,* **14:** 155–158 (1949).

tent often is important in determining the properties of a clay material, and also a knowledge of clay-mineral–organic relations might throw much light on some important geologic processes. It might improve our understanding of the origin of petroleum, since the clay minerals may well have acted as catalysts in the transformation of the parent organic matter into hydrocarbons.[11]

Exchangeable ions and soluble salts. Some clay materials contain water-soluble salts which may have been entrained in the clay at the time of accumulation or may have developed subsequently as a consequence of weathering or alteration processes, as in the oxidation of pyrite to produce sulfates. It is frequently necessary to wash out the soluble salts before other attributes of the material are studied. Some salts may act to flocculate the clay, so that it cannot be dispersed for particle-size analysis or for fractionation preliminary to clay-mineral analysis until the salts are washed out. Common water-soluble salts found in clay materials are chlorides, sulfates, and carbonates of alkalies, alkaline earths, aluminum, and iron.

The clay minerals and some of the organic material found in clay materials have significant ion-exchange capacity. The ion-exchange capacity of the clay minerals and the organic components, as well as the identity and relative abundance of the exchangeable ions which are present, are extremely important attributes of clay materials. To characterize a clay material completely, the relative abundance of both the exchangeable anions and cations should be determined. It is difficult to distinguish sometimes between exchangeable ions and those present in a moderately soluble compound, so that determinations of ion-exchange characteristics are difficult in a material containing appreciable water-soluble salts. This whole matter, together with analytical procedures, is discussed in detail in Chap. **7**.

Texture. The textural factor refers to the particle-size distribution of the constituent particles, the shape of the particles, the orientation of the particles in space and with respect to each other, and the forces tending to bind the particles together.

Some knowledge of the particle-size distribution of the coarser grains can be obtained quickly by microscopic examinations, and detailed determinations can be made by sieving and/or wet sedimentation methods. Fine-grained particles require wet methods, and this applies to the clay-mineral fraction. It must be remembered that wet methods are likely to reflect only the degree to which clay-mineral units or aggregates have been cleaved or

[11] Grim, R. E., Relation of Clay Mineralogy to the Origin and Recovery of Petroleum, *Bull. Am. Assoc. Petrol. Geologists,* **31**: 1491–1499 (1947).

broken down in the process of making the analysis rather than any inherent attribute of the natural material. In dispersing clays in water for analysis, the material is usually agitated, which splits and cleaves natural particles. The particle-size distribution records the amount of agitation applied. There are clay-mineral materials from which literally any particle-size distribution can be obtained by relatively slight variations of the preparation procedure. In general the particle-size distribution of clay materials composed of smectite, vermiculite, and the attapulgite-sepiolite clay minerals would be more affected by analytical procedures than clay materials composed of the other clay minerals.

The use of chemical dispersing agents almost certainly will alter the base-exchange composition of the material, and consequently such agents must not be used or at least used only with great caution, if exchangeable ions are to be determined. It is generally essential to determine the exchangeable ions on the "as received" material, since any mixing in water or washing is likely to cause a significant change. Also such chemicals are likely to yield salts in the resulting fractions which complicate the identification of any clay minerals therein.

The dewatering of the fine clay grades may well yield a material in a hydration state different from that of the original material, and such dehydration, if it is complete, may tend to conceal some of the clay-mineral components. Thus, completely collapsed smectite from which all adsorbed water has been removed is easily mistakable for illite. If clay-mineral determinations are to be made on clay fractionations, it is essential they be only air-dried, not oven-dried.

It is obvious that the particle-size-grade analysis of clay materials is difficult, and care must be taken to devise a tailor-made procedure best suited to the material at hand and to the objectives of the investigation if pertinent, reproducible, and comparable data are to be obtained.

The shape of the finest particles is revealed best by electron-microscope studies. Such investigations have shown the hexagonal outline of the flake-shaped units of kaolinite, the elongate tubular shape of the halloysite minerals, the irregular flake shape of the illite, chlorite, vermiculite, and most smectite mineral particles, and the elongate lath or fiber shape of some of the smectite minerals and of attapulgite-sepiolite-palygorskite. Information on the thickness as well as areal dimensions can frequently be obtained from electron micrographs; kaolinite particles that have been studied show a ratio of areal diameter to thickness of 2–25:1, whereas for smectite it is 100–300:1. In the application of the electron beam in electron microscopy, considerable heat is developed in the specimen so that some concern has been felt as to whether or not some of the observed results are due to this heat and the possible resulting dehydration rather than to the natural mineral.

A microscope using ordinary light can, of course, be used to study the coarser particles. The lower limit for the study of the shape of particles by ordinary microscopic methods is about 5 μ.

Some information regarding the orientation of extremely fine particles can sometimes be obtained from the study of thin sections. In the absence of appreciable amounts of nonclay-mineral components, aggregate parallel orientation of the anisotropic clay-mineral particles as compared with random orientation is shown by uniform extinction and birefringence characteristics. Thin-section studies appear to have distinct limitations. The thickness of the sections is many times that of the individual clay-mineral components, so that many individuals lie on top of each other. The presence of even small amounts of organic material or free ferric iron oxide or hydroxide will mask the individual components and distort the optical values. Also the clay material must be dried in preparation for the cutting of the section, so that the texture observed may be not quite that of the original material. Even with these deficiencies of thin-section study, it is usually worthwhile to cut sections and study them in any clay-material investigation. Such studies,[12] for example, have revealed particularly pertinent data on the paragenesis of hydrothermal clay minerals in wall-rock alteration associated with ore bodies.

Some few attempts have been made to devise new methods of studying the texture of clays in their natural state. Information on the nature of the aggregation of the clay minerals may be obtained by diffraction techniques in which sections of the rock cut in definite directions are mounted in the X-ray beam (Odom[13]). Electron-microscope studies of replicas of actual clay-material surfaces have provided textural data on the arrangement of the clay minerals. Much remains to be done, and the texture of clay materials is an important and promising field for research that should attract able investigators.

So little is known in detail about the forces binding the particles together in clay materials that the possible types of binding forces can merely be enumerated.

1. Forces due to the attraction of the mass of one clay-mineral particle for the mass of another particle.

2. Intermolecular forces resulting from the nearness of one particle to another with the overlap of fields of force of molecules in the surface layers of adjacent particles.

[12] Sales, R. H., and C. Meyer, Wall Rock Alteration at Butte, Montana, *Am. Inst. Mining Met. Engrs., Tech. Publ.* 2400 (1948).

[13] Odom, I. E., Clay Mineralogy and Clay Mineral Orientation of Shales and Claystones Overlying Coal Seams in Illinois, Ph.D. thesis, University of Illinois (1963).

3. Electrostatic forces due to charges on the lattice resulting from unbalanced substitution within the lattice, broken bonds on edges of the lattice, and the attractive force of certain ions adsorbed on clay-mineral surfaces. Examples are to be found in the bonding action ·of K^+ between mica layers and of multivalent ions, with one valence tied to one particle and another valence tied to a second particle.

4. The bonding action of adsorbed polar molecules. Oriented water molecules (see Chap. 8) between two clay-mineral surfaces may form a bridge of considerable strength if only a few molecules thick and of no strength if more than a few molecules thick. Similarly, adsorbed polar organic molecules could serve as a bond between clay-mineral particles.

In any given clay material all the bond forces probably are at work, and they are interrelated. Thus the nature of the adsorbed ion will itself influence bonding and also affect the development of oriented adsorbed water, which in turn is related to bonding.

Recent work by colloid chemists and others has begun to shed light on the interparticle forces in clay-mineral assemblages. Van Olphen,[14] for example, emphasized the probable difference and importance of the bonds found at the edges of clay-mineral particles as compared with those on the large flake surfaces. Fripiat[15] considered in detail the nature of the clay-mineral surfaces on structural grounds and the probable resulting bonding forces.

The matter of the bonding force in clay materials is of particular importance to soil-mechanics investigations and construction engineers, since it largely determines the sensitivity and strength of soil materials. Construction failures have occurred because the strength properties of a soil that developed during construction could not be predicted adequately from empirical laboratory testing data. Without fundamental data on how and why clay materials are held together, it is impossible always to predict safely from any empirical data how a clay material will act when load is applied, when the water table is altered, or when other conditions are changed.

Additional references

Atterberg, A., Die Plastizität der Tone, *Intern. Mitt. Bodenk.*, pp. 10–43 (1911).
Baver, L. D., "Soil Physics," Wiley, New York (1940).

[14] Van Olphen, H., "An Introduction to Clay Colloid Chemistry," Interscience, New York (1963).

[15] Fripiat, J. J., Surface Properties of Alumino-Silicates, Proceedings of the 12th National Clay Conference, pp. 327–358, Pergamon Press, New York (1964).

Edwards, D. G., A. M. Posner, and J. P. Quirk, Repulsion of Chloride Ions by Negatively Charged Clay Surfaces, *Trans. Faraday Soc.*, **61**: 2808–2815, 2815–2820, 2820–2823 (1965).

Glossop, R., and A. W. Skempton, Particle Size in Silts and Sands, *J. Inst. Civil Engrs. (London)*, no. 5492, pp. 81–105 (1945).

Grim, R. E., Modern Concepts of Clay Materials, *J. Geol.*, **50**: 225–275 (1950).

Jenny, H., "Factors of Soil Formation," McGraw-Hill, New York (1941).

Joffe, H., "Pedology," Rutgers University Press, New Brunswick, N.J., (1949).

Knight, H. G., New Size Limit of Clay-Silt, *Soil Sci. Soc. Am. Proc.*, **2**: 592 (1937).

Krumbein, W. C., and F. J. Pettijohn, "Manual of Sedimentary Petrography," Appleton-Century-Crofts, New York (1938).

Marshall, C. E., "Physical Chemistry and Mineralogy of Soils," Wiley, New York (1964).

Michaels, A. S., and J. C. Bolger, Particle Interactions in Aqueous Kaolinite Dispersions, *Ind. Eng. Chem., Fundamentals,* **3**(1): 14–20 (1964).

Oden, S., General Introduction to the Chemistry and Physical Chemistry of Clays, *Bull. Geol. Inst. Univ. Upsala,* **15**: 175–194 (1916).

Popov, I. V., and G. G. Zubkovich, Microstructure of Clays, *Akad. Nauk SSSR, Lab. Gidrogeol. Probl.*, pp. 21–34 (1963).

Ries, H., "Clays, Occurrence, Properties and Uses," 3d ed., Wiley, New York (1927).

Twenhofel, W. H., "Principles of Sedimentation," McGraw-Hill, New York (1950).

Von Moos, A., and F. de Quervain, "Technische Gesteinkunde," Birkhauser, Basel (1948).

Two
Concepts of the Composition of Clay Materials

Old concepts

Because of the importance of clay materials in ceramics and other industries, in agriculture, in geology, and elsewhere, their investigation goes back far into antiquity. Many people have devoted most of their lives to the study of clay materials. From the first, investigators learned that clays and soils had widely varying properties. Even soils and clays which had the same color and general appearance and the same texture were found to differ widely in other characteristics. As soon as procedures for the ultimate analysis of clay materials were worked out, it was learned that clay materials differed widely in their chemical composition. The finest fractions of clay materials, which were thought to be the essence of the material, showed wide variations in the amounts of alumina, silica, alkalies, and alkali earths that they contained. It was also found that clays of the same ultimate chemical composition frequently had very different physical attributes, and that clays with substantially the same physical properties might have very different chemical compositions.

It is obvious from the foregoing statements that there must be variations not only in the amounts of the ultimate chemical constituents, but also in the way in which they are combined, or in the manner in which they are present in various clay materials. A review of the older literature shows that a considerable number of concepts were suggested to portray the fundamental and essential components of all clay materials and to explain their variation in properties. These concepts essentially present ideas of the nature of the way in which the alumina, silica, etc., are made up into the fundamental building blocks of clay materials.

Until very recent years there have been no adequate analytical tools to determine with any degree of certainty the exact nature of the fundamental building blocks of most clay materials. It is understandable, therefore, that many different concepts were suggested and that there was no general agreement among the workers in this field.

It is desired to present here very brief statements of the older concepts to serve as a background for the later consideration of the development of the present, generally accepted clay-mineral concept. For a more detailed discussion of these early ideas of the nature of clays and soils, reference should be made to the works of Blanck,[1] Stremme,[2,3] Oden,[4] Bradfield,[5] Marshall,[6] and Kelley.[7] There is no definite sequence in the development of these older concepts; in general they existed contemporaneously in the minds of various investigators.

A very old idea is that there is a single pure clay substance and that this pure clay substance is the mineral kaolinite or something substantially similar to it. Clay materials, according to this concept, are composed of kaolinite, frequently with varying amounts of other materials considered to be impurities. Differences in the chemical composition between kaolinite and natural clays were explained by the presence of these impurities. Kaolinite, however, was thought to be the essence of clays. This concept of the general prevalence of kaolinite was held widely by geologists and, indeed, unfortunately persists in the thinking and writings of some present-day members of this profession.

[1] Blanck, E., Anorganische Bestandteile des Bodens, "Handbuch der Bodenlehre," vol. 7, pp. 1–60, Springer, Berlin (1933).

[2] Stremme, H., Die Chemie des Kaolins, *Fortschr. Mineral. Krist. Petrog.*, **2**: 87–128, Dresden (1912).

[3] Stremme, H., Allgemeines über die wasserhaltigen Aluminumsilicate, "Handbuch der Mineralchemie," vol. 2, pt. 2, pp. 30–94, 130–134, Doelter, ed., Steinkopff, Dresden (1917).

[4] Oden, S., General Introduction to the Chemistry and Physical Chemistry of Clays, *Bull. Geol. Inst. Univ. Upsala,* **15**: pp. 175–194 (1916).

[5] Bradfield, R., The Chemical Nature of Colloidal Clay, *Missouri, Univ., Agr. Expt. Sta., Res. Bull.* 60 (1923).

[6] Marshall, C. E., "The Colloid Chemistry of the Silicate Minerals," Academic, New York (1949).

[7] Kelley, W. P., "Cation Exchange in Soils," Reinhold, New York (1948).

Some clays are composed almost wholly of kaolinite, and in some of such clays, the particles of kaolinite are large enough to be seen and identified positively by a microscope with relatively low magnification. Such clays are the rare exceptions in which the fundamental building blocks could be identified definitely in the years prior to the development of modern research tools for studying extremely small particles. Also such kaolinite clays are of particular importance in the ceramic art and to geologists; hence they were among the first to be studied in detail. It was a simple matter to extend the findings of the study of these kaolinite clays to all clay materials. It is now established, of course, that there are many clay materials in which there is no kaolinite present. Merrill[8] and Ries[9] early emphasized the error of considering that kaolinite was the base of all clays, but the error tended to persist.

Another concept very widely held, particularly by soil investigators, was that the essential component of all clay materials was a colloid complex. Particularly in early days, all colloidal material was thought to be amorphous, and the colloidal complex in clays was believed to be amorphous. The complex was thought to be partly inorganic and partly organic when the clay material contained some organic material.

In general there were two more or less clearly defined ideas concerning the character of the colloidal complex. One of the ideas, with which the names of Van Bemmelen[10] and Stremme[11] are particularly associated, regarded the complex not as a definite compound but as a loose mixture of the oxides of silicon, aluminum, and iron. In later years, the researches of Thugutt,[12] Bradfield,[13] and many others showed that clay materials generally did not contain a colloidal mixture of oxides. The other idea regarded the complex as a compound or a mixture of compounds. The compounds were generally thought of as salts of weak ferroaluminosiliceous acids. In some cases these compounds were considered definitely to be amorphous, but mostly there was no real concept of the structure of the colloidal complex.

Way[14] in his early work on the exchange reaction in soils concluded

[8] Merrill, G. P., "Non-metallic Minerals," Wiley, New York (1904).

[9] Ries, H., "Clays, Occurrence, Properties and Uses," 3d ed., Wiley, New York (1927).

[10] Van Bemmelen, J. M., Die Absorptionverbindung und das Absorptionvermögens der Ackererde, *Landw. Vers. Sta.*, **35:** 69–136 (1888).

[11] Stremme, H., On Allophane, Halloysite and Montmorillonite, *Centr. Mineral. Geol.*, pp. 205–211 (1911).

[12] Thugutt, H., Are Allophane, Montmorillonite, and Halloysite Units or Are They Mixtures of Alumina and Silica Gels?, *Centr. Mineral Geol.*, pp. 97–103 (1911).

[13] Bradfield, R., The Colloidal Chemistry of the Soil, "Colloid Chemistry," J. Alexander, ed., vol. III, pp. 569–590, Reinhold, New York (1928).

[14] Way, J. T., On the Power of Soils to Absorb Manure, *J. Roy. Agr. Soc. (Engl.)*, **13:** 123–143 (1852).

that the exchange complex in soils was a hydrous aluminum silicate quite similar to artificial precipitates produced in the laboratory.

Van Bemmelen[10] and later Stremme[11] divided their colloidal fraction into two parts. One part, which was soluble in hydrochloric acid, they called allophaneton, and a second part not soluble in hydrochloric acid but soluble in hot concentrated sulfuric acid came to be called kaolinton. The allophaneton was thought to be highly colloidal of very varying composition and largely responsible for the plastic and adsorptive properties of clay materials. Kaolinton was thought to be largely amorphous but at times containing also some crystalline material. Its composition showed relatively little variation and usually approached that of the mineral kaolinite. Attempts were made to classify clay materials on the basis of their kaolinton and allophaneton content.

Mellor[15] and Searle[16] also developed the idea that there were two essential components of clay. One, which was called *clayite,* was thought to be the true clay substance in kaolins and was considered to be an amorphous substance with about the same chemical composition as the mineral kaolinite. The other, for which the name *pelinite* was suggested, was the true clay substance in clay materials other than the kaolins. The latter was thought of as an amorphous material of varying composition, but of generally higher silica content than clayite and also with appreciable alkalies and/or alkaline earths.

Wiegner[17] in his extensive studies of cation exchange viewed the exchange material as made up of three parts: (1) a kernel, (2) a layer of adsorbed anions external to the kernel but lying in contact with it, and (3) exchangeable cations attracted to the particle by the adsorbed anions. The kernel was considered to be a hydrous compound chiefly of alumina and silica of variable composition and of unknown structural attributes.

In the extensive studies of cation exchange in soils by Gedroiz,[18] this investigator considered the complex as zeolitic material, but not as zeolitic in the mineralogical sense. In other words the complexes had certain of the properties of zeolites but were not considered to have their precise composition or structure. The nature of their structure was not known.

Another slight variation of this same concept, developed extensively in

[15] Mellor, J. W., Nomenclature of Clays, *Trans. Ceram. Soc. (Engl.),* **8:** 23 (1908).

[16] Searle, A. B., Clay and Clay Products, *Brit. Assoc. Advan. Sci. Rept., pp.* 113–154 (1920).

[17] Wiegner, G., Ionenumtaush and Struktur, *Trans. 3rd Intern. Congr. Soil Sci., Oxford,* **3:** 5–28 (1936).

[18] Gedroiz, K. K., On the Absorptive Power of Soil, Commissariat of Agriculture U.S.S.R., Petrograd (1922). Translated by S. A. Waksman and distribution by U.S. Department of Agriculture.

the relatively recent work of Mattson,[19,20] is that the colloid complex is made up of a relatively inert framework of silica, iron, and aluminous materials encased in an active amorphous envelope of a varying compound of silica, alumina, and iron with alkalies and alkaline earths. Mattson considered the latter compound to be an amorphous isoelectric precipitate of hydrated sesquioxides and silicic acid. In the light of advances in clay mineralogy and the finding of the general crystalline nature of the components of clay materials, Mattson[21] somewhat modified his concept by postulating the colloidal complex as a crystalline kernel covered with an amorphous heterogeneous coating which lacks a definite composition and is not identical with the nucleus. According to Mattson, X-ray-diffraction analysis would reveal only the character of the crystalline nucleus and not of the heterogeneous coating which is the essence of the complex. Mattson's concepts were criticized by Kelley[22] and Marshall,[6] and there is no doubt that in many clay materials, X-ray analyses have shown that substantially all the components are definite crystalline compounds.

In a more recent work, Puri[23] considered soils to be composed essentially of ferroaluminosilicates of varying composition but all composed of the same framework. According to him, when soils from different localities are subject to treatment by mild acids, a framework residue is obtained which in every case behaves in the same manner. Studies of the structures of the clay minerals have, of course, shown that there are important and significant differences in the structure of the various components of the finest fractions of soils.

Asch[24] and Byers[25] and his colleagues in the U.S. Department of Agriculture considered that the essential components of soils were a number of substances rather than a single compound. They viewed these substances as aluminosilicic acids or salts of such acids with definite compositions and with definite structures. This concept approaches the present clay-mineral concept,

[19] Mattson, S., The Laws of Soil Colloidal Behavior, III, Isoelectric Precipitates, *Soil Sci.,* **30:** 459–495 (1930).

[20] Mattson, S., The Laws of Soil Colloidal Behavior, IX, Amphoteric Reactions and Isoelectric Weathering, *Soil Sci.,* **34:** 209–239 (1932).

[21] Mattson, S., The Constitution of the Pedosphere, *Ann. Roy. Agr. Coll. Sweden,* **5:** 261–276 (1938).

[22] Kelley, W. P., Mattson's Papers "The Laws of Soil Colloidal Behavior," Review and Comments, *Soil Sci.,* **56:** 443–456 (1943).

[23] Puri, A. N., "Soils: Their Physics and Chemistry," Reinhold, New York (1949).

[24] Asch, W., and D. Asch, "The Silicates of Chemistry and Commerce," Constable, London (1914).

[25] Byers, H. G., L. T. Alexander, and R. S. Holmes, The Composition and Constitution of the Colloids of Certain of the Great Soil Groups, *U.S. Dept. Agr., Tech. Bull.* **484** (1935).

and, in fact, Byers et al. suggested clay-mineral names, e.g., montmorillonitic, halloysitic, for their postulated acids.

It has long been known by mineralogists that the zeolite minerals are silicate compounds that possess the property of cation exchange. When Way[14] and his successors showed that soil materials had cation-exchange capacity and that it resided in the silicate complex, an understandable step was to postulate that soil materials contained zeolites. Lemberg[26] in 1876 particularly developed the concept of the presence of zeolites in soil materials. Later, when the general idea was that the colloidal complex was amorphous, it was postulated, notably by Gans,[27] that the complex was zeolitic. That is, the complex was an amorphous counterpart of the crystalline zeolite minerals. Even in relatively recent work the colloidal complex is sometimes referred to as zeolitic, although the work of Gedroiz[28] and many others has shown wide differences between the properties of mineral zeolites and the finest fractions of clay materials and indicated that the exchange complex can be considered zeolitic only in the sense that it possesses cation-exchange capacity. Modern X-ray analyses have revealed instances when zeolites occur in bentonite clays, but such minerals are not general and significant components of clay materials.

It is generally recognized that the small size of the particles in clay materials is one of the reasons for their special attributes. It was suggested, notably by Oden,[4] that particle size is the major factor and that, in fact, clays can be composed of almost any minerals if they are fine enough—about 1 μ was considered the upper size limit. According to Oden, clays are composed of a heterogeneous array of extremely small particles of crystalline and amorphous components. Some clays, especially those of glacial origin, may contain an unusually large variety of minerals in extremely small particle sizes. Present data indicate that certain minerals, i.e., the clay minerals, must be present in appreciable amounts if the clays are to have the plastic properties associated with the term clay. The shape of such particles, their adsorptive and surface properties, in addition to their small size, are essential if a material is to have the characteristics of clay.

Clay-mineral concept

For many years some students of clay materials suggested that such materials are composed of extremely small particles of a limited number of crystalline

[26] Lemberg, J., Ueber Silicatumwandlungen, *Z. Deut. Geol. Ges.*, **28**: 519–621 (1876).

[27] Gans, R., Ueber die chemische oder physikalische Natur der kolloidalen wasserhaltigen Tonerdesilikate, II, *Centre. Mineral. Geol.*, pp. 728–741 (1913).

[28] Gedroiz, K. K., Die Lehre, vom Adsorptionvermögens der Bodens, *Kolloidchem. Beih.*, **33**: 317–448 (1931). Translated by H. Kuron.

minerals. For example, Le Châtelier[29] and Lowenstein[30] arrived at this conclusion in 1887 and 1909, respectively. This is the clay-mineral concept, but prior to about 1920 to 1925 there were no adequate research tools to provide positive evidence for it. The clay-mineral concept, therefore, is not new; rather it has been well established and generally accepted in recent years.

In 1923, Hadding[31] in Sweden and in 1924 Rinne[32] in Germany, working quite independently, published the first X-ray-diffraction analyses of clay materials. Both these investigators found crystalline material in the finest fractions of a series of clays and also found that all the samples studied seemed to be composed of particles of the same small group of minerals. They did not find a large heterogeneous array of minerals of a wide variety of types in the fine fractions of the samples studied.

In the early years of the present century, careful researches by some soil scientists were leading many of them to the idea that soils generally were composed essentially of definite compounds and that there were a limited number of such compounds in soils. Previous mention has been made of the work of Byers[25] and his colleagues of the U.S. Department of Agriculture which led them to postulate a few definite compounds, such as halloysitic acid, montmorillonitic acid, or their salts, as the essential constituents of all soils. Another example of this trend on the part of students of soils is the chemical work of Bradfield[33] on the fine fractions of the Putnam soil from Missouri, which showed that the clay fractions below a certain size resembled each other very strongly but that none of them behaved like mixed gels of hydrous oxides.

About 1924, Ross and some colleagues[34,35] of the U.S. Geological Survey began a study of the mineral composition of clays that led to a series of monumental papers on the subject. Working particularly with bentonites at first, but within a few years with a variety of clays used in industry

[29] Le Châtelier, H., De l'action de la chaleur sur les argiles, *Bull. Soc. Fran. Mineral.*, **10**: 204–211 (1887).

[30] Lowenstein, E., Ueber Hydrate deren Dampfspannung sich kontinuerlich mit der Zusammensetzung andert, *Z. Anorg. Chem.*, **63**: 69–139 (1909).

[31] Hadding, A., Eine röntgenographische Methode kristalline und kryptokristalline Substanzen zu identifizieren, *Z. Krist.*, **58**: 108–112 (1923).

[32] Rinne, F., Röntgenographische Untersuchungen an einigen feinzerteilten Mineralien Kunsprodukten und dichten Gesteinen, *Z. Krist.*, **60**: 55–69 (1924).

[33] Bradfield, R., The Nature of the Acidity of Colloidal Clay of Acid Soils, *J. Am. Chem. Soc.*, **45**: 2669–2678 (1923).

[34] Ross, C. S., and E. V. Shannon, The Chemical Composition and Optical Properties of Beidellite, *J. Wash. Acad. Sci.*, **15**: 467–468 (1925).

[35] Ross, C. S., and E. V. Shannon, Minerals of Bentonite and Related Clays, and Their Physical Properties, *J. Am. Ceram. Soc.*, **9**: 77–96 (1926).

and a variety of soils, it was shown, on the basis of extremely careful and painstaking optical work with a petrographic microscope supplemented by excellent chemical data, that the components of clay materials were largely essentially crystalline and that there was a limited number of such crystalline components, to which the name *clay minerals* was applied. A classification of the clay minerals was suggested.[36] Ross's[37] work corrected the erroneous notion, still held in some quarters, that microscopic studies are of no value in clay researches. Ross[38] and his colleagues later added X-ray analysis to their investigations; in general it substantiated their earlier findings.

About 1926, Marshall[39,40] began a study of the optical characteristics of clay-water suspensions when they were placed in an electric field. He devised a quantitative method for measuring the birefringence resulting from the aggregate orientation that developed in the electric field. Marshall's work showed the crystalline nature of the finest fraction of the soils that he studied. Marshall also showed that the birefringence exhibited a measurable variation with a variation in the nature of the exchange cation, indicating, at least in the soils studied, that the sites of the exchange cations were internal and related to the anisotropy of the crystal in some definite way. This latter finding did not hold for kaolinite clays, whose ionic exchanges could then be ascribed to external surfaces only.

Hendricks and Fry[41] in 1930 and Kelley, Dore, and Brown[42] in 1931 presented separate papers from independent work showing, chiefly on the basis of X-ray-diffraction analyses, that soil materials, even in their finest size fractions, are composed of crystalline particles and that the number of different crystalline minerals likely to be found is limited.

By the early 1930s what has come to be known as the clay-mineral concept became firmly established in the minds of a great many people ac-

---

[36] Ross, C. S., The Mineralogy of Clays, *First Intern. Congr. Soil Sci.,* **4:** 555–556 (1928).

[37] Ross, C. S., Altered Paleozoic Volcanic Materials and Their Recognition, *Bull. Am. Assoc. Petrol. Geologists,* **12:** 143–164 (1928).

[38] Ross, C. S., and P. F. Kerr, The Kaolin Minerals, *U.S. Geol. Surv., Profess. Paper* 165E, pp. 151–175 (1931).

[39] Marshall, C. E., The Orientation of Anisotropic Particles in an Electric Field, *Trans. Faraday Soc.,* **26:** 173–189 (1930).

[40] Marshall, C. E., Clays as Minerals and Colloids, *Trans. Ceram. Soc. (Engl.),* **30:** 81–96 (1931).

[41] Hendricks, S. B., and W. H. Fry, The Results of X-ray and Microscopical Examinations of Soil Colloids, *Soil Sci.,* **29:** 457–478 (1930).

[42] Kelley, W. P., W. H. Dore, and S. M. Brown, The Nature of the Base-Exchange Materials of Bentonites, Soils, and Zeolites as Revealed by Chemical and X-ray Analyses, *Soil. Sci.,* **31:** 25–45 (1931).

tively studying clay materials. Ross and Kerr[43] in 1931, Endell, Hofmann, and Wilm[44] in 1933, and Correns[45] in 1936 published particularly concise statements of this concept. At the present time it has come to be accepted by almost all students of clays. A few investigators, notably Mattson,[21] have clung to the notion that the essence of substantially all clays is an amorphous, extremely variable material which cannot be revealed by X-ray-diffraction analyses.

According to the clay-mineral concept, clays generally are essentially composed of extremely small crystalline particles of one or more members of a small group of minerals which have come to be known as the clay minerals. The clay minerals are essentially hydrous aluminum silicates, with magnesium or iron proxying wholly or in part for the aluminum in some minerals and with alkalies or alkaline earths present as essential constituents in some of them. Some clays are composed of a single clay mineral, but in many there is a mixture of them. In addition to the clay minerals, some clay materials contain varying amounts of so-called nonclay minerals, of which quartz, calcite, feldspar, and pyrite are important examples. Also many clay materials contain organic matter and water-soluble salts (see page 7).

According to the clay-mineral concept, the crystalline clay minerals are the essential constituents of nearly all clays and, therefore, the components that largely determine their properties. As noted before, the nonclay minerals and some other factors (see page 5) also influence properties if they are present in appreciable amounts. In the early years of the acceptance of the clay-mineral concept, it was thought that amorphous material was substantially completely absent in almost all clay materials. Some amorphous material had been found, but it was thought to be limited to a few unique clays, e.g., in association with the halloysite in the so-called *indianaite* from Indiana.[46] Subsequent work[47] indicated that extremely poorly crystalline material that appears in some cases to be actually amorphous to X-ray diffraction is not so rare as it was believed to be earlier (Chukhrov,[48] Ginzburg and

[43] Ross, C. S., and P. F. Kerr, The Clay Minerals and Their Identity, *J. Sediment. Petrol.,* **1**: 55–65 (1931).

[44] Endell, K., U. Hofmann, and D. Wilm, Ueber die Natur der keramischen Tone, *Ber. Deut. Keram. Ges.,* **14**: 407–438 (1933).

[45] Correns, C. W., Petrographie der Tone, *Naturwissenschaften,* **24**: 117–124 (1936).

[46] Ross, C. S., and P. F. Kerr, Halloysite and Allophane, *U.S. Geol. Surv., Profess. Paper* 185G, pp. 135–148 (1934).

[47] Grim, R. E., Some Factors Influencing the Properties of Soil Materials, *Second Intern. Congr. Soil Mech.,* **3**: 8–12 (1948).

[48] Chukhrov, F. V., Colloids in the Earth's Crust, Academy of Sciences of the USSR, 672 pp. (1955).

Yashina;[49] Beutelspacher and Van der Marel[50]).   It does not, however, appear to be a very common component, and the great majority of clay materials appear to be entirely crystalline.   Certainly amorphous material is not present in all or even in most clay materials, and it cannot be considered a universal constituent responsible for clay properties.   The presence of poorly crystalline material is revealed by X-ray-diffraction data, but the presence of definitely amorphous material is usually hard to establish.   Its presence is usually suggested when the analytical data do not indicate the crystalline constituents to be present in sufficient quantities to add up to 100 percent, or by chemical treatment processes indicating easily removable components.

Since about 1930 and the general acceptance of the clay-mineral concept, there has been intense interest in the study of clay materials, and a very voluminous literature has developed.   Workers have approached the study of clay mineralogy from many different fields—mineralogy, geology, chemistry, physics, agronomy, etc.   Also a tremendous amount of work has been done on applied clay mineralogy by ceramists, engineers, etc., in a host of university, commercial, and other laboratories.   The clay-mineral literature appears in an extremely wide variety of publications—in chemical, physical, mineralogical, ceramic, and other journals—as is to be expected because of the wide range of backgrounds and approaches of the persons working in the field.

In the following statements an attempt will be made to indicate some of the more important early contributions which appeared in the development of clay mineralogy.   It is impossible to mention all or even most of those who have contributed significantly, and an attempt can be made merely to list some of those who pioneered in the study of the clay minerals.

Reference has already been made to the monumental work of Ross and his colleagues, which did much to establish the clay-mineral concept.   The work of Ross and Kerr on kaolinite[38] and halloysite[46] and later of Ross and Hendricks[51] on montmorillonite provided fundamental data on the properties of these clay minerals essential to their determination in clay materials. Hendricks, working alone and with a series of colleagues from the U.S. Department of Agriculture, produced a series of outstanding papers on the structure of the clay minerals and on many of their physical attributes.   Among the many contributions of Hendricks of particular importance, in addition to his struc-

[49] Ginzburg, I. I., and R. S. Yashina, Displacement of Bases by Salts in Some Clay Minerals, Their Electrodialysis and Hydrolysis, *Dokl. k Sobraniyu Mezhdunar, Komis. po Izuch. Glin, Akad. Nauk SSSR*, pp. 59–67 (1960).

[50] Beutelspacher, H., and H. W. Van der Marel, Über die Amorphen Stoffe in der Tonen verschiedener Boden, *Acta Univ. Carolinae, Geol. Suppl.*, **1:** 97–114 (1961).

[51] Ross, C. S., and S. B. Hendricks, Minerals of the Montmorillonite Group, *U.S. Geol. Surv., Profess. Paper* 205B, pp. 23–79 (1945).

tural studies,[52-54] are his works on cation exchange,[55] on the reaction of organic ions and the clay minerals,[56] on the hydration characteristics of certain of the clay minerals,[57] and on the nature of the water adsorbed on the surface of the clay-mineral particle.[58]

Since about 1930, Kelley and his colleague[59,60] at the University of California have contributed immensely to our knowledge of the distribution of the clay minerals in various soil types and the soil-forming conditions under which various clay minerals form and are stable. Jenny[61] of this school made outstanding contributions to concepts of cation exchange, and Barshad,[62] also of this school, published valuable data on the vermiculite and chlorite clay minerals.

Beginning about 1931, Grim, Bradley, and their colleagues[63-65] at the Illinois State Geological Survey and the University of Illinois have studied the

[52] Hendricks, S. B., On the Structure of the Dickite, Halloysite and Hydrated Halloysite, *Am. Mineralogist,* **23**: 275–300 (1938).

[53] Hendricks, S. B., Polymorphism of the Micas with Optical Measurements by M. E. Jefferson, *Am. Mineralogist,* **24**: 729–771 (1939).

[54] Hendricks, S. B., Lattice Structure of Clay Minerals and Some Properties of Clays, *J. Geol.,* **50**: 276–290 (1942).

[55] Hendricks, S. B., Base-Exchange in the Crystalline Silicates, *Ind. Eng. Chem.,* **37**: 625–630 (1945).

[56] Hendricks, S. B., Base Exchange of the Clay Mineral Montmorillonite for Organic Cations and Its Dependence upon Adsorption Due to van der Waals Forces, *J. Phys. Chem.,* **45**: 65–81 (1941).

[57] Hendricks, S. B., R. A. Nelson, and L. T. Alexander, Hydration Mechanism of the Clay Mineral Montmorillonite Saturated with Various Cations, *J. Am. Chem. Soc.,* **62**: 1457–1464 (1936).

[58] Hendricks, S. B., and M. E. Jefferson, Structures of Kaolin and Talc-Pyrophyllite Hydrates and Their Bearing on Water Sorption of the Clays, *Am. Mineralogist,* **23**: 863–875 (1938).

[59] Kelley, W. P., W. H. Dore, and A. O. Woodford, The Colloidal Constituents of California Soils, *Soil Sci.,* **48**: 201–255 (1939).

[60] Kelley, W. P., W. H. Dore, and J. B. Page, The Colloidal Constituents of American Alkali Soils, *Soil Sci.,* **51**: 101–124 (1941).

[61] Jenny, H., Studies of the Mechanism of Ionic Exchange in Colloidal Aluminum Silicates, *J. Phys. Chem.,* **36**: 2217–2258 (1932).

[62] Barshad, I., The Effect of Interlayer Cations on the Expansion of the Mica Type of Crystal Lattice, *Am. Mineralogist,* **35**: 275–238 (1950).

[63] Grim, R. E., R. H. Bray, and W. F. Bradley, The Mica in Argillaceous Sediments, *Am. Mineralogist,* **22**: 813–829 (1937).

[64] Grim, R. E., Relation of the Composition to the Properties of Clays, *J. Am. Ceram. Soc.,* **22**:141–151 (1939).

[65] Grim, R. E., and F. L. Cuthbert, The Bonding Action of Clays, *Illinois State Geol. Surv., Repts. Invest.* 102, 110 (1946).

illite clay minerals and the composition of many clays and shales.   They have been particularly interested in the relation of the clay-mineral composition to the plastic, burning, strength, and other properties of clay materials which determine their utility in ceramics, soil mechanics, oil-well drilling, and other applied fields.   They have also worked with the development of the differential thermal procedure[66] for the analysis of clay materials and have studied the composition of recent marine sediments.[67]   Bradley,[68] working independently, determined the structure of attapulgite, which provided for the first time an insight into the structure of some of the fibrous clay minerals.

The classical investigation of the structure of the layer silicates by Pauling[69] provided the basic ideas permitting the elaboration of the structure of the layer clay minerals.   Following Pauling's original ideas, Gruner worked out a structure for kaolinite[70] and vermiculite.[71]   The structural concepts of the latter mineral largely led the way to an understanding of interlayered mixtures of clay minerals.   Hendricks and Teller[72] later worked out the theory of X-ray diffraction for such interlayered mixtures, which has permitted their detailed study and evaluation.

Contributions by Gieseking[73] greatly enhanced our knowledge of the adsorption of organic ions and their influence on the structure of smectite. Bradley[74] pointed out that certain nonionic polar organic molecules are also adsorbed, and he indicated the nature of the bond between such molecules and the clay-mineral surfaces.   Later Jordan[75] in some extremely interesting and important work showed that multiple adsorption of organic molecules

[66] Grim, R. E., and R. A. Rowland, Differential Thermal Analysis of Clay Minerals and Other Hydrous Materials, *Am. Mineralogist,* **27:** 746–761 (1942).

[67] Grim, R. E., W. F. Bradley, and R. S. Dietz, Clay Mineral Composition of Some Sediments from the Pacific Ocean off the California Coast and the Gulf of California, *Bull. Geol. Soc. Am.,* **60:** 1785–1805 (1949).

[68] Bradley, W. F., The Structural Scheme of Attapulgite, *Am. Mineralogist,* **25:** 405–410 (1940).

[69] Pauling, L., The Structure of Micas and Related Minerals, *Proc. Natl. Acad. Sci. U.S.,* **16:** 123–129 (1930).

[70] Gruner, J. W., The Crystal Structure of Kaolinite, *Z. Krist.,* **83:** 75–88 (1932).

[71] Gruner, J. W., The Structure of Vermiculites and Their Collapse by Dehydration, *Am. Mineralogist,* **19:** 557–574 (1934).

[72] Hendricks, S. B., and E. Teller, X-ray Interference in Partially Ordered Layer Lattices, *J. Chem. Phys.,* **10:** 147–167 (1942).

[73] Gieseking, J. E., Mechanism of Cation Exchange in the Montmorillonite-Beidellite-Nontronite Type of Clay Mineral, *Soil Sci.,* **47:** 1–14 (1939).

[74] Bradley, W. F., Molecular Associations between Montmorillonite and Some Polyfunctional Organic Liquids, *J. Am. Chem. Soc.,* **67:** 975–981 (1945).

[75] Jordan, J. W., Organophilic Bentonites, *J. Phys. Colloid Chem.,* **53:** 294–306 (1949).

by smectite was possible and described organophilic smectite-organic complexes which formed gels in organic liquids.

At Pennsylvania State University, a group led by Henry[76] and Bates[77] made notable contributions to clay mineralogy. Bates contributed particularly to the knowledge of the halloysite minerals.

In Germany beginning about 1931 several schools of investigators began to study the clay minerals. At the Technische Hochschule in Berlin, Hofmann, Endell, and Wilm[78] began a study of smectite which resulted in a suggested structure for this mineral which is now widely accepted in its broader outlines. This same team of workers[79,80] investigated many German clays and were among the first to show the relationship between the properties of the individual clay minerals and the larger ceramic and other properties of the clays themselves.

Correns[81] and his associates at the University of Rostock published valuable data on the X-ray, optical, and chemical specifications of the various clay minerals. Mehmel[82] of this group was the first to indicate clearly the two forms of halloysite and to show their relations to each other. They also showed that the indices of refraction of some of the clay minerals may vary with the nature of the index liquid used to measure them. Correns,[83] studying the marine-bottom samples collected by the *Meteor,* was among the first to investigate recent marine sediments by modern analytical techniques.

At Hanover in Germany, Noll[84] began about 1933 the study of the laboratory synthesis of the clay minerals and produced some of the best data yet available on this subject.

[76] Henry, E. C., and A. C. Siefert, Plasticity and Drying Properties of Certain Clays as Influenced by Electrolyte Content, *J. Am. Ceram. Soc.,* **24:** 281–285 (1941).

[77] Bates, T. F., F. A. Hildebrand, and A. Swineford, Morphology and Structure of Endellite and Halloysite, *Am. Mineralogist,* **35:** 463–484 (1950).

[78] Hofmann, U., K. Endell, and D. Wilm, Kristallstruktur und Quellung von Montmorillonit, *Z. Krist.,* **86:** 340–348 (1933).

[79] Hofmann, U., K. Endell, and D. Wilm, Röntgenographische und kolloidchemische Untersuchungen über Ton, *Angew. Chem.,* **47:** 539–547 (1934).

[80] Endell, K., U. Hofmann, and E. Maegdefrau, Ueber die Natur des Tonanteils in Rohstoffen der deutschen Zementindustrie, *Zement,* **24:** 625–632 (1935).

[81] Correns, C. W., and M. Mehmel, Ueber den optischen und röntgenographischen Nachweis von Kaolinit, Halloysit, und Montmorillonit, *Z. Krist.,* **94:** 337–348 (1936).

[82] Mehmel, M., Ueber die Struktur von Halloysit und Metahalloysit, *Z. Krist.,* **90:** 35–43 (1935).

[83] Correns, C. W., et al., Die Sedimente des äquatorialen atlantischen Ozeans, Deutsche atlantische Exped. *Meteor* 1925–1927, *Wiss. Ergeb.,* **3:** Teil 3 (1937).

[84] Noll, W., Mineralbildung im System Al_2O_3-SiO_2-H_2O, *Neues Jahrb. Mineral. Geol. Beil. Bd. A,* **70:** 65–115 (1935).

In the Netherlands at the Agricultural College in Wageningen, Edelman and his colleagues[85] began an active study of clay mineralogy in the early 1930s. They developed a structural concept[86] of smectite and halloysite, somewhat different from that suggested by Hofmann et al.,[78] which still claims considerable attention. They[87] also determined the clay-mineral composition of many types of soil.

In 1927, Orcel[88] in Paris first applied the differential thermal procedure in its modern form to the study of clay minerals, and its general use today is due in no small part to Orcel's efforts. In association with Orcel[89] at first and later independently, Henin and Caillere[90] actively pursued a variety of clay-mineral investigations. They contributed particularly to our knowledge of the sepiolite-palygorskite clay minerals.[91] Other French investigators, notably Longchambon[92] and De Lapparent,[93] also contributed important information regarding this group of clay minerals.

More recently a second group of French students of clays, led by Mering[94,95] at the National Chemical Laboratory in Paris, produced extremely important contributions to the subject. Mering[96] contributed to the X-ray analysis of the interlayer mixtures of clay minerals, extending the earlier

[85] Edelman, C. H., Relation between the Crystal Structure of Minerals and Their Base-Exchange Capacity, *Trans. 3rd Intern. Congr. Soil Sci., Oxford,* 3: 97–99 (1936).

[86] Edelman, C. H., and J. C. L. Favejee, On the Crystal Structure of Montmorillonite and Halloysite, *Z. Krist.,* 102: 417–431 (1940).

[87] Edelman, C. H., J. C. L. Favejee, and F. A. Van Baren, General Discussion of the Mineralogical Composition of Clays and Quantitative X-ray Analyses of Dutch Clays, *Overdruk uit Mededeel. Landbouwhoogeschool,* 43: 1–39 (1939).

[88] Orcel, J., Researches on the Chemical Composition of the Chlorites, *Bull. Soc. Franc. Mineral.,* 50: 75–450 (1927).

[89] Orcel, J., and S. Caillere, L'Analyse thermique différentielle des argiles à montmorillonite (bentonite), *Compt. Rend.,* 197: 774–777 (1933).

[90] Caillere, S., and S. Henin, Application de l'analyse thermique différentielle de l'etude des argiles des sols, *Ann. Agron.,* 50 pp. (1947).

[91] Caillere, S., Observation sur les composition chemique des palygorskites, *Compt. Rend.,* 199: 1795–1798 (1934).

[92] Longchambon, H., Sur certain caractéristiques de la sepiolite d'Ampandandrava et le formule les sepiolitès, *Bull. Soc. Franc. Mineral.,* 60: 1–45 (1937).

[93] De Lapparent, J., Les Argiles des terres à foulon, *Congr. Intern. Mines Met. Geol. Appl.,* 1: 381–387 (1935).

[94] Mering, J., Reactions des montmorillonit, *Bull. Soc. Chim. France,* pp. 218–223 (1949).

[95] Mering, J., The Hydration of Montmorillonite, *Trans. Faraday Soc.,* 42B: 205–219 (1946).

[96] Mering, J., L'Interférence des rayons X dans les systèmes à stratification disordonnée, *Acta Cryst.,* 3: 371–377 (1949).

work of Hendricks and Teller.[72] He also enhanced our knowledge of the hydration of smectite and of the catalytic activity of the same mineral.   His colleagues[97,98] contributed to the adsorption phenomena of the clay minerals and their texture as revealed by the electron microscope.

Among the early contributors to clay mineralogy in Great Britain were Nagelschmidt and Brammall. Nagelschmidt[99,100] published clay-mineral analyses of a variety of soils and was among the first to suggest that some of the clay minerals had an elongate fibrous form. Brammall[101,102] contributed particularly to knowledge of the illite clay minerals and of the possible variations in the composition of some of the other clay minerals.

More recently a very great interest in clay mineralogy has developed in Great Britain. MacEwan[103] at the Rothamsted Agricultural Experiment Station studied in detail the reaction of the clay minerals and organic compounds. He presented the analysis of many soil materials[104] and contributed to X-ray techniques[105] and the structure of the clay minerals. Brindley[106,107] at the University of Leeds studied the kaolinite minerals in detail and showed the possible variations in the degree of crystallinity of these minerals. At the Macaulay Agricultural Research Institute in Aberdeen, Scotland, Walker[108]

[97] Glaeser, R., On the Mechanism of Formation of Montmorillonite-Acetone Complexes, *Compt. Rend.,* **222:** 935–938 (1948).

[98] Mering, J., A. Mathieu-Sicaud, and I. Perrin-Bonnet, Observation au microscope électronique de montmorillonite saturée par différent cations, *Trans. 4th Intern. Congr. Soil Sci.,* **3:** 29–32 (1950).

[99] Nagelschmidt, G., Rod-shaped Clay Particles, *Nature,* **142:** 114–115 (1938).

[100] Nagelschmidt, G., Identification of Minerals in Soil Colloids, *J. Agr. Sci.,* **29:** 477–501 (1939).

[101] Brammall, A., J. G. Leech, and F. A. Bannister, The Paragenesis of Cookeite and Hydromuscovite, *Mineral. Mag.,* **24:** 507–520 (1937).

[102] Brammall, A., and J. G. Leech, Base-Exchange and Its Problems, *Sci. J. Roy. Coll. Sci.,* **7:** 69–78 (1937).

[103] MacEwan, D. M. C., Complexes of Clays with Organic Compounds, *Trans. Faraday Soc.* **44:** 349–367 (1948).

[104] MacEwan, D. M. C., Les Minéraux argileux de quelques sols écossais, *Verre Silicates Ind.,* **12:** 3–7 (1947).

[105] MacEwan, D. M. C., Some Notes on the Recording and Interpretation of X-ray Diagrams of Soil Clays, *J. Soil Sci.,* **1:** 90–103 (1949).

[106] Brindley, G. W., and K. Robinson, The Structure of Kaolinite, *Mineral. Mag.,* **27:** 242–253 (1946).

[107] Brindley, G. W., and K. Robinson, Randomness in the Structures of Kaolinitic Clay Minerals, *Trans. Faraday Soc.,* **42B:** 198–205 (1946).

[108] Walker, G. F., Trioctahedral Minerals in the Soil Clays of Northeast Scotland, *Mineral. Mag.,* **29:** 72–84 (1950).

and Mackenzie[109] studied the composition of Scottish soils and contributed to our knowledge of the biotite-like and vermiculite-like clay minerals.

Considerable investigation of clay minerals has been made in the U.S.S.R. by Belyankin,[110] Sedletsky,[111] and others.[112]   Unfortunately much of the early work of the Russian investigators is not available outside of that country, so that neither the magnitude nor the worth of it can be adequately evaluated.

In other countries, notably in Sweden with the work of Forslind,[113] in Australia with the work of Hosking[114] and others,[115] in India with the work of Mukherjee,[116] Chatterjee,[117] and others, in Italy with the work of Gallitelli,[118] and in Spain with the work of Hoyos,[119] notable early contributions to the knowledge of clay mineralogy were made.

The foregoing statements record only the early work which served to establish the clay-mineral concept and some of the properties of the clay minerals themselves.   In the past 20 years a very large number of investigators have produced such a voluminous literature that it is impossible to review it individually or even by countries.   An illustration of the growth of clay-mineral studies is the formation of clay-mineral organizations in many countries, e.g., The Clay Minerals Society of the United States, The Clay Minerals Group of the Mineralogical Society of Great Britain, etc., which hold meetings devoted solely to clay-mineral subjects.   These organizations now existing in about a dozen countries have formed an International Association for the

[109] Mackenzie, R. C., G. F. Walker, and R. Hart, Illite in Decomposed Granite at Ballater, Aberdeenshire, *Mineral. Mag.*, **28**: 704–714 (1949).

[110] Belyankin, D. S., Characteristics of the Mineral "Monothermite," *Compt. Rend. Acad. Sci. URSS*, **18**: 673–676 (1938).

[111] Sedletsky, I., Mineralogy of Dispersed Colloids, *Acad. Sci. URSS*, 114 pp. (1945).

[112] Pinsker, Z. G., Electronographic Determination of the Elementary Cell and Space Group of Kaolinite, *Dokl. Akad. Nauk SSSR*, **73**: 107–110 (1950).

[113] Forslind, E., The Clay Water System, *Swed. Concrete Cement Res. Inst. Bull.* 11 (1948).

[114] Hosking, J., The Clay Mineralogy of Some Australian Soils Developed on Granitic and Basaltic Parent Material, *J. Council Sci. Ind. Res.*, **13**: 206–216 (1941).

[115] Shearer, J., X-ray Powder Analysis and Its Application to Soil Colloids, *Australian J. Sci.*, **5**: 43–47 (1942).

[116] Mukherjee, J. N., and R. P. Mitra, Some Aspects of the Electrochemistry of Clays, *J. Colloid Sci.*, **1**: 141–159 (1946).

[117] Chatterjee, B., The Role of Aluminum in the Interaction of Clays and Clay Minerals with Neutral Salts and Bases, *J. Indian Chem. Soc.*, **12**: 81–99 (1949).

[118] Gallitelli, P., Uno Sguardo ad alcuni nuori Aspetti del Problema "Argilla," *Soc. Ital. Progr. Sci.*, **42**: 1–10 (1951).

[119] Hoyos de Castro, A., and M. Delgado, Origin of the Bed of Kaolin of Carataunas, *Anales Edafol. Fisiol. Vegetal. (Madrid)*, **8**: 703–785 (1949).

Study of Clays (Association Internationale Pour l'Etude des Argile; A.I.P.E.A.) which holds international conferences devoted to clay mineralogy.

Additional references

Old concepts

Ashley, H. E., The Colloidal Matter of Clay and Its Measurement, *U.S. Geol. Surv., Bull.* 388, pp. 1–62 (1909).

Byers, H. G., and M. S. Anderson, The Composition of Soil Colloids in Relation to Soil Classification, *J. Phys. Chem.*, **36**: 348–366 (1932).

Calsow, G., Kaolin und Tone, *Chem. Erde*, **2**: 415–441 (1926).

Davis, N. B., The Plasticity of Clay and Its Relation to Mode of Origin, *Trans. Am. Inst. Mining Met. Engrs.*, **51**: 451–480 (1915).

Dufrenoy, A., "Traité de minéralogie," 2d ed., vol. III, Paris (1856).

Harrassowitz, H., Fossile Verwitterungsdecken, "Handbuch der Bodenlehre," vol. 4 pp. 225–302, Springer, Berlin (1930).

Hissink, D. J., Base-Exchange in Soils, *Trans. Faraday Soc.*, **20**: 551–566 (1924).

Koettgen, P., Die Zusammensetzung der Silikatkomplexe einiger dialytischer Pelite, *Jahrb. Preuss. Geol. Landesanstalt (Berlin)*, **42**: 626–656 (1921).

Linck, G., Ueber den Chemismus der toniger Sedimente, *Geol. Rundschau*, **4**: 289–311 (1913).

Merrill, G. P., "Rock-Weathering and Soils," Macmillan, New York (1904).

Robinson, G. W., "Soils: Their Origin, Constitution and Classification," 2d ed., Murby, London (1936).

Roburgh, R. H. J., and H. Kolkmeyer, Ueber die Struktur des Adsorptionkomplexes der Tone, *Z. Krist.*, **94**: 74–79 (1936).

Savelly, T. F., Kaolins and Refractory Clays in Italy, *Corriere Ceram.*, **11**: 299–303 (1930).

Schloessing, T., Sur la constitution des argiles, *Compt. Rend.*, **79**: 376–380, 473–477 (1874).

Van Bemmelen, J. M., Beiträge zur Kenntnis der Verwitterungsprodukte des Silikate in Ton-Vulkanischen und Laterit-Boden, *Z. Anorg. Chem.*, **42**: 265–314 (1904).

Clay-mineral concepts

Alexander, L. T., S. B. Hendricks, and R. A. Nelson, Minerals Present in Soil Colloids, II, Estimation in Some Representative Soils, *Soil Sci.*, **48**: 273–279 (1939).

Caillere, S., and S. Henin, "Minéralogie des argiles," Masson et Cie, Paris (1963).

Chukhrov, F. V., "Colloids in the Earth's Crust," Academy of Sciences of Moscow (1955).

Ginzburg, I. I., Immediate Aims in the Solution of Important Problems in the Mineralogy of Clays, *Inst. Geol. Ore Deposts. Petrog., Mineral. Geochem., Moscow; Issled. i Ispol'z. Glin, L'vov. Gos. Univ., Materialy Soveshch., Lvov*, pp. 7–33 (1957).

Grim, R. E., Petrography of the Fuller's Earth Deposit, Olmstead, Illinois, with a Brief Study of Some Non-Illinois Earths, *Econ. Geol.*, **28**: 344–363 (1933).

Grim, R. E., "Applied Clay Mineralogy," McGraw-Hill, New York (1962).

Hardon, H. J., and J. C. L. Favejee, Mineralogical Investigations of Clays and Clay Minerals, III, Quantitative X-ray Analysis of the Clay Fraction of the Principal Soil Types of Java, *Mededeel. Landbouwhoogeschool, Wageningen*, **43**: 55–59 (1939).

Jasmund, K., "Die silikatischen Tonminerale," Verlag Chemie GmbH, Weinheim, Germany (1955).

Kelley, W. P., The Evidence as to the Crystallinity of Soil Colloids, *Trans. 3rd Intern. Congr. Soil Sci., Oxford,* **3:** 88–91 (1936).

Le Châtelier, H., Ueber die Konstitution der Tone, *Z. Physik. Chem.,* **1:** 396–402 (1887).

Marshall, C. E., Mineralogical Methods for the Study of Silts and Clays, *Z. Krist.,* **90:** 8–34 (1935).

Millot, G., "Géologie des argiles," Masson et Cie, Paris (1963).

Sudo, T., "Mineralogical Study of Clays of Japan," Maruzen Co. Ltd., Tokyo (1959).

Three
Classification and Nomenclature of the Clay Minerals

Classification of the clay minerals

The nomenclature and classification of clay minerals have been discussed both nationally and internationally for many years. At the present time, there is still considerable difference of opinion regarding the proper basis for a satisfactory classification and the mineral names themselves. It has not always been appreciated that terms which are too narrowly defined or require characterizations not generally determinable because of the characteristics of the clay minerals will be unusable or at best usable only in specially favorable circumstances. For example, it frequently is impossible at the present time to determine whether a given clay mineral which may occur intimately mixed with another mineral is dioctahedral or trioctahedral. Similarly, the polymorphic form of the mica constituent in a complex clay-mineral assemblage frequently cannot be determined.

Many different classifications have been suggested; only a few examples will be mentioned here. Thus, a committee on nomenclature of the Clay Minerals Group of the Mineralogical Society of Great Britain[1] suggested that

[1] Brown, G., Report of the Clay Minerals Group Sub-Committee on Nomenclature of Clay Minerals, *Clay Minerals Bull.*, **2** (13): 294–302 (1955).

the crystalline clay minerals be divided into chain and layer structures and that the layer structures be divided into 2:1 and 1:1 families, with the names triphormic and diphormic applied to these families. They suggested that a further division be made on the basis of the dioctahedral or trioctahedral character of the minerals (see Chap. 4). They proposed the names kandites for the kaolinite minerals, including halloysite, and smectite for the montmorillonite minerals.

In the first edition of this volume, the author suggested the classification shown in Table 3-1, which based distinctions on the shape of the clay minerals

Table 3-1 Classification of the clay minerals

I. Amorphous
   Allophane group
II. Crystalline
   A. Two-layer type (sheet structures composed of units of one layer of silica tetrahedrons and one layer of alumina octahedrons)
      1. Equidimensional
         Kaolinite group
            Kaolinite, nacrite, etc.
      2. Elongate
         Halloysite group
   B. Three-layer types (sheet structures composed of two layers of silica tetrahedrons and one central dioctahedral or trioctahedral layer)
      1. Expanding lattice
         a. Equidimensional
            Montmorillonite group
               Montmorillonite, sauconite, etc.
            Vermiculite
         b. Elongate
            Montmorillonite group
               Nontronite, saponite, hectorite
      2. Nonexpanding lattice
         Illite group
   C. Regular mixed-layer types (ordered stacking of alternate layers of different types)
      Chlorite group
   D. Chain-structure types (hornblende-like chains of silica tetrahedrons linked together by octahedral groups of oxygens and hydroxyls containing Al and Mg atoms)
      Attapulgite
      Sepiolite
      Palygorskite

and the expandable or nonexpandable character of the 2:1 and 1:1 layer silicates. The distinction between dioctahedral and trioctahedral minerals was not used as a distinguishing characteristic because of the difficulty in determining this character in many complex clay-mineral mixtures. The expandable characteristic was emphasized because of the ease with which it can be determined.   Weaver,[2] however, concluded that the importance of the expandability in the 2:1 clay minerals had been overemphasized, and he proposed that the 2:1 minerals be grouped together, with the primary division based on the relative amount of replacement in the tetrahedral and octahedral positions, and that the expandable or nonexpandable character be used as secondary consideration.

Examples of other classifications of clay minerals that have been presented can be found in the literature, for example, Frank-Kamenetskii,[3] Jasmund,[4] and Caillere and Henin.[5]   The latter authors, in their book "Minéralogie des argiles," have divided the layer silicates into 1:1, 2:1, and 2:1:1 classes, with further division into dioctahedral and trioctahedral, and then still further into those with fixed versus variable c-axis dimensions.   Recently Pedro[6] summarized the problems involved in the classification of the clay minerals and brought together in one publication most of the classifications recorded in the literature.

In the United States, a Nomenclature Committee was set up in 1961 by the Clay Minerals Committee of the National Research Council, under the chairmanship of G. W. Brindley.   This committee was continued under the Clay Minerals Society.   Earlier the Association Internationale pour L'Etude des Argiles (A.I.P.E.A.) established a subcommittee on nomenclature and classification.   Following a meeting of this subcommittee at the International Clay Conference in Stockholm in 1963 a summary of the discussions, recommendations, and unresolved problems resulting from the deliberations of the committee was circulated to persons interested in clay mineralogy in 31 countries.   On the basis of their replies, the proposed classification scheme for the phyllosilicates shown in Table 3-2 seems to be the nearest approach the clay mineralogists throughout the world have yet achieved toward an agreed nomenclature.   There is still a strong difference of opinion concerning the use of montmorillonite-saponite versus smectite as a group name. Further, there is still substantial difference of opinion as to whether illite should be relegated to a footnote or appear as a group name.   The A.I.P.E.A. recommended the terms halloysite and metahalloysite and that the name endellite be dropped.   A vigorous exception was taken to this by some clay mineralogists in the United States.   In this volume, the clay minerals are

[2] Weaver, C. E., A Classification of the 2:1 Clay Minerals, *Am. Mineralogist,* **38**: 698–706 (1953).

[3] Frank-Kamenetskii, V. A., A Crystallochemical Classification of Clay and Mixed-layer Clay Minerals, *Vestn. Leningr. Univ.* 16(12), *Ser. Geol. Geograf.* no. 2, pp. 5–17 (1961).

[4] Jasmund, K., "Die silikatischen Tonminerale," Verlag Chemie GmbH, Weinheim, Germany (1955).

[5] Caillere, S., and S. Henin, "Minéralogie des argiles," Masson et Cie, Paris (1963).

[6] Pedro, G., La Classification des minéraux argileux, *Ann. Agron.,* **16**: 108 pp. (1965).

Table 3-2   Proposed classification scheme for the phyllosilicates,
including layer-lattice clay minerals
(As submitted by the AIPEA Nomenclature Committee to the
International Mineralogical Association)

Type	Group (x = layer charge)	Subgroup	Species*
	Pyrophyllite-talc	Pyrophyllites	Pyrophyllite
	$x \sim 0$	Talcs	Talc
	Smectite *or* montmorillonite-	Dioctahedral smectites *or* mont-morillonites	Montmorillonite, beidellite, nontronite
	saponite $x \sim 0.5 - 1$	Trioctahedral smectites or saponites	Saponite, hectorite, sauconite
	Vermiculite	Dioctahedral vermiculite	Dioctahedral vermiculite
2:1	$x \sim 1 - 1.5$	Trioctahedral vermiculite	Trioctahedral vermiculite
	Mica†	Dioctahedral micas	Muscovite, paragonite
	$x \sim 2$	Trioctahedral micas	Biotite, phlogopite
	Brittle mica	Dioctahedral brittle micas	Margarite
	$x \sim 4$	Trioctahedral brittle micas	Seybertite, xanthophyllite, brandisite
	Chlorite	Dioctahedral chlorites	
2:1:1	x variable	Trioctahedral chlorites	Pennine, clinochlore, prochlorite
	Kaolinite-serpentine	Kaolinites	Kaolinite, halloysite
1:1	$x \sim 0$	Serpentines	Chrysotile, lizardite, antigorite

* Only a few examples given.

† The status of *illite* (or *hydromica*), *sericite*, etc., must at present be left open since it is not clear whether or at what level they would enter the table; many materials so designated may be interstratified.

discussed under headings that correspond approximately to the group headings, with the exception that halloysite is considered separately from kaolinite.

Nomenclature of the clay minerals

The name allophane was first applied by Stromeyer and Hausmann[7] in 1816 to material lining cavities in marl. The term is derived from Greek words meaning "to appear" and "other," in allusion to its frequent change, on standing, from a glassy material to one with an earthy appearance because of the loss of water. Since the work of Stromeyer and Hausmann a consider-

[7] Stromeyer, P., and J. F. L. Hausmann, *Göttingische Gelehrte Anziegen*, **2**: 125 (1816).

able number of materials have been described as allophane or classed with it. In general all such material was thought to be amorphous, and hence the name allophane came to be associated with amorphous constituents of clay. The study of much of this material, particularly by Ross and Kerr,[8] showed that some of it was actually crystalline but that much of it was really amorphous to X-ray diffraction. They suggested that the term be used for all amorphous clay-mineral materials regardless of their composition. These investigators pointed out that the known examples of amorphous material in clays and soils show considerable variation in composition. Recently electron diffraction has demonstrated considerable organization in some material amorphous to X-ray diffraction because of very small particle size or a relatively low degree of order.

As has been indicated (page 16), the word allophaneton has been used for the portion of a clay soluble in hydrochloric acid. At the time the term was suggested, such material was thought to be amorphous. It is now known that some of the crystalline clay minerals, e.g., montmorillonite, are fairly soluble in this acid and that others are somewhat soluble if present in extremely small particles. The term allophaneton has about disappeared from use.

Ross and Kerr[8] discussed in detail the nomenclature of the kaolin minerals. According to them,

> . . . by kaolin is understood the rock mass which is composed essentially of a clay material that is low in iron and usually white or nearly white in color. The kaolin-forming clays are hydrous aluminum silicates of approximately the composition $2H_2O \; Al_2O_3 \; 2SiO_2$ and it is believed that other bases if present represent impurities or adsorbed materials. Kaolinite is the mineral that characterizes most kaolins.

The name kaolin is a corruption of the Chinese "kauling" meaning "high ridge," the name of a hill near Jauchau Fu, China, where the material was obtained centuries ago. Occurrences in many parts of the world are well known today. Johnson and Blake[9] in 1867 appear to have first clearly intended the name kaolinite for the "mineral of kaolin." Ross and Kerr[8] showed that the kaolin minerals cannot be assigned to a single species, i.e., clays of this character are not composed of a single mineral species, and

[8] Ross, C. S., and P. F. Kerr, The Kaolin Minerals, *U.S. Geol. Surv., Profess. Paper* 165E, pp. 151–175 (1931).

[9] Johnson, S. W., and J. M. Blake, On Kaolinite and Pholerite, *Am. J. Sci., ser.* 2, **43**: 351–361 (1867).

minerals of that composition also are not all the same species. They concluded that three distinct species are represented, namely, kaolinite, nacrite, and dickite. Nacrite was proposed by Brongniart[10] in 1807. Later Des Cloizeaux,[11] and much later Dick,[12] described a mineral called nacrite from mines in Saxony with sufficient analytical data to differentiate it from the "Dick mineral" but not from the "mineral of kaolin." Mellor[13] accepted nacrite as a distinct mineral, and Ross and Kerr[8] finally established its identity.

Dick[12] in 1908 described a mineral from the island of Anglesey in Wales, without giving it any specific name, which was referred to with other clay minerals as a "mineral of kaolin." Ross and Kerr[8] showed that the "Dick mineral" was a distinct species and first applied the name dickite.

Anauxite was proposed as a mineral name by Breithaupt[14] in 1838 for a mineral from Bilin, Czechoslovakia. Smirnoff[15] in 1907 gave a detailed description of it and showed photomicrographs of characteristic "worm-like" structures resembling those of kaolinite. The mineral from Bilin had a higher silica-to-alumina ratio than that usually found in the "mineral of kaolin," and Ross and Kerr[8] proposed that anauxite be defined as a mineral with essentially the same attributes as kaolinite but with a higher silica-alumina molecular ratio, frequently approaching 3.

It is difficult to reconcile the high silica content of anauxite with the kaolinite structure. The substitution of Al for Si as suggested by Gruner[16] or vacancies in octahedral positions as suggested by Hendricks[17] do not seem feasible. Later Hendricks[18] suggested an irregular interlayer mixture of a double silica layer and a two-layer type of sheet (see pages 68 and 70). The dimensions of both layers are substantially the same, and so the structure would yield X-ray and optical data similar to kaolinite. Hendricks suggested

[10] Brongniart, A., "Traité élémentaire de minéralogie," vol. 1, p. 506, Paris (1807).

[11] Des Cloizeaux, A., "Supplement to Manual of Mineralogy," vol. 1, pp. 548–549, Paris (1862).

[12] Dick, A. B., Supplementary Note on the Mineral Kaolinite, *Mineral. Mag.*, **15**: 127 (1908).

[13] Mellor, J. W., Do Fireclays Contain Halloysite or Clayite? *Trans. Ceram. Soc. (Engl.)*, **16**: 83 (1916).

[14] Breithaupt, A., Bestimmung neuer Mineralien, *J. Prakt. Chem.*, **15**: 325 (1838).

[15] Smirnoff, W. P., Ueber ein Verwitterungsprodukt des Augits, *Z. Krist.*, **43**: 338–346 (1907).

[16] Gruner, G. W., Structure of Kaolinite, *Z. Krist.*, **83**: 75–88 (1932).

[17] Hendricks, S. B., Crystal Structure of Kaolinite and the Composition of Anauxite, *Z. Krist.*, **95**: 247–252 (1936).

[18] Hendricks, S. B., Lattice Structure of Clay Minerals and Some Properties of Clays, *J. Geol.*, **50**: 276–290 (1942).

structure as a possible explanation although direct proof is nearly impossible. However, it is probable that much material described in the literature as anauxite is actually a discrete mixture of kaolinite and some form of silica that has not been detected.

The minerals of this group have come to be described as the kaolinite minerals. The suggestion that kandites be used as a general group name (Brown[1]) has not been accepted generally.

It has been shown, particularly by Brindley and Robinson,[19] that there is considerable variation in the perfection of stacking and possibly also in the precise positions of the aluminums in the octahedral sheet of members of the kaolinite group. In the usual mineral the stacking is regular, whereas in some specimens random variations in certain directions are to be found. Brindley[20] at first suggested the name mellorite for the less well-crystallized material, thinking that there was a specific degree of such disorder. More recent work by Brindley[21] himself suggested that there is a considerable range of disorder in the poorly crystallized kaolinites and that no new specific mineral name is warranted at this time. Zvyagin[22] and Bailey[23] have investigated in detail the various possible polymorphic forms of the kaolinite minerals and qualifying symbols; for example, 1Tc, D, 2M$_2$ have been suggested to designate the forms.

The name halloysite was given by Berthier[24] in 1826 for material found in pockets in Carboniferous limestone near Liége, Belgium, in a district of old zinc and iron mines. It was named in honor of Omalius d'Halloy, who had observed the mineral several years previously. In the years prior to the development of X-ray-diffraction techniques many materials were described as halloysite. Dana[25] listed under halloysite 16 mineral names that he considered to be synonymous with it. Ross and Kerr[26] studied much of this material by modern methods and in addition obtained samples from

[19] Brindley, G. W., and K. Robinson, Randomness in the Structures of the Kaolinitic Clay Minerals, *Trans. Faraday Soc.*, **42B:** 198–205 (1946).

[20] Brindley, G. W., Structural Relationships in the Kaolin Group of Minerals, presented before the International Geological Congress, London (1948).

[21] Brindley, G. W., The Kaolin Minerals, "X-ray Identification and Structures of Clay Minerals," chap. III, pp. 32–75, Mineralogical Society of Great Britain Monograph (1951).

[22] Zvyagin, B. B., Polymorphism of Double-layer Minerals of the Kaolin Type, *Kristallografiya*, **7:** 51–65 (1962).

[23] Bailey, S. W., Polymorphism of the Kaolin Minerals, *Am. Mineralogist,* **48:** 1196–1209 (1963).

[24] Berthier, P., Analyse de l'halloysite, *Ann. Chim. Phys.*, **32:** 332–335 (1826).

[25] Dana, J. D., "System of Mineralogy," 6th ed., Wiley, New York (1914).

[26] Ross, C. S., and P. F. Kerr, Halloysite and Allophane, *U.S. Geol. Surv., Profess. Paper* 185G, pp. 135–148 (1934).

the mineralogical collections of the University of Liége which are probably as nearly representative of the type material as can be obtained at the present time. They showed that halloysite is crystalline and stated that it is closely related to but distinct from kaolinite. It is now generally accepted that halloysite is distinct from kaolinite and warrants a separate specific name.

Ross and Kerr[26] in 1934 pointed out that "there are two types of halloysite—one that is usually white or light-colored, porous, friable, or almost cottony in texture; and another that is dense, nonporous and porcelainlike." In the same year Hofmann, Endell, and Wilm[27] pointed out that there were two forms of the mineral and also that one form was more hydrous than the other. They found that the more hydrous form had a larger c-axis spacing than kaolinite and, when dried at 105°C, experienced a structural change accompanying the dehydration to a material similar to kaolinite. Mehmel[28] in 1935 and Correns and Mehmel[29] clearly distinguished two forms of halloysite: one with the same chemical composition as kaolinite, and the other, more hydrated form with a composition differing from that of kaolinite by having an added $2H_2O$. They showed that the transition from the higher hydrated form to the lower hydrated form was not reversible, that it could take place at temperatures at least as low as 60°C, and that the lower hydrated form resembled but was not identical with kaolinite. Mehmel[28] suggested the name halloysite for the highly hydrated form and metahalloysite for the lower hydration form.

Alexander et al.[30] in 1943 suggested that the name halloysite be applied to the lower hydration form and the new name endellite be given to the higher hydration form. They made this suggestion because of their belief that the original material described by Berthier[24] was the lower hydration form. MacEwan[31] disputed this point, claiming that the original material was the higher hydration form. MacEwan suggested that halloysite be used as a general term for all naturally occurring specimens of the mineral regardless of their state of hydration. He pointed out that natural materials may be in an intermediate state of hydration, and he would use the adjectives hydrated, nonhydrated, and intermediate when the state of hydration is known and when it is important to describe it.

[27] Hofmann, U., K. Endell, and D. Wilm, Röntgenographische und kolloidchemische Unteruchungen über Ton, *Angew. Chem.*, **47**: 539–547 (1934).

[28] Mehmel, M., Ueber die Struktur von Halloysit und Metahalloysit, *Z. Krist.*, **90**: 35–43 (1935).

[29] Correns, C. W., and M. Mehmel, Ueber die optischen und röntgenographischen Nachweis von Kaolinit, Halloysit, und Montmorillonit, *Z. Krist.*, **94**: 337–348 (1936).

[30] Alexander, L. T., G. Faust, S. B. Hendricks, and H. Insley, Relationship of the Clay Minerals Halloysite and Endellite, *Am. Mineralogist,* **28**: 1–18 (1943).

[31] MacEwan, D. M. C., Halloysite Nomenclature, *Mineral. Mag.*, **28**: 36–44 (1947).

Unfortunately there is no agreement among clay mineralogists at the present time regarding the proper nomenclature for the halloysite mineral. The original locality is no longer accessible, and it seems impossible to establish, to the satisfaction of everyone, the state of hydration of the original material described by Berthier.[24] This is particularly true because of the fact that the transition and dehydration may take place at room temperature over long periods of time.

At the International Congress of Soil Science held in Amsterdam in 1950 a special meeting was held of those interested in clay mineralogy to discuss the nomenclature of the clay minerals, particularly the problem of the halloysites. It was the general feeling of the meeting that great simplification and clarification would result if the word halloysite were employed for all forms of the mineral and, when necessary, additional self-explanatory qualifications be used, such as fully hydrated, partially hydrated, dehydrated, halloysite–7-Å, etc.,[32] and this usage will be followed in the present volume. The discussion of the nomenclature of the halloysite minerals has continued to the present time. According to the recent canvass made by Brindley,[33] most workers outside the United States favor halloysite and metahalloysite, whereas in the United States some investigators favor endellite and halloysite.

Damour and Salvetat[34] proposed the name montmorillonite in 1847 for a mineral from Montmorillon, France, which is a hydrous aluminum silicate with a silica-to-R_2O_3 ratio equal to about 4 and with a small content of alkalies and alkali earths. Le Châtelier[35] later studied the material and presented the formula $4SiO_2 \cdot Al_2O_3 \cdot H_2O +$ aq. for montmorillonite; this was accepted by Dana.[25] Dana listed a considerable number of names of minerals thought to be similar to montmorillonite wholly or in part.

Ross and his colleagues in a series of classical studies published from about 1926[36] to 1945[37] established the identity of montmorillonite as a valid clay-mineral group. They also indicated the variations in composition that are to be found in members of this group. For example, they showed the possible variation in the ratio of silica to R_2O_3 and the possible complete

---

[32] Brindley, G. W., et al., The Nomenclature of the Clay Minerals, *Am. Mineralogist,* **36**: 370–371 (1951).

[33] Brindley, G. W., personal communication.

[34] Damour, A. A., and D. Salvetat, Et analyses sur un hydrosilicate d'alumine trouvé à Montmorillon, *Ann. Chim. Phys. ser.* 3, **21**: 376–383 (1847).

[35] Le Châtelier, H., De l'action de la chaleur sur les argiles, *Bull. Soc. Franc. Mineral.,* **10**: 204–211 (1887).

[36] Ross, C. S., and E. V. Shannon, Minerals of Bentonite and Related Clays and Their Physical Properties, *J. Am. Ceram. Soc.,* **9**: 77–96 (1926).

[37] Ross, C. S., and S. B. Hendricks, Minerals of the Montmorillonite Group, *U.S. Geol. Surv., Profess. Paper* **205B**, pp. 23–79, (1945).

replacement of aluminum by iron and magnesium. They also emphasized the very frequent presence of magnesium in relatively small amounts in many specimens, apparently as an essential ingredient.

Hofmann, Endell, and Wilm[38] in 1933 published a structure for montmorillonite showing the expanding-lattice characteristics of the mineral, and this attribute is now generally considered to be an essential characteristic of the group. In 1935 Gruner[39] and Marshall[40] pointed out possible replacements within the montmorillonite structure and emphasized their importance.

Cronstedt[41] in 1788 described a material called "smectis" which seems to be the same as montmorillonite. Kerr[42] in 1932 showed that certain clay materials which have been described as smectite are actually montmorillonite. The name smectite, therefore, is earlier than montmorillonite. Kerr concluded, "in view of the large amount of modern literature on montmorillonite it seems in the best interests of science to continue the use of montmorillonite and drop that of smectite."

The earliest use of the name saponite is difficult to establish. The name, derived from "sapo," meaning soap, was used by Svanberg in 1840,[43] and in 1842[44] he published chemical analyses showing the material to be essentially a hydrous magnesium silicate. As in the case of many of the other clay minerals, prior to the development of modern analytical techniques, the mineral could not be well characterized, and the early literature includes a considerable variety of materials under this name. Ross and Kerr[45] in 1931 identified saponite as a member of the montmorillonite group with a high content of MgO. Ross and Hendricks[37] in 1945 defined saponite as a member of the montmorillonite group in which the replacement of Al^{3+} by Mg^{++} is essentially complete and with some replacement of Si^{4+} by Al^{3+}.

Berthier[46] proposed the name nontronite for a material associated with manganese ore in the Arrondissement of Nontron near the village of Saint

[38] Hofmann, U., K. Endell, and D. Wilm, Kristallstruktur und Quellung von Montmorillonit, *Z. Krist.*, **86:** 340–347 (1933).

[39] Gruner, J. W., Structural Relations of Nontronites and Montmorillonites, *Am. Mineralogist,* **20:** 475–483 (1935).

[40] Marshall, C. E., Layer Lattices and Base-Exchange Clays, *Z. Krist.*, **91:** 433–449 (1935).

[41] Cronstedt, A., "Mineralogie," Stockholm (1758). English translation by Magellan, John Hyacinth, London (1788).

[42] Kerr, P. F., Montmorillonite or Smectite as Constituents of Fuller's Earth and Bentonite, *Am. Mineralogist,* **17:** 192–198 (1932).

[43] Svanberg, A., Saponit, *Akad. Handl. Stockholm,* p. 153 (1840).

[44] Svanberg, A., Saponit, *Ann. Physik Chem. (Poggendorff),* **57:** 165–170 (1842).

[45] Ross, C. S., and P. F. Kerr, The Clay Minerals and Their Identity, *J. Sediment. Petrol.,* **1:** 55–65 (1931).

[46] Berthier, P., Nontronite nouveau minéral, *Ann. Chim. Phys.,* **36:** 22–27 (1827).

Pardoux in France. A chemical analysis given by Berthier shows the mineral to be a hydrous ferric iron silicate. Collins[47] in 1877 appears to have been the first to recognize the association between nontronite and montmorillonite. The similarity of nontronite and montmorillonite was established by Larsen and Steiger,[48] Ross and Kerr,[45] and Gruner.[39] Nontronite is now the name generally applied to the iron-rich end member of the montmorillonite group.

The name montmorillonite is used currently both as a group name for all clay minerals with an expanding lattice, except vermiculite, and also as a specific mineral name. Specifically it indicates a high-alumina end member of the montmorillonite group with some slight replacement of Al^{3+} by Mg^{++} and substantially no replacement of Si^{4+} by Al^{3+}. MacEwan[49] suggested the term montmorillonoid for the group name to avoid confusion with montmorillonite as a specific mineral name, and Correns[50] suggested montmorin as the group name. Neither of these names has found favor. The name smectite suggested as a group name by the Clay Minerals Group of the Mineralogical Society of Great Britain (Brown[1]) at the outset met strong opposition, particularly by many American mineralogists, but it is becoming widely accepted, and it will be used in this volume, with montmorillonite the species name as just defined.

The name vermiculite from the Latin "vermiculari, to breed worms," was first used by Webb[51] in 1824 for some material from Millbury, Massachusetts. Vermiculites have been classed with the micas and have frequently been listed as alteration products, chiefly of biotite and phlogopite. They have also been considered as closely related to the micas and were stated to show a considerable range in chemical composition. Until the work on the structure of vermiculites by Gruner[52] and later by Hendricks and Jefferson,[53] it was not known whether or not vermiculite was a distinct mineral.

[47] Collins, J. H., Remarks on Gramenite from Smallcombe, and the Chloropal Group of Minerals, *Mineral. Mag.*, 1: 67–82 (1877).

[48] Larsen, E. S., and G. Steiger, Dehydration and Optical Studies of Alunogen, Nontronite, and Griffithite, *Am. J. Sci.*, ser. 5, 15: 1 (1928).

[49] MacEwan, D. M. C., The Montmorillonite Minerals, "X-ray Identification and Structures of Clay Minerals," chap. IV, pp. 86–187, Mineralogical Society of Great Britain Monograph (1951).

[50] Correns, C. W., Objection to the Name Montmorillonoid, *Clay Minerals Bull.*, 1 (6): 194–195 (1950).

[51] Webb, T. H., Miscellaneous Localities of Minerals, *Am. J. Sci.*, 7: 55–61 (1824).

[52] Gruner, J. W., Vermiculite and Hydrobiotite Structures, *Am. Mineralogist*, 19: 557–575 (1934).

[53] Hendricks, S. B., and M. E. Jefferson, Crystal Structure of Vermiculites and Mixed Vermiculite-Chlorites, *Am. Mineralogist*, 23: 851–862 (1938).

More recently[54] the occurrence of vermiculite in small particles as a clay-mineral constituent of clay materials has been recognized.

Vermiculites have an expanding lattice, differing from smectite in that the expansion can take place only to a limited degree. The possible range of composition of vermiculites is not known, but the natural materials always seem to contain considerable magnesium and iron (some of it ferrous), and Mg^{++} seems to be the characteristic exchangeable cation. The vermiculites have high cation-exchange capacity, in general higher than smectites, as a consequence of a layer charge of 1 to 1.5 as compared with 0.5 for smectites. The composition of vermiculites can be about the same as that of some smectites, in which case the only difference would be in the large particle size of the vermiculites and higher layer charge.

The term illite was proposed by Grim, Bray, and Bradley[55] in 1937 as a general term, not as a specific clay-mineral name, for the mica-like clay minerals. The name was derived from the abbreviation for the state of Illinois. Prior to 1937 the widespread occurrence of a mica-like clay mineral had been recognized by many investigators, and many names had been suggested, e.g., potash-bearing clay mineral,[45] sericite-like mineral,[56] and glimmerton.[57] Grim et al.[55] pointed out objections to these earlier names, and the term illite has now been widely accepted for a mica-type clay mineral with a 10-Å c-axis spacing which shows substantially no expanding-lattice characteristics.

Grim, Bray, and Bradley[55] gave the general formula for illites as $(OH)_4K_y(Si_{8-y} \cdot Al_y)(Al_4 \cdot Fe_4 \cdot Mg_4 \cdot Mg_6)O_{20}$. In muscovite y is equal to 2, whereas in illite y is less than 2 and frequently equal to 1 to 1.5. According to the formula, illites would include both trioctahedral and dioctahedral types, and no attempt was made to differentiate between biotite and muscovite types of crystallization. At the present time the name illite is generally used, and will be used herein, for clay-mineral micas of both dioctahedral and trioctahedral types and of muscovite and biotite crystallizations.

Recent investigations have shown that some material described in the literature as illite is actually a mixed-layer complex. However, it has also been shown (Grim, Gaudette, and Eades[58]) that a mineral with the character-

[54] MacEwan, D. M. C., Chlorites and Vermiculites in Soil Clays, *Verre Silicates Ind.,* **13**: 41–46 (1948).

[55] Grim, R. E., R. H. Bray, and W. F. Bradley, The Mica in Argillaceous Sediments, *Am. Mineralogist,* **22**: 813–829 (1937).

[56] Grim, R. E., Petrology of Pennsylvania Shales and Noncalcareous Underclays Associated with Illinois Coals, *Bull. Am. Ceram. Sec.,* **14**: 113–119, 129–134, 170–176 (1935).

[57] Endell, K., U. Hofmann, and E. Maegdefrau, Ueber die Natur des Tonanteils in Rohstoffen der deutschen Zementindustrie, *Zement,* **24**: 625–632 (1935).

[58] Grim, R. E., H. E. Gaudette, and J. L. Eades, The Nature of Illite, Proceedings of the 13th National Clay Conference, pp. 33–48, Pergamon Press, New York (1966).

istics described by Grim et al.[55] that is not a mixed-layer complex also is a component of clays. Further, it is frequently impossible to characterize such a component further than that it has a 10-Å periodicity. The name illite has come to be used almost universally for such material.

It has been suggested, chiefly by Ross[37] and his colleagues, that the name illite should be replaced by bravaisite on the basis that bravaisite is an earlier name and is a clay-mineral mica. Actually type bravaisite is a mixture[59] of montmorillonite and a clay-mineral mica and not a specific mineral and therefore has no standing as a mineral species. Sarospatite has also been suggested by Hofmann et al.[60] as a substitute for illite, but the same objection[61] can be raised, namely, that the type material from Sárospatak, Hungary, is a mixture of clay minerals. It is believed that experience has shown the desirability of using a new name for this clay-mineral group rather than attempting a redefinition of an old name, particularly if the old name originally described a mixture of minerals.

Werner[62] appears to have first used the name chlorite about 1800. It has been used for a group of green hydrous silicates in which ferrous iron is prominent and which are closely related to the micas. A large variety of materials have been described as chlorites, and there has been much confusion regarding the identity and validity of species belonging to the group.

The work of Pauling[63] and McMurchy[64] showed the general structural attributes of the chlorites, and more recent works by Barshad[65] and Brindley[66] provided detailed information on the relation of the chlorites to other micas and on the specific variations in structure between members of the group. Structurally the chlorites are regular interstratifications of single biotite mica layers and brucite layers. The chlorites provided an early example of inter-

[59] Grim, R. E., and R. A. Rowland, Differential Thermal Analyses of Clay Minerals and Other Hydrous Minerals, *Am. Mineralogist,* **27**: 746–761 (1942).

[60] Hofmann, U., J. Endell, and E. Maegdefrau, Specific Identity of the Micaceous Clay Mineral Sarospatite, *Ber. Deut. Keram. Ges.,* **24**: 339–344 (1943).

[61] Grim, R. E., W. F. Bradley, and G. Brown, The Mica Clay Minerals, "X-ray Identification and Structures of Clay Minerals," chap. V, pp. 138–172, Mineralogical Society of Great Britain Monograph (1951).

[62] Werner, A. G. See Dana,[25] p. 643.

[63] Pauling, L., Structure of Chlorites, *Proc. Natl. Acad. Sci. U.S.,* **16**: 578–582 (1930).

[64] McMurchy, R. C., Structure of Chlorites, *Z. Krist.,* **88**: 420–432 (1934).

[65] Barshad, I., Vermiculite and Its Relation to Biotite as Revealed by Base-Exchange Reactions, X-ray Analyses, Differential Thermal Curves, and Water Content, *Am. Mineralogist,* **33**: 655–678 (1948).

[66] Brindley, G. W., and K. Robinson, The Chlorite Minerals, "X-ray Identification and Structure of Clay Minerals," chap. VI, pp. 173–198, Mineralogical Society of Great Britain Monograph (1951).

layering of units into more complex structures, a phenomenon which is now known to be fairly common in clay materials.

More recently chlorites have been recognized as important constituents of many clay materials, and it is likely that this clay mineral is much more abundant in clay materials than is now recognized. It is frequently very difficult to detect small amounts of chlorite when it is mixed with other clay minerals. The identification of chlorite is particularly difficult when kaolinite is also a constituent of the clay material being studied.

Work by several investigators, e.g., Brindley, Oughton, and Robinson,[67] Nelson and Roy,[68] Steinfink,[69] and Bailey and Brown,[70] has shown the occurrence of various polymorphic forms of chlorite. In clays, the mineral is commonly present in mixtures, and analytical data are adequate to determine only the presence of mineral with a 14-Å periodicity, and not its polymorphic form.

Sepiolite and meerschaum have long been considered as synonymous by mineralogists. The name meerschaum appears to have been first applied by Werner[71] in 1789 and is German for "sea froth," alluding to the lightness and color of the material. The term sepiolite was first applied in 1847 by Glocker[72] and is derived from the Greek for "cuttlefish," the bone of which is light and porous. The material is reported by Dana[25] to vary from compact with a smooth feel to earthy or fibrous. He gave the formula $2H_2O\ 2MgO\ 3SiO_2$ for the mineral. Later Schaller[73] stated the formula to be $4H_2O\ 2MgO\ 3SiO_2$.

The name palygorskite was first applied by Fersman[74] to a family of fibrous hydrous siliceous minerals forming an isomorphous series between two end members: an aluminum end member called paramontmorillonite because of its resemblance to montmorillonite in all ways except the fibrous character,

[67] Brindley, G. W., B. M. Oughton, and K. Robinson, Polymorphism of the Chlorites, *Acta Cryst.*, **3**: 408–416 (1950).

[68] Nelson, B. W., and R. Roy, Composition and Identification of Chlorites, *Natl. Acad. Sci.*, Publ. 327, pp. 335–348 (1954).

[69] Steinfink, H., Crystal Structure of a Monoclinic Chlorite, *Acta Cryst.*, **11**: 191–195 (1958).

[70] Bailey, S. W., and B. E. Brown, Chlorite Polytypism, I, Regular and Semi-random One-layer Structures, *Am. Mineralogist,* **46**: 819–850 (1962).

[71] Werner, A. G., "Letztes mineral System." Notes by Hofmann, *Bergm. J.,* **1**: 377 (1789).

[72] Glocker, E. F., "Generum et Specierum Mineralium secundum Ordines Naturales digestorum Synopsis," Halle (1847).

[73] Schaller, W. T., The Chemical Composition of Sepiolite (Meerschaum), *Am. Mineralogist,* **21**: 202–210 (1936).

[74] Fersman, A., Studies in Magnesian Silicates, *Mem. Acad. Sci. (St. Petersburg), ser.* **8, 32**: 323–365 (1913).

and a magnesium end member called sepiolite.   Longchambon,[75] Migeon,[76] and De Lapparent[77] studied these materials in considerable detail, but much additional work is required before the minerals can be well understood.   The term palygorskite has also been used in a more specific sense for specimens thought to be like sepiolite except for some replacement of magnesium by aluminum.   Longchambon first suggested that the palygorskite-sepiolite minerals have an amphibole-like structure composed of double chains of silica tetrahedrons, whereas De Lapparent believed them to have micaceous similarities.

The name attapulgite was first applied by De Lapparent[78] in 1935 to a clay mineral he encountered in fuller's earth from Attapulgus, Georgia, Quincy, Florida, and Mormoiron, France.   Bradley[79] determined the structure of the mineral, showing silica chains similar to those in amphibole to be essential components of its structure.   Bradley gave the ideal formula of attapulgite as $(OH_2)_4(OH)_2Mg_5Si_8\text{-}O_{20}\cdot4H_2O$, in which there is considerable replacement of magnesium by aluminum.

Much material described as palygorskite is clearly the same as attapulgite.   However, the descriptions of palygorskite are frequently vague, and the exact relation of the mineral to either sepiolite or attapulgite cannot always be determined.

As in the case of chlorite, the sepiolite and attapulgite-palygorskite minerals may well be much more abundant and widespread than is now considered to be the case.   These minerals are easily missed in clay-mineral analysis. They are very soluble in acids, and they frequently occur in calcareous material.   They would be destroyed if solution of the carbonates by acids preceded attempts at clay-mineral identification.

As will be shown (see pages 100 and 121), there are clay-mineral mixtures in which there is a regular interlayering of different units.   These regular mixed-layer structures have definite periodicities, and specific names have been suggested for some of them.   Thus, allevardite has been suggested for a regular interstratification of two mica layers and one or more water layers

---

[75] Longchambon, H., Sur certaines caractéristiques de la sepiolite d'Ampandandrava et la formule des sepiolites, *Bull. Soc. Franc. Mineral.,* **60:** 232–276 (1937).

[76] Migeon, G., Contribution à l'étude de la définition des sepiolites, *Bull. Soc. Franc. Mineral.,* **59:** 6–133 (1936).

[77] De Lapparent, J., The Relation of the Sepiolite-Attapulgite Series of Phyllitic Silicates of the Mica Type, *Compt. Rend.,* **203:** 482–484 (1936).

[78] De Lapparent, J., Formula and Structure of Attapulgite, *Compt. Rend.,* **202:** 1728–1731 (1936).

[79] Bradley, W. F., The Structural Scheme of Attapulgite, *Am. Mineralogist,* **25:** 405–410 (1940).

(Caillere, Mathieu-Sicaud, and Henin,[80] and Brindley[81]), and corrensite for a regular interstratification of a regular chlorite layer and one that swells in a liquid (Lippmann[82]).

Clay-mineral mixtures in which there is a random interlayering of components do not warrant the application of specific names. They must be described on the basis of the identity and abundance of the component layers (see pages 121 and 154).

Questionable and discredited clay-mineral names

The literature contains a vast number of mineral names applied to materials which are mainly hydrous silicates of aluminum and iron and have clay-like properties. These materials were originally believed to be specific minerals. Many of these names have now been discredited, because the type material was shown to be a mixture of specific minerals or identical with another species whose name had priority. Most of the discredited names were suggested prior to the development of X-ray-diffraction and differential thermal techniques when there were no adequate analytical methods for differentiating the very fine-grained minerals. These early names were suggested on the basis of optical and chemical data, and recent work has shown that these data are likely to be inadequate to differentiate and characterize the clay minerals (see Chap. 11).

Kerr and Hamilton[83] published a "Glossary of Clay Mineral Names" in which are listed most of the names that have been given to minerals which can be classed as clay minerals, together with their original description. Kerr and Hamilton also recorded the work which established or discredited these names. In 1963 Caillere and Henin[5] published a list of about 400 mineral and rock names which have been used to describe clay materials. The original reference and definitions for many of the various names are also given. No attempt will be made here to discuss all the suggested clay-mineral species mentioned in the literature. Consideration will be restricted to those which

[80] Caillere, S., A. Mathieu-Sicaud, and S. Henin, Nouvel essai d'identification du minéral de la table prés Allevard, l'allevardite, *Bull. Soc. Franc. Mineral Crist.*, **73**: 193–201 (1950).

[81] Brindley, G. W., Allevardite, a Swelling Double-layer Mica Mineral, *Am. Mineralogist*, **41**: 91–103 (1956).

[82] Lippmann, F., Keuper Clay From Zaisersweiher-Heidelberg, *Beitr. Mineral*, **4**: 130–134 (1954).

[83] Kerr, P. F., and P. K. Hamilton, "Glossary of Clay Mineral Names," Rept. 1, American Petroleum Institute Project 49, Columbia University, New York (1948).

have received some attention and whose discredited or questionable status has not yet been implanted in the thinking of mineralogists, geologists, agronomists, and others.

Leverrierite was first described by Termier[84] in 1890 from material in black carbonaceous shales near St.-Etienne, France. It was named after the mining engineer LeVerrier. Dana[25] in 1892 recognized the identity of leverrierite with kaolinite, as did Cayeux[85] some what later (1916). Ross and Kerr[8] studied the original type material in 1931 and concluded it was kaolinite. De Lapparent[86] in 1934 restudied more of the type material and found a mixture of alternating plates of kaolinite and muscovite. There seems no doubt, therefore, that the name should be discarded. However, the name persists in some more recently published textbooks on mineralogy—in one instance listed as a member of the smectite group.

Beidellite was first used by Larsen and Wherry[87] in 1925 for a clay mineral occurring in a gouge clay in a mine at Beidell, Colorado. These authors[88] first described this same material as leverrierite in 1917 but in their later work suggested that it was a distinct species. The original material was thought to be a member of the smectite group which differed from the type montmorillonite in having a lower silica-to-alumina molecular ratio (about 3 instead of 4) and somewhat higher indices of refraction.

Larsen and Wherry did their work before the development of X-ray and differential thermal techniques, and their identification was based on optical, chemical, and dehydration data. Grim and Rowland[59] concluded, on the basis of X-ray-diffraction and thermal data, that the type material was a mixture of clay minerals and showed that the other samples listed as beidellite in the U.S. National Museum collections were also mixtures. For example, a sample from Namiquipa, Mexico, listed as beidellite, which had apparently the appropriate optical properties, proved to be composed of a mixture of halloysite, limonite, and a small amount of illite.

In early clay-mineral studies in the 1930s a large amount of material was identified as beidellite on the basis of analytical data now known to be inadequate. Many of these identifications were based on X-ray-diffraction patterns which did not include low-angle reflections, which are now known

[84] Termier, P., Étude sur a leverrierite, *Ann. Mines,* **17:** 372–398 (1890).

[85] Cayeux, L., Introduction à l'étude petrographique des roches sédimentaires, *Mem. Serv. Carte Geol. France,* pp. 230–232 (1916).

[86] De Lapparent, J., Constitution et origine de la leverrierite, *Compt. Rend.,* **198:** 669–671 (1934).

[87] Larsen, E. S., and E. T. Wherry, Beidellite—A New Mineral Name, *J. Wash. Acad. Sci.,* **15:** 465–466 (1925).

[88] Larsen, E. S., and E. T. Wherry, Leverrierite from Colorado, *J. Wash. Acad. Sci.,* **7:** 208–217 (1917).

to be most critical, and/or on optical measurements, which are now also known to be inadequate. Also such identifications were made before the development of the glycol technique for the detection of small amounts of smectite in mixtures with other clay minerals. Recent studies show that a very large amount of this early so-called beidellite is actually a mixture, in many cases an interlayered mixture of illite and smectite. Many clay-mineral investigators feel, therefore, that the term beidellite should be dropped from clay-mineral terminology, and the author subscribes to their view. Ross and Hendricks[37] published a definition of beidellite as a member of the smectite group in which there is substantially no magnesium or iron present and in which replacement of Si^{4+} by Al^{3+} accounts for the cation-exchange capacity. MacEwan[89] and Caillere and Henin[5] followed this usage. This is substantially in accord with the original definition of Larsen and Wherry.[87] However, because of the past use of the term, it is believed desirable to drop it entirely in order to avoid confusion.

Monothermite was suggested by Belyankin[90] in 1938 for a material from Chassov-Yar, U.S.S.R., that showed a single high-temperature thermal reaction at about 550°C and had a composition $2RO\ AL_2O_3\ 3SiO_2\ 1.5H_2O\ 0.5Ag$. The name has not been used outside the U.S.S.R. and only recently has come to be used in that country by others besides Belyankin. Sedletsky[91] stated that it occurs in the weathered products of some slates from the Don Basin. It is impossible to determine positively from the published analytical data whether or not the mineral is a valid species and, if so, its proper classification. However, the data strongly suggest that the mineral is a mixture in which illite and kaolinite are important components.

Gedroizite was first suggested as a mineral name by Sedletsky[92] in 1939 and later[93] (1941) described in more detail for a material believed to be a specific mineral and characteristic of many alkali soils. The name is after Gedroiz, the famous Russian soil scientist. Sedletsky gave the formula $6(K \cdot Na)_2O\ 5Al_2O_3\ 14SiO_2\ 12H_2O$ for the mineral and stated that it belongs to the vermiculite group. He considered it to be a vermiculite in which the magnesium is replaced by potassium and sodium. As in the case of monother-

[89] MacEwan, D. M. C., Montmorillonite Minerals, "The X-ray Identification and Crystal Structures of Clay Minerals," chap. IV, pp. 143–207, Mineralogical Society of Great Britain Monograph (1961).

[90] Belyankin, D. S., On the Characteristics of the Minerals Monothermite, *Compt. Rend. Acad. Sci. URSS,* **18**: 673–676 (1938).

[91] Sedletsky, I. D., Mineralogy of White Clays of the Rostov Regions, *Dokl. Akad. Nauk. SSSR,* **69**: 69–72 (1949).

[92] Sedletsky, I. D., Gedroizite in the Alkali-Soils, *Compt. Rend. Acad. Sci. URSS,* **23**: 565–568 (1939).

[93] Sedletsky, I. D., Characteristics of the Mineral Gedroizite, *Compt. Rend. Acad. Sci. URSS,* **27**: 308–411 (1941).

mite, material called gedroizite has not been studied outside the U.S.S.R., and the validity or possible classification of the mineral cannot be determined.

Pholerite was originally used by Guillemin[94] in 1825 for a mineral from Rive-de-Gier in France. The original descriptions are somewhat vague, but the same material was described later in more detail by Des Cloizeaux[95] in 1862, and his descriptions apply to kaolinite. Ross and Kerr[8] in 1931 showed definitely that the mineral is identical with kaolinite, or one of the other kaolin minerals, and pointed out that the name should be abandoned.

Morencite was first described from Morenci, Arizona, by Lindgren and Hillebrand[96] in 1904 as a brownish-yellow hydrous silicate of ferric iron with magnesium, calcium, and aluminum. Gruner[39] showed the mineral to be structurally the same as nontronite.

Celadonite was proposed in 1847 by Glocker[97] for a soft gray-green mineral that is a hydrous silicate of iron, magnesium, and potassium. The name, meaning "sea-green" in French, refers to the color of the mineral. Similar material appears to have been described earlier as *terra verti* by DeLish[98] in 1783 and as *grunerde* by Hofmann[99] in 1788. Hendricks and Ross[100] in 1941 showed that celadonite and glauconite have a very similar structure. The original celadonite came from amygdaloidal fillings, and Kerr and Hamilton[83] suggested that it might be desirable to retain the name celadonite because of its different origin from that of glauconite.

Faratsihite was first described by Lacroix[101] in 1914 from Faratsiho, Madagascar. The original material was pale yellow in color and was described as a hydrous silicate of aluminum and iron. On the basis of X-ray-diffraction data Gruner[39] showed that some specimens of the mineral were identical with nontronite, and Hendricks[102] showed that other specimens are mixtures of kaolinite and nontronite.

[94] Guillemin, A., Sur la pholerite, *Ann. Mines,* **11**: 489 (1825).

[95] Des Cloizeaux, A., "Manuel de minéralogie," vol. 1, p. 190, Paris (1862).

[96] Lindgren, W., and W. F. Hillebrand, Minerals from the Clifton, Morenci District, Arizona, *Am. J. Sci., ser.* 4, **18**: 448–460 (1904).

[97] Glocker, E. F., "Generum et Specierum Mineralium Secundum ordines Naturales Digestorium Synopsis," p. 193, Halle (1847).

[98] DeLish, R., "Cristallographie ou description des formes propres à tous les corps du regne minéral," vol. 2, p. 502, Paris (1783).

[99] Hofmann, C., *Berfmannisches J.,* p. 519 (1788).

[100] Hendricks, S. B., and C. S. Ross, Chemical Composition and Genesis of Glauconite and Celadonite, *Am. Mineralogist,* **26**: 683–708 (1941).

[101] Lacroix, A. A., Sur l'opale et sur une nouvelle espice minérale (faratsihite) de Faratsiho, Madagascar, *Bull. Soc. Franc. Mineral.,* **37**: 231–236 (1914).

[102] Hendricks, S. B., Random Structures of Layer Minerals as Illustrated by Cronstedite $(2FeO \cdot Fe_2O \cdot SiO_2 \cdot 2H_2O)$. Possible Iron Content of Kaolin, *Am. Mineralogist,* **24**: 329–539 (1939).

Sedletsky and Yusupova[103] in 1940 described a mineral from Jurassic beds of Tashkent, U.S.S.R., as *ablykite*. The material was said to resemble halloysite in its dehydration characteristics but to differ from it in its X-ray-diffraction properties. Analytical data are insufficient to determine the validity or possible classification of the material.

Chloropal, meaning "green opal," was first described by Bernhardi and Brandes[104] in 1822 for material from Unghwar, Hungary. Gruner[39] and Ross and Hendricks[37] showed that chloropal and the smectite rich in iron are the same. The name chloropal has priority over nontronite, but nontronite has come to be the accepted name for the mineral. Further, the name chloropal is unsatisfactory, since it erroneously implies that the mineral is an opal.

---

[103] Sedletsky, I. D., and S. Yusupova, Argillaceous Minerals Closely Approaching Halloysite, *Compt. Rend. Acad. Sci. URSS,* **26:** 944–947 (1940).

[104] Bernhardi, J. J., and R. Brandes, Mineralogische-chemische Untersuchungen zwei neuer ungarisches Mineralien des Muschligen und des erdigen Chloropals, *J. Chem. Physik,* **5:** 29 (1882).

Four
Structure of the Clay Minerals

The atomic structures of the common clay minerals have been determined in considerable detail by numerous investigators, based on the generalizations of Pauling[1] for the structure of the micas and related layer minerals.

Two structural units are involved in the atomic lattices of most of the clay minerals. One unit consists of two sheets of closely packed oxygens or hydroxyls in which aluminum, iron, or magnesium atoms are embedded in octahedral coordination, so that they are equidistant from six oxygens or hydroxyls (Fig. 4-1). When aluminum is present, only two-thirds of the possible positions are filled to balance the structure, which is the gibbsite structure and has the formula $Al_2(OH)_6$. When magnesium is present, all the positions are filled to balance the structure, which is the brucite structure and has the formula $Mg_3(OH)_6$. The normal O-to-O distance is 2.60 Å, and a common OH-to-OH distance is about 3 Å, but in this structural unit the OH-to-OH distance is 2.94 Å, and the space available for the ion in octahedral

[1] Pauling, L., The Structure of Micas and Related Minerals, *Proc. Natl. Acad. Sci. U.S.*, **16**: 123–129 (1930).

coordination is about 0.61 Å. The theoretical thickness of the undistorted unit is 5.05 Å in clay-mineral structures.

The second unit is built of silica tetrahedrons which can be described as follows, assuming no distortions: In each tetrahedron a silicon atom is equidistant from four oxygens, or hydroxyls if needed to balance the structure, arranged in the form of a tetrahedron with a silicon atom at the center. The silica tetrahedral groups are arranged to form a hexagonal network, which is repeated indefinitely to form a sheet of composition $Si_4O_6(OH)_4$ (Fig.

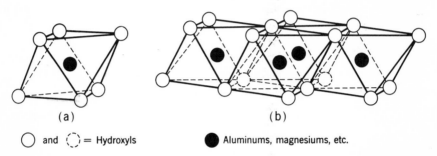

○ and ◌ = Hydroxyls   ● Aluminums, magnesiums, etc.

Fig. 4-1. Diagrammatic sketch showing (*a*) a single octahedral unit and (*b*) the sheet structure of the octahedral units.

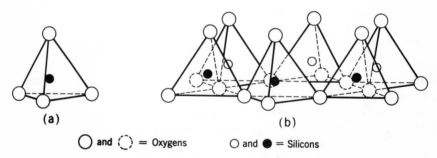

○ and ◌ = Oxygens   ○ and ● = Silicons

Fig. 4-2. Diagrammatic sketch showing (*a*) a single silica tetrahedron and (*b*) the sheet structure of silica tetrahedrons arranged in a hexagonal network.

4-2). The tetrahedrons are arranged so that the tips of all of them point in the same direction, and the bases of all tetrahedrons are in the same plane. The structure can be considered as made of a perforated plane of oxygens which is the plane of the base of the tetrahedral groups; a plane of silicon atoms with each silicon in the cavity at the junction of three oxygen atoms and therefore forming a hexagonal network; and a plane of hydroxyl atoms with each hydroxyl directly above the silicon at the tip of the tetrahedrons. The open hexagonal network can be considered as composed of three strings of oxygen atoms intersecting at angles of 120°. The O-O distance in the silica tetrahedral sheet is 2.55 Å, and the space available for the ion in tetrahedral coordination is about 0.55 Å. The thickness of the undistorted unit is 4.65 Å in clay-mineral structures. Each of these units presents a center-to-center height of about 2.1 Å.

As shown, particularly by Newnham,[2] Newnham and Brindley,[3] Brindley,[4] and Radoslovich and Norrish,[5] substantial distortion of these units in clay-mineral structures must be considered in order to fit them into determined unit-cell dimensions of the minerals. Thus, according to Brindley, if the Si-O hexagonal network is referred to a rectangular cell with side lengths of a and b and with $b/a = \sqrt{3}$, and if the value of 1.6 Å is taken for the Si-O distance (Smith[6] in a later paper Smith and Bailey[6a] gave values of 1.62 and 1.77 Å, respectively, for S-O and Al-O tetrahedral distances), then the calculated value for b is 9.05 Å. This is close to the observed value for many layer silicates which range from about 8.95 Å for kaolinite to about 9.2 to 9.3 Å for micas and chlorites. An increase in b above the calculated value could arise from substitution of Al for Si ions as in the micas, but a decrease in b can arise only from deformation of the Si-O network, or of the individual tetrahedrons, or both these causes. If the sum of ionic radii is taken for the Al-O distance, namely, $1.32 + 0.57 = 1.89$ Å, then the O-O distance for a regular octahedron is 2.67 Å, which gives a value of 8.01 Å for the b parameter. This is far short of observed values, and deformations arising from the shared octahedral edges must play an important role. It is evident that very significant departures from ideal geometry are necessary to fit the silica and alumina layers together in the various clay minerals.

Radoslovich and Norrish[5] pointed out that, for most of the layer-silicate structures now known in some detail, the network of silica tetrahedrons—ideally hexagonal—is distorted to a ditrigonal surface symmetry by the opposing rotation of alternate tetrahedrons. The amount of this rotation varies from a few degrees to near the theoretical maximum of 30°. The tetrahedral

[2] Newnham, R. E., Crystal Structure of the Mineral Dickite, Ph.D. thesis, Pennsylvania State University (1956).

[3] Newnham, R. E., and G. W. Brindley, Structure of Dickite, *Acta Cryst.*, **9**: 759–764 (1956); see also corrections, *Acta Cryst.*, **10**: 88 (1957).

[4] Brindley, G. W., Kaolin, Serpentine, and Kindred Minerals, "The X-ray Identification and Crystal Structures of Clay Minerals," chap. II, pp. 51–131, Mineralogical Society of Great Britain Monograph (1961).

[5] Radoslovich, E. W., and K. Norrish, The Cell Dimensions and Symmetry of Layer-Lattice Silicates, I, Some Structural Considerations, *Am. Mineralogist,* **47**: 599–616 (1962).

[6] Smith, J. V., Review of Si-O and Al-O Distances, *Acta Cryst.*, **7**: 479–481 (1954).

[6a] Smith, J. V., and S. W. Bailey, Second Review of Al-O and Si-O Tetrahedral Distances, *Acta Cryst.,* **16**: 801–811 (1963).

rotation is generally accepted as due to the misfit of a larger tetrahedral layer onto a smaller octahedral layer. The strain between these two layers is supposedly relieved by expansion of the octahedral layer with an accompanying contraction of thickness and partially by contraction of the tetrahedral layer due to the rotation of the basal triads. These authors showed how a value for average tetrahedral rotation from hexagonal symmetry may be computed from the observed b axis and known Al-for-Si substitutions tetrahedrally. They gave observed and computed values for kaolinite of 10.9° (10°48′) and 9° (9°18′).

Veitch and Radoslovich[7] concluded that for micas, kaolinites, and montmorillonites, at least, the dioctahedral layers are noticeably stretched and thinned but that the corresponding trioctahedral layers are more nearly regular. They pointed out the relation of the thickness of the octahedral layer to the b dimension and to the size of the octahedral hole. The octahedral layer departs from complete geometrical regularity by rotation of the upper and lower equilateral triads of anions around the octahedral site; for example, Drits and Kashaev[8] observed experimentally that in kaolinite the upper and lower triads around aluminum sites rotate plus 6.5° and minus 4°, respectively.

Radoslovich[9] deduced that, in general, forces within the octahedral layers control major features of the layer-silicate structures, that these forces tend to produce ordering of the octahedral cations, and that individual octahedrons cannot be geometrically regular. The importance of bonds between interlayer cations and surface oxygens is greater than is usually recognized. The cell dimensions of an octahedral layer correspond to an equilibrium between three kinds of forces: (1) cation-cation repulsion across shared octahedral edges, (2) anion-anion repulsion along shared edges, and (3) cation-anion bonds within octahedrons. On all available evidence these forces result in severe deformation of all octahedral layers except for minerals in which they are opposed by additional and strong external forces. The deformation can be pictured simply if an octahedron in such a layer is viewed as an upper and lower triad of oxygens around the cation, with neighboring octahedrons exhibiting counterrotation of these triads.

Some of the clay minerals are fibrous and are composed of different structural units from those noted above. These minerals resemble the amphiboles in their structural characteristics, and the basic structural unit is composed of silica tetrahedrons arranged in a double chain of composition Si_4O_{11}, as shown in Fig. 4-3. The structure is similar to that of the sheet of silica

[7] Veitch, L. G., and E. W. Radoslovich, The Cell Dimensions and Symmetry of Layer Lattice Silicates, III, Octahedral Ordering, *Am. Mineralogist,* **48:** 62–75 (1963).

[8] Drits, V. A., and A. L. Kashaev, Structural Peculiarities of Kaolinite Minerals, *Dokl. k Sobraniyu Mezhdunar. Komis. po Izuchen. Glin, Akad. Nauk SSSR,* pp. 15–18 (1960).

[9] Radoslovich, E. W., The Cell Dimensions and Symmetry of Layer-Lattice Silicates, IV, Interatomic Forces, *Am. Mineralogist,* **48:** 76–99 (1963).

tetrahedrons in the layer minerals except that it is continuous in only one direction.   In the other direction it is restricted to a width of about 11.5 Å.

The chains are bound together by atoms of aluminum and/or magnesium placed so that each such atom is surrounded by six "active" oxygen atoms. The active oxygens are those with only one link to silicon and hence are those at the edges of the chains and at the tips of the tetrahedrons.   Detailed structural data permitting specific statements are not available, but it appears likely that some distortions similar to those described above are present also in these structural units.

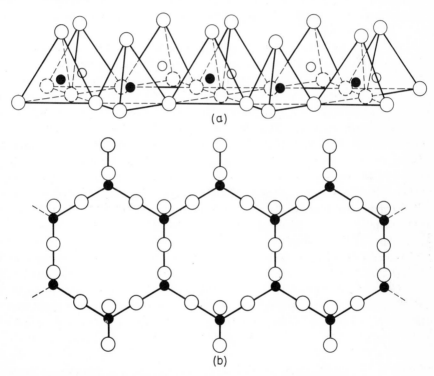

Fig. 4-3.   Diagrammatic sketch of double chains of silica tetrahedrons, as in the amphibole structural type of clay minerals: (*a*) in perspective; (*b*) projected on the plane of the base of the tetrahedrons.

Allophane minerals

As defined by Ross and Kerr,[10] the allophane clay minerals are amorphous to X-ray diffraction.   Some materials give substantially no diffraction effects,

[10] Ross, C. S., and P. F. Kerr, Halloysite and Allophane, *U.S. Geol. Surv., Profess. Paper* 185G, pp. 135-148 (1934).

and some give broad bands rather than ordered reflections. White[11] described material from Indiana which showed four diffraction bands at 3.5 to 3.0 Å, 2.26 to 2.08 Å, 1.45 to 1.27 Å, and 1.22 to 1.12 Å, from which he concluded that the material had a more ordered structure than glass. Allophanes may be considered as silicon in tetrahedral coordination and metallic ions in octahedral coordination with occasional other units such as phosphate tetrahedrons without substantial symmetry. In at least some samples there is a crude order of these structural units—enough to give diffraction bands. Further, it seems probable, on the basis of compositional factors and mineral associations, that some of these crude organizations resemble halloysite whereas others resemble montmorillonite. Egawa[11a] determined both 4- and 6-coordinated Al in allophanes.

It might be expected that the composition of allophanes would vary widely, and this appears to be the case even though it is often difficult or impossible to get pure material for analysis. In the purer materials that have been described, the molecular ratio of SiO_2 to R_2O_3 is commonly less than in the crystalline clay minerals, and the content of alkalies and alkaline earths is extremely low. Thus Birrell and Fieldes[12] described allophanes from New Zealand with SiO_2-to-R_2O_3 ratios ranging from 1.55 to 0.82 and with only a trace of MgO. Some of these samples from New Zealand soils had up to 18.40 percent Fe_2O_3. White's[11] materials from Indiana had SiO_2-to-Al_2O_3 ratios in the range of 1.23 to 1.01 and extremely small amounts of MgO (maximum 0.04 percent); two samples contained 9.51 and 10.57 percent P_2O_5, respectively. Allophanes frequently contain substantial phosphate, and Chukhrov et al.[12a] showed that in some cases, at least, the phosphate is not present in distinct compounds. Slivko[13] described an allophane with a SiO_2-to-Al_2O_3 molecular ratio of 1.5 and a MgO content of 0.47 percent.

Aomine and Wada[14] pointed out that in the allophane from Japan studied by them the silica and alumina content is considerably less than that for the ideal composition of kaolinite and that the hydrogen content is considerably larger. They suggested that the excess hydrogen occurs as $(H_4)^{4+}$ in

[11] White, W. A., Allophanes from Lawrence County, Indiana, *Am. Mineralogist,* **38**: 634–642 (1953).

[11a] Egawa, T., Coordination Number of Al in Allophane, *Clay Sci. (Tokyo),* **2**: 1–7 (1964).

[12] Birrell, K. S., and M. Fieldes, Allophane in Volcanic Ash Soils, *J. Soil Sci.,* **3**: 156–166 (1952).

[12a] Chukhrov, F. V., E. S. Rudnitskaya, V. A. Moleva, and L. P. Ermilova, Phosphate-Allophanes, *Izv. Akad. Nauk SSSR, Ser. Geol.,* **30**(3): 51–57 (1965).

[13] Slivko, M. M., Allophane from the Vyshkovo Region in Transcarpathia, *Mineralog. Sb., L'vovsk. Geol. Obshchestvo,* **7**: 187–190 (1953).

[14] Aomine, S., and K. Wada, Differential Weathering of Volcanic Ash and Pumice, Resulting in Formation of Hydrated Halloysite, *Am. Mineralogist,* **47**: 1024–1048 (1962).

tetrahedral units, and they believe that there is a continuous chemical transition from allophane to halloysite to kaolinite which can be regarded as a continuous process of dehydroxylation through condensation replacement with SiO_4 and which in turn results in higher crystallinity of these minerals.

It is difficult and at times impossible to prove unequivocally the presence of small or even modest amounts of amorphous material in some clays and soils. This is particularly true in complex mixtures and when the clay-mineral components themselves are rather poorly crystallized. Commonly, the only evidence for amorphous material is negative evidence: For example, the diffraction effects are not adequate for 100 percent crystalline material, and the solubility and exchange properties do not completely fit those of the completely crystalline clay minerals. It is not surprising, therefore, that there is a difference of opinion among investigators as to how widespread and abundant allophane material is in clays and soils. Beutelspacher and Van der Marel[15] concluded that allophane is very widespread in many soils, where it occurs as a coating of crystalline components. They believe that the allophane component is an important factor in accounting for the physical properties of soils. It is the author's opinion that allophane is rather uncommon in soils and clays and that it is rarely an important factor in accounting for the physical properties of such clay materials.

Kaolinite minerals

The structure of kaolinite was first suggested in general outlines by Pauling.[16] It was worked out in some detail by Gruner[17] and later revised by Brindley and his colleagues.[18,19] More recently, the structure of kaolinite has been studied in detail by Newnham,[2,20] Brindley and Nakahira,[21] Zvyagin,[22] Drits

[15] Beutelspacher, H., and H. W. Van der Marel, Der elektronenoptische Nachweis von amorphem Mineral als störendem Faktor bei der quantitativen Analyse von Tonmineralien in Boden, *Second Conf. Clay Mineral. and Petrog.*, Prague (1961).

[16] Pauling, L., The Structure of the Chlorites, *Proc. Natl. Acad. Sci. U.S.*, **16**: 578–582 (1930).

[17] Gruner, J. W., The Crystal Structure of Kaolinite, *Z. Krist.*, **83**: 75–88 (1932).

[18] Brindley, G. W., and K. Robinson, The Structure of Kaolinite, *Mineral. Mag.*, **27**: 242–253 (1946).

[19] Brindley, G. W., The Kaolin Minerals, "X-ray Identification and Structures of Clay Minerals," chap. II, pp. 32–75, Mineralogical Society of Great Britain Monograph (1951).

[20] Newnham, R. E., A Refinement of the Dickite Structure and Some Remarks on Polymorphism of the Kaolin Minerals, *Mineral. Mag.*, **32**: 683–704 (1961).

[21] Brindley, G. W., and M. Nakahira, Further Consideration of the Crystal Structure of Kaolinite, *Mineral. Mag.*, **31**: 781–786 (1958).

[22] Zvyagin, B. B., Electron-diffraction Determination of the Structure of Kaolinite, *Kristallografiya*, **5**: 40–50 (1960).

and Kashaev,[8] Brindley,[4] Radoslovich,[9] and Bailey,[23] and the following statements are taken from their work which has shown that the structure diagrammatically illustrated in Fig. 4-4 is modified substantially by distortions in the lattice

The structure is composed of a single silica tetrahedral sheet and a single alumina octahedral sheet combined in a unit so that the tips of the silica tetrahedrons and one of the layers of the octahedral sheet form a common layer (Fig. 4-4). All the tips of the silica tetrahedrons point in the same direction and toward the center of the unit made by the silica and octahedral sheets. The dimensions of the sheets of tetrahedral units and of the octa-

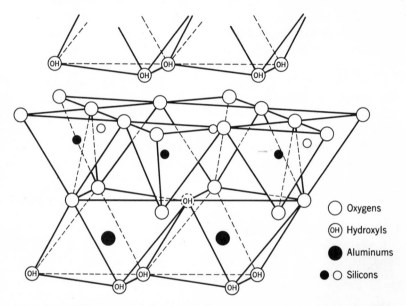

Oxygens
Hydroxyls
Aluminums
Silicons

Fig. 4-4. Diagrammatic sketch of the structure of the kaolinite layer. (After Gruner.[17])

hedral units are sufficiently similar in their *a* and *b* dimensions so that composite octahedral-tetrahedral layers are formed.

In the layer common to the octahedral and tetrahedral groups, two-thirds of the atoms are shared by the silicon and aluminum, and then they become O instead of OH. Only two-thirds of the possible positions for aluminum in the octahedral sheet are filled, and there are three possible plans of regular population of the octahedral layer with aluminums. The aluminum atoms are considered to be so placed that two aluminums are separated by an OH

[23] Bailey, S. W., Polymorphism of the Kaolin Mineral, *Am. Mineralogist*, pp. 1196–1209 (1963).

above and below, thus making a hexagonal distribution in a single plane in the center of the octahedral sheet. The OH groups are placed so that each OH is directly below the perforation of the hexagonal net of oxygens in the tetrahedral sheet.

The charge distribution in the layers is as follows:

$6O^{--}$ $12-$

$4Si^{4+}$ $16+$

$4O^{--} + 2(OH)^-$ $10-$ (Layer common to tetrahedral and octahedral sheets)

$4Al^{3+}$ $12+$ *Al₄ (Si₄ O₁₀) (OH)₈*

$6(OH)^-$ $6-$

The charges within the structural unit are balanced. The structural formula is $(OH)_8Si_4Al_4O_{10}$, and the theoretical composition expressed in oxides is SiO_2, 46.54 percent; Al_2O_2, 39.50 percent; H_2O, 13.96 percent. The analyses[24] of many samples of kaolinite minerals have shown that there is very little substitution within the lattice. In a few instances, the evidence suggests a very small amount of substitution of iron and/or titanium for aluminum in the relatively poorly crystalline variety (see page 66).

The minerals of the kaolinite group consist of sheet units of the type just described continuous in the a and b directions and stacked one above the other in the c direction. The variation between members of this group consists in the way in which the unit layers are stacked above each other and possibly in the position of the aluminum atoms in the possible positions open to them in the octahedral layer.

Table 4-1   Lattice parameters of kaolinite

	Newnham[2]	Brindley and Robinson[18]
a	5.139 ± 0.014 Å	5.15 Å
b	8.392 ± 0.016 Å	8.95 Å
c	7.371 ± 0.019 Å	7.39 Å
α	$91.6° \pm 0.2°$	$91.8°$
β	$104.8° \pm 0.2°$	$104.5–105°$
γ	$89.9° \pm 0.1°$	$90°$

The mineral is triclinic, with the lattice parameters given in Table 4-1. Newnham[2] recomputed the parameters by a systematic least-squares method, employing only those reflections having single indices. His results are in substantial agreement with those published by Brindley and Robinson[18] for

[24] Ross, C. S., and P. F. Kerr, The Kaolin Minerals, *U.S. Geol. Surv., Profess. Paper* 165E, pp. 151–176 (1931).

a triclinic unit cell which has $d(001) = c(1 - \cos^2 \alpha - \cos^2 \beta)^{1/2} = 7.15 \, \text{Å}$. It contains only a single structural layer, and analysis resolves itself into finding the correct orientation of the successive layers with respect to a and b axes. There are six ways of arranging the structural layers to conform with the cell dimensions, but the relation of adjacent layers to each other illustrated in Fig. 4-5, which shows a projection on (001) of the Si-O network of one layer and the hydroxyl sheet of the adjacent layer, gives the best agreement between calculated and observed intensities. Oxygen and hydroxyl ions are grouped in pairs, and this suggests that the layers are bound together

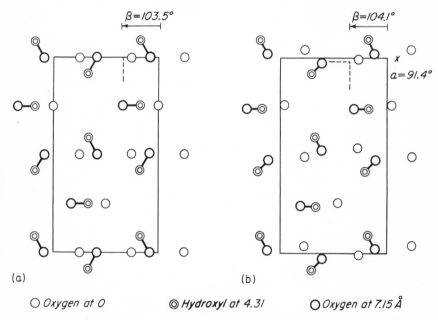

O Oxygen at 0 $\circledcirc$ Hydroxyl at 4.31 O Oxygen at 7.15 Å

Fig. 4-5. Relation between adjacent oxygens and hydroxyls seen in projection on (001). (After Brindley.[4])

by hydrogen- or hydroxyl-type bonds. It is reasonable to expect that the layers will be stacked with respect to one another so that the O-OH distances between adjacent layers will be equal or at least approximately equal. However, when one takes the triclinic parameters $\beta = 104.8$, and $\alpha = 91.6$ (Table 4-1) and still assumes the ideal layer structure, then there is no longer a stacking arrangement in which the O-OH distances are equal. An explanation for this must be sought in the departure of the layer structure from ideal geometry. Since the bonds joining atoms within the layer structure are considerably stronger than the O—OH bonds between layers, it can reasonably be assumed that the layer structure is essentially the same in all the kaolin minerals so that the accurately analyzed structure of dickite can be applied with some justification in the case of kaolinite (Brindley and Nakahira[21]). Newnham and Brindley's[3] analysis of dickite showed that the departure from ideal geometrical arrangements takes the form of distortions, mainly of rota-

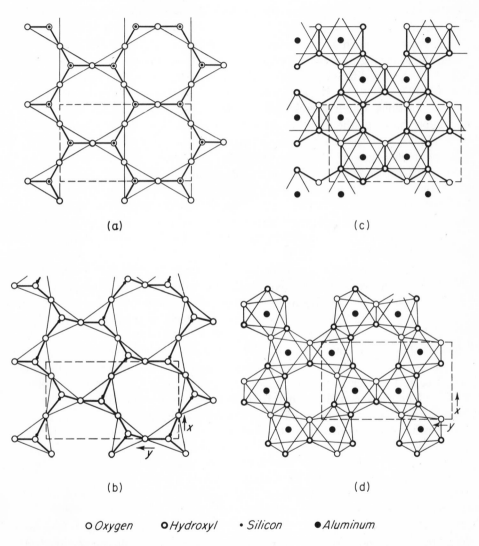

(a)

(c)

(b)

(d)

o *Oxygen*   O *Hydroxyl*   • *Silicon*   ● *Aluminum*

Fig. 4-6.   **Tetrahedral and octahedral configurations in kaolin minerals.   Broken lines show the unit-cell boundary.   (After Brindley.[4])**

tions of Si-O tetrahedrons which are alternately left-handed and right-handed rotations, and a shortening of the shared edges of Al-O (OH) octahedrons (Fig. 4-6).   Brindley[4] pointed out that, if similar distortions are assumed for the kaolinite unit layer, there is better agreement between observed and calculated X-ray intensities, and it leads naturally to an explanation of the α and β angles of kaolinite. If successive layers are so placed as to equalize O-OH distances (Fig. 4-5) then the a-axis displacement must be slightly greater than $-a/3$ to give $\beta = 104.1°$, and a small b-axis displacement is also required to give $\alpha = 91.4°$.

Zvyagin[22] studied the structure of kaolinite by electron diffraction and found the following parameters, which are in good agreement with those determined by X-ray diffraction: $a = 5.13$ Å, $b = 8.89$ Å, $c = 7.25$ Å, $\alpha = 91.40°$, $\beta = 104.40°$, $\gamma = 90°$. He also reached the following conclusions: Relative to the positions which correspond to closest packing of anions, the lower and upper bases of the octahedrons are rotated through angles corresponding to 3 and 5°, and the bases of the tetrahedrons, on an average, through 20°. The common edges of the octahedrons are shortened, and the octahedrons as a whole are somewhat flattened. The aluminum atoms are displaced toward the lower OH bases and the silicon atoms toward the bases of the tetrahedrons; hence the atoms of the bases of the polyhedrons do not lie in one plane and are of several different z coordinates. The closest atoms in successive layers are grouped in pairs O-OH which, however, differ somewhat in their length.

Drits and Kashaev[8] made a single-crystal analysis of kaolinite and found that the tetrahedrons are rotated in such a way that their bases form a ditrigonal pattern, with the average angle of rotation equal to 21°. The top and bottom ends of the octahedrons also form a ditrigonal array but the angles are much smaller, being 6.5 and 4°, respectively. The common edges of the octahedrons are shorter than the others. The successive layers are stacked in such a way that the oxygen atoms and OH groups in adjacent layers come together in pairs: the distances vary somewhat, the mean being 2.93 Å. The authors pointed out that their results do not differ essentially from Zvyagin's conclusions based on electron diffraction.

Newham[20] concluded that the silica tetrahedrons are rotated in opposite directions by about 7.5° and that another feature of the silica sheet is a puckering in the basal oxygen plane. One of the three oxygen atoms is approximately 0.17 Å higher than the other two, thus causing corrugations along (110) and ($\bar{1}$10) in alternate kaolinite layers. Corrugations arise as the result of the shortened shared edges of the alumina octahedrons.

In the octahedral part of the structure the hydroxyl layer also deviates from regularity because one of the hydroxyls shares a shorter octahedral edge with another hydroxyl group, whereas two other hydroxyls are associated with the apex oxygens of the tetrahedral group. The hydroxyl sharing the shorter octahedral edge with another hydroxyl is slightly elevated. Newnham[20] pointed out that the best interlayer linkage is obtained by matching the elevated oxygens and hydroxyls in successive layers. He pointed out further that the suggested kaolinite structures do not satisfy the puckering criterion as do those of dickite or nacrite, which may explain why kaolinite crystals are seldom as large as those of dickite or nacrite.

Radoslovich[9] discussed in considerable detail the causes of distortions in the unit layers of kaolin minerals, particularly emphasizing the importance of the octahedral part of the structure. He extended and amended at some points structural detail proposed by Newnham[20] and presented a slightly different concept of the puckering in the basal surface layers.

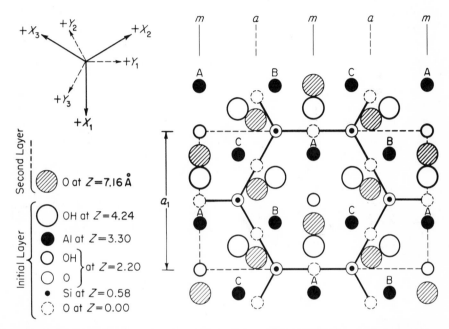

Fig. 4-7. Normal projection onto (001) of an undistorted 7-Å layer of space group *Cm.* The three possible octahedral sites, only two of which are occupied in kaolins, are labeled A, B, and C. The second layer has been shifted by $-1/3a_1$, as in kaolinite and dickite, to provide long hydrogen bonds between the paired OH and O atoms at the layer interface. (After Bailey.[23])

Bailey[23] studied the stacking sequence of the kaolin minerals and concluded that kaolinite and dickite have the same interlayer shift of $-\frac{1}{3}a_1$, when referred to a standard layer organization (Fig. 4-7). The two structures differ only in regard to the distribution of the vacant cation sites in successive octahedral sheets and in the consequence of this distribution in terms of symmetry, layer distortion, and Z-axis periodicity.

Figure 4-8 illustrates Bailey's concept of the pattern of vacant octahedral sites in successive layers of kaolinite and dickite. In well-crystallized kaolinite each layer is identical and has octahedral site C (or B) vacant. In dickite the vacant site alternates between C and B in successive layers to create a two-layer monoclinic superstructure. The alternation of vacant sites in dickite tends to balance the stress distribution of the two layers so that the cell shape remains monoclinic. The pattern of vacant sites also creates c and n glide planes parallel to (010) and changes the space group to Cc;

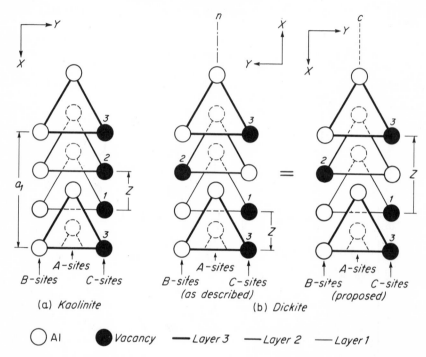

Fig. 4-8. Normal projection onto (001) of the octahedral portions of three layers (labeled 1, 2, 3) of the kaolinite and dickite structures, showing the distribution of cations and vacancies over the A, B, and C octahedral sites. In both structures each layer is shifted by $-1/3a_1$ relative to the layer below. The projected Z-axis vector is shown as a solid-line arrow. For dickite the cation distribution may be interpreted as related by an n glide plane or by a c glide plane, depending on the definition of the Z-axis vector. Two sets of octahedral positions, separated by a_1, are shown in layer 3 to illustrate the two choices for the Z-axis vector in dickite. (After Bailey.[23])

thus, dickite can be considered as a regular alternation of right- and left-handed kaolinite layers in one sense, or a superstructure of the ideal 1M polytype due to a particular ordering pattern of octahedral cations and vacancies.

Krstanovic and Radosevic[25] described a "monoclinic kaolinite" from

[25] Krstanovic, I., and S. Radosevic, Monoclinic Kaolinite from Kocevjo Mine, Yugoslavia, *Am. Mineralogist,* **46:** 1198 (1961).

Yugoslavia. Bailey[23] pointed out that a one-layer structure analogous to kaolinite, but with octahedral site A vacant, would belong to the monoclinic space group *Cm* as noted by Gruner.[17] He stated that an alternative possible monoclinic structure would be an average structure in which the vacant site is distributed at random over the three possible octahedral positions.

The diversity in detail of structural interpretation of kaolinite diffraction data reflects the difficulty of finding and handling suitable crystals. When made, an authoritative detailed structure will be known, but it will still be true that kaolins in bulk are composed of variously deformed individuals.

Numerous investigators[26-28] of clays have reported the finding of a kaolinite mineral of lower crystallinity than that of the well-crystallized material just noted. Brindley and his colleagues[19,26] investigated in detail some examples of rather poorly crystallized kaolinite and showed that their examples from English fireclays contained fewer reflections than normal kaolinite. The feature most commonly observed is that reflections with the k index not a multiple of 3 tend to be weak or missing while reflections with $k = 3n$ tend to be largely unaffected. The results are readily interpreted in terms of random layer displacements parallel to the b axis of $nb/3$. In the idealized layer structure, the OH ions in the wholly hydroxyl layer lie in lines parallel to the b axis at the intervals of $b/3$. Therefore, the structure layers can be displaced parallel to b by $3nb/3$ without altering the OH—O bonds between the adjacent layers. The net result of such displacements is that the Al and Si atoms occupy a number of positions statistically in the average unit cell. Layer displacements of $b/3$ cause phase changes of $k \times 120°$ so that, if k is not a multiple of 3, phase changes of 0, 120, and 240° occur randomly and the corresponding hkl reflections are cut out. When k is a multiple of 3, the phase changes are multiples of 360°, and the corresponding hkl reflections are unaffected. However, when three-dimensional coherence of the scattered X-rays does not occur, two-dimensional coherence is still possible, and in place of sharp hkl reflections one may obtain hk bands of scattering, provided they are sufficiently strong to be visible. Usually only one hk band is seen, having indices 02, 11, or $1\bar{1}$; this band occurs in the range of 2θ values where the $02l$ and $11l$ reflections occur for a well-crystallized kaolinite.

Bailey[23] pointed out that in poorly crystallized kaolinite it is quite conceivable that the vacancy does not always occur in the same octahedral site in each layer as in well-crystallized kaolinite, or it alternates regularly between two sites in successive layers as in dickite. Earlier Newnham[20] sug-

[26] Brindley, G. W., and K. Robinson, Randomness in the Structures of Kaolinitic Clay Minerals, *Trans. Faraday Soc.*, **42B**: 198–205 (1946).

[27] Grimshaw, R. W., E. Heaton, and A. L. Roberts, Constitution of Refractory Clays, *Trans. Brit. Ceram. Soc.*, **44**: 69–92 (1945).

[28] Grim, R. E., Differential Thermal Curves of Prepared Mixtures of Clay Minerals, *Am. Mineralogist*, **32**: 493–501 (1947).

gested that the diffraction effects shown by certain fireclays could be interpreted as a random mixing of right- and left-hand kaolinite crystals, i.e., a random choice of C or B as the vacant site in different layers (Fig. 4-8).

The first-order spacings are slightly higher (7.15 to 7.20 Å) for the poorly crystalline than for the well-crystallized mineral, suggesting some occasional interlayer water between the kaolinite units. Dehydration data tend to confirm the presence of such water (see page 298). X-ray-diffraction characteristics of kaolinite ranging from well-crystallized to *b*-axis disordered mineral, after Murray and Lyons,[29] are given in Fig. 4-9. Brindley[30] suggested mellorite as a name for such poorly crystallized kaolinite, and Roberts[31] suggested livisite for similar material. As there are probably all gradations from well-crystallized kaolinite to that of complete randomness in the *b* direction, and in the population of aluminum positions, it appears doubtful that a specific name should be applied. Currently, the names disordered kaolinite or poorly crystalline kaolinite are commonly used.

Most available data suggest that there is only slight substitution of titanium or iron for aluminum in kaolinite and that such substitutions are restricted to the poorly crystallized kaolinite. It is not necessarily established that any of the titanium is in the structure, as anatase is an almost universal minor component of kaolins.

Youell[32] concluded that isomorphous replacement of iron and aluminum occurs between chamosite and kaolinite over a wide range of composition.

Dickite[33] and nacrite[34] have structures somewhat similar to that of kaolinite and are usually listed as clay minerals, although they are rarely found in clay materials. They are made up of unit layers of an alumina octahedral sheet and a silica tetrahedral sheet like that in kaolinite but differ in the stacking of the layers and in the arrangement of aluminum ions in octahedral positions. In Dickite[33,35,36] the unit cell is made up of two unit layers. The

[29] Murray, H. H., and S. C. Lyons, Degree of Crystal Perfection of Kaolinite, *Natl. Acad. Sci., Publ.* **456**, pp. 31–40 (1956).

[30] Brindley, G. W., Structural Relationships in the Kaolin Group of Minerals, presented before International Geological Congress, London (1948).

[31] Roberts, A. L., and R. W. Grimshaw, The Principal Clay Mineral in Certain Refractory and Bond Clays, presented before International Geological Congress, London (1948).

[32] Youell, R. F., Isomorphous Replacement in the Kaolin Group of Minerals, *Nature,* **181:** 557–558 (1958).

[33] Gruner, J. W., Crystal Structure of Dickite, *Z. Krist.,* **83:** 394–404 (1932).

[34] Gruner, J. W., The Crystal Structure of Nacrite and a Comparison of Certain Optical Properties of the Kaolin Group with Its Structure, *Z. Krist.,* **85:** 345–354 (1933).

[35] Ksanda, C. J., and T. F. W. Barth, Note on the Structure of Dickite and Other Clay Minerals, *Am. Mineralogist,* **20:** 631–637 (1935).

[36] Hendricks, S. B., On the Structure of the Clay Minerals: Dickite, Halloysite, and Hydrated Halloysite, *Am. Mineralogist,* **23:** 295–301 (1938).

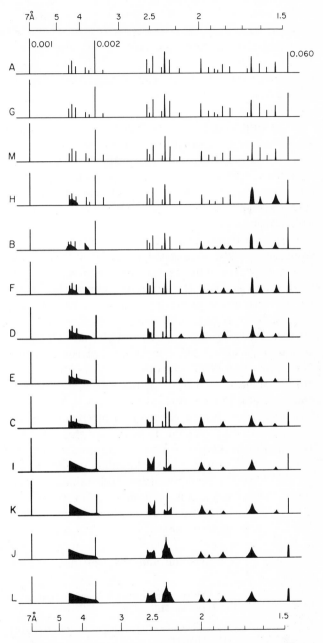

Fig. 4-9. X-ray powder diagrams of samples of kaolinite arranged in order of crystallinity. (After Murray and Lyons.[29])

exact stacking of successive layers is not well established. The parameters of the mineral are best accounted for by shifts along the a axis of $n = 1$ in $na_0/6$ and along the b axis of $m = 3$ in $mb_0/6$, with the shift being positive and negative in successive layers. A shift of $m = 1$ is also a possibility but seems less likely. However, shifts of $m = 1$ or 3 are not completely in accord with the diffraction data. According to Gruner,[33] the mineral is monoclinic $a = 5.15$ Å, $b = 8.96$ Å, $c = 14.45$ Å, $\beta = 96°50'$, and the space group is $C_s{}^4–Cc$.

Newnham and Brindley[3] conducted single-crystal analyses of dickite and confirmed the earlier measurements and the space group. Their analysis shows the departures of the tetrahedral and octahedral sheets from ideal geometric arrangements as illustrated in Fig. 4-6. The distortions take the form mainly of rotations of Si-O tetrahedrons, which are alternately left- and right-handed rotations, and a shortening of shared edges of Al-O (OH) octahedrons.

In nacrite, the unit cell was first described as made up of six unit layers, each having the arrangement described for kaolinite. Thus, according to Hendricks[37] the unit dimensions are $a = 5.15$ Å, $b = 8.96$ Å, $c = 43$ Å, and $\beta = 90°0'$, so that the structure approaches rhombohedral symmetry. In this structure, successive layers are stacked one above the other so that the c axis is about perpendicular to the ab plane; that is, n and m are equal to zero, and the space group is $C_s{}^4–Cc$. Threadgold[38] studied the structure of nacrite in detail and concluded that rotation of the tetrahedrons of about 7.8° causes significant distortion of the structure. He concluded further that the distribution of vacant octahedral sites is such that a two-layer unit cell can be selected to describe the structure.

Bailey[23] analyzed the polymorphism of the kaolin minerals in terms of the direction and amount of interlayer shift and the location of vacant octahedral sites in successive layers. He presented a modified system of polytype notation for describing the kaolin minerals.

Occasionally kaolinite-type clay minerals appear to have considerably higher silica-to-alumina molecular ratio than kaolinite, often approaching 3. Such silica-rich kaolinites are called *anauxite*. It is now believed that much of such material is a discrete mixture of kaolinite and silica. However, examples have been described which seem to be monomineral species on the basis of X-ray-diffraction and chemical data. Thus, Orcel et al.[39] described

[37] Hendricks, S. B., Crystal Structure of Nacrite and the Polymorphism of the Kaolin Minerals, *Z. Krist.*, **100**: 509–518 (1938).

[38] Threadgold, I. M., The Crystal Structures of Hellyrite and Nacrite, *Trans. Brit. Ceram. Soc.*, **63**: 60A (1964).

[39] Orcel, J., S. Henin, and S. Caillere, The Presence of Anauxite in France, *Bull. Soc. Franc. Mineral. Crist.*, **79**: 435–448 (1956).

a specimen from France which had a silica-to-alumina molecular ratio of 2.7 following treatment with a caustic sodium solution which was believed to have removed all colloidal free silica.

The structure of anauxite has been the subject of considerable controversy, but the suggestion of Hendricks[40] seems most plausible. According to Hendricks, anauxite consists of kaolinite unit layers between which units composed of double silica tetrahedral sheets are interlayered randomly. In the double silica units, the tetrahedrons of each sheet point toward the center, and a common oxygen forms the tip of the tetrahedrons in both sheets (Fig. 4-10). The charge distribution is as follows:

$6O^{--}$ $12-$
$4Si^{4+}$ $16+$
$4O^{--}$ $8-$ (Sheet common to both tetrahedral sheets)
$4Si^{4+}$ $16+$
$6O^{--}$ $12-$

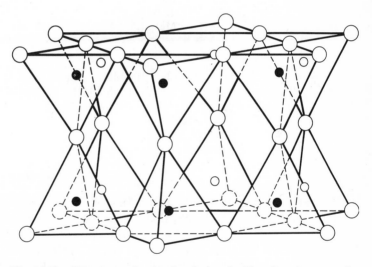

Fig. 4-10. Diagrammatic sketch of the double silica layer in the anauxite structure. (After Hendricks.[40])

The dimensions and characteristics of the double silica sheet are about those of kaolinite; hence an interlamination with kaolinite sheets is possible without changing materially the diffraction characteristics. Also the optical properties of the two units would not be very different. The silica-to-alumina molecular ratio of anauxites is known to vary; this may be accounted for

[40] Hendricks, S. B., Lattice Structure of Clay Minerals and Some Properties of Clays, *J. Geol.,* **50**: 276–290 (1942).

by variations in the relative abundance of the interlayering of the double silica sheets.

A number of minerals are structurally similar to the kaolinite minerals but are chemically distinct. The most important of them are listed in Table 4-2 with the cation composition of octahedral and tetrahedral layers indicated. These minerals do not usually occur in clay materials and are not

Table 4-2   Compositions of kaolin minerals and structurally similar minerals
(*After Brindley*[4])

Dioctahedral minerals			
Kaolinite and polymorphic varieties	Al_2	Si_2	$O_5(OH)_4$
Trioctahedral minerals			
Serpentine minerals	Mg_3	Si_2	$O_5(OH)_4$
	$(Mg_{3-x}Al_x)$	$(Si_{2-x}Al_x)$	$O_5(OH)_4$
Amesite	(Mg_2Al)	$(SiAl)$	$O_5(OH)_4$
Cronstedtite	$(Fe^{++}Fe^{3+})$	$(SiFe^{3+})$	$O_5(OH)_4$
Greenalite	$(Fe^{++}_{2.2}Fe^{3+}_{0.5})$	Si_2	$O_5(OH)_4$
Chamosite	$(Fe^{++},Mg)_{2.2}(Fe^{3+},Al)_{0.7}$	$(Si_{1.4}Al_{0.6})$	$O_5(OH)_4$

normally classed as clay minerals. Brindley[4] summarized the structural attributes of these minerals, and his work should be consulted for information concerning them.

Halloysite minerals

As indicated earlier, there are two forms of halloysite (see page 38), one with the composition $(OH)_8Si_4Al_4O_{10}$ and the other with the composition $(OH)_8Si_4Al_4O_{10} \cdot 4H_2O$. The latter form dehydrates to the former irreversibly at relatively low temperatures. Various structures for the halloysite minerals have been suggested by Mehmel,[41] Edelman and Favejee,[42] Stout,[43] and Hendricks.[36]

Hendricks showed that the earlier suggested structures are not in accord with the observed intensities of the basal reflections or with the very easy

[41] Mehmel, M., Ueber die Struktur von Halloysit und Metahalloysit, *Z. Krist.*, **90:** 35–43 (1935).

[42] Edelman, C. H., and J. C. L. Favejee, On the Crystal Structure of Montmorillonite and Halloysite, *Z. Krist.*, **102:** 417–431 (1940).

[43] Stout, P. P., Alterations in the Crystal Structure of the Clay Minerals as a Result of Phosphate Fixation, *Soil Sci. Soc. Am. Proc.*, **4:** 177–182 (1939).

dehydration of the mineral. The basal spacing of the dehydrated form is about 7.2 Å or about the thickness of the kaolinite layer, and the basal spacing of the hydrated form is about 10.1 Å. The difference, 2.9 Å, is about the thickness of a single molecular sheet of water molecules. Hendricks suggested therefore that the highly hydrated form consists of kaolinite layers separated from each other by single molecular layers of water. On this basis he ex-

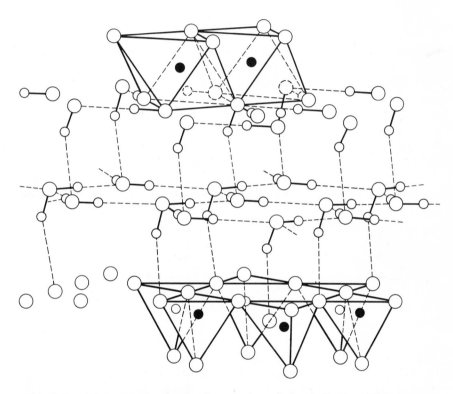

Fig. 4-11. Diagrammatic sketch of a portion of the halloysite $4H_2O$ structure, showing a single layer of water molecules with the configuration suggested by Hendricks and Jefferson.[44]

plained a strong 001 and 003 reflection and absence of the 002 reflection, the easy dehydration without change of layer structure, and the additional $2H_2O$ in the highly hydrated form. Later Hendricks and Jefferson[44] suggested that the water molecule in this layer had a definite configuration (Fig. 4-11).

[44] Hendricks, S. B., and M. E. Jefferson, Structures of Kaolin and Talc-Pyrophyllite Hydrates and Their Bearing on Water Sorption of the Clays, *Am. Mineralogist,* **23**: 863–875 (1938).

Diffraction data for halloysite are not suited for detailed structural study, but the intensities of the basal reflections are in accord with Hendricks' suggested structure. The transition to the dehydrated form is due to the loss of the interlayer of water molecules.

Halloysites dried at 110°C have strong reflections for spacing at about 7.2 and 3.6 Å, corresponding to the 001 and 002 reflections from kaolinite. However, as compared with kaolinite, these reflections usually show considerable broadening. In addition there are broad diffraction bands which can be indexed as hk diffraction. The absence of sharp hkl reflections could be explained by an irregular stacking sequence. On this basis this form of halloysite consists of layers similar to those in kaolinite but displaced in a highly random manner parallel to both a and b directions. The displacements are thought to be simple fractions of the cell dimensions $na_0/6$ and $mb_0/6$. As will be seen presently, halloysite appears to consist of curved units, and the curvature could only account for the hk *bands*.

Brindley and his colleagues[45-47] studied the transition of the hydrated to the dehydrated form of the mineral in great detail. According to them, only partial dehydration takes place at low temperatures (60 to 75°C), and temperatures of the order of 400°C are necessary for complete removal of the interlayer water and the development of the 7.2-Å spacing. At temperatures of 60 to 75°C or at a lower temperature for longer time, a partially dehydrated form develops which tends to persist and has, therefore, considerable stability. The partially hydrated form has a basal spacing of 7.36 to 7.9 Å, corresponding to a hydration of about 0.5 to $1.5H_2O$. The partially hydrated form consists of a statistical distribution of hydrated layers and nonhydrated layers; the maximum value would correspond to a little more than one water layer for every four kaolinite layers. Brindley[19] stated that examples with hydrations ranging between $1.5H_2O$ and about $3H_2O$ have not been observed. Equilibrium studies by Bates[48] and by Roy and Osborn[49] indicate only two stable forms so that natural halloysites with 2.2 to $2.3H_2O$ must be regarded as very persistent metastable forms.

[45] Brindley, G. W., and K. Robinson, X-ray Studies of Halloysite and Metahalloysite, I, The Structure of Metahalloysite, *Mineral. Mag.,* **28:** 393–407 (1948).

[46] Brindley, G. W., and J. Goodyear, X-ray Studies of Halloysite and Metahalloysite, II, The Transition of Halloysite to Metahalloysite in Relation to Relative Humidity, *Mineral. Mag.,* **28:** 407–422 (1948).

[47] Brindley, G. W., K. Robinson, and J. Goodyear, X-ray Studies of Halloysite and Metahalloysite, III, Effect of Temperature and Pressure on the Transition from Halloysite to Metahalloysite, *Mineral. Mag.,* **28:** 423–428 (1948).

[48] Bates, T. F., Structure and Genesis in the Kaolinite Group, Problems of Clay and Laterite Genesis, pp. 144–153, American Institute of Mining, and Metallurgical Engineers, (1952).

[49] Roy, R., and E. F. Osborn, The System $Al_2O_3\text{-}SiO_2\text{-}H_2O$, *Am. Mineralogist,* **39:** 853–885 (1954).

Bates et al.[50] showed the tubular nature (see Chap. 6) of the halloysite minerals from electron micrographs and suggested that the $4H_2O$ form consists of tubes made up of overlapping, curved sheets of the kaolinite type, with the c axis for any point on the tube nearly perpendicular to a plane tangent to the tube at that point. The axis of the tube may be parallel to either the a or b axis, or possibly to any intermediate crystallographic direction in the plane of the sheets. In the dehydration to the $2H_2O$ form the tubes frequently collapse, split, or unroll.

The difference in effective size of the O associated with the silicon and of the OH associated with the aluminum has been stated previously (pages 53 and 62). Bates[50] stated that the a dimension in the O plane is 5.14Å and in the OH plane is 5.06 Å and that the b dimension in the O plane is 8.93 Å and in the OH plane is 8.62 Å. According to him, this difference in the dimensions of the upper and lower plane of the silica-alumina layer causes a curvature of the layers, with a radius in accordance with the dimensions of the observed tubes (see Fig. 4-12). The interlayer water molecules, which cause a weak bond between successive layers, would favor the development of curvature in hydrated halloysite. The regular stacking and close spacing of layers cause a relatively strong bond between successive layers in kaolinite, and curvature would not be favored.

Gastuche and others[51] subjected kaolinite to boiling nitrobenzene for 305 hr; they reported that they had broken the hydrogen bonds between the unit layers and that following this, the kaolin layers had rolled into tubes. Oberlin and Tchoubar[52] subjected kaolinite particles to successive cycles of wetting and drying and observed the curling of the edges of thin flakes.

Although electron micrographs of halloysite seem satisfactorily to document Bates's morphology, and recent electron-diffraction data (see below) are in accord, there are disturbing characteristics which are difficult to understand. For example, Pundsack[53] very carefully determined the density of the hydrated form and concluded that these data were not in accord with tubular forms. Further, some examples have an angular rather than a rounded cross section (Hofmann et al.[54]), and electron micrographs of some elongate units show saw-toothed edges suggesting twinning.

[50] Bates, T. F., F. A. Hildebrand, and A. Swineford, Morphology and Structure of Endellite and Halloysite, *Am. Mineralogist,* **35**: 463–484 (1950).

[51] Gastuche, M. C., J. Delvigne, and J. J. Fripiat, Altération chimique des kaolinites, *Compt. Rend. Congrs. Intern. Sci. du Sol,* II, pp. 439–456 (1954).

[52] Oberlin, A., and C. Tchoubar, Examination of Certain Minerals of the Kaolinite Group by Electron Microdiffraction, *Compt. Rend.,* **248**: 3184–3186 (1959).

[53] Pundsack, F. L., Density and Structure of Endellite, *Natl. Acad. Sci. Publ.* **566**, pp. 129–135, New York (1958).

[54] Hofmann, U., S. Morcos, and F. W. Schembra, The Very Strange Clay Mineral—Halloysite, *Ber. Deut. Keram. Ges.,* **39**: 475–482 (1962).

Waser[55] analyzed mathematically the scattering effects of tubular objects and obtained results which correlated with experimental data obtained by electron diffraction from single crystals (or tubes) of halloysite. He concluded that the halloysite tube was composed of curved sheets of a layered structure and that the layers had random displacement with respect to each other.

Single-crystal electron-diffraction studies of halloysites have been made

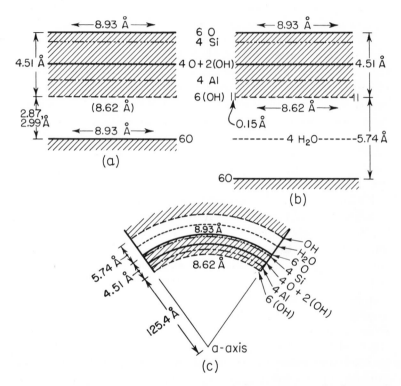

Fig. 4-12. Diagrammatic representation of the structure of kaolinite and halloysite $4H_2O$ after Bates et al.[50]   (*a*) Arrangement of layers in kaolinite; (*b*) arrangement of layers in halloysite according to Hendricks;[36] (*c*) proposed arrangement of halloysite layers.

also by Honjo and coworkers,[56] with results suggesting that kaolinite sheets may be rolled about different axes to build up tubes. Some of the "single crystal" diagrams obtained by Honjo and his coworkers indicate some three-dimensional order "higher than that appropriate for halloysite."

[55] Waser, J., The Scattering of Tubular Objects, *Natl. Acad. Sci., Publ.* 395, p. 65 (1955).

[56] Honjo, G., N. Kitamura, and K. Mihama, Study of Clay Minerals by Single-crystal Electron Diffraction Diagrams—the Structure of Tubular Kaolin, *Clay Minerals Bull.,* **2**(12): 133–141 (1954).

Bates[57] concluded, on the basis of a detailed analysis of the chemical composition of a considerable number of samples of kaolinite and halloysite, that the only definitely significant difference is that halloysite contains more H_2O+ than kaolinite. He also concluded that in some way the excess water causes the interlayer bonds to become too weak to overcome the misfit and that curved layers or tubes are to be expected. In kaolinite, the interlayer bonds are presumably strong enough to stretch the gibbsite sheet of one layer to fit the Si-O sheet of the adjacent layer, thereby overcoming the misfit and producing platey crystals of limited size. Bates suggested that the transition from kaolinite to halloysite ($2H_2O$) is simply the first part of a series leading to halloysite ($4H_2O$) and allophane as the H_2O content is increased. Radoslovich[58] disagreed with Bates et al.[50] and Bates[57] that the curvature found in halloysite is a consequence of the supposed misfit between a larger tetrahedral layer and a smaller octahedral layer. Radoslovich pointed out that, if the tetrahedral layer of a layer silicate would on its own accord exceed the dimensions of the neighboring octahedral layer, then the former may contract very readily and quite markedly simply by tetrahedral rotations, leading to ditrigonal rather than hexagonal surface symmetry so that either there would be no mismatch with adjacent octahedral layers or at least the stresses due to any mismatch would be of very secondary importance and inadequate to explain the tubular morphology. As Radoslovich pointed out, nevertheless halloysites form tubes and this implies unequal stresses at different levels in the kaolinite layers. According to him, it appears likely that the unbalanced stresses are the result of expansion due to Al—Al repulsion across shared octahedral edges, and secondly to a contraction of surface hydroxyls within the layer, brought on probably by OH—OH bonds in the hydroxyl triads around vacant octahedral sites. He concluded that halloysite has clearly the possibility of some OH—OH distances being shortened around unoccupied sites and that this could happen on only one side of the octahedral layer. The net result should be an unbalanced pair of forces leading to a tubular morphology. According to Radoslovich, the wide variations in morphology with respect to crystallinity are not in the least surprising since the pattern of OH—OH forces may well be systematically related to the octahedral network in some specimens so that one direction, for example, the *b* axis, is the preferred tubular axis.

Various investigators seem to have evidence for a tubular mineral with a high degree of crystalline order. Thus De Keyser and Degueldre[59] described

[57] Bates, T. F., Morphology and Crystal Chemistry of 1:1 Layer Lattice Silicates, *Am. Mineralogist*, **44**: 78–114 (1959).

[58] Radoslovich, E. W., The Cell Dimensions and Symmetry of Layer-Silicates, VI, Serpentine and Kaolin Morphology, *Am. Mineralogist*, **48**: 368–378 (1963).

[59] De Keyser, W., and L. Degueldre, Relations Between the Morphology and Structure of Kaolins and Halloysites, *Bull. Soc. Belge Geol., Paleontol., Hydrol.*, **63**: 100–110 (1954).

a material from Les Eyzies, France, which electron micrographs show to be composed of tubes but X-ray-diffraction diagrams indicate to be a kaolinite with a high degree of crystallinity. Visconti and collaborators[60] described kaolin clays from Brazil which show tubular forms in electron micrographs but give sharp X-ray-diffraction patterns. Brindley[4] offered two possible explanations for this apparently contradictory situation. As he had shown earlier (Brindley and Comer[61]), a synthetic mixture of equal amounts of kaolinite and tubular halloysite gave an X-ray powder diagram scarcely distinguishable from that of pure kaolinite but an electron micrograph in which tubular particles were very noticeable. In other words, tubular particles may be very obvious in samples in an electron micrograph but have relatively little influence on an X-ray-diffraction pattern. The second possible explanation based on the work of Honjo et al.[56] is that some tubular particles may consist of kaolinite sheets arranged in an orderly fashion with respect to the crystallographic axis parallel to the tube length which then could give diffraction characteristics similar to those for well-crystallized kaolinite. Santos and Brindley[62] described a tubular mineral from Brazil for which selected-area electron diffraction of single fibers shows some degree of regularity in the layer structure.

Von Engelhardt[63] described a kaolin which is intermediate in structure between halloysite and poorly crystallized kaolinite. It has shifts parallel to both the a and b axes, but not as random as in halloysite. Electron micrographs show no tubes and good hexagonal flakes.

Drits and Kashaev[8] concluded, on the basis of X-ray-diffraction studies, that there is a continuous series of structural transitions from perfect kaolinite to imperfect kaolinite and that the nature of the imperfect structure is determined by the imperfect structure of the individual layers and the irregularities of superposition of the layers.

Brindley et al.[64] also presented X-ray-diffraction data of monomineralic samples, indicating that a continuous sequence exists with increasing amounts of layer-stacking disorder from kaolinite to halloysite. These authors showed

[60] Visconti, Y. S., B. N. F. Nicot, and E. G. de Andrade, Tubular Morphology of Some Brazilian Kaolins, *Am. Mineralogist*, **41**: 67–75 (1956).

[61] Brindley, G. W., and J. J. Comer, Structure and Morphology of Kaolin Clay from Les Eyzies, *Natl. Acad. Sci., Publ.* 456, pp. 61–66 (1956).

[62] Santos, P. de Souza, G. W. Brindley, and H. de Souza Santos, Mineralogical Studies of Kaolinite-Halloysite clays, III, *Am. Mineralogist*, **50**: 619–628 (1965).

[53] Von Engelhardt, W., Ein Tonminerale der Kaolinit-Halloysitgruppe von Provins (Frankreich), *Heidelberger Beitr. Mineral Petrog.*, **4**: 319–324 (1954).

[64] Brindley, G. W., P. de Souza Santos, and H. de Souza Santos, Mineralogical Studies of Kaolinite-Halloysite Clays, I, Identification Problems, *Am. Mineralogist*, **48**: 897–910 (1963).

that a sequence of morphological forms from platey through lath-like and curved forms to fully rolled and tubular forms also exists. They suggested that the name halloysite should be applied to materials that have a "significant" degree of rolling or curling. They recognized that this distinction may not always be satisfactory because platey forms may exist with a high degree of disorder with respect to both a and b axes; consequently, these forces would have X-ray-diffraction characteristics of most halloysites with the morphology of kaolinite.

Most chemical analyses reported for halloysite minerals suggest very little or no substitutions within the lattice. However, Von Engelhardt[63] reported a sample containing 3.5 percent Fe_2O_3, and Chukhrov et al.[65] a sample with 5.09 percent CuO; in both cases the metals are believed to be in the lattice. As in the case of kaolinite, possible limits of substitution within the halloysite structure are not yet known.

In summary it is important to note that the whole body of detailed and somewhat discordant descriptions related to kaolinites and halloysites in the literature cited does not obviate the fact that an essentially invariant $OH_{10}Al_4Si_4O_8$ unit actually characterizes all the deduced or postulated combinations. As will be indicated later (see Chap. 12), infrared-absorption data cannot be fully reconciled with the postulated kaolinite structures. There is, therefore, much yet to be learned concerning these structures.

Smectite minerals

In the first edition of this volume, montmorillonite was used both as a group name and for the aluminous member of the group. In this edition smectite is used as the group name. The smectite minerals occur only in extremely small particles so that precise and detailed diffraction data are very difficult to obtain. The first structural concepts were deduced from powder data and inferences from better-known structures. In recent years electron- and X-ray-diffraction studies of selected areas and the use of oriented aggregates have added substantial detailed structural data. Currently the generally accepted structure for the smectite minerals follows the original suggestion made in 1933 by Hofmann, Endell and Wilm[66] modified by later suggestions of

[65] Chukhrov, F. V., S. I. Berkhin, and V. A. Moleva, Cuprous Clay Minerals, *Dokl. k Sobraniyu Mezhdunar. Komiss. po Izuchen. Glin, Akad. Nauk SSSR,* pp. 29–44 (1960).

[66] Hofmann, U., K. Endell, and D. Wilm, Kristallstruktur und Quellung von Montmorillonit, *Z. Krist.,* **86:** 340–348 (1933).

Maegdefrau and Hofmann,[67] Marshall,[68] and Hendricks.[40]   According to this concept, smectite is composed of units made up of two silica tetrahedral sheets with a central alumina octahedral sheet.   All the tips of the tetrahedrons point in the same direction and toward the center of the unit.   The tetrahedral and octahedral sheets are combined so that the tips of the tetrahedrons of each silica sheet and one of the hydroxyl layers of the octahedral sheet form a common layer.   The atoms common to both the tetrahedral and octahedral layer become O instead of OH.   The layers are continuous in the *a* and *b* directions and are stacked one above the other in the *c* direction.

In the stacking of the silica-alumina-silica units, O layers of each unit are adjacent to O layers of the neighboring units, with the consequence that there is a very weak bond and an excellent cleavage between them.   The outstanding feature of the smectite structure is that water and other polar molecules, such as certain organic molecules (see Chaps. 8 and 10), can enter between the unit layers, causing the lattice to expand in the *c* direction.   The *c*-axis dimension of smectite is, therefore, not fixed but varies from about 9.6 Å, when no polar molecules are between the unit layers, to substantially complete separation of the individual layers in some cases.   Figure 4-13 is a diagrammatic sketch of this structure of smectite.

X-ray powder diagrams given by smectites are made up of two distinct types of reflections: (1) basal 00*l* reflections exhibiting an integral series of orders varying in position according to the separation of the structural layers and (2) general reflections from the *hk*-band system which are characteristic of the mineral and, in general, are the same for all smectites, the differences being in the details of spacing and relative intensities.   MacEwan[69] described diffraction techniques whereby the two types of reflections may be separately emphasized and their resolution enhanced.

The smectite minerals show the structural distortions common to the layer silicates.   According to Radoslovich,[70] the degree of twist is relatively small, and he gave values of 6 to 13°, with nontronites rather smaller than montmorillonites.   Cowley and Goswami[71] stated that electron-diffraction data indicate bent crystal sheets of smectite in which the tetrahedral oxygen atoms are rotated from their idealized positions.

[67] Maegdefrau, E., and U. Hofmann, Die Kristallstruktur des Montmorillonits, *Z. Krist.*, **98**: 299–323 (1937).

[68] Marshall, C. E., Layer Lattices and Base-Exchange Clays, *Z. Krist.*, **91**: 433–449 (1935).

[69] MacEwan, D. M. C., Montmorillonite Minerals, "The X-ray Identification and Crystal Structures of Clay Minerals," chap. IV, pp. 143–207, Mineralogical Society of Great Britain Monograph (1961).

[70] Radoslovich, E. W., The Cell Dimensions and Symmetry of Layer Lattice Silicates, II, Regression Relations, *Am. Mineralogist,* **47**: 617–636 (1962).

[71] Cowley, J. M., and A. Goswami, Electron Diffraction Patterns from Montmorillonite, *Acta Cryst.,* **14**: 1071–1079 (1961).

Exchangeable cations occur between the silicate layers, and the *c*-axis spacing of completely dehydrated smectite depends somewhat on the size of

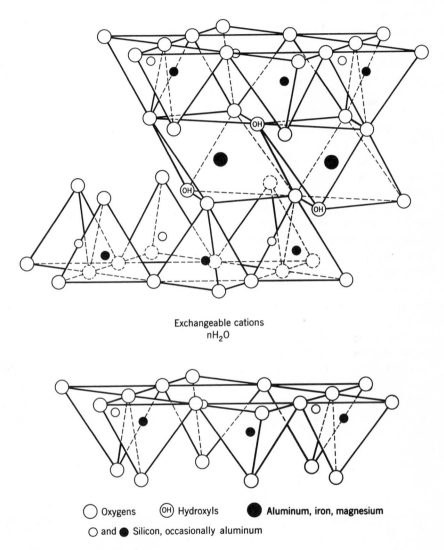

Exchangeable cations
nH$_2$O

○ Oxygens   (OH) Hydroxyls   ● Aluminum, iron, magnesium

○ and ● Silicon, occasionally aluminum

Fig. 4-13. Diagrammatic sketch of the structure of smectite according to Hofmann, Endell, and Wilm,[66] Marshall,[68] and Hendricks.[40]

the interlayer cation, being larger the larger the cation. In the case of adsorption of polar organic molecules between the silicate layers, the *c*-axis dimension also varies with the size and geometry of the organic molecule.

The thickness of the water layers between the silicate units depends on the nature of the exchangeable cation at a given water-vapor pressure (see Chap. 7). Under ordinary conditions a smectite with Na^+ as the exchange ion frequently has one molecular water layer and a c-axis spacing of about 12.5 Å; with Ca^{++} there are frequently two molecular water layers and a c-axis spacing from about 14.5 to 15.5 Å. The expansion properties are reversible. However, when the structure is completely collapsed by removal of all the interlayer polar molecules, reexpansion may proceed with difficulty.

Experiments by Mering[72] and others[73] with montmorillonite in the presence of large quantities of water suggest that with certain adsorbed cations, for example, Na^+, the unit layers completely separate but that with other cations, for example, Ca^{++} and H^+, the separation is not complete.

Numerous studies (see Weiss[74] for a general summary) have been made of the factors determining the swelling of the layer silicates. Thus, Barshad[75] found that the expansion is greater, the larger the dipole moment of the reacting liquid, but found no clear-cut relation with the dielectric constant. Foster[76] investigated the swelling of sodium-saturated montmorillonites from 12 different localities and found that the degree of swelling showed no correlation with cation-exchange capacity or with tetrahedral charge on the lattice but did show a correlation with octahedral substitution, in the direction of a decrease in swelling with an increase in octahedral substitutions. The substitution of ferrous iron had a greater depressing effect than that of ferric iron, which in turn appeared to have about the same depressing effect as magnesium. Hofmann and others[77] showed that in the absence of any charge on the lattice, as in the case of talc and pyrophyllite, there is no intercrystalline swelling. Below about 0.55 charge equivalent per unit formula, there is a very intensive intercrystalline expansion in water. Between about 0.55 and 0.65 equivalent units, even with alkali ions, the intercrystalline swelling in water remains within the limit of about 14 to 16 Å; above 0.65 equivalent

[72] Mering, J., On the Hydration of Montmorillonite, *Trans. Faraday Soc.*, **42B:** 205–219 (1946).

[73] Bradley, W. F., and R. E. Grim, Colloidal Properties of Layer Silicates, *J. Phys. Colloid Chem.*, **52:** 1404–1413 (1948).

[74] Weiss, Armin, Die innerkristalline Quellung als allgemeines Modell für Quellungsvorgänge, *Chem. Ber.*, **91:** 487–502 (1958).

[75] Barshad, I., Factors Affecting the Interlayer Expansion of Vermiculite and Montmorillonite, *Soil Sci. Soc. Am. Proc.*, **16:** 176–182 (1952).

[76] Foster, M. D., Geochemical Studies of Clay Minerals, II, Relation Between Ironic Substitution and Swelling in Montmorillonites, *Am. Mineralogist*, **38:** 994–1006 (1953).

[77] Hofmann, U., A. Weiss, G. Koch, A. Mehler, and A. Scholz, Intracrystalline Swelling, Cation Exchange, and Anion Exchange of Minerals of the Montmorillonite Group and of Kaolinite, *Natl. Acad. Sci., Publ.* **465,** pp. 273–287 (1956).

units, in the presence of potassium ions, the basal spacing in water remains at 10.3 Å, the same as in the dry state. Hight et al.[78] showed that, on the basis of small-angle X-ray-scattering investigations, in dilute aqueous suspensions, sodium montmorillonite occurred in independent layers 10 Å thick. With the addition of chloride salts of various cations, aggregation took place at varying critical concentrations.

It appears from the work of Bradley, Grim, and Clark[79] that the thickness of the water layers between successive silicate layers is an integral number of molecules. That is, the water layer is one, two, three, etc., molecular water layers thick. A natural smectite may have a regular ordering of a single thickness of water layers, or it may be a random mixture of different "hydrates." Roth[80] showed that important physical characteristics of clays composed of smectite are related to the regularity or randomness of the inter-layer water layers.

Mering[72] demonstrated that the diffraction effects of Ca montmorillonite prepared at low relative humidities (50%±) can be explained by a mixing of hydrates with thicknesses of 14 and 15 Å. Similar clays prepared at 90 percent relative humidity show only a 15-Å reflection, 15 Å being the thickness of the double-layer hydrate described by Bradley et al.[79]

Investigations of the uptake of water as a function of vapor pressure by Mooney, Keenan, and Wood[81] and Cornet[82] show clear evidence of layer formation of the water molecules.

Norrish[83] investigated in detail the sequence of hydration states of mont-morillonite carrying various exchangeable cations. In the case of the common ions, he showed that for sodium montmorillonite there is at first a stepwise increase in c-axis spacing and then (above about 40 Å) complete dissociation of the individual layers. Calcium has a considerable range of stability with two molecular layers of water with a c-axis spacing of about 15.4 Å, and it has substantially no expansion beyond 18 Å. For montmorillonites with potassium and magnesium there is only very limited expansion.

Hofmann, Endell, and Wilm[66] described the smectite structure as generally

[78] Hight, R., Jr., W. T. Higdon, H. C. H. Darley, and P. W. Schmidt, Small Angle X-ray Scattering from Montmorillonite Clay Suspensions, II, *J. Chem. Phys.*, **37**: 502–510 (1962).

[79] Bradley, W. F., R. E. Grim, and G. L. Clark, A Study of the Behavior of Montmoril-lonite on Wetting, *Z. Krist.*, **97**: 216–222 (1937).

[80] Roth, R. S., The Structure of Montmorillonite in Relation to the Properties of Certain Bentonites, Ph.D. thesis, University of Illinois (1951).

[81] Mooney, R. W., A. G. Keenan, and L. A. Wood, Adsorption of Water Vapour by Montmorillonite, *Z. Krist.*, **93**: 481–487 (1952).

[82] Cornet, I., Expansion of Montmorillonite, *J. Chem. Phys.*, **18**: 623–626 (1950).

[83] Norrish, K., Swelling of Montmorillonite, *Trans. Faraday Soc.*, **18**: 120–134 (1954).

similar to that of pyrophyllite except for the expanding-lattice characteristic. They postulated a stacking of layers to give an orthorhombic unit. Later Maegdefrau and Hofmann[67] concluded that generally the unit layers are stacked without any fixed periodical arrangement in the a and b directions.

Zvyagin and Pinsker[84] concluded, on the basis of electron-diffraction data of an oriented sample, that the mineral was monoclinic with β equal to $99°54'$ and space group C_2/m.

Later Zvyagin[85] investigated a variety of montmorillonites, beidellites, and nontronites by electron diffraction and found that in every case the ratio of the pseudohexagonal axes $b_0:a_0$ was approximately $\sqrt{3:1}$ and that the angle β of the monoclinic crystallographic cell for normal montmorillonites and nontronites varied from $100°30'$ to $101°15'$, and for beidellites it was about $105°$.

Mering and Glaeser[86] deduced that the symmetry of the layers was C_2 rather than C_2/m. Nakahira[87] concluded that detailed X-ray-diffraction bands may best be explained by supposing that there are regions of local ordering with random displacements of multiples of $b/3$.

Weiss, Koch, and Hofmann[88] described a saponite which showed definite signs of hkl reflection. These authors stated that, in expandable clay minerals, hkl interferences are likely to be observed in any mineral in which the number of exchangeable cations per unit-cell layer is greater than 1.1. Brown and Stephen[89] and Weir[90] also described smectites showing considerable stacking order. In general, it may be concluded that some montmorillonites show considerable stacking order whereas others exhibit substantially none.

Oberlin and Mering[91] showed, on the basis of selected-area electron micrography and diffraction, that the montmorillonite in Wyoming bentonites

[84] Zvyagin, B. B., and Z. G. Pinsker, Structure of Montmorillonite, *Compt. Rend. Acad. Sci. URSS,* 68: 65–67, 505–508 (1949).

[85] Zvyagin, B. B., Electronographic Study of Minerals of the Montmorillonite Group, *Dokl. Akad. Nauk SSSR,* 86: 149–152 (1952).

[86] Mering, J., and R. Glaeser, Cations échangeables dans montmorillonite, *Bull. Groupe Franc. Argiles,* 5: 61–72 (1953).

[87] Nakahira, M., Crystal Structure of Montmorillonite, *J. Sci. Res. Inst. (Tokyo),* 46: 268–287 (1952).

[88] Weiss, A., G. Koch, and U. Hofmann, Saponite, *Ber. Deut. Keram. Ges.,* 32: 12–17 (1955).

[89] Brown, G., and I. Stephen, Expanding-lattice Minerals, *Mineral. Mag.* 32: 251–253 (1959).

[90] Weir, A. H., Beidellite, Ph.D. thesis, University of London (1960).

[91] Oberlin, A., and J. Mering, Observations en microscopie et microdiffraction electroniques sur la montmorillonite Na, *J. Microscopie,* 1: 107–120 (1962).

is composed of large flakes randomly stacked and that the montmorillonite from Camp Berteaux in Algeria is composed of flakes which in turn are made up of a mozaic of books with some order in their stacking and with fairly distinct geometrical outlines. A possible explanation is that in the former case there is random population of cations in octahedral positions, whereas in the latter case there is some regularity in similar positions in the individual mosaic units.

The theoretical charge distribution, without considering lattice substitutions within the layer, is as follows:

$6O^{--}$ $12-$
$4Si^{4+}$ $16+$ $+ ↵$
$4O^{--}2(OH)^-$ $10-$ (Layer common to tetrahedral and octahedral sheets)
$4Al^{3+}$ $12+$
$4O^{--}2(OH)^-$ $10-$ (Layer common to tetrahedral and octahedral sheets)
$4Si^{4+}$ $16+$
$6O^{--}$ $12-$

Interlayer layer of H_2O or other polar molecules

The theoretical formula without considering lattice substitutions is $(OH)_4Si_8Al_4O_{20} \cdot nH_2O$ (interlayer), and the theoretical composition without the interlayer material is SiO_2, 66.7 percent; Al_2O_3, 28.3 percent; H_2O, 5 percent.

As first emphasized by Marshall[68] and Hendricks,[40] smectite *always* differs from the above theoretical formula because of substitution within the lattice of aluminum and possibly phosphorus for silicon in tetrahedral coordination and/or magnesium, iron, zinc, nickel, lithium, etc., for aluminum in the octahedral sheet. In the tetrahedral sheet the substitution of Al^{3+} for Si^{4+} appears to be limited to less than about 15 percent. In the formula noted above, only two-thirds of the possible positions in the octahedral sheet are filled. The substitution of Mg^{++} for Al^{3+} can be one for one, or three Mg^{++} for two Al^{3+}, with all possible octahedral positions being filled in the latter case. Substitutions within the octahedral sheet may vary from few to complete. Total replacement of $2Al^{3+}$ by $3Mg^{++}$ yields the mineral saponite; replacement of aluminum by iron yields nontronite; by chromium, volkhonskoite; by zinc, sauconite.

The Mg^{++} ion of diameter 0.65 Å and the Fe^{3+} ion of diameter 0.67 Å are somewhat large to fit into the lattice, and as a consequence montmorillonite minerals with large substitutions of these ions are subject to a directional strain which results in elongate lath- or needle-shaped units (see Chap. 6).

Layer minerals of this general type in which all the possible octahedral positions are filled are called octaphyllites, or trioctahedral, and those in which only two-thirds of the possible positions are filled are called heptaphyllites, or dioctahedral. Numerous analyses of smectite have shown that the substitutions within the octahedral sheet are such that the mineral is either almost exactly trioctahedral or dioctahedral and not intermediate.

Radoslovich[70] suggested that there is some ordering in the way magnesium and ferric iron enter the octahedral sites of dioctahedral smectites, and Grim and Kulbicki[92] reached a similar conclusion for some but not all such minerals. According to the latter authors, this is an important factor explaining the variation in physical properties of smectite clays.

Ross and Hendricks[93] computed the structural fit for a large number of chemical analyses of smectites. Unfortunately there are no accompanying X-ray data, and some of the samples may be mixtures of clay minerals. However, the number of analyses is so great that their conclusion seems well established, namely, that the number of ions in 6 coordination, i.e., in the octahedral positions, lie in two distinct ranges, 4.00 to 4.44 and 5.76 to 6.00. Further, it appears that, if the mineral is dioctahedral, there may be considerable variation in the exact position of the aluminums or other atoms in the possible octahedral positions.

Another way in which smectite *always* differs from the theoretical formula, is that the lattice is *always* unbalanced by the substitutions noted above, that is, Mg^{++} for Al^{3+}, Al^{3+} for Si^{4+}, etc. The unbalancing may result from substitution of ions of different valence in the tetrahedral or octahedral sheet or both. Unbalancing in one sheet may be compensated for in part, but only in part, by substitution in the other sheets of the unit layer. Thus, the substitutions of Al^{3+} for Si^{4+} may in part be compensated by filling slightly more than two-thirds of the octahedral positions. Compensation also may be by substitution of OH for O in the octahedral layer. It is significant that the substitutions in the smectite lattice, with the internal compensating substitutions, always result in about the same net charge on the lattice. Many analyses have shown this to be about 0.66 — per unit cell. This net-charge deficiency is balanced by exchangeable cations adsorbed between the unit layers and around their edges (see Chap. 7, Ion Exchange). This charge deficiency corresponds to about two-thirds unit per unit cell. It would require the substitution of one Mg^{++} for every sixth Al^{3+}, for example, or about one out of every six Si^{4+} by Al^{3+}.

Greene-Kelly[94] found that the lithium ion enters the vacant octahedral position in dioctahedral montmorillonites on treating with the lithium salt

---

[92] Grim, R. E., and G. Kulbicki, Montmorillonite: High Temperature Reaction and Classification, *Am. Mineralogist,* **46:** 1329–1369 (1961).

[93] Ross, C. S., and S. B. Hendricks, Minerals of the Montmorillonite Group, *U.S. Geol. Surv., Profess. Paper* 205B, pp. 23–47 (1945).

[94] Greene-Kelly, R., Dehydration of the Montmorillonite Minerals, *Mineral. Mag.,* **30:** 604–615 (1955).

at a moderately elevated temperature (300°C), thereby balancing any charge deficiency in the octahedral layer with a consequent reduction in intercrystalline swelling. Similar treatment of trioctahedral smectite had no effect on the swelling of the mineral. White[95] also was able to place lithium in vacant octahedral positions in dioctahedral montmorillonites.

Mering and Glaeser[96] pointed out that the cations compensating for lattice unbalance would tend to take a position as close to the negative charge as possible. Thus, in the case of monovalent cations, for example, sodium, they would go into the hexagonal hole, whereas in the case of divalent calcium balancing two charges the cation would go into the hole nearest to one negative charge. Unlike the sodium ion, which does not hydrate, the calcium ion in the presence of water tends to leave the hexagonal hole and become hydrated if it is present in amounts greater than the charge due to the substitution of alumina for silica. With lesser amounts of Ca^{++}, it remains in the hole and does not affect solvation. On the basis of electron-diffraction data, Pezerat and Mering[97] confirmed the findings that sodium tends to remain in the hole even with sufficient water present to develop a one-layer hydrated structure, whereas under similar conditions the calcium ion reaches the water layer.

Smectite has been synthesized from pure hydrous mixtures of magnesia and silica, in which case the charge deficiency cannot be due to lattice substitutions. It must be the result of vacancies in the lattice, and such vacancies probably also occur in natural minerals.

Ross and Hendricks[93] computed the structural formula from the chemical composition of many smectites, and these have shown something of the range of substitutions within the lattice and the partial compensating substitutions within the structure. The nomenclature of the members of the smectite group depends on the substitution within the lattice, and Ross and Hendricks[93] assigned the names in Table 4-3 to smectites with the compositions indicated. The names conform closely to general usage within the exception of beidellite, which many investigators consider to be discredited. Arrows are placed under the group having the charge deficiency, which requires the addition of a cation external to the silicate layer to balance the structure. In each case the balancing external cation has been indicated as Na^+ for convenience. The water or other polar molecules between the silicate layers are omitted from the formulas.

[95] White, J. L., Reaction of Molten Salts with Layer-Lattice Silicates, *Natl. Acad. Sci., Publ.* 456, pp. 133–146 (1955).

[96] Mering, J., and R. Glaeser, Sur le rôle de la valence des cations échangeables dans la montmorillonite, *Bull. Soc. Franc. Mineral. Crist.*, **77**: 519–530 (1954).

[97] Pezerat, H., and J. Mering, Detection of the Exchangeable Cations of Montmorillonite with the Help of Differential Series, *Bull. Groupe Franc. Argiles*, **10**: 25–26 (1958).

Greene-Kelly[94] suggested that the term beidellite be applied to minerals in which the structural charge arises mainly from the tetrahedral layers and that montmorillonite be applied to minerals in which structural charge arises mainly from the octahedral layer. On the basis of this concept, therefore, there would be a complete gradation from nontronite to beidellite. The trioctahedral smectites that have been described are primarily those in which there is complete substitution of magnesia for alumina. In saponite with

Table 4-3   *Formulas for some members of the smectite group*
(Suggested by Ross and Hendricks[93])

Dioctahedral (heptaphyllitic) smectite

Montmorillonite

$$(OH)_4 Si_8 (Al_{3.34} \cdot Mg_{0.66}) O_{20}$$
$$\downarrow$$
$$Na\ 0.66$$

Beidellite

$$(OH)_4 (Si_{6.34} \cdot Al_{1.66}) Al_{4.34} O_{20}$$
$$\downarrow$$
$$Na\ 0.66$$

or
Beidellite

$$(OH)_4 (Si_6 \cdot Al_2) Al_{4.44} O_{20}$$
$$\downarrow$$
$$Na\ 0.66$$

Nontronite

$$(OH)_4 (Si_{7.34} \cdot Al_{0.66}) Fe_4^{3+} O_{20}$$
$$\downarrow$$
$$Na\ 0.66$$

or
Nontronite (aluminian)

$$(OH)_4 (Si_{6.34} \cdot Al_{1.66}) Fe_{4.34}^{3+} O_{20}$$
$$\downarrow$$
$$Na\ 0.66$$

Trioctahedral (octaphyllitic)

Hectorite

$$(OH)_4 (Si_8 (Mg_{5.34} \cdot Li_{0.66}) O_{20}$$
$$\downarrow$$
$$Na\ 0.66$$

Saponite

$$(OH)_4 (Si_{7.34} \cdot Al_{0.66}) Mg_6 O_{20}$$
$$\downarrow$$
$$Na\ 0.66$$

or
Saponite (aluminian)

$$(OH)_4 (Si_{6.66} \cdot Al_{1.34}) (Mg_{5.34} \cdot Al_{0.66}) O_{20}$$
$$\downarrow$$
$$Na\ 0.66$$

the structural charge arising primarily from the substitution of aluminum for silicon in the tetrahedral layer the theoretical formula would be as follows:

$$(Si_{7.33} Al_{0.67})^{IV} (Mg_6)^{VI} O_{20} (OH)_4$$
$$\downarrow$$
$$Mg_{0.67}^+$$

In hectorite, which has no aluminum, the structural charge results from a substitution of lithium for magnesium. The mineral stevensite, a substantially pure trioctahedral hydrous magnesium silicate with a structural charge due to deficiencies in the octahedral positions, has been described by Strese and Hofmann[98] and Faust and Murata.[99] Brindley[100] interpreted X-ray-diffraction data for stevensite as indicating a mixed-layer mineral. A careful reexamination of a number of samples by Faust, Hathaway, and Millot[101] suggested a defect structure due to a deficiency of octahedral cations, and since the deficiency is large the mineral is best grouped with the smectites. The literature contains the names of a considerable number of dioctahedral smectites with a wide range of composition, for example, the chromiferous variety volkhonskoite, the zinc variety sauconite, and the nickel variety pimelite (see MacEwan[69]). In some cases the monomineralic character of the specimens is not established.

MacEwan[102] considered that the relation of the a and b dimensions of the smectite unit is due to variations in its chemical composition and offered a formula for computing b_0 from which a_0, of course, can also readily be computed. Brindley and MacEwan[103] suggested a modified semiempirical formula for such computations for mica- and kaolin-type minerals but stated that it has questionable value for the beidellite-montmorillonite minerals. Radoslovich and Norrish[5] concluded that the modified formula gives good results when applied to some minerals, but not to others, and that the results of recent structural analyses indicate that a new b-axis formula is required for general application to the layer silicates. These authors pointed out that the following factors were omitted by Brindley and MacEwan[103] and that more recent structural data indicate that they are important:

1. No account was taken of the effect of interlayer cation.

2. No factor was introduced for varying octahedral layer thicknesses (Bradley[104]).

[98] Strese, H., and U. Hofmann, Synthese von Mg-Silikatgelen, *Z. Anorg. Chem.*, **247:** 65–95 (1941).

[99] Faust, G. T., and K. J. Murata, Stevensite Redefined, *Am. Mineralogist*, **38:** 973–987 (1953).

[100] Brindley, G. W., Stevensite, a Mixed-layer Mineral, *Am. Mineralogist*, **40:** 239–247 (1955).

[101] Faust, G. T., J. C. Hathaway, and G. Millot, Restudy of Stevensite, *Am. Mineralogist*, **44:** 342–370 (1959).

[102] MacEwan, D. M. C., The Montmorillonite Minerals (Montmorillonoids), "X-ray Identification and Structures of Clay Minerals," chap. IV, pp. 86–137, Mineralogical Society of Great Britain Monograph (1951).

[103] Brindley, G. W., and D. M. C. MacEwan, Structural Aspects of the Mineralogy of Clays, Ceramics, A Symposium, pp. 15–59, British Ceramics Society (1953).

[104] Bradley, W. F., Current Progress in Silicate Structures, Proceedings of the 6th National Clay Conference, pp. 18–25, Pergamon Press, New York (1959).

3. Some corrections may be required because the charge balance for the interlayer cation is sometimes in the tetrahedral and sometimes in the octahedral site.

4. The expansion due to increased ionic size is computed by comparing, for example, dioctahedral $Al(OH)_3$, which has two trivalent cations, with trioctahedral $Mg(OH)_2$, which has three divalent cations, whereas some minerals either are intermediate in the number of octahedral cations or differ in octahedral charge or both.

5. The three octahedral sites (per one-layer cell) are treated as similar, whereas the accepted space groups imply that they are crystallographically distinct for the common mica polymorphs.

These authors developed a new b-axis formula, using regression analysis of b against composition. For details on this matter, the works of Radoslovich and his colleagues should be consulted.

The molecular silica-to-alumina ratio in smectites ranges from about 3:2 corresponding to no substitutions for aluminum in the octahedral layer, and substitutions of aluminum for silicon in the tetrahedral layer adequate to account for the cation-exchange capacity, to 4.8:1 corresponding to no substitutions in the tetrahedral layer and adequate substitutions for aluminum in the octahedral layer to account for the cation-exchange capacity. There appears to be a continuous range between these extremes. This ratio may be extended somewhat, particularly by increasing the population of the octahedral positions slightly above four and partially balancing the positive charges by increasing the aluminum-for-silicon substitution. The foregoing statements apply to dioctahedral varieties and to those species where there is limited small substitution of ferric iron for aluminum.

Iron can apparently proxy for aluminum in all positions in the octahedral layer and not at all in the tetrahedral layer. Iron-rich varieties, i.e., nontronite, on the basis of currently available analyses, show little replacement of Fe^{3+} by Mg^{++}, and the charge deficiency seems to result mainly from substitutions of Al^{3+} for Si^{4+}. In the trioctahedral smectites the charge deficiency results mainly from substitutions of Al^{3+} for Si^{4+}. It would seem from data of Ross and Hendricks[93] that up to about one atom of aluminum or iron per unit cell may be present in the octahedral layer of trioctahedral smectites. The excess charge in the octahedral layer is balanced by deficiencies in the tetrahedral layer due to larger replacements of Al^{3+} for Si^{4+}.

The opinion has frequently been expressed that the structure suggested by Hofmann et al.,[66] Marshall,[68] and Hendricks[40] does not adequately account for all the properties of montmorillonite, notably its ion-exchange capacity. Edelman and Favejee[42] suggested an alternative structure that, it is claimed, explains these properties more adequately. This structure differs from that of Hofmann et al.[66] in that every other silica tetrahedron in both silica sheets is inverted, so that half of them point in the opposite direction. Those that point away from the silicate sheet would have the tip O replaced by OH. In this structure the silicon atoms are not all in a single plane in the silica tetrahedral sheets, and there must be some substitutions of OH for O in the octahedral layers to balance the structure (Figs. 4-14 and 4-15). The charge distribution is as follows:

$2(OH)^-$	$2-$
$2Si^{4+}$	$8+$
$6O^{--}$	$12-$
$2Si^{4+}$	$8+$
$2O^{--}4(OH)^-$	$8-$
$4Al^{3+}$	$12+$
$2O^{--}4(OH)^-$	$8-$
$2Si^{4+}$	$8+$
$6O^{--}$	$12-$
$2Si^{4+}$	$8+$
$2(OH)^-$	$2-$

Interlayer H_2O or other polar groups

The structural formula is $(OH)_{12}Si_8Al_4O_{16} \cdot H_2O$ (interlayer). In this structure, no substitution is necessary within the lattice to account for the exchange capacity. The lattice may be completely balanced. The exchange reaction is believed due primarily to replacement of the H of the OH of the projecting tetrahedrons. As originally postulated, there are far more available OH groups than required by the exchange capacity. Edelman and Favejee[42] postulated that only some of them were available for exchange. Evidence from X-ray-diffraction data based on Fourier syntheses,[105] chemical data indicating lattice substitution, and careful dehydration studies all are strongly against the original Edelman and Favejee[42] structure. Edelman[106] suggested a modified structure in which only about 20 percent of the tetrahedrons are inverted, which would satisfy the discrepancy with the exchange capacity. It is questionable whether or not X-ray data would be adequate to provide evidence for or against the modified structure. However, the modified structure still does not seem to fit with chemical data indicating lattice replacements and with the geometry of the adsorption of organic molecules. Studies of the methylation of certain organics during their adsorption by montmorillonite have led some workers (Berger,[107] Duell[108]) to favor this

[105] Brown, G., A Fourier Investigation of Montmorillonite, *Clay Minerals Bull.*, **4**: 109–111 (1950).

[106] Edelman, C. H., Relations entre les propriétés et la structure de quelques minéraux argileux, *Verre Silicates Ind.*, **12**: 3–6 (1947).

[107] Berger, G., De Struktuur van Montmorilloniet, *Chem. Weekblad*, **38**: 42–43 (1941).

[108] Duell, H., G. Huber, and R. Iberg, Organische Derivate von Tonmineralien, *Helv. Chim. Acta*, **33**: 1229–1232 (1950).

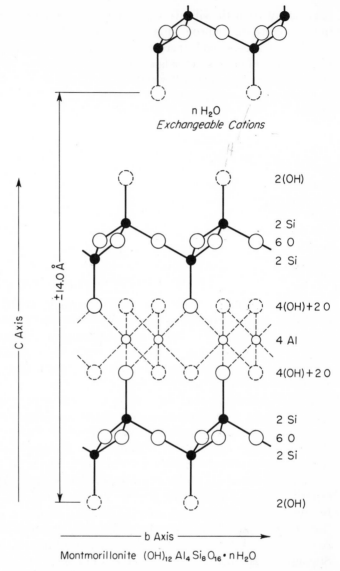

Fig. 4-14. Schematic presentation of the structure of smectite, suggested by Edelman and Favejee.[42]

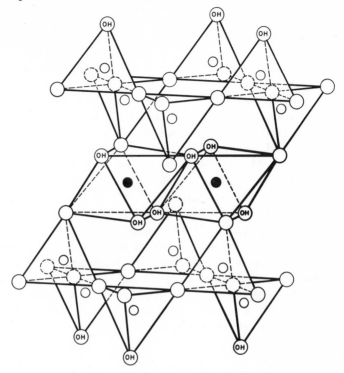

Exchangeable Cations
n H₂O

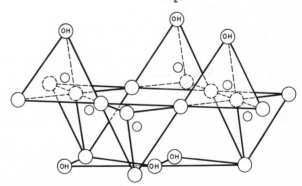

Fig. 4-15. Diagrammatic sketch of the structure of smectite, suggested by Edelman and Favejee.[42]

structure, since more OH groups seem to be called for on the surface of the montmorillonite silicate unit than are provided for by the structure postulated by Hofmann et al.[66]   Further studies are necessary before the possible structural implications of such organic-montmorillonite reactions become clear.

McConnell[109] suggested a revision in the structure postulated by Hofmann et al.[66] whereby some of the silica tetrahedrons are replaced by $(OH)_4$ tetrahedrons.   This is essentially a hole in the silicon position with adjacent balancing by OH for O.   This would provide the needed quantity of surface OH required by certain organic-adsorption characteristics and, according to McConnell, is in accord with the dehydration of the mineral. Further investigations are necessary before the feasibility of McConnell's suggestion can be evaluated.

Primarily on the basis of structural changes taking place when smectites are heated, Franzen, Muller-Hesse, and Schwiete[110] suggested the following structure.   Instead of tetrahedral and octahedral layers alternating in the layered package of smectite, two tetrahedral layers, tied together, as in α-quartz, alternate with an octahedral layer.   The water molecules lie between the octahedral layer of one package and the silica layer of the next.   The corresponding structural formula is $Al_2Si_4O_9(OH_4)$.   No diffraction evidence for this structure seems to have been presented.

Illite minerals

The structure of the micas, the variations in their composition, and their polymorphic variations have been worked out in considerable detail by Pauling,[1] Mauguin,[111,112] Jackson and West,[113,114] Winchell,[115] Hendricks and

[109] McConnell, D., The Crystal Chemistry of Montmorillonite, *Am. Mineralogist,* **35:** 166–172 (1950).

[110] Franzen, G., H. Muller-Hesse, and H. E. Schwiete, The Structure of Montmorillonite, *Naturwissenschaften,* **42:** 176 (1955).

[111] Mauguin, C. H., Etude des micas au moyens des rayons X, *Compt. Rend.,* **185:** 288–291 (1927).

[112] Mauguin, C. H., Etude des micas au moyens des rayons X, *Bull. Soc. Franc. Mineral.,* **51:** 285–332 (1928).

[113] Jackson, W. W., and J. West, The Crystal Structure of Muscovite, *Z. Krist.,* **76:** 211–227 (1930).

[114] Jackson, W. W., and J. West, The Crystal Structure of Muscovite, *Z. Krist.,* **85:** 160–164 (1933).

[115] Winchell, A. N., Studies in the Mica Group, *Am. J. Sci.,* ser. 5, **9:** 309–327, 415–430 (1925).

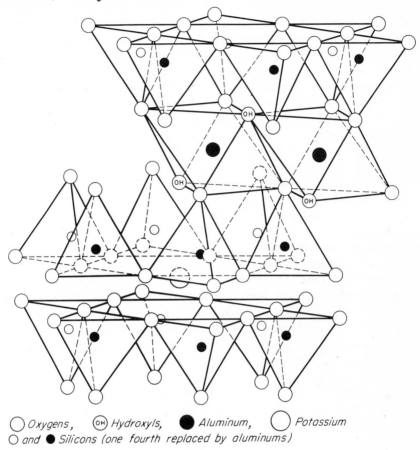

○ *Oxygens,*   ⊙ *Hydroxyls,*   ● *Aluminum,*   ◯ *Potassium*
○ *and* ● *Silicons (one fourth replaced by aluminums)*

Fig. 4-16.   Diagrammatic sketch of the structure of muscovite.

Jefferson,[116] Smith and Yoder,[117] and Radoslovich,[118]   The basic structural unit is a layer composed of two silica tetrahedral sheets with a central octahedral sheet.   The tips of the tetrahedrons in each silica sheet point toward the center of the unit and are combined with the octahedral sheet in a single layer with suitable replacement of OH by O.   The unit is the same as that for montmorillonite *except* that some of the silicons are always replaced by

[116] Hendricks, S. B., and M. Jefferson, Polymorphism of the Micas with Optical Measurements, *Am. Mineralogist,* **24:** 729–771 (1939).

[117] Smith, J. V., and H. S. Yoder, Studies of Mica Polymorphs, *Mineral. Mag.,* **31:** 209–235 (1956).

[118] Radoslovich, E. W., The Structure of Muscovite, KAl₂(Si₃Al)O₁₀(OH)₂, *Acta Cryst.,* **13:** 919–930 (1960).

aluminums and the resultant charge deficiency is balanced by potassium ions. In many of the well-crystallized micas one-fourth of the silicons are replaced by aluminums so that the charge deficiency is about 2 per unit cell. The unit layers extend indefinitely in the a and b directions and are stacked in the c direction. The potassium ions occur between unit layers (Fig. 4-16).

Muscovite is dioctahedral; i.e., only two-thirds of the possible octahedral positions are filled, and the octahedral sheet is populated only by aluminums. The structural formula is $(OH)_4K_2(Si_6 \cdot Al_2)Al_4O_{20}$, and the theoretical composition is K_2O, 11.8 percent; SiO_2, 45.2 percent; Al_2O_3, 38.5 percent; H_2O, 4.5 percent.

The biotite micas are trioctahedral with the octahedral positions populated mostly by Mg^{++}, Fe^{++}, and/or Fe^{3+}. Examples are biotite, $(OH)_4K_2(Si_6Al_2)(Mg \cdot Fe)_6O_{20}$, with the relative abundance of the iron and magnesium varying widely, and phlogopite, $(OH)_4K_2(Si_6Al_2)Mg_6O_{20}$. Hendricks and Jefferson[116] showed the existence of at least six polymorphic variations of biotite, due to variations in the number of silica-alumina-silica units per unit cell and in the manner of stacking of the unit cells. Unit cells composed of 1, 2, 3, 6, and 24 silica-alumina-silica units are known, with stacking yielding monoclinic, rhombohedral, or triclinic forms. Later Smith and Yoder[117] studied the various mica polymorphs in detail and introduced a convenient system of short designations as 1M, 2M₁, and 2M₂, 3T, etc., for one-layer monoclinic, two kinds of two-layer monoclinic, three-layer trigonal, etc., structures.

It is important that, in the large well-crystallized micas with no imperfections in the regularity of stacking, there seem to be no isomorphous gradations from dioctahedral to trioctahedral types.

The charge distribution within the layers of the well-crystallized micas is as follows:

K^+	$1+$
$6O^{--}$	$12-$
$3Si^{4+}1Al^{3+}$	$15+$
$4O^{--}2(OH)^-$	$10-$
$4Al^{3+}$ (Dioctahedral) or	
$\quad$ 6R (Trioctahedral, R = Mg^{++}, Fe^{++}, Fe^{3+}, Li^+, Ti^{3+})	$12+$
$4O^{--}2(OH)^-$	$10-$
$3Si^{4+}1Al^{3+}$	$15+$
$6O^{--}$	$12-$
K^+	$1+$

Radoslovich and his associates[5,7,9,70] studied in great detail the structure of the micas, and the following statements are taken from their works. These publications should be consulted for details concerning the symmetry and cell dimensions of the various micas.

The tetrahedral layers play a secondary role in determining the b axis, rather than the dominant role previously assumed (Smith and Yoder[117]). The cell dimensions of most micas appear to be controlled largely by the octahedral layer and the interlayer cation. The surface-oxygen triads rotate until some (probably one-half) of the cation-oxygen bonds have normal bond lengths, i.e., until one-half of the oxygens "lock" onto the interlayer cations. If the octahedral layer of a mica tends to be much smaller than the tetrahedral layer, the tetrahedrons may rotate beyond this point; normal bond lengths from the surface oxygens to the interlayer cation are then attained by the latter being held with its center slightly above the top of the oxygen layers; that is, the oxygen surfaces are no longer in contact.[118]

The calculated value for the rotation of the silica tetrahedrons in biotites ranges from 7 to $9\frac{1}{2}°$, and the calculated value for the separation of the basal planes of successive layers lies between 2.5 and 2.9 Å, suggesting that successive layers are generally in contact. Similar values for muscovite are 14°41′ and 3.49 Å, indicating that successive layers are not in contact.

In the ditrigonal structure of the silica layer of muscovite, there are K—O bonds only to the six nearest oxygens, which are approximately octahedrally arranged around it. In detail, the K^+ occupies an equilibrium position determined by a complex balance system of interlocking strong bonds reaching through the adjacent layers to the K's at the next level above and below. This system of bonds is the direct consequence of $2M_1$ muscovite being a dioctahedral mineral with $2Al^{3+}$ octahedrally and with an order and arrangement of $2Si$ and $2Si_{1/2}Al_{1/2}$ tetrahedrally. This view of the role of potassium in muscovite is far removed from the earlier concepts of an ion of the right charge fitting into a hole of comfortable size. Gatineau[119] recently arrived at a somewhat different concept of the cation distribution in the muscovite structure.

The illite clay minerals[120] differ from the well-crystallized micas in several possible ways, all of which may be exhibited by a given sample; these are as follows: There is less substitution of Al^{3+} for Si^{4+}; in the well-crystallized micas one-fourth of the Si^{4+} are replaced, whereas in the illites frequently only about one-sixth are replaced. As a consequence of this smaller substitution, the silica-to-alumina molecular ratio of the illites is higher than that of the well-crystallized micas, and the net-unbalanced-charge deficiency is reduced from 2 per unit cell to about 1.3 per unit cell. The potassium ions between the unit layers may be partially replaced by other cations, possibly Ca^{++}, Mg^{++}, H^+. There is some randomness in the stacking of the layers

---

[119] Gatineau, L., Localization of Isomorphous Replacements in Muscovite, *Compt. Rend.*, **256**(22): 4648–4649 (1963).

[120] Grim, R. E., R. M. Bray, and W. F. Bradley, The Mica in Argillaceous Sediments, *Am. Mineralogist*, **22**: 813–829 (1937).

in the *c* direction, and the size of the illite particles occurring naturally is very small, of the order of 1 to 2 μ or less.

Because of the foregoing variations, illite clay minerals show differences in their characteristics from those of the well-crystallized micas.[121]   The 10-Å diffraction line of the micas is shown by illite usually as a characteristic diffraction effect, but it is often modified into a band which tends to tail off slightly toward the low-angle region as a consequence of the small particle size, variation in the interlayer cation, and occasional slight interlayer hydration.   The potassium ions may be replaced by $(H_3O)^+$, accounting, at least partially, for the reduced potassium content and also for the interlayer hydration.   Weaver[121a] showed that the K_2O content for illites is commonly 8 to 10 percent.   The occasional hydration layers may simply be the nontypical features associated with any particle surface.   Also because of these defects *hkl* reflections are weak, or absent in the case of those normally weak in the well-crystallized micas.   Most of the illite clay minerals so far described are dioctahedral, but some are trioctahedral with unit cells composed of one, two, or three silica-alumina-silica layers.[121,122]   Careful X-ray-diffraction data (see page 144) permit the identification of the polymorphic form of illite if the sample is reasonably pure and if the crystallinity is not too poor. In some cases, it is not possible with currently available analytical techniques to identify the polymorphic forms of the illites.   Levinson[123] showed that three-layer trigonal (3T), two-layer monoclinic (2M), one-layer monoclinic (1M), and one-layer monoclinic disordered (1Md) structures are to be found in specimens described in the literature as illite or hydrous mica.

Bohor[124] studied five samples of illite and in each case identified 1Md, 1M, and 2M polymorphs, with the relative amount of the polymorphs varying from sample to sample.

There are dioctahedral illites in which the aluminum is replaced by some iron and magnesium.   Whether or not there is any isomorphous relationship between the dioctahedral and trioctahedral illites is not definitely known, but none has been established thus far.

[121] Grim, R. E., W. F. Bradley, and G. Brown, The Mica Clay Minerals, "X-ray Identification and Structures of Clay Minerals," chap. V, pp. 138–172, Mineralogical Society of Great Britain Monograph (1951).

[121a] Weaver, C. E., Potassium Content of Illite, *Geol. Soc. Am., Program (Miami)*, p. 217 (1964).

[122] Walker, G. F., Trioctahedral Minerals in Soil-Clays of Northeast Scotland, *Mineral. Mag.*, **29**: 72–84 (1950).

[123] Levinson, A. A., Polymorphism Among Illites and Hydrous Micas, *Am. Mineralogist*, **40**: 41–49 (1955).

[124] Bohor, B. F., Characterization of Illite and Its Associated Mixed Layers, Ph.D. thesis, University of Illinois (1959).

Burst[125] found that material described in the literature as glauconite ranges from substantially monomineral material, with little or no mixed layering, to structures in which mixed layering is dominant. The monomineral examples are single-layer monoclinic dioctahedral micas with varying degrees of disorder, so that they would be classified from 1M to 1Md, and with a relatively high iron content ($20\% \pm$ as Fe_2O_3). The mixed-layer structures have varying amounts of expandable material and can be described as interlayering of the foregoing type of mica and smectite. Burst pointed out that the well-ordered micaceous materials consistently contain more potassium than do the disordered structures. When the potassium atom equivalent falls below approximately 1.4 per unit cell, ordered stacking becomes less apparent. Evidently disorder begins when fewer than two out of three possible potassium positions are filled. Burst also indicated that there is a transition from the low-potassium to the high-potassium varieties, which suggests that the potassium fixation is limited not by the capacity of this structure but by the capacity of the depositional medium to supply potassium to the structure. Earlier Hendricks and Ross[126] concluded that the dominant component of glauconite is a dioctahedral illite with considerable replacement of Al^{3+} by Fe^{3+}, Fe^{++}, and Mg^{++}, frequently with even less than two-thirds of the possible octahedral positions filled. There is consequently a charge deficiency in the octahedral sheet as well as in the tetrahedral sheets, and the interlayer cations seem to balance both these charges. Often Ca^{++} and Na^{++} as well as K^+ are the interlayer cations.

Hower[127] showed for a series of glauconites that the octahedral cation, as well as the interlayer potassium, decreases with increasing amounts of expandable layers. He also showed that potassium increases with octahedral charge and is independent of tetrahedral charge.

The usefulness of 00*l* diffraction-intensity sequences for characterizing micas was first outlined by Nagelschmidt.[128] He observed that the 5-Å reflection was much stronger for muscovite than biotite-phlogopite micas. In a more recent extension of this principle, Brown[129] tabulated sets of partial sums for the amplitudes of the first three orders of the basal reflections through pertinent composition ranges for iron- and magnesium-bearing compositions.

[125] Burst, J. F., Glauconite Pellets, Their Mineral Nature and Application to Stratigraphic Interpretations, *Bull. Am. Assoc. Petrol. Geologists,* **42:** 310–327 (1958).

[126] Hendricks, S. B., and C. S. Ross, The Chemical Composition and Genesis of Glauconite and Celadonite, *Am. Mineralogist,* **26:** 683–708 (1941).

[127] Hower, J., Some Factors Concerning the Nature and Origin of Glauconite, *Am. Mineralogist,* **46:** 313–334 (1961).

[128] Nagelschmidt, G., X-ray Investigations on Clays, *Z. Krist.,* **97:** 514–521 (1937).

[129] Brown, G., Intensities of 00*l* Reflections of Mica- and Chlorite-type Structures, *Mineral. Mag.,* **30:** 657–665 (1955).

Applications to the macrocrystalline micas are straightforward. Extension to illites is less rigorous, but it is frequently possible to infer essentially aluminian or ferroian compositions from the relative prominence or essential absence, respectively, of a scattering maximum near 5.0 Å.

Bradley and Grim[130] stated that illites most frequently encountered in sediments have b parameters near 9.0 Å; glauconites are frequently about 9.1 Å; and examples related to the biotites are about 9.2 Å.

Grim, Bradley, and Brown[121] presented an empirical formula relating the b_0 dimension to the chemical composition of the micas. The formula did not provide uniformly reliable results but did serve to show the relative influence of various octahedral and tetrahedral components on this parameter. As indicated earlier, Radoslovich et al.[5,70] discussed this matter in detail and presented a new formula.

The structure of illite differs from that of smectite in several important ways as follows: The charge deficiency due to substitutions per unit-cell layer is about 1.30 to 1.50 for illite and 0.65 for smectite. The seat of this charge deficiency in illite is largely in the silica sheet and therefore close to the surface of the unit layer, whereas in smectite it is frequently, perhaps chiefly, in the octahedral sheet at the center of the unit layer. Also in the case of illite the balancing cation between the unit layers is chiefly or entirely potassium. Because of these differences the illite structural unit layers are relatively fixed in position, so that polar ions cannot enter readily between them and cause expansion. Also the interlayer balancing cations are not easily exchangeable (see Chap. 7).

It is conceivable that all gradations can exist between illite and well-crystallized muscovites and biotites on the one hand and smectite on the other hand. Data are not yet available to settle this point positively, but it is noteworthy that the illites so far studied in detail have shown similar and distinctive attributes rather than complete gradational variations with muscovite or biotite. With regard to gradations with smectite the data are less clear. Many clays seem to be gradational between illite and smectite but are actually composed of interlayered mixtures of true illite and smectite (see mixed-layered minerals, pages 121 and 122). It is possible, even likely, that clay minerals are to be found with a stacking of layers differing slightly from illite in having less unbalanced charge and with substantial replacement of K^+ by another cation like Na^+. It would be expected that in such specimens some water could penetrate between the unit layers with some accompanying expansion. In such specimens there might well be some variation in the unbalancing between different layers, and the nature of the balancing cations might vary from layer to layer as well as between the same layer. In such

[130] Bradley, W. F., and R. E. Grim, Mica Clay Minerals, "The X-ray Identification and Crystal Structures of Clay Minerals," chap. V, pp. 208–241, Mineralogical Society of Great Britain Monograph (1961).

material, expansion between all layers would not be the same and might even vary within a single interface between adjacent units.   The classification of such material would be difficult, and the separation of illite and smectite must frequently be purely arbitrary, with, in general, definitely expanding material being called smectite and nonexpanding being called illite.

Many materials described in the literature as illites are undoubtedly mixed-layer structures of micas and expandable components, and it has been considered (see Radoslovich[130a] and Yoder[131]) that all materials showing the diffractional and compositional properties assigned to illites are mixed-layer structures.   Recently, Gaudette, Eades, and Grim[132] showed that there are clay-mineral specimens that are deficient in potassium as compared with macrocrystalline micas and with the other diffraction and composition attributes assigned to illites which are not mixed-layer structures but monomineral species.

Chlorite minerals

The structure of chlorite was first suggested by Pauling.[16] Later Mauguin[133] and McMurchy[134] examined the mineral in great detail, verifying Pauling's suggested structure and providing additional information on its symmetry and dimensions.   More recently many investigators, but particularly Brindley and his colleagues,[135,136] Steinfink,[137] and Brown and Bailey,[138] further studied the mineral and considerably amplified knowledge of the general chlorite structure and the variations of polymorphic forms.

[130a] Radoslovich, E. W., The Cell Dimensions and Symmetry of Layer-Lattice Silicates, V, Composition Limits, *Am. Mineralogist,* **48:** 348–367 (1963).

[131] Yoder, H. S., Experimental Studies on Micas: a Review, Proceedings of the 6th National Clay Conference, pp. 42–60, Pergamon Press, New York (1959).

[132] Gaudette, H. E., J. L. Eades, and R. E. Grim, The Nature of Illite, Proceedings of the 13th National Clay Conference, pp. 33–48, Pergamon Press, New York (1966).

[133] Mauguin, C. H., Unit Cell of Chlorites, *Bull. Soc. Franc. Mineral.,* **53:** 279–300 (1930).

[134] McMurchy, R. C., Crystal Structure of Chlorites, *Z. Krist.,* **88:** 420–432 (1934).

[135] Robinson, K., and G. W. Brindley, Structure of Chlorites, *Proc. Leeds Phil. Lit. Soc. Sci. Sect.,* **5:** 102–108 (1948).

[136] Brindley, G. W., and K. Robinson, The Chlorite Minerals, "X-ray Identification and Structures of Clay Minerals," chap. VI, pp. 172–198, Mineralogical Society of Great Britain Monograph (1951).

[137] Steinfink, H., Crystal Structure of a Monoclinic Chlorite, *Acta Cryst.,* **11:** 191–195 (1958); Crystal Structure of a Triclinic Chlorite, *Acta Cryst.,* **11:** 195–198 (1958).

[138] Brown, B. E., and S. W. Bailey, Chlorite Polytypism, I, Regular and Semirandom One-layer Structures, *Am. Mineralogist,* **47:** 819–850 (1962).

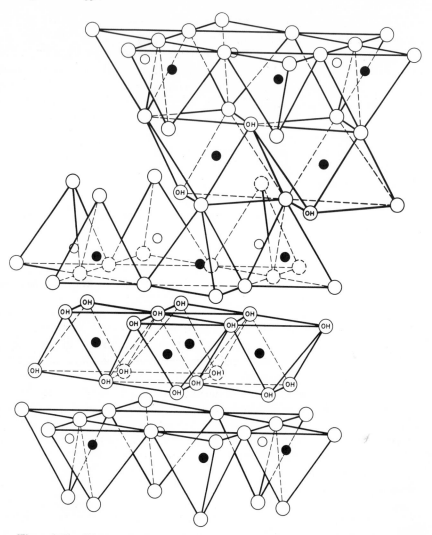

Fig. 4-17. Diagrammatic sketch of the structure of chlorite. (After McMurchy.[134])

All true chlorites have the same general structural framework. The structure consists of alternate mica-like and brucite-like layers. The layers are continuous in the a and b dimensions and are stacked in the c direction with basal cleavage between the layers. The mica-like layers are triocta-hedral with the general composition $(OH)_4(SiAl)_8(Mg \cdot Fe)_6O_{20}$. The brucite-like layer has the general composition $(Mg \cdot Al)_6(OH)_{12}$. The mica layer is unbalanced by substitution of Al^{3+} for Si^{4+}, and this de-ficiency of charge is balanced by an excess charge in the brucite sheet as

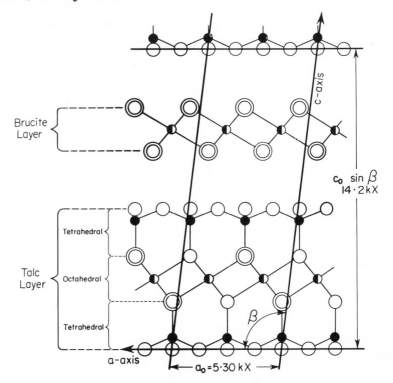

Fig. 4-18. Projection of the chlorite structure on the *ac* plane. (After McMurchy[134] from Brindley and Robinson.[136])

a consequence of substitution of Al^{3+} for Mg^{++}. The general chlorite structure is illustrated in Figs. 4-17 and 4-18. The charge distribution in the layers is as follows, where $X = 1$ or 2:

$6(OH)^-$	$6-$
$(6 - 2X)Mg^{++} \cdot 2X Al^{3+}$	$[2(6 - 2X) + 3(2X)]+$
$6(OH)^-$	$6-$
$6O^{--}$	$12-$
$(4 - X)Si^{4+} \cdot X Al^{3+}$	$[4(4 - X) + 3X] +$
$4O^{--}2(OH)^{--}$	$10-$
$6R(Trioctahedral, R = Mg^{++}, Mn^{++}, Fe^{++},$	
$Fe^{3+}, Cr^{3+}, Ti^{4+})$	$12+$
$4O^{--}2(OH)^{--}$	$10-$
$(4 - X)Si^{4+} \cdot X Al^{3+}$	$[4(4 - X) + 3X] +$
$6O^{--}$	$12-$

On the basis of single-crystal analyses, Steinfink[137] found that the tetrahedral network was deformed by the rotation of the tetrahedrons alternating in opposite directions. Radoslovich[70] found that the rotation varied substantially to a maximum of about 14°.

The bonding between the layers is partly electrostatic in character as the consequence of substitutions within the lattice. A second important bonding mechanism arises from adjacent sheets of oxygen and hydroxyl ions where the principal bonding mechanism arises from pairs of adjacent oxygen and hydroxyl ions. This bonding mechanism is similar to that found in the kaolin mineral.

Various members of the chlorite group differ from each other in the kind and amount of substitutions within the brucite layer and the tetrahedral and octahedral positions of the mica layer. They also differ in the detailed orientation of successive octahedral and tetrahedral layers, in the relation of the mica to brucite layers, and in the stacking of successive chlorite units. Specimens may be monoclinic or triclinic.

Brown and Bailey[138] concluded that most chlorites have fixed relations between talc and brucite sheets within each layer but semirandom stacking of successive layers, with shifts of plus or minus $b/3$. These investigators and numerous others in the references given on page 99 studied the stacking arrangements in the chlorite minerals and the resulting symmetries. For details these works should be consulted.

Substitutions within the tetrahedral sheets vary from about Si_3Al to Si_2Al_2, and substitutions within the octahedral layers from Mg_5Al to Mg_4Al_2, with Fe^{++} and Mn^{++} partially replacing Mg^{++} and Fe^{3+} or Cr^{3+} partially replacing Al^{3+}. The substitutions of Al^{3+} for Si^{4+} in the tetrahedral sheets expand them sufficiently to accommodate the somewhat larger octahedral layers between them.

Brindley and MacEwan[103] published a formula for computing the b parameter of chlorites from the chemical composition. Various authors (Von Engelhardt,[139] Hey,[140] and Shirozu[141]) found an approximately linear relation between the b parameter and the iron content of chlorites. The most recent formula given by Shirozu is $b = 9.210 + 0.037$ (Fe^{2+}, Mn) when (Fe^{2+}, Mn) is the number of atoms per six octahedral positions. On this basis it appears that the ferrous iron content can be estimated reasonably well from an accurate determination of the b parameter.

Radoslovich[70] stated that magnesium and ferric iron in octahedral coordination did not significantly influence the b parameter. This author considered in detail the relation of composition to the parameters of the chlorite minerals.

[139] Von Engelhardt, W., Structures of Thuringite, Bavalite and Chamosite, *Z. Krist.*, **104**: 142–159 (1942).

[140] Hey, M. H., New Review of Chlorites, *Mineral Mag.*, **30**: 277–292 (1954).

[141] Shirozu, H., X-ray Patterns and Cell Dimensions of Chlorites, *Mineral. J. (Tokyo)*, **2**: 209–223 (1958).

Brindley[141a] pointed out that the basal spacing of chlorites varied from about 14.1 to 14.35 Å as the layer charge per unit cell decreased from 1.5 to about 0.5. He also summarized the work of numerous investigators on the variation of the basal spacing with composition. Schoen[142] studied the influence of substitution in octahedral sites of chlorite on the intensities of basal reflections.

By detailed study of single crystals Brindley[136] and his colleagues showed that chlorites exhibit polymorphic forms similar to the biotite micas. They found structures consisting of several different stacking arrangements of the chlorite layers. Units with three, six, and nine chlorite layers in the orthohexagonal cell have been described. For details of these polymorphic forms and other details of the chlorite structure the original work of Brindley et al.[136] should be consulted.

Fine-grained chlorites are found in some clay materials. The range of polymorphic forms of the chlorite minerals and the variations in structure between clay-mineral chlorite and well-crystallized chlorite occurring in large units have not been established. Available data seem to suggest that clay-mineral chlorites differ from well-crystallized material in a somewhat random stacking of layers and perhaps in some hydration. Unfortunately chlorite in clay material is generally found intimately mixed with other clay minerals so that diffraction data permit only the identification of a 14-Å nonexpanding mineral and not the polymorphic form.

An equivalent dioctahedral chlorite seems to be theoretically possible, containing pyrophyllite-like gibbsite-like layers. Sudo[143,144] and Brown and Jackson[145] presented substantial evidence for the existence of such a mineral.

Various authors have described a mineral with a 14-Å spacing which does not collapse on heating and which expands slightly following treatment with glycol (to 16 to 18 Å) as "swelling chlorite"; thus, Honeyborne[146] and Stephen and MacEwan[147] described such material from the Keuper marl in

[141a] Brindley, G. W., Chlorite Minerals, "The X-ray Identification and Crystal Structures of Clay Minerals," chap. VI, pp. 242–296, Mineralogical Society of Great Britain Monograph (1961).

[142] Schoen, R., Semiquantitative Analysis of Chlorites by X-ray Diffraction, *Am. Mineralogist,* 47: 1384–1492 (1962).

[143] Sudo, T., H. Takahashi, and H. Matsui, Fireclay from Kurata Mine, *Japan J. Geol. Geography,* 24: 71–85 (1954)

[144] Sudo, T., and H. Kodama, Aluminian Mixed-layer Montmorillonite-Chlorite, *Z. Krist.,* 109: 379–387 (1957).

[145] Brown, B. E., and M. L. Jackson, Clay Mineral Distribution in Soils of Northern Wisconsin, *Natl. Acad. Sci., Publ.* 566, pp. 213–226 (1958).

[146] Honeyborne, D. B., Clay Minerals in the Keuper Marl, *Clay Minerals Bull.* 1(5): 150–155 (1951).

[147] Stephen, I., and D. M. C. MacEwan, Swelling Chlorites, *Geotechnique (London),* 2: 82–83 (1950).

central England. A similar component has been described in mixed-layer structures; thus, Lippmann[148] described a mineral from the Keuper clay in Germany with a natural spacing of 28.3 Å which expanded to 32.5 Å with glycerol but which after heat treatment to 550°C for $\frac{1}{2}$ hr gave a weak and broad 28-Å reflection and a similar 13- to 14-Å reflection. He called the mineral corrensite and considered it a mixed-layer structure of chlorite and "swelling chlorite." Such "swelling chlorites" are probably imperfect structures in which the brucite layer is discontinuous and can be viewed as occurring in islands between the silicate layers. The islands would prevent the collapse of the mineral on heating but would be insufficient to prevent expansion on glycol treatment.

Bradley[149] described a somewhat similar structure as a regular mixed-layer assemblage consisting of alternations of one pyrophyllite unit and one vermiculite unit, which was designated as a rectorite, and Bradley and Weaver[150] described a mixed-layer structure consisting of the regular interstratification of chlorite and vermiculite units which is comparable to Lippmann's corrensite.

Vermiculite minerals

The structure of vermiculite was first worked out by Gruner.[151,152] Later Hendricks and Jefferson[153,44] confirmed Gruner's general conclusions but changed certain structural details. More recently Barshad,[154-156]

[148] Lippmann, F., Keuper Clay from Zaisersweiher, *Heidelberger-Beitr. Mineral. Petrog.,* 4: 130–134 (1954).

[149] Bradley, W. F., The Alternating Layer Sequence of Rectorite, *Am. Mineralogist,* 35: 590–595 (1950).

[150] Bradley, W. F., and C. E. Weaver, Chlorite-Vermiculite, *Am. Mineralogist,* 41: 497–504 (1956).

[151] Gruner, J. W., Vermiculite and Hydrobiotite Structures, *Am. Mineralogist* 19: 557–575 (1934).

[152] Gruner, J. W., Water Layers in Vermiculite, *Am. Mineralogist,* 24: 428–433 (1939).

[153] Hendricks, S. B., and M. E. Jefferson, Crystal Structure of Vermiculites and Mixed Vermiculite-Chlorites, *Am. Mineralogist,* 23: 851–863 (1938).

[154] Barshad, I., Vermiculite and Its Relation to Biotite, *Am. Mineralogist,* 33: 655–678 (1948).

[155] Barshad, I., The Nature of Lattice Expansion and Its Relation to Hydration in Montmorillonite and Vermiculite, *Am. Mineralogist,* 34: 675–684 (1949).

[156] Barshad, I., The Effect of Interlayer Cations on the Expansion of the Mica Type Crystal Lattice, *Am. Mineralogist,* 35: 225–238 (1950).

Walker,[157-159] and Grudemo[160] independently added much to our knowledge of the structure of the mineral. Gruner[151] also showed that many materials classed as vermiculite are mixed-layer mica-vermiculite structures.

Gruner[151] showed that the structure consists of sheets of trioctahedral mica or talc separated by layers of water molecules occupying a definite space (4.98 Å) which is about the thickness of two water molecules. In its natural state, therefore, the mineral consists of an alternation of mica and double water layers. Gruner assigned vermiculite to the space group C_{2h}^6–$C2/c$. Megascopic vermiculite is invariably trioctahedral, whereas dioctahedral forms are also found in clays.

The structure is unbalanced chiefly by substitutions of Al^{3+} for Si^{4+}. These substitutions may be partially balanced by other substitutions within the mica lattice, but there is always a residual net-charge deficiency of 1 to 1.4 per unit cell. The charge deficiency is satisfied by cations which occur chiefly between the mica layers and are largely exchangeable. In the natural mineral, which has the same cation-exchange capacity as smectite, or somewhat higher, the balancing cation is Mg^{++}, sometimes with a small amount of Ca^{++} also present.

A general formula for natural vermiculite is $(OH)_4(Mg \cdot Ca)_X(Si_{8-X} \cdot Al_X)$ $(Mg \cdot Fe)_6 O_{20} \cdot yH_{20}$ with $X = 1$ to 1.4 and $y =$ about 8. The Mg^{++} and Ca^{++} are the balancing and largely exchangeable cations. The charge distribution in the layers is as follows:

yH_2O double water layers

$X(Mg^{++} \cdot Ca^{++})$	$X+$
$6O^{--}$	$12-$
$(4-X)Si^{4+} \cdot XAl^{3+}$	$(16-X)+$
$4O^{--}2(OH)^-$	$10-$
$6(Mg \cdot Fe)^{++}$	$12-$
$4O^{--}2(OH)^-$	$10-$
$(4-X)Si^{4+} \cdot XAl^{3+}$	$(16-X)+$
$6O^{--}$	$12-$
$X(Mg^{++} \cdot Ca^{++})$	$X+$

yH_2O double water layers

[157] Walker, G. F., Water Layers in Vermiculite, *Nature,* **163:** 726 (1949).

[158] Walker, G. F., Vermiculite and Some Related Mixed-layer Minerals, "X-ray Identification and Structures of Clay Minerals," chap. VII, pp. 199–223, Mineralogical Society of Great Britain Monograph (1951).

[159] Walker, G. F., Vermiculite Minerals, "The X-ray Identification and Crystal Structures of Clay Minerals," chap. VII, pp. 297–324, Mineralogical Society of Great Britain Monograph (1961).

[160] Grudemo, A., Structure of Vermiculite, *Handl. Svensk. Forskn. Inst. Cement,* no. 22, 1954.

On heating vermiculite to temperatures as high as 500°C, the water is driven out from between the mica layers, but the mineral quickly rehydrates on exposure to moisture at room temperature. The mineral, therefore, has an expanding lattice, but the expansion is restricted to about 4.98 Å, or two water layers. If the mineral is heated to 700°C, there is no expansion again. In such material the 14-Å line and higher orders of it disappear, and a new line at 9.3 Å with other new mica lines appears on the diffraction pattern.

Hendricks and Jefferson[153] assigned the following crystallographic constants, based on single-crystal data, to the mineral: $c = 28.91$ Å, $b = 9.20$ Å, $a = 5.34$ Å, $\beta = 93°15'$. They showed that vermiculites have the same types of shifts along the a axis as muscovite, talc, and pyrophyllite and have a partially random displacement of structural layers parallel to the b axis. They gave the space group as C_c–$C_s{}^4$ and pointed out the pseudo nature of the space group because of the random displacement parallel to the b axis.

Grudemo[160] gave the unit-cell constants, on the basis of single-cell studies, as $a = 5.347$ Å, $b = \sqrt{3}a$, $c = 14.44$ Å. Mathieson and Walker[161] gave the following constants: $a = 5.33$ Å, $b = 9.18$ Å, $c = 28.9$ Å, $\beta = 97°$, space group Cc. A two-dimensional Fourier synthesis projected on (010) disclosed the distortion of the surface oxygen sheet of the silicate layers, representing a displacement of the oxygen from regular hexagonal sites, in general, comparable to the rotation of the silica tetrahedrons suggested by Radoslovich.[9] Their data indicated that the unit cell contained two silicate and magnesium water layers and that there were two stacking sequences of the silicate layers of equal probability.

The variation in the cell constants reported by different investigators is probably explained by inherent variations in the mineral.

There has been considerable discussion regarding the exact structure of the water layers in vermiculite. Thus, Gruner[152] suggested that the water might occur as charged hydronium $(H_3O)^+$ groups. Hendricks and Jefferson[44] postulated an extended hexagonal network of water molecules (Fig. 4-19) (see Chap. 8 for a further discussion). Bradley and Serratosa[162] concluded that chemical, thermogravimetric, infrared, and cation-exchange data together with X-ray-diffraction interpretations of Walker (see below) for vermiculite can be reconciled with the essential feature of water network described by Hendricks and Jefferson; namely, that each water molecule is in an approximate tetrahedral environment. It was pointed out by Barshad[155,156] and

[161] Mathieson, A. McL., and G. F. Walker, Structure of Mg-Vermiculite, *Am. Mineralogist,* **39:** 231–255 (1954).

[162] Bradley, W. F., and J. M. Serratosa, A Discussion of the Water Content of Vermiculite, Proceedings of the 7th National Clay Conference, pp. 260–270, Pergamon Press, New York (1960).

Walker[157,158] independently that the nature of the exchangeable cations must influence the state of the interlayer water since they occur between the mica layers. These investigators showed this to be the case (Table 4-4).

Table 4-4 Thickness of water layers in vermiculite saturated with various cations

(*After Walker*[158])

Cation	$d(002)$, Å	Thickness of water layer, Å
Mg^{++}	14.39	5.11
Ca^{++}	15.0	5.75
Sr^{++}	15.0	5.75
Ba^{++}	12.3	3.04
Li^+	12.2	2.94
Na^+	14.8	5.55
K^+	10.6 diffuse	About 1.34
NH_4^+	10.8 diffuse	About 1.54

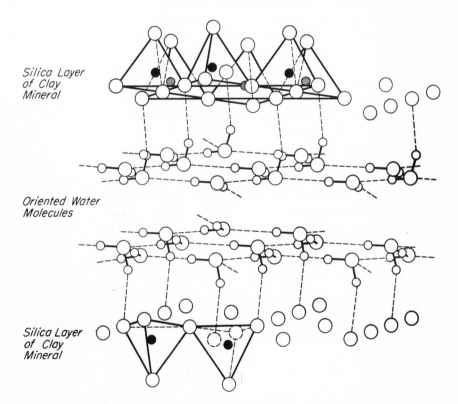

Silica Layer
of Clay
Mineral

Oriented Water
Molecules

Silica Layer
of Clay
Mineral

Fig. 4-19. Diagrammatic sketch of the vermiculite structure, showing layers of water. (After Hendricks and Jefferson.[44])

Grudemo's[160] data demonstrated that the interlayer molecules and cations occupied definite sites, but his data did not permit the deduction of their relationships to the silicate layers. Mathieson and Walker[161] were able to locate the sites of the interlayer water molecules and cations of their magnesium vermiculite. Using two-dimensional Fourier methods, they showed that the water network, consisting of two sheets of water-molecule sites, was arranged in such a way as to provide octahedral coordination for the exchangeable magnesium ions, which are midway between the silicate layers (Fig. 4-20). The water-molecule sites within a sheet are arranged in hexagonal array, and each site is equivalently related to a single oxygen in the silicate-layer surface. From the chemical data, about two-thirds of the available water-molecule sites and one-ninth of the exchangeable-cation sites were found to be occupied in the specimen examined. Their evidence suggested a slight distortion of the water network from their mean position. These investigators assumed a tendency for the interlayer cations to distribute themselves fairly evenly throughout the available sites and for the interlayer water molecules to group themselves around the cation.

Mathieson[163] compared the structural features of vermiculite and chlorites and suggested that the essential structure difference between the groups lies in the sites of the interlayer magnesium ion.

Walker[164] found that when small magnesium-vermiculite crystals are placed in water there is a gradual increase in the basal spacing from 14.36 to 14.81 Å in the course of several days. Swelling of the lattice normal to the plane of the layers is regular throughout the crystal and is accompanied by an increase in interlayer water content, but the reaction is not shown by all specimens. All or nearly all the interlayer water-molecule sites are occupied in the 14.81-Å lattice. It was suggested that, in accommodating further water molecules, the interlayer water network adopts the undistorted hexagonal form but that the structure in other respects resembles that of the normal 14.36-Å phase.

Progressive removal of interlayer water from vermiculite leads to the development of a series of less hydrated phases (Walker[164]). On the dehydration of the normal 14.36-Å lattice, a 13.82-Å phase is first formed corresponding to a structure containing double sheets of interlayer water with an arrangement different from that in the 14.36-Å lattice. On further dehydration an 11.59-Å phase is formed in which a single sheet of water molecules is interleaved with a silicate layer; with still further dehydration a 20.6-Å phase consisting of approximately regular alternations of 11.59-Å and 9.02-Å com-

[163] Mathieson, A. McL., Structure of Mg-Vermiculite, *Am. Mineralogist,* **43**: 216–227 (1958).

[164] Walker, G. F., Mechanism of Dehydration of Mg-Vermiculite, *Natl. Acad. Sci., Publ.* 456, pp. 101–115 (1956).

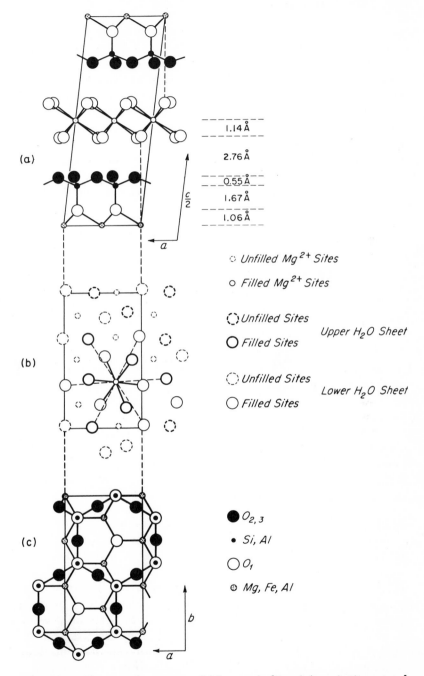

Fig. 4-20. The crystal structure of Mg vermiculite: (*a*) projection normal to the *ac* plane, (*b*) projection normal to the *ab* plane, showing the interlayer region; (*c*) projection normal to the *ab* plane showing one-half of a silicate layer (*z* = 0 to *c*/8). (After Mathieson and Walker.[161])

ponents is developed, and finally a 9.02-Å phase is formed from which all interlayer water has been excluded. In Fig. 4-21 the observed silicate-layer relationships at various stages of hydration are shown. It should be noted that the β angle changes progressively as the water is removed from the lattice. The two intermediate states at 13.82 and 20.6 Å have no stability fields and do not appear in equilibrium hydration studies (Van Olphen[164a]).

Starting from the fully hydrated 14.81-Å phase, which contains approximately 16 water molecules per cation, removal of water first produces a regular contraction of the lattice along the c axis without lateral displacement of the silicate layers with respect to each other. During this initial stage the gradual introduction of vacant sites causes the remaining water molecules

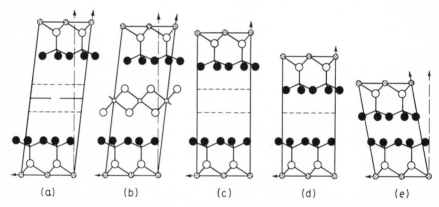

Fig. 4-21. Projections normal to the *ac* plane in Mg vermiculite at various stages of hydration, showing the silicate-layer relationships: (*a*) 14.81-Å phase; (*b*) 14.36-Å phase; (*c*) 13.82-Å phase; (*d*) 11.59-Å phase; (*e*) 9.02-Å phase. (Key as for Fig. 4-20.) (After Walker.[159])

to be displaced increasingly from their regular hexagonal sites so that a distortion of the network develops. When the lattice spacing reaches 14.36 Å, contraction ceases temporarily while the number of water molecules per cation falls from about 12 to 9. At this point there is an abrupt contraction of the lattice to the 13.82-Å phase during which the silicate stacking sequence of the silicate layers changes (Fig. 4-21). The location of the cations and water molecules in the 13.82-Å phase has not been fully established, but it appears that the cations are displaced from their central positions and occupy a less perfect octahedral environment consisting of three interlayer water molecules and three oxygens of the silicate-layer surface.

[164a] Van Olphen, H., Composition of Clay Sediments in the Range of Molecular Particle Distances, Proceedings of the 11th National Clay Conference, pp. 178–187, Pergamon Press, New York (1963).

A further abrupt contraction to 11.59 Å develops when the water-molecule-to-cation ratio is reduced to about 8:1. In this transition, the silicate-layer relationships are preserved, as in the 13.82-Å phase, and it is probable that interlayer cation sites remain unchanged also. Further withdrawal of water molecules proceeds without accompanying structure changes until about three water molecules per cation remain. The gradual development of an approximately regular alternation of 11.59-Å and 9.02-Å components is then observed as an intermediate stage before complete contraction of the lattice to 9.02 Å takes place. A further lateral displacement of silicate layers with respect to one another is associated with the 11.59- to 9.02-Å contraction. Walker[164] also studied the diffusion of water molecules into interlayer regions of dehydrated samples and found that phase transitions generally similar to those developing during dehydration take place in a highly regular manner. He also supplied data regarding conditions of temperature and humidity at which partially hydrated phases of magnesium vermiculite are stable, and he showed further that, for macroscopic flakes of magnesium vermiculite, completely dehydrated samples do not rehydrate readily.

In the case of vermiculite prepared with exchangeable Ba^{++} and Li^+, there seems to be a single water layer between the mica layers. For vermiculites with K^+ and NH_4^+, the space is smaller than that required for a single molecular water layer, and Barshad[155] considered the mineral to be dehydrated. Walker,[158] however, considered that some water is present with the K^+ and NH_4^+ and that the reduced space is due to a partial embedding in the silicon-oxygen sheets. He pointed out that, when K^+ saturated vermiculite is heated, the basal spacing is further reduced to 10.1 Å.

In general, however, it may be said that the basal reflection of about 14 Å should always be more intense than the subsequent lower orders if the mineral is vermiculite rather than chlorite. Vermiculite and chlorite can also be distinguished by the persistence of the reflection corresponding to the 7-Å periodicity when chlorite is heated at 110°C. Caution must be used, however, because of the very rapid rehydration of the 11.6-Å phase of magnesium vermiculite in air. This is particularly true for clay-size particles.

Since vermiculite has an expanding structure and an unbalanced lattice with high cation-exchange capacity, it is important to consider its relation to smectite, which also has these characteristics. Vermiculite differs from smectite in that the expansion with water is limited to about 4.98 Å. Vermiculites adsorb certain organic molecules (see Chap. 10) between the mica layers but differ from smectite in that the adsorbed layer is thinner and less variable. These characteristics may be the result of the relatively larger particle size of the vermiculite layers and possibly also of the fact that there is considerably less randomness in the stacking of the vermiculite layers. All these differences may be due to the fact that the unbalancing in the vermiculite is mainly and perhaps essentially in the tetrahedral layer, whereas in smectite it is mainly and perhaps essentially in the octahedral layer.

Available evidence indicates that vermiculites always have greater layer charge than smectites. Weiss, Koch, and Hofmann,[88] and Hofmann et al.[77] favored the idea of a continuous gradation between the minerals and proposed that a distinction should be made on the basis of a layer-charge value. They suggested that, with the layer-charge value greater than 0.55 equivalent corresponding to a cation-exchange capacity of approximately 115 meq/100 g of air-dried samples, the material be designated as vermiculite. With a lower value the designation would be smectite. Their suggestion resulted largely from their observation that expanding-lattice three-layer minerals with a capacity less than this value show only *hk* reflections, whereas those in excess of this value show *hkl* reflections.

Walker[165] suggested that the threshold values should be slightly higher, as otherwise certain samples of saponite and beidellite would be included with the vermiculites. Walker pointed out that significant cation-exchange-capacity values are difficult to obtain, particularly for complex mixtures, and that the nature of the interlayer cation and the amount of interlayer water may also influence the values obtained. According to Walker and Cole,[166] chemical differences between trioctahedral smectites and macroscopic vermiculites center largely on the greater aluminum-for-silicon substitution of the latter, but they pointed out that comparable data for pure clay-mineral vermiculites are not available. Walker pointed out that the mineralogical complexity of many natural clays would render it difficult or frequently impracticable to differentiate the minerals on the basis of chemical composition.

Walker[167] considered that a test based on relative swelling ability offers the best prospects of success, since there is a general tendency even for vermiculites reduced to clay size to expand less than smectites. Further failure of the hydrated magnesium lattice to expand beyond 14.5 Å on treatment with glycerol appears to be characteristic of vermiculites, whereas all magnesium smectites so far tested expand to about 17.8 Å. For vermiculites containing other interlayer cations, this distinction may not be valid.

Various investigators have described minerals in clays and soils with characteristics mostly similar to those generally given for vermiculite, but with certain differences. Thus, Hathaway[168] described a dioctahedral clay mineral found in various soils with the basal spacing of about 14 Å, which

[165] Walker, G. F., Reactions of Expanding-lattice Minerals with Glycerol and Ethylene Glycol, *Clay Minerals Bull.,* **3**(20): 302–313 (1958).

[166] Walker, G. F., and W. F. Cole, The Vermiculite Minerals, "The Differential Thermal Investigation of Clays," chap. VII, pp. 191–206, Mineralogical Society, London (1957).

[167] Walker, G. F., Differentiation of Vermiculites and Smectites in Clays, *Clay Minerals Bull.,* **3**(17): 154–163 (1957).

[168] Hathaway, J. C., Some Vermiculite-type Clay Minerals, *Natl. Acad. Sci., Publ.* 395, pp. 74–86 (1955).

does not expand further on treatment with glycerol or ethylene glycol but collapses by moderate heating; he described this material as vermiculite. Sawhney[169] described the formation of aluminum interlayers in expandable clay minerals with the consequent loss of the expansion characteristic. He suggested the term chloritized montmorillonite and chloritized vermiculite for such clay minerals. The common difference between such materials and common vermiculite probably resides in the cation population between the silicate layers.

The variation, if any, between clay-mineral vermiculite and large flakes of the mineral is not known. Any such differences would probably be small and would likely be the same sort of variation as suggested for well-crystallized chlorite and clay-mineral chlorite.

Sepiolite—palygorskite—attapulgite minerals

The general attributes of structure and composition of these minerals are not very well known. According to Fersman[170] there is a complete series of minerals from a magnesium end member, sepiolite, to an aluminum end member, which he called paramontmorillonite. Fersman called the whole series palygorskites. Longchambon,[171] Migeon,[172] and Caillere[173-175] studied intensively many of these minerals, which show a considerable range in relative amounts of aluminum and magnesium but which have substantially the same general X-ray-diffraction and dehydration characteristics.

These authors concluded that the structurally important element in these minerals is the amphibole double silica chain oriented with its long direction

[169] Sawhney, B. L., Aluminum Interlayers in Soil Clay Minerals, Montmorillonite and Vermiculite, *Nature,* **182:** 1595–1596 (1958).

[170] Fersman, A., Recherches sur les silicates de magnésie, *Mem. Russian Acad. Sci.* **32:** 377–392 (1913).

[171] Longchambon, H., Sur certaines caractéristiques de la sepiolite d'Ampandandrava et la formule des sepiolites, *Bull. Soc. Franc. Minerals.,* **60:** 232–276 (1937).

[172] Migeon, G., Contribution à l'étude de la définition des sepiolites, *Bull. Soc. Franc. Mineral.,* **59:** 6–133 (1936).

[173] Caillere, S., Etude de quelques silicates magnésiens à facies asbestiforme ou papyrace n'appartenant pas du groupe de l'antigorite, *Bull. Soc. Franc. Mineral.,* **59:** 353–386 (1936).

[174] Caillere, S., Sepiolite, "X-ray Identification and Structures of Clay Minerals," chap. VIII, pp. 224–233, Mineralogical Society of Great Britain Monograph (1951).

[175] Caillere, S., and S. Henin, Palygorskite-Attapulgite, "X-ray Identification and Structures of Clay Minerals," chap. IX, pp. 234–243, Mineralogical Society of Great Britain Monograph (1951).

parallel to the c axis. Nagelschmidt[176] independently concluded that there were clay-mineral types which were fibrous and composed of silica chains. Longchambon[171] suggested the formula $Si_4O_{11}(Mg \cdot H_2)_3H_2O \, 2(H_2O)$ for sepiolite, which fits the sample that he studied and includes the Si_4O_{11} double chain. It appears that the magnesium can be replaced by iron or aluminum to a considerable extent. Migeon[172] and Longchambon[171] pointed out that the minerals of this group contain "zeolitic water," i.e., water loosely held in the lattice and lost at low temperatures (300°C), as well as water lost at higher temperatures. They did not present a structural concept of the water lost at low temperatures. The deployment of these chains and the water, which adequately explained the diffraction data, was deduced by Bradley.[177]

It is obvious from the above statements that there has been much confusion regarding the composition and character of these minerals. Recent work has shown that there are two distinct fibrous clay minerals: one type that has been called palygorskite, attapulgite, and/or pilolite; and a second type similar to sepiolite.

The structure of attapulgite was first studied by De Lapparent[178,179] and later in greater detail by Bradley.[177] The structure as worked out by Bradley is now generally accepted and is as follows: Attapulgite consists of double silica chains running parallel to the c axis with the chains linked together through oxygens at their longitudinal edges. The apexes of the tetrahedrons in successive chains point in opposite directions. The linked chains, therefore, form a kind of double-ribbed sheet, with two rows of tetrahedral apexes at alternate intervals in the top and bottom of the sheets. The ribbed sheets are arranged so that the apexes of successive sheets point together, and the sheets are held together by aluminum and/or magnesium in octahedral coordination between the apex oxygens of successive sheets (Fig. 4-22). The octahedral layer is similar to that in the layer clay minerals, but it is continuous in only one direction. The octahedral layer is completed by central OH groups and by hydroxyls at the open sides. The mineral has good cleavage parallel to (110) caused by the weak link through O of the silica chains in the ribbed layer. Chains of water molecules running parallel to c fill the interstices between the amphibole chains. The cavities will accommodate four molecules of water per unit cell, and this water would account for the dehydration loss at low temperatures. The dimensions of the cell suggested by Bradley[177] are $a_0 \sin \alpha = 12.9$ Å, $b_0 = 18$ Å, and $c_0 = 5.2$ Å. There are

[176] Nagelschmidt, G., Rod-shaped Clay Particles, *Nature*, **142**: 114–115 (1938).

[177] Bradley, W. F., The Structural Scheme of Attapulgite, *Am. Mineralogist,* **25**: 405–410 (1940).

[178] De Lapparent, J., Défense de l'attapulgite, *Bull. Soc. Franc. Mineral.,* **61**: 253–283 (1938).

[179] De Lapparent, J., Argile attapulgitique, *Compt. Rend.,* **212**: 971–974 (1941).

H 2 0

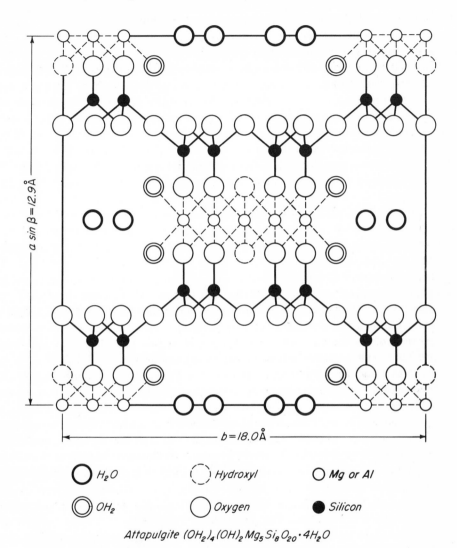

H_2O

OH₂

Hydroxyl

Oxygen

Mg or Al

Silicon

Attapulgite $(OH_2)_4(OH)_2Mg_5Si_8O_{20}\cdot 4H_2O$

Fig. 4-22. Schematic presentation of the structure of attapulgite. (After Bradley.[177])

two molecules in the unit cell, and the space group is probably $C_{2h}{}^3$–$C2/m$. The structure is balanced, and the composition of the ideal cell is $(OH_2)_4(OH)_2Mg_5Si_8O_{20}\cdot 4H_2O$, in which trivalent cations are considered equivalent to $1.5Mg^{++}$. Some substitution of Al^{3+} for Si^{4+} is considered probable. Bradley pointed out that substitution of Al^{3+} for either Mg^{++} or Si^{4+} or both should weaken the structure, so that it appears doubtful that extensive replacement takes place, and an aluminum end member would not be expected.

Some samples of these minerals have a fibrous texture and a cardboard or paper-like appearance due to the tangling of fibers. Other samples, such as the material from Attapulgus, Georgia, are compact and cryptocrystalline. They all give similar diffraction data. According to Huggins et al.,[180] attapulgite and palygorskite are structurally the same; attapulgite is simply short-fibered palygorskite. They obtained X-ray-diffraction data which were indexed on an orthorhombic unit cell rather than a monoclinic cell as done by Bradley. They obtained slightly different unit-cell constants and stated that the variations are probably due to variations in the chemical composition of the samples investigated. Their calculated density value was **2.327**, and their observed density was **2.272**. Serdyuchenko and Shashkina[181] have provided considerable chemical and X-ray data on palygorskite. Their data on the unit-cell parameters are in excellent agreement with those determined earlier by Bradley. Zvyagin et al.[181a] published the following structural data for palygorskite: a, 5.22 Å, b, 18.06 Å; c, 12.75 Å; β, 95°50′; space group $P2/a$.

Caillere and Henin[182] presented chemical analyses for several palygorskite samples, showing contents of Al_2O_3 from 6.82 to 15.44 percent and of Fe_2O_3 from 0.87 to 3.8 percent. Palygorskite is commonly associated with smectite, and it cannot be concluded, at least in every case, that these constituents are part of the palygorskite; however, it is certain that iron and aluminum occur in the palygorskite lattice in significant amounts.

Nagy and Bradley[183] and Brauner and Preisinger[184] presented structures for sepiolite which are generally similar but differ in detail. The scheme of Nagy and Bradley is that two pyroxene chains are linked to form an amphibole chain with an extra silica tetrahedron added at regular intervals on each side, as shown in Fig. 4-23. In contrast, in the structure postulated by Brauner and Preisinger, three pyroxene chains are linked to form two continuous amphibolic chains (Fig. 4-24). In both structures the linked Si-O

[180] Huggins, C. W., M. V. Denny, and H. R. Shell, Properties of Palygorskite, an Asbestiform Mineral, *U.S. Bur. Mines, Rept. Invest.* 6071, 17 pp. (1962).

[181] Serdyuchenko, D. P., and V. P. Shashkina, Sepiolite, Palygorskite, and Attapulgite, *Mineralog. Sb. L'vovsk. Geol. Obshchestvo*, **12**: 396–404 (1958).

[181a] Zvyagin, B. B., K. S. Mishchenko, and V. A. Shitov, Electron Diffraction Data on the Structures of Sepiolite and Palygorskite, *Kristallografiya*, **8**(2): 201–206 (1963).

[182] Caillere, S., and S. Henin, Palygorskite, "The X-ray Identification and Crystal Structures of Clay Minerals," chap. IX, pp. 343–353, Mineralogical Society of Great Britain Monograph (1961).

[183] Nagy, B., and W. F. Bradley, Structure of Sepiolite, *Am. Mineralogist,* **40**: 885–892 (1955).

[184] Brauner, K., and A. Preisinger, Structure of Sepiolite, *Mineral. Petrog. Mitt.,* **6**: 120–140 (1956).

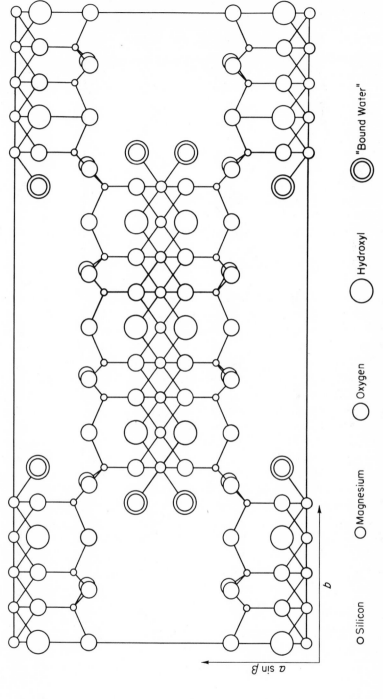

○ Silicon   ○ Magnesium   ○ Oxygen   ○ Hydroxyl   ◎ "Bound Water"

Fig. 4-23. Projection on (001) of a unit cell of sepiolite. (After Nagy and Bradley.[183])

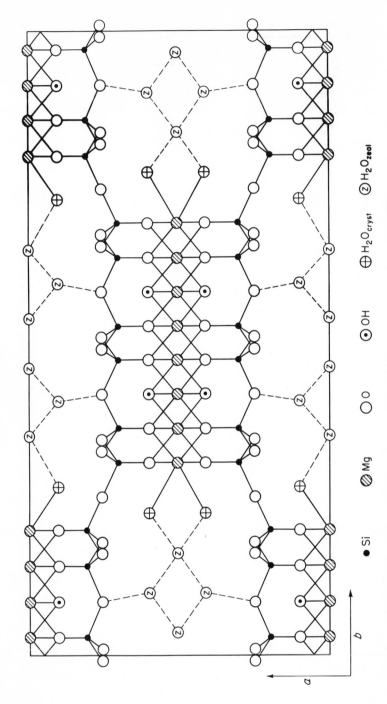

Fig. 4-24. Projection on (001) of a unit cell of sepiolite. (After Brauner and Preisinger[184])

● Si   ⊗ Mg   ○○ O   ⊙ OH   ⊕ H_2O_{cryst} Ⓩ H_2O_{zeol}

chains face each other between the adjacent continuous oxygen sheets. The sheets formed by the apexes of the tetrahedrons are completed by hydroxyls and magnesiums in octahedral coordination which tie the sheets together. A single ribbon is similar in structure to a layer of talc. Both arrangements leave channels in the structure between and parallel to the ribbons. At the edges of the ribbons, exposed hydroxyls are neutralized by protons, giving rise to bound water designated as OH_2. Water molecules, designated as zeolitic water, may enter the channels.

While the difference in arrangement of the silicons does not modify the size of the crystal motif, it does lead to a difference in composition of the octahedrally coordinated part of the structure. The Nagy-Bradley model has one extra hydroxyl in each of the discontinuous oxygen sheets and one magnesium less than the Brauner-Preisinger model in the octahedral cation sheet. Nagy and Bradley were able to index the $hk0$ reflections on the basis of the monoclinic cell with $a \sin \beta = 13.4$ Å, $b = 27$ Å, $c = 5.3$ Å, and suggested $C2/m$ as the possible space group. Brauner and Preisinger favored an orthorhombic cell with $a = 13.4$ Å, $b = 26.8$ Å, and $c = 5.28$ Å, with a space group of *Pncn*. The latter authors used more reflections in deducing their structure and, according to Caillere and Henin,[184a] chemical and dehydration data favor the structure suggested by Brauner and Preisinger. Zvyagin et al.[181a] reported the following structural data for sepiolite: a, 5.24 Å; b, 27.2 Å; c, 13.4 Å; $\beta = 90°$; space group *Pnan*.

Brindley[185] studied a number of sepiolites by electron diffraction and found marked variations in their crystallinity. He found that his data could be indexed on the basis of an orthorhombic cell.

The two structural models imply different ideal formulas for the mineral. The half-unit-cell contents for the Nagy-Bradley model should be $(Si_{12})(Mg_9)O_{30}(OH)_6(OH_2)_46H_2O$, and $(Si_{12})(Mg_8)O_{30}(OH)_4(OH_2)_88H_2O$ for the Brauner-Preisinger model. It is extremely difficult to obtain relatively pure sepiolite and consequently to determine which structure is favored by chemical data. Chemical analyses of various sepiolites suggest that the mineral may contain small but significant amounts of aluminum, ferrous iron, nickel, and manganese (Caillere and Henin[184a]). It appears that substitutions within the lattice do not leave it unbalanced. Rogers, Quirk, and Norrish[186] also described an aluminous sepiolite.

[184a] Caillere, S., and S. Henin, Sepiolite, "The X-ray Identification and Crystal Structures of Clay Minerals," chap. VIII, pp. 325–343, Mineralogical Society of Great Britain Monograph (1961).

[185] Brindley, G. W., X-ray and Electron Diffraction Data for Sepiolite, *Am. Mineralogist,* **44:** 495–500 (1959).

[186] Rogers, L. E., J. P. Quirk, and K. Norrish, Aluminum Sepiolite, *J. Soil Sci.,* **7:** 177–183 (1956).

In the structure presented by Nagy and Bradley, the b dimension is one and a half times that for attapulgite. Attapulgite has amphibole-type chains composed of two pyroxene-type chains, whereas in sepiolite the ribbons contain three instead of two pyroxene chain components and, therefore, are not amphibole-like in character but rather present a new silicate structure type. The ideal density is 2.2, comparing favorably with determined values ranging from 2.0 to 2.1.

Chemical analyses indicate that fewer than nine octahedral sites are actually occupied, especially if allowance is made for alumina ions in tetrahedral position, for the uncertainty of the actual valence state of the minor constituents, and for the possibility that minor constituents may actually be accessories. Neutrality is maintained with the presence of sufficient additional protons and/or exchange ions to balance any residual charge deficiency.

A practical rationalization formula is, therefore, written as $H_6Mg_8Si_{12}O_{30}(OH)_{10}6H_2O$. This formula allows for water of four separate bonding energies, namely (1) molecular water, (2) water bound on brucite ribbon edges, (3) hydroxyls analogous with those of various layer mineral structures, and (4) balancing protons analogous with exchange ions. These categories are in the ratio of 6:4:3:2. It should be noted that oxygen ions at positions which join chains in this structural scheme are double links as compared with single-link oxygens in the attapulgite. It is quite possible that both structures are actually alike in that an improper choice of models was made in one case or the other. If the difference is real, however, it can be speculated that this is the factor which renders the sepiolite modification described by Longchambon[171] more stable than comparable high-temperature modifications of attapulgite.

Martin-Vivaldi and Cano-Ruiz[187] suggested that the palygorskite-sepiolite minerals occupy a region of discontinuity between dioctahedral and trioctahedral minerals and that there is a continuous series, with the two extremes having a planar structure and the intermediate members a fibrous structure. The change in structure from lamellar to fibrous takes place as the number of vacancies increases progressively. Among the minerals of fibrous structure, the structural change from sepiolite to attapulgite occurs in such a way as to allow no more than one vacancy per structural-fiber half cell, as if this vacancy were not distributed at random but symmetrically among the octahedral positions. The greater the number of vacant octahedral positions, the smaller is the extension of the octahedral layer. The margin of variation in the number of octahedral positions per structural fiber is greater in sepiolite than in attapulgite, probably because of the greater width of the fiber in the former mineral.

[187] Martin-Vivaldi, J. L., and J. Cano-Ruiz, Effect of Heat on Sepiolite, *Natl. Acad. Sci., Publ.* 456, pp. 170–180 (1956).

Martin-Vivaldi and Gonzalez[188] described a fibrous clay mineral occurring in cracks in a bentonite near Almeria, Spain, which seems to be intermediate between attapulgite and sepiolite on the basis of X-ray-diffraction and thermal analytical data. Further work is necessary to evaluate the structural concept of these investigators.

Mixed-layer minerals

Many clay materials are composed of more than one clay mineral, and the clay minerals may be mixed in several ways. The mixture may be of discrete clay-mineral particles in which there is no preferred geometric orientation of one particle with respect to its neighboring clay-mineral particles.

Another type of mixing is the interstratification of the layer clay minerals in which the individual layers are of the order of a single or a few alumino-silicate sheets. These so-called mixed-layer structures are a consequence of the fact that the layers of the different layer clay minerals are very similar, all being composed of silica tetrahedral-hexagonal layers and closely packed octahedral layers of oxygens and hydroxyl groups. Mixed-layer structures as stable as those composed of a single kind of layer are therefore possible.

Mixed-layer structures are of two different types. The interstratification may be regular; i.e., the stacking along the *c* axis is a regular repetition of the different layers. In such cases, the resulting structure has distinctive characteristics; the unit cell is equivalent to the sum of the component layers, and regular (001) reflections are obtained. An example of a regular mixed-layer mineral is chlorite composed of a regular alternation of mica and brucite layers (see pages 43 and 100).

Other examples of regular mixed-layer clay-mineral structures are allevardite (Caillere, Mathieu-Sicaud, and Henin[189]) a mica-type mineral with layers of water molecules separating double mica layers; rectorite (Bradley[149]), a continuous pyrophyllite layer and vermiculite layer; and corrensite (Lippmann[148]), a regular interstratification of chlorite and swelling chlorite, which is perhaps better described as a regular interstratification of chlorite and vermiculite.

Another kind of mixed-layer structure is due to a random irregular interstratification of layers in which there is no uniform repetition of layers. The importance of such random mixing was pointed out by Gruner[151] and later

[188] Martin-Vivaldi, J. L., and J. L. Gonzalez, A Random Intergrowth of Sepiolite and Attapulgite, Proceedings of the 9th National Clay Conference, pp. 592–602, Pergamon Press, New York (1962).

[189] Caillere, S., A. Mathieu-Sicaud, and S. Henin, Allevardite, *Bull. Soc. Franc. Mineral.*, **73:** 193–201 (1950).

elaborated by Hendricks and Teller[190] and Bradley.[191] Weaver[192] emphasized the importance of such structures by stating that over 70 percent of about 6,000 samples, ranging in age from Cambrian to Recent, studied by him contained some variety of mixed-layer clay minerals. Mixtures of the various three-layer clay minerals are most common, but occurrences of clay minerals which are probably mixtures of illite and kaolinite[193] and chlorite and kaolinite (Brindley and Gillery[194]) have also been described.

The study of mixed-layer structures is rather difficult, and in many cases, particularly in early investigations, their presence was not recognized. Recently, X-ray-diffraction procedures have been developed which simplify their identification and make easier the evaluation of the kind and abundance of layers that are mixed.

As pointed out previously (pages 46–50), the literature contains the names of many discredited or questionable clay-mineral species. Probably a great many of these "species" are mixed-layer combinations. Careful X-ray-diffraction techniques (page 154) are required to detect the occurrence of mixed-layer minerals. Sometimes differential thermal procedure will aid in their detection. Very frequently mixed layering cannot be detected solely on the basis of optical measurements, and the widespread occurrence of mixed-layer minerals is an important reason why clay-mineral identifications based solely on optical measurements must be made with great caution.

Since random mixed-layer minerals have an inherent variability, they cannot be given specific names. They can only be designated as mixtures of the layers involved.

Additional references

Kaolinite minerals

Brindley, G. W., and K. Robinson, X-ray Study of Some Kaolinitic Fireclays, *Trans. Brit. Ceram. Soc.,* **46:** 49–62 (1947).

De Keyser, W. L., and L. Degueldre, Relation between Morphology and Structure of Kaolinite and Halloysite, *Bull. Soc. Belge Geol.,* **63:** 100–110 (1954).

[190] Hendricks, S. B., and E. Teller, X-ray Interference in Partially Ordered Layer Lattices, *J. Chem. Phys.,* **10:** 147–167 (1942).

[191] Bradley, W. F., Diagnostic Criteria for Clay Minerals, *Am. Mineralogist,* **30:** 704–713 (1945).

[192] Weaver, C. E., Mixed-layer Clays in Sedimentary Rocks, *Am. Mineralogist,* **41:** 202–221 (1956).

[193] De Lapparent, J., Constitution et origine de la leverrierite, *Compt. Rend.,* **198:** 669–671 (1934).

[194] Brindley, G. W., and F. H. Gillery, Mixed-layer Kaolin-Chlorite, *Natl. Acad. Sci., Publ. 327,* pp. 349–353 (1954).

Gruner, J. W., Density and Structural Relations of Kaolinites and Anauxites, *Am. Mineralogist,* **22**: 855–860 (1937).

Hendricks, S. B., Concerning the Crystal Structure of Kaolinite, $Al_2O_3 \cdot 2SiO_2 \cdot 2H_2O$, and the Composition of Anauxite, *Z. Krist.,* **95**: 247–252 (1936).

Whittaker, E. J. W., Diffraction of X-rays by a Cylindrical Lattice, I–IV, *Acta Cryst.,* **7**: 827–832; **8**: 261–265; **8**: 265–271; **8**: 726–729 (1954–1955).

Smectite minerals

Faust, G. T., The Relation Between Lattice Parameters and Composition for Montmorillonite-group Minerals, *J. Wash. Acad. Sci.,* **47**: 146–147 (1957).

Forslind, E., The Clay-Water System, I, Crystal Structure and Water Adsorption of the Clay Minerals, *Swed. Cement Concrete Inst., Res. Bull.* 11 (1948).

Ginsburg, I. I., Nontronites from the Southern Urals, *Compt. Rend. Acad. Sci. URSS,* pp. 41–46 (1946).

Gruner, J. W., The Structural Relationships of Nontronites and Montmorillonites, *Am. Mineralogist,* **20**: 475–483 (1935).

Hendricks, S. B., and C. S. Ross, Lattice Limitations of Montmorillonite, *Z. Krist.,* **100**: 251–264 (1938).

Hofmann, U., and W. Bilke, Ueber die innerkristalline Quellung und das Basenaustauschvermögens des Montmorillonits, *Kolloid-Z.,* **77**: 239–253 (1936).

Hofmann, U., and A. Hausdorf, Kristallstruktur und innerkristalline Quellung von Montmorillonit, *Z. Krist.,* **104**: 265–293 (1942).

Jonas, E. C., Experimental Structure Factor Curves of Montmorillonites, *Natl. Acad. Sci. Publ.,* **566**: 295–307, Pergamon Press, New York (1959).

McAtee, J. L., Random Interstratification in Montmorillonite, *Am. Mineralogist,* **41**: 627–631 (1956).

Mering, J., and A. Oberlin, A Microscope and Electron-microdiffraction Study of Hectorite, Montmorillonite, and Nontronite, *Bull. Groupe Franc. Argiles,* **14**(9): 147 (1964).

Nagelschmidt, G., On the Atomic Arrangement and Variability of the Members of the Montmorillonite Group, *Mineral. Mag.,* **25**: 140–155 (1938).

Nakahira, M., Crystal Structure of Montmorillonite, *J. Sci. Res. Inst. (Tokyo),* **46**: 268–287 (1952).

Noll, W., Zur Kenntniss des Nontronits, *Chem. Erde,* **5**: 373–384 (1930).

Sedletsky, I. D., Kolloidno-Despersnaja Mineralogia, *Izv. Akad. Nauk SSSR* (1945).

Stresse, H., and U. Hofmann, Synthese von Magnesium-Silikatgelen mit Zweidimensional Regelmässiger Struktur, *Z. Anorg. Allgem. Chem.,* **247**: 65–95 (1941).

Tettenhorst, R., and W. D. Johns, Interstratifiation in Montmorillonite, Proceedings of the 13th National Clay Conference, pp. 85–94, Pergamon Press, New York (1966).

Winkler, H. G. F., Kristallstruktur von Montmorillonit, *Z. Krist.,* **105**: 291–303 (1943).

Illite Minerals

Andreatta, C., A New Type of Illite-Hydromica in Hydrothermal Deposit, *Clay Minerals Bull.,* No. 3, pp. 96–99 (1949).

Bannister, F. A., Brammallite (Sodium Illite): A New Mineral from Llandebie, South Wales, *Mineral. Mag.,* **26**: 304–307 (1943).

Dell'Anna, L., Glauconite in Cretaceous Limestone of the Salentino Peninsula, *Periodico Mineral. (Rome),* **33**(2/3): 521–548 (1964).

Gatineau, L., Real Structure of Muscovite—Distribution of Isomorphous Substitutions, *Bull. Groupe Franc. Argiles,* **16**(11): 3–10 (1965).

Gaudette, H. E., Illite from Fond du Lac County, Wisconsin, *Am. Mineralogist,* **50**: 411–417 (1965).

Grim, R. E., Petrology of the Pennsylvanian Shales and Noncalcareous Underclays Associated with Illinois Coals, *Bull. Am. Ceram. Soc.,* **14:** 18–25 (1934).

Gruner, J. W., The Structural Relationship of Glauconite and Mica, *Am. Mineralogist,* **20:** 699–714 (1935).

Hellman, N. N., D. G. Aldrich, and M. L. Jackson, Further Note on an X-ray Diffraction Procedure for the Positive Differentiation of Montmorillonite and Hydrous Mica, *Soil Sci. Soc. Am. Proc.,* **7:** 197–199 (1943).

Hofmann, U., J. Endell, and E. Maegdefrau, Specific Identity of the Clay Mineral, Sarospatite, *Ber. Deut. Keram. Ges.,* **24:** 339–344 (1942).

Hofmann, U., K. Endell, and D. Wilm, Röntgenographische Untersuchungen an Tonen, *Angew. Chem.,* **47:** 539–547 (1934).

Maegdefrau, E., and U. Hofmann, Glimmerartige Mineralien als Tonsubstanzen, *Z. Krist.,* **90:** 31–59 (1937).

Smith, J. V., and H. S. Yoder, Studies of Mica Polymorphs, *Mineral. Mag.,* **31:** 209–235 (1956).

Smulikowski, K., The Problem of Glauconite, *Arch. Mineral.,* **18:** pt. I, 21–120 (1954).

Yoder, H. S., Experimental Studies on Micas: A Review, Proceedings of the 6th National Clay Conference, pp. 42–60, Pergamon Press, New York (1959).

Yoder, H. S., and H. P. Eugster, Synthetic and Natural Muscovites, *Geochim. Cosmochim. Acta,* **8:** 225–280 (1955).

Halloysite minerals

Alexander, L. T., G. Faust, S. B. Hendricks, and H. Insley, Relationship of the Clay Minerals Halloysite and Endellite, *Am. Mineralogist,* **28:** 1–18 (1943).

Correns, C. W., and M. Mehmel, Ueber die Optischen und Röntgenographischen Nachweis von Kaolinit, Halloysit und Montmorillonit, *Z. Krist.,* **94:** 337–348 (1936).

Nagelschmidt, G., Röntgenographische Untersuchungen an Tonen, *Z. Krist.,* **87:** 120–146 (1934).

Ross, C. S., Minerals and Mineral Relationships of the Clay Minerals, *J. Am. Ceram. Soc.,* **28:** 173–183 (1945).

Chlorite minerals

Bradley, W. F., X-ray Diffraction Criteria of Chloritic Material, *Natl. Acad. Sci., Publ.* 327, pp. 324–334 (1954).

Bradley, W. F., Structural Irregularities in Hydrous Magnesium Silicates, *Natl. Acad. Sci., Publ.* 395, pp. 94–104 (1955).

Brindley, G. W., and F. H. Gillery, X-ray Identification of Chlorite Species, *Am. Mineralogist,* **41:** 169–196 (1956).

Brindley, G. W., B. M. Oughton, and K. Robinson, Polymorphism of the Chlorites, I, Ordered Structures, *Acta Cryst.,* **3:** 408–416 (1950).

Foster, M. D., Interpretation of the Composition and a Classification of Chlorites, *U.S. Geol. Surv., Profess. Paper* 414-A (1962).

Nelson, B. W., and R. Roy, Composition and Identification of Chlorites, *Natl. Acad. Sci., Publ.* 327, pp. 335–348 (1954).

Orcel, J., Chemical Composition of the Chlorites, *Bull. Soc. Franc. Mineral.,* **50:** 70–454 (1927).

Serdyuchenko, D. P., Chemical Constitution of the Chlorites, *Dokl. Akad. Nauk SSSR,* **60:** 1561–1564 (1948).

Serdyuchenko, D. P., Chlorites, Their Chemical Constitution and Classification, *Akad. Nauk SSSR, Inst. Geol. Nauk* (1953).

Shirozu, H., and S. W. Bailey, Chlorite Polytypism, III, Crystal Structure of an Ortho-hexagonal Iron Chlorite, *Am. Mineralogist,* **50**: 868–886 (1965).

Stephen, I., and D. M. C. MacEwan, Some Chloritic Clay Minerals of Unusual Type, *Clay Minerals Bull.,* **1**(5): 157–161 (1951).

Sudo, T., "Mineralogical Study on Clays of Japan," Maruzen Co. Ltd., Tokyo (1959).

Vermiculite minerals

Barshad, I., Vermiculite in Soil Clays, *Soil Sci.,* **61**: 423–442 (1946).

Barshad, I., Adsorptive and Swelling Properties of Clay-Water System, *Calif. Dept. Nat. Resources, Div. Mines, Bull.* 169, pp. 70–77 (1955).

Caillere, S., and S. Henin, Transformation of Minerals of the Montmorillonite Family into 10 Å Micas, *Mineral. Mag.,* **28**: 606–611 (1949).

Foster, M. D., Interpretation of the Composition of Vermiculites and Hydrobiotites, Proceedings of the 10th National Clay Conference, pp. 70–89, Pergamon Press, New York (1961).

Hathaway, J. C., Some Vermiculite-type Clay Minerals, *Natl. Acad. Sci., Publ.* 395, pp. 74–86 (1955).

Kazantzev, V. P., Structure of Vermiculite, *Mem. Soc. Russe Mineral., ser.* 2, **63**: 464–480 (1934).

Wager, L. R., Hydrobiotite, *Proc. Yorkshire Geol. Soc.,* **25**: 366–372 (1944).

Sepiolite-Palygorskite-Attapulgite minerals

Caillere, S., and S. Henin, The Sepiolite and Palygorskite Minerals, "The Differential Thermal Investigation of Clays," chap. IX, pp. 231–247, Mineralogical Society, London (1957).

Caldwell, O. G., and C. E. Marshall, A Study of Some Chemical and Physical Properties of the Clay Minerals Nontronite, Attapulgite and Saponite, *Univ. Missouri, Coll. Agr., Res. Bull.* 354 (1942).

De Lapparent, J., Formula and Structure of Attapulgite, *Compt. Rend.,* **202**: 1728–1731 (1935).

De Lapparent, J., The Constituents of Fuller's Earths, *Ann. Office Natl. Combustibles Liquides,* pp. 863–943 (1936).

De Lapparent, J., A propos de l'attapulgite, *Z. Krist.,* **97**: 237–248 (1937).

De Lapparent, J., Formules structurales et classification des argiles, *Z. Krist.,* **98**: 233–258 (1938).

Longchambon, H., Sur les caractéristiques des palygorskites, *Compt. Rend.,* **204**: 44–58 (1937).

Longchambon, H., and G. Migeon, Sur les sepiolites, *Compt. Rend.,* **204**: 431–434 (1936).

Marshall, C. E., and O. G. Caldwell, The Colloid Chemistry of the Clay Mineral Attapulgite, *J. Phys. Chem.,* **51**: 311–320 (1947).

Rogers, L. E., A. E. Martin, and K. Norrish, Palygorskite from Queensland, *Mineral. Mag.,* **30**: 534–540 (1954).

Stephen, I., Palygorskite from Shetland, *Mineral. Mag.,* **30**: 471–480 (1954).

Urbain, P., Classification of Hydrated Aluminum Silicates, *Compt. Rend. Soc. Geol. Franc.,* pp. 147–149 (1936).

Five
Diffraction Data

X-ray diffraction

General statement. In this chapter X-ray-diffraction data for the clay minerals are given, together with a discussion of the application of these data in their identification. The diffraction characteristics of many of the clay minerals have considerable similarity so that identification based solely on diffraction data as a "fingerprint" cannot be made safely; i.e., some knowledge of the cause of the diffraction characteristics is necessary.

In this chapter and in the tables to follow X-ray-diffraction data are given in angstrom units. In the first edition of this volume the data were given in kX units; i.e., they were based on comparisons with the values for calcite. Recent determinations of Avogadro's number, on which the calculated value of the absolute calcite spacings depends, show that values expressed in kX units are converted to angstroms by multiplying by 1.00202.[1-3] The difference between kX and angstrom units is significant only when measurements are accurate to 0.5 percent.

[1] Bearden, J. A., Evaluation of N and *e* by X-rays, *J. Appl. Phys.*, **12**: 395–403 (1941).

[2] Siegbahn, J., X-ray Wavelengths, *Nature*, **151**: 502 (1943).

[3] Bragg, W. L., Conversion of kX to A units, *J. Sci. Instr.*, **24**: 27 (1947).

The diffraction data presented herein are taken from the second edition of "The X-ray Identification and Crystal Structures of Clay Minerals," edited by G. Brown and published as a monograph of the Mineralogical Society of Great Britain, by which permission was kindly granted.

Clay minerals exist for the most part only in very fine particles, and some form of the powder method usually must be used to obtain the diffraction data. Because of certain inherent characteristics of the clay minerals, special cameras and special techniques are frequently required for clay-mineral work. Cameras and techniques entirely adequate for other materials have been found to be unsatisfactory for clay-mineral investigations. Brindley[4] discussed X-ray methods as applied to clay-mineral research, and the following is largely taken from his work. For a consideration of X-ray methods in general, the standard works of Bragg,[5,6] Buerger,[7] James,[8] Wilson,[9] Bragg et al.,[10] Azaroff and Buerger,[11] and Klug and Alexander[12] should be consulted.

In the clay minerals few, if any, reflections are obtained from spacings less than about 1 Å, and therefore with copper $K\alpha$ radiation, it is unnecessary to extend either the film or the camera to values of 2θ greater than about 90 or 100°. The specimen, therefore, can be brought nearer the focus of the X-ray tube and the exposure time shortened by omitting that part of the usual cylindrical camera corresponding to $90° < 2\theta < 180°$.

The first-order basal reflections of the clay minerals are frequently their most important reflections. Such reflections may correspond to spacings as large as 20 to 30 Å, and it is to obtain clear recordings of such low-angle reflections that special cameras and special techniques have been devised. Further, it may be necessary to study the exact shape or profile of such a reflection, and for this purpose cameras capable of recording spacings of

[4] Brindley, G. W., Experimental Methods, "The X-ray Identification and Crystal Structures of Clay Minerals," chap. I, pp. 1–50, Mineralogical Society of Great Britain Monograph (1961).

[5] Bragg, W. L., "The Crystalline State," G. Bell, London (1933).

[6] Bragg, W. L., "Atomic Structure of Minerals," Oxford University Press, Fair Lawn, N.J. (1937).

[7] Buerger, M. L., "X-ray Crystallography," Wiley, New York (1942).

[8] James, R. W., Optical Principles of the Diffraction of X-rays, "The Crystalline State," vol. 2, G. Bell, London (1948).

[9] Wilson, A. J. C., "X-ray Optics," Methuen, London (1949).

[10] Bragg, W. H., M. Laue, and C. Hermann, eds., "International Tables for the Determination of Crystal Structure," ref. ed., I, Bell, London (1944).

[11] Azaroff, L. V., and M. J. Buerger, "The Powder Method in X-ray Crystallography," McGraw-Hill, New York (1958).

[12] Klug, H. P., and L. E. Alexander, "X-ray Diffraction Procedures for Polycrystalline and Amorphous Substances," Wiley, New York (1954).

the order of 50 Å are required.[13,14]   Cameras with diameters of 9 to 20 cm are commonly employed in clay-mineral work.

Most of the clay minerals occur in flake-shaped units, and consequently when a powder is packed, an aggregate orientation of the flakes develops readily. Thus, care must be taken in the preparation and mounting of clay samples, or enhanced basal reflections will be obtained from an aggregate arrangement. For some types of work where basal reflections are particularly desirable, as in the study of mixed-layered structures, advantages may be taken of this tendency to develop oriented aggregates. Bradley, Grim, and Clark[15] described a method of forming well-oriented aggregates and using them in X-ray-diffraction studies. Such aggregates yielding up to nine basal orders have been prepared in the author's laboratory, using a centrifuge technique modified after that of Kinter and Diamond.[16]

In the smectite and vermiculite clay minerals the basal spacings vary with the humidity of the atmosphere to which they are subjected. It may, therefore, be necessary to control the humidity of the atmosphere within the camera. Techniques using clay-mineral–organic complexes[17,18] have been devised for these minerals (see pages 78 and 423) and are superior for identification purposes.

Imperfections of crystals affect their diffraction characteristics, and they are particularly important for the clay minerals. The imperfections may be of several kinds, as follows:

Because of the highly symmetrical arrangement of the atoms in the various layers and the relatively weak binding force between them, the layers can be displaced with respect to one another. The geometrical relation between adjacent sheets of atoms in contact may remain the same, although the relation between more distant atoms is changed. The result of such imperfections is the absence of certain types of reflections. Two-dimensional diffraction effects become the major feature when the layers are displaced randomly in two directions. If, for example, displacements occur along both the a and b axes, the only hkl reflections which are possible are the basal

[13] MacEwan, D. M. C., The Identification and Estimation of the Montmorillonite Group of Minerals, *J. Soc. Chem. Ind.* (*London*), **65**: 298–306 (1946).

[14] MacEwan, D. M. C., Some Notes on the Recording and Interpretation of X-ray Diagrams of Soil Clays, *J. Soil Sci.*, **1**: 90–103 (1949).

[15] Bradley, W. F., R. E. Grim, and G. F. Clark, X-ray Study of Montmorillonite, *Z. Krist.*, **97**: 216–222 (1937).

[16] Kinter, E. B., and S. Diamond, A New Method for Preparation and Treatment of Oriented Aggregate Specimens of Soil Clays for X-ray Diffraction Analysis, *Soil Sci.*, **81**: 111–120 (1956).

[17] MacEwan, D. M. C., Identification of Montmorillonite, *Nature,* **154**: 577 (1944).

[18] Bradley, W. F., Molecular Associations between Montmorillonite and Some Polyfunctional Organic Liquids, *J. Am. Chem. Soc.*, **67**: 975–981 (1945).

(00*l*) reflections.   Two-dimensional diffraction effects then arise from the regularity within the layers and require only two indices, namely, *hk*.   They give broad diffraction bands rather than sharp lines and are generally easily recognized.

Another type of imperfection is found in the mixed-layered structures when layers of different types are interstratified.   Because of the fact that many clay-mineral layers are very similar, being composed by silica tetrahedral sheets and closely packed layers of oxygen and hydroxyls, they can interstratify to build structures as stable or almost as stable as structures composed of a single layer.   When the alternation of different layers is regular, the result is equivalent to a single unit of larger unit-cell dimensions, e.g., chlorite.   When the alternation of different layers is irregular, new diffraction effects arise, in particular a nonintegral series of reflections from the basal planes (see pages 121 and 154).

The shape and size of the crystal particles exposed to X-ray influence the diffraction effects.   Thus, if the crystal particles have a distinct shape e.g., plate-like or rod-like, certain reflections may be broader than others. Theoretically a study of the diffuseness of the reflections and their profile should provide information on the shape and size of the crystals.   However, because the lattice imperfections of the clay minerals cause somewhat similar diffraction effects, only qualitative results could be hoped for at best, and much more useful results on size and shape are likely to be revealed by the electron microscope.

The aforementioned characteristics of the clay minerals make quantitative determinations of the components of complex mixtures particularly difficult.   Numerous investigators (Johns et al.;[19] Von Engelhardt,[20] Nicolas and Legrand,[21] Bertrand and Loisel,[22] Norrish and Taylor,[23] Drits,[24] and Gibbs[24a]) have presented a variety of techniques and procedures designed to obtain

[19] Johns, W. D., R. E. Grim, and W. F. Bradley, Quantitative Estimations of Clay Minerals, *J. Sediment. Petrol.*, **24:** 242–251 (1954).

[20] Von Engelhardt, W., Ueber die Moglichkeit der quantitativen Plasenanalyse von Tonen mit Röntgenstrahlen, *Z. Krist.*, **106:** 430–459 (1955).

[21] Nicolas, J., and C. Legrand, Contribution de la diffraction des rayon X au dosage des constituants des argiles: le quartz et la kaolinite, *Bull. Soc. Franc. Ceram.*, **42:** 113–116 (1959).

[22] Bertrand, A., and M. Loisel, Quantitative Analysis of Constituents of Clay Minerals by X-ray Diffraction, *Bull. Soc. Franc. Ceram.*, **50:** 53–61 (1961).

[23] Norrish, K., and R. M. Taylor, Quantitative Analysis by X-ray Diffraction, *Clay Minerals Bull.*, **5**(28): 98–109 (1962).

[24] Drits, V. A., The Quantitative X-ray Phase Analysis of Clay Minerals, *Soviet Phys.-Cryst. (English Transl.)*, **6:** 423–427 (1962).

[24a] Gibbs, R. J., Error Due to Segregation in Quantitative Clay Mineral X-ray Diffraction Mounting Techniques, *Am. Mineralogist,* **50:** 741–751 (1965).

quantitative data. Jasmund[25] and Melka[26] discussed the application of the Guinier camera to clay-mineral analysis.

Kaolinite and halloysite minerals. Diffraction data for well-crystallized kaolinite are given in Table 5-1; for *b*-axis disordered kaolinite in Table 5-2; for dickite in Table 5-3; for nacrite in Table 5-4; and for halloysite minerals in Tables 5-5 and 5-6. In general the determination of the kaolinite group by X-ray diffraction is simple, but the identification of the particular members of the group may be more difficult. The prominent basal reflections at about 7.16 Å (001) and 3.57 Å (002) are usually adequate for identification.

The chloritic clay minerals may be confused with kaolinite minerals, especially if the 14-Å chlorite reflection is not pronounced. However, usually the third order at 4.7 Å is seen to indicate the presence of chlorite. Chlorites rich in iron frequently give weak first- and third-order reflections, and differentiation from kaolinite is particularly difficult. The slight difference in the *c*-axis dimension of the two minerals may show as a doublet at 3.50 to 3.57 Å, which may serve to differentiate them. However, further tests may be necessary for precise determinations. Such a further test may be treatment with warm HCl to take advantage of the great solubility of chlorite in this acid. Also, on heating to 600°C, kaolinite tends to lose its crystalline character, whereas chlorite at this temperature is only partially dehydrated, causing increased intensity of the 14-Å reflection. Frequently other analytical methods, such as differential thermal analyses, are necessary to supplement the X-ray data to establish the presence of chlorite.

Nacrite, dickite, and kaolinite can be differentiated from each other on the basis of X-ray-diffraction data if the technique used provides adequate resolution (Tables 5-1 to 5-4).

Diffraction characteristics of well-crystallized versus poorly crystallized kaolinite are given in Fig. 4-9 and Tables 5-1 and 5-2. The transition from well-crystallized to poorly crystallized kaolinite is shown by a broadening and weakening of the reflections with the complete elimination of the weaker ones. There is a tendency for adjacent reflections to fuse into one. The basal reflection also increases from 7.14 to as much as 7.20 Å. The group of lines from (020) ($d = 4.46$) to (002) ($d = 3.57$) particularly reflect the change to lower crystallinity. In this region the clearly resolved doublet (11$\bar{1}$) and (1$\bar{1}$1) of well-crystallized material gives way to a single band, affording good evidence for the decrease in crystallinity. It is frequently difficult to deter-

[25] Jasmund, K., Erfahrungen bei röntgenographischen Untersuchungen von Tonmineralien mit einer Guinier-Kamera, *Geol. Foren. Stockholm Forh., Bd.,* **78**(1): 156–170 (1956).

[26] Melka, K., The Application of the Guinier Camera According to P. M. de Wolff for the X-ray Study of Clay Minerals, *Acta Univ. Carolinae, Geol.,* no. 1–2, pp. **3–24** (1959).

Table 5-1* *X-ray powder data for kaolinite
(After Brindley and Robinson[27]*)*

d (Å)	I	d (Calc.)	hkl
7.16	10+	7.15	001
4.46	4	4.469	020
4.36	5	4.370	1$\bar{1}$0
		4.332	110
4.18	5	4.172	1$\bar{1}$1
4.13	3	4.125	1$\bar{1}\bar{1}$
3.845	4	3.849	02$\bar{1}$
3.741	2	3.736	021
3.573	10+	3.573	002
		3.423	1$\bar{1}$1
3.372	4	3.370	111
3.144	3	3.148	11$\bar{2}$
3.097	3	3.098	1$\bar{1}\bar{2}$
		2.838	0$\bar{2}$2
2.753	3	2.748	022
2.558	6	2.566	1$\bar{3}$0
		2.563	20$\bar{1}$
		2.548	130
2.526	4	2.530	13$\bar{1}$
		2.520	1$\bar{1}$2
2.491	8	2.500	13$\bar{1}$
		2.490	200
		2.483	112
2.379	6	2.383	003
2.338	9	2.342	20$\bar{2}$
		2.341	1$\bar{3}$1
		2.335	11$\bar{3}$
2.288	8	2.301	13$\bar{1}$
		2.288	131
2.247	2	2.248	132
		2.234	040
2.186	3	2.182	2$\bar{2}$0+
2.131	3	2.130	02$\bar{3}$+
2.061	2	2.063	2$\bar{2}\bar{2}$+
1.989	6	1.994	20$\bar{3}$
		1.989	1$\bar{3}$2
1.939	4	1.936	132+
1.896	3	1.897	13$\bar{3}$+
1.869	2	1.869	042
1.839	4	1.845	1$\bar{3}$3
		1.836	202
		1.835	22$\bar{3}$
1.809	2	1.810	11$\bar{4}$+
1.781	4	1.786	004

Table 5-1   X-ray powder data for kaolinite
(Continued)

(After Brindley and Robinson[27])

d (Å)	I	d (Calc.)	hkl
1.707	2	1.711	$2\bar{2}2$
1.685	2	1.687	$24\bar{1}$
1.662	7	$\begin{cases}1.666 \\ 1.662\end{cases}$	$20\bar{4}+$ $1\bar{3}3+$
1.619	6	1.617	$133+$
1.584	4	1.587	$13\bar{4}+$
1.542	5B	1.542	$1\bar{3}4+$
1.489	8	$\begin{cases}1.490 \\ 1.487 \\ 1.486\end{cases}$	060 $33\bar{1}$ $33\bar{1}$
1.467	2	$\begin{cases}1.469 \\ 1.467\end{cases}$	$06\bar{1}$ $33\bar{2}+$
1.452	4B	1.455	$3\bar{3}0+$
1.429	4	1.430	005
1.403	2	1.404	$20\bar{5}$
1.390	2	1.391	$33\bar{3}+$
1.371	2		
1.338	4		
1.305	6B		
1.292	2		
1.282	5		
1.264	3		
1.246	3		
1.235	3		
1.217	1		
1.200	3		
1.190	3	1.190	006
1.168	2		
1.124	1		
1.094	3		
1.082	2		
1.057	1		
1.049	2		
1.039	2		
1.021	2	1.021	007
1.013	2		

d (Calc.) based on triclinic cell: $a = 5.15$ Å, $b = 8.95$ Å, $c = 7.39$ Å, $\alpha = 91.8°$, $\beta = 104.5°$, $\gamma = 90°$

+ signifies other possible indices.

B signifies broad reflections.

[27] Brindley, G. W., and K. Robinson, The Structure of Kaolinite, *Mineral. Mag.*, **27:** 242–253 (1946).

Table 5-2   *X-ray powder data for b-axis disordered kaolinites*

(After Brindley[30])

d (R., B., M.[28])	I	Line profile	d (Calc.)	hkl	d (E., G.[29])
7.18	10	sharp	7.159	001	7.30
4.48	8	band ↓	4.466	02l	4.42
3.584	10+	sharp	3.579	002	3.56
2.565	8	sharp	2.565	$20\bar{1}$	2.57
			2.557	130	
2.502	8	sharp	2.512	$13\bar{1}$	2.51
			2.496	200	
2.386	8	sharp	2.387	003	2.374
2.341	9	broad ↓	2.341	$20\bar{2}$	2.34
			2.316	131	
2.206	1	broad ↓	2.219	$13\bar{2}$	
			2.192	201	
1.989	4	broad ↓	1.992	$20\bar{3}$	1.99
			1.965	132	
			1.869	$13\bar{3}$	
			1.843	202	
1.789	4	sharp	1.789	004	
1.666	5	broad ↓	1.665	$20\bar{4}$	1.68
			1.642	133	
			1.563	$13\bar{4}$	
1.541	1	broad ↑	1.543	203	
1.488	10	sharp	1.489	060	1.49
			1.488	$33\bar{1}$	
			1.462	$33\bar{2}$	
1.458	3	broad ↕	1.457	061	
			1.453	330	
1.432	2	sharp	1.432	005	
			1.382	$33\bar{3}$	
1.375	1	broad ↕	1.375	062	
			1.367	331	
1.339	1		1.324	$13\bar{5}$	
1.310	1	broad ↓	1.308	204	
1.287	3	sharp	1.287	$26\bar{1}$	
			1.285	$40\bar{1}$	
			1.272	$33\bar{4}$	
1.265	1	broad ↕	1.263	063	

Table 5-2   X-ray powder data for b-axis disordered kaolinites (Continued)

(*After Brindley*[30])

d (R., B., M.[28])	I	Line profile	d (Calc.)	hkl	d (E., G.[29])
			1.256	26$\bar{2}$	
			1.254	332	
1.250	$\frac{1}{2}$		1.248	400	
			1.241	40$\bar{3}$	
1.237	3	broad↓	1.231	261	
1.194	1		1.193	006	

d (R., B., M.): Observed data by Robertson, Brindley, and Mackenzie[28] for kaolin from Tanganyika.

d (E., G.): Observed data by Von Engelhardt and Goldschmidt[29] for kaolin from Provins (France).

d (Calc.): Based on $a = 5.157$ Å, $b = 8.933$ Å, $c = 7.394$ Å, $\beta = 104.5°$.

Line profile indicates whether line is sharp or diffuse, and arrows indicate the direction in which the lines are diffused.

Spacings in angstroms.

[28] Robertson, R. H. S., G. W. Brindley, and R. C. Mackenzie, Kaolin Clays from Pugu, Tanganyika, *Am. Mineralogist*, **39:** 118–139 (1954).

[29] Von Engelhardt, W., and H. Goldschmidt, A Clay Mineral from Provins, France, *Heidelberger Beitr. Mineral. Petrog.*, **4:** 319–324 (1954).

[30] Brindley, G. W., Kaolin, Serpentine, and Kindred Minerals, "The X-ray Identification and Crystal Structures of Clay Minerals," chap. II, pp. 51–131, Mineralogical Society of Great Britain Monograph (1961).

mine precisely the crystallinity of very fine kaolinite since the small size would also cause a broadening and weakening of reflections. Johns and Murray[40] and Hinckley[41] published methods of expressing the crystallinity of kaolinites based on the ratio of the intensity of various reflections.

In the case of halloysite (2H_2O) the broadening of reflections, the development of bands replacing adjacent lines, and the elimination of reflections are carried to the stage (Fig. 5-1, page 143) where differentiation is usually not difficult.

Halloysite (4H_2O) gives a basal reflection to about 10.1 Å, a second-order reflection at about 5 Å, which is usually unobservable, and a third-order reflection almost coinciding with the second order for halloysite (2H_2O). Except for these few basal reflections, the patterns of the two forms of halloysite

[40] Johns, W. D., and H. H. Murray, Empirical Crystallinity Index for Kaolinite, Program Annual Meetings, *Geol., Soc. Am. Abstr.* (1959).

[41] Hinckley, D. N., Mineralogical and Chemical Variations in the Kaolin Deposits of the Coastal Plain of Georgia and South Carolina, *Am. Mineralogist*, **50:** 1865–1883 (1965).

[41a] Brindley, G. W., and K. Robinson, Randomness in the Structures of Kaolinitic Clay Minerals, *Trans. Faraday Soc.*, **42B:** 198–205 (1946).

Table 5-3   *X-ray powder data for dickite*
(*After Brindley*[30])

d (N., B.[31])	I	d (Calc.)	hkl	d (S., B.[32])	d (B., R.)
7.16	10	7.153	002	7.153	7.17
4.462	$\frac{1}{2}$	4.475	020⎫	4.451	4.452
4.439	4	4.439	110⎭		
4.370	4	4.369	11$\bar{1}$	4.366	4.374
4.270	3	4.271	021	4.254	4.273
4.131	7	4.122	111	4.118	4.129
3.950	2	3.959	11$\bar{2}$	3.953	3.962
3.795	6	3.794	022	3.790	3.798
		3.610	112		
3.587	10	3.579	004	3.578	3.582
3.427	3	3.431	11$\bar{3}$	3.428	3.432
3.272	2	3.264	023	3.262	3.253
3.101	2	3.096	113	3.094	3.103
2.938	2	2.938	11$\bar{4}$	2.936	2.937
2.794	2	2.795	024	2.794	2.796
		2.656	114	2.650	2.656
2.560	4	⎰2.562	13$\bar{1}$⎫	2.558	2.565
		⎱2.556	200⎭		
		2.527	11$\bar{5}$	2.524	2.523
2.510	5	⎰2.510	131⎫	2.503	2.506
		⎱2.503	20$\bar{2}$⎭		
2.400	1	2.386	006⎫	2.383	2.388
2.376	2	2.380	132⎭		
2.322	9	⎰2.326	13$\bar{3}$⎫	2.322	2.323
		⎱2.322	202⎭		
2.212	2	2.213	133⎫	2.210	2.211
		2.207	20$\bar{4}$⎭		
2.106	1	2.106	026	2.105	2.103
2.025	$\frac{1}{2}$	2.026	043		
1.975	5	⎰1.975	13$\bar{5}$⎫	1.974	
		⎱1.973	204⎭		
1.937	1	⎰1.938	223⎫	1.935	
		⎱1.935	11$\bar{7}$⎭		
1.898	2	1.897	044	1.896	
1.859	3	⎰1.861	135⎫	1.862	
		⎱1.857	20$\bar{6}$⎭		
		1.849	22$\bar{5}$	1.850	
1.805	1	⎰1.805	13$\bar{6}$⎫	1.803	
		⎱1.805	224⎭		
1.785	1	⎰1.790	008⎫	1.789	
		⎱1.789	117⎭		
1.762	$\frac{1}{2}$	1.763	045	1.761	
1.720	1	⎰1.722	11$\bar{8}$⎫	1.717	
		⎱1.715	22$\bar{6}$⎭		
1.686	1	1.687	24$\bar{1}$ +	1.686	
		1.670	151 +	1.669	

Table 5-3 **X-ray powder data for dickite** (Continued)

(*After Brindley*[30])

d (N., B.)[31]	I	d (Calc.)	hkl	d (S., B.)[32]
1.652	5	{ 1.651	13$\bar{7}$	1.651
		{ 1.650	206	
		1.641	311+	1.644
		1.628	24$\bar{3}$	1.626
1.613	1	{ 1.612	15$\bar{3}$ }	1.610
		{ 1.611	242 }	
1.586	1	1.589	31$\bar{4}$+	1.589
		1.573	153	1.574
1.555	4	{ 1.557	137 }	1.558
		{ 1.555	20$\bar{8}$ }	
		1.548	226+	1.548
		1.525	313 }	1.525
		1.524	31$\bar{5}$ }	
1.508	$\frac{1}{2}$	1.510	047	1.509
		1.503	24$\bar{5}$+	1.502
1.489	5	{ 1.491	060 }	1.488
		{ 1.488	33$\bar{1}$ }	
		1.469	22$\bar{8}$	1.469
1.458	3	{ 1.460	062 }	1.457
		{ 1.456	33$\bar{3}$+ }	
		{ 1.457	$\bar{2}$31 }	
1.429	2	{ 1.432	0,0,10 }	1.433
		{ 1.430	138 }	
1.395	1	1.391	208+	1.392
1.374	3	{ 1.377	064 }	1.376
		{ 1.374	333 }	
1.318	5		065+	1.319
				1.296
1.287	2			1.289
				1.279
1.263	1			1.264
1.253	1			1.254
1.236	2			1.238

d (N., B.): Experimental data, Newnham and Brindley[31].

d (S., B.): Experimental data, Smithson and Brown.[32]

d (B., R.): Experimental data, Brindley and Robinson.[30]

d (Calc.): Newnham and Brindley,[31] based on a = 5.15 Å, b = 8.95 Å, c = 14.42 Å, β = 96.8°.

+ signifies other possible indices.

Spacings in angstroms.

[31] Newnham, R. E., and G. W. Brindley, Structure of Dickite, *Acta Cryst.*, **9:** 759–764 (1956); see also corrections, *Acta Cryst.*, **10:** 88 (1957).

[32] Smithson, F., and G. Brown, Dickite in Northern England and North Wales, *Mineral. Mag.*, **31:** 381–391 (1957).

Table 5-4   X-ray powder data for nacrite
(*After Brindley*[30])

1 d (Å)	2 I	3 d (Å)	4 I	5 d (Å)	6 I	7 d (Å)	8 I
7.17	10	7.23	10	7.09	10	7.17	10
4.414	7	4.38	7	4.413	8	4.413	8
4.123	3	4.12	6	4.134	3	4.17	6
		3.96	1				
3.577	10+	3.59	9	3.586	9	3.60	10
		3.44	2B				
3.061	1	3.07	3	3.048	1B	3.08	2
2.917	½	2.93	1				
2.578 ⎫	3B	2.59	2	2.542	1B	2.54	6
2.502 ⎭		2.52	3				
2.438	7	2.43	6	2.423	10	2.423	10
2.397	7			2.398	1-2		
2.330	2	2.34	1			2.310	1
		2.29	1				
2.237	½	2.26	1	2.267	1B		
2.099	1B	2.09	2B	2.073	1B	2.104	6
1.982	½	1.95	2				
1.921	2B	1.93	2	1.906	2B	1.926	6
		1.897	1				
1.818	2	1.800	1	1.799	1	1.801	6
1.792	3	1.772	1				
		1.735	1				
		1.685	2	1.678	2	1.679	2
1.668	3B	1.660	1				
1.617	½	1.619	2B	1.619	2B	1.621	2
		1.584	1B				
1.488	5	1.489	8	1.489	8	1.492	8
1.462	1	1.463	4	1.458	4B	1.469	2
1.435	½	1.435	2	1.437	1		
		1.417	1				
		1.369	3				
		1.340	1				
1.319	½	1.312	1	1.317	½		
1.285	½	1.283	2	1.287	1	1.284	1
1.266	½	1.268	3	1.265	3	1.272	2
1.228	½	1.237	2B	1.232	2B		
		1.210	2	1.210	1		
		1.195	2				

1, 2: Nacrite from Hirvivaara, Finland; camera radius 10.0 cm. Von Knorring, Brindley, and Hunter.[33]

3, 4: Nacrite from Leicestershire, England; camera radius 3.0 cm. Claringbull.[34]

5, 6: Nacrite from Brand, Saxony; camera radius 2.88 cm. Gruner.[35]

7, 8: Nacrite from Freiberg, Saxony; camera radius 2.88 cm. Nagelschmidt.[36]
I = visually estimated intensity; B = broad line.

[33] Von Knorring, O., G. W. Brindley, and K. Hunter, Nacrite from Hirvivaara, Finland, *Mineral. Mag.*, **29:** 963–972 (1952).

[34] Claringbull, G. F., Nacrite from Groby, Leicestershire, *Mineral. Mag.*, **29:** 973 (1952).

[35] Gruner, J. W., Structure of Nacrite and Optical Properties of Kaolin Group Minerals, *Z. Krist.*, **85:** 345–354 (1933).

[36] Nagelschmidt, G., X-ray Studies on Clays, *Z. Krist.*, **87:** 120–145 (1934).

Table 5-5   X-ray powder data for halloysite 2H₂O

Table 5-5   X-ray powder data for halloysite $2H_2O$
(After Brindley[30])

d (1)	I	d (Calc.)	Indices	d (2)	d (3)
7.41	6	7.21	001	7.70	7.56
4.432	10	4.460	02	4.43	4.43
3.603	4	3.007	002	3.585	3.67
2.562	4	2.575	20	2.564	2.56
2.493	?				
2.405	1	2.405	003	2.408	
2.340	?				2.36
2.222	$\frac{1}{2}$	2.229	04	2.222	2.23
1.805	$\frac{1}{2}$*	1.803	004		
1.680	2	1.685	24	1.681	1.681
1.484	5	1.487	06	1.484	1.484
1.283	$1\frac{1}{2}$	1.287	40	1.283	1.285
1.233	$1\frac{1}{2}$	1.236	42	1.233	1.234
1.203	$\frac{1}{2}$*	1.202	006		
1.110	$\frac{1}{2}$	1.114	08		1.107
1.023	$\frac{1}{2}$	1.023	28		1.023
0.970	$\frac{1}{2}$	0.973	46		0.970 / 0.960
0.858	$\frac{1}{2}$	0.859	60		0.857
0.842	$\frac{1}{2}$	0.843	2,10		0.840

Experimental values:

d (1): Lawrence County, Missouri, data by Brindley and Robinson.[37]

d (2): Simla, India, data by Brindley and Robinson.[37]

d (3): Hungary, data by D. M. C. MacEwan.[38]

d (Calc.): Based on $a = 5.15$ Å, $b = 8.92$ Å, $d(001) = 7.214$ Å.

* Observed only after heat treatment at 300°C.

? Doubtful metahalloysite lines.

Spacings in angstroms.

[37] Brindley, G. W., and K. Robinson, Structure of Metahalloysite, *Mineral. Mag.*, **28**: 393–406 (1948).

[38] MacEwan, D. M. C. See MacEwan, D. M. C., and J. L. Amoros, Röntgenographic Investigation of Clays, *Anales Edafol. Fisiol. Vegetal (Madrid)*, **9**: 363–379 (1950).

consist of identical bands. Brindley[41b] and Brown and MacEwan[42] showed in detail the diffraction effect resulting from mixed layering of $4H_2O$ and $2H_2O$ forms when the interlayer water is gradually eliminated. The 10.1-Å

[41b] Brindley, G. W., The Kaolin Minerals, "X-ray Identification and Structures of Clay Minerals," chap. II, pp. 32–75, Mineralogical Society of Great Britain Monograph (1951).

[42] Brown, G., and D. M. C. MacEwan, X-ray Diffraction by Structures with Random Interstratification, "X-ray Identification and Structures of Clay Minerals," chap. XI, pp. 266–284, Mineralogical Society of Great Britain Monograph (1951).

Table 5-6   X-ray powder data for halloysite 4H₂O
(*Mehmel[39]*)

d (Å)	I	Indices	d (Å)	I	Indices
10.1	10	001	2.23	3	04
4.46	8	02	1.67	3	24
3.40	5	003	1.48	5	06
2.56	5	20	1.28	1	40
2.37	3	?	1.23	1	42

[39] Mehmel, M., Structure of Halloysite and Metahalloysite, *Z. Krist.*, **90:** 35–43 (1935).

spacing of halloysite (4H₂O) occurs at about the same place as the (001) basal spacings of the illites, but other reflections usually serve to differentiate these minerals easily. Also the reflection can be shifted in the halloysite mineral by drying at low temperature, and the halloysite will adsorb certain organic molecules (see Chap. 10), so that auxiliary procedures are available to identify the mineral.

Smectite minerals. The diffractions shown by powder diagrams of the smectite minerals can be placed in two categories. One class consists of basal reflections, which vary with the state of hydration of the mineral, i.e., with the thickness and regularity of the water layers between the silicate sheets. The *c*-axis spacing, the diffuseness of the reflections, and the number of orders shown vary from sample to sample, depending on the thickness of the water layers and their regularity, which factors in turn are dependent on the exchangeable cation present and the conditions, e.g., water-vapor pressure, under which the sample has been prepared. Because of the variability of the basal spacings no table can be given for them.

The second class of diffraction consists of general diffractions which are characteristic of the structure of the smectite layers themselves and are not dependent on the interlayer hydration. These are *hk* bands, which in general are the same for all smectites. The differences shown by various smectites are in details of spacings and relative intensities. Table 5-7 (after MacEwan[43]) gives data for the general diffractions of some smectite minerals.

By using oriented aggregates, as MacEwan[43] emphasized, patterns which contain either the (00*l*) or the *hk* lines can be obtained by suitably mounting the specimen in front of the X-ray beam.

As will be shown elsewhere (see Chap. 10), smectites possess the property of adsorbing certain organic molecules between the individual silicate layers

[43] MacEwan, D. M. C., The Montmorillonite Minerals, "The X-ray Identification and Crystal Structures of Clay Minerals," chap. IV, pp. 143–207, Mineralogical Society of Great Britain Monograph (1961).

Table 5-7 *Spacings of hk reflections of smectites*
(*After MacEwan*[43])

Indices	d(Å) calc.	Montmorillonite 1 d(Å)	1 I	Montmorillonite 2 d(Å)	2 I	2 S	Beidellite 3 d(Å)	3 I	Nontronite 4 d(Å)	4 I	Volkhonskoite 5 d(Å)	5 I	5 S	Saponite 6 d(Å)	6 I	Hectorite 7 d(Å)	7 I	Sauconite 8 d(Å)	8 I	8 S	Pimelite 9 d(Å)	9 I	9 S	Cardenite 10 d(Å)	10 I	Lembergite 11 d(Å)	11 I
11; 02	4.50	4.51	10	4.61	10	vbr	4.46	10	4.56	10	4.461	3		4.52; (4.07); (3.02)	8; 6; 6	4.58	10	4.62	6		4.21; 4.13	8	br	4.50; 4.43; (3.04)	7; 5	4.56	10
13; 20	2.60	2.62; 2.56; 2.42	0	2.56	8	vbr	2.61; 2.50; 2.39	10	2.63; 2.57	8	2.575	5	br	2.55; 2.46	8; 4	2.66; 2.49	8; 6	2.67; 2.584	8; 6		2.425	10	vbr	2.65	5	2.62; 2.47	8
22; 04	2.25	2.24; 2.159; (1.909)	3	2.221	3	br	2.244; 2.168; 2.094; (1.894)	4			2.24	½				2.289	3							2.298	1		
31; 15; 24	1.706	1.711; 1.685	6	1.692	6	br	1.697; 1.657	3; 6	1.719; 1.671	3	1.706	½		1.74	6	1.721; 1.692	6	1.757	3		1.691	3	br	1.739; 1.682	1	1.70	5
33; 06	1.503	1.500	10	1.492	10	vsh	1.491	10	1.522	10	1.506; (1.483)	6; 1		1.52	8	1.530	10	1.547	8		1.513	8	sh	1.536; (1.483)	6; 1	1.536	10
26; 40	1.301	1.295	6	1.289	6	br	1.287	6	1.319; 1.303	6	1.302	2	br	1.32	6	1.323; 1.304	6	1.330	5		1.307	5	br	1.326	1	1.326	5
35; 17; 42	1.252	1.250	6	1.244	6	sh	1.244	6	1.269; 1.249	3	1.267; 1.254	½; 1	br	1.29	6	1.270	2	1.288	2	br							
08; 44	1.127			1.120	3	sh								1.157	4			1.136	2	br							
37; 28; 51	1.036			1.029	3	br								1.052	4			1.069	2	br							
19; 53; 46	0.984	0.972	3	0.9765	4	br	0.972									0.992		1.008	4	br							
0,10; 55	0.902																	0.893	2	br							
39; 60	0.8683	0.865	3	0.8642	4	sh	0.864	3								0.880	4	0.846	1	br							
2,10; 48; 62																		0.823	2	br							
1,11; 57; 64																											
a	5.21 Å	5.18 Å		5.17 Å			5.15 Å		5.24 Å		5.18 Å			5.31 Å		5.25 Å		5.35 Å			5.24 Å			5.32 Å		5.32 Å	
b	9.02 Å	9.00 Å		8.95 Å			8.95 Å		9.13 Å		9.02 Å			9.20 Å		9.18 Å		9.28 Å			9.08 Å			9.22 Å		9.22 Å	

1. Unter-Rupsroth, Rhön; [Si$_{7.48}$Al$_{0.52}$]^{IV}[Al$_{3.54}$Fe^{3+}$_{0.06}$Mg$_{0.40}$]^{VI}O$_{20}$(OH)$_4$Ca$_{0.32}$Na$_{0.14}$Mg$_{0.03}$, spacings and formula from Nagelschmidt.[44]

2. From Wyoming bentonite; typical formula of such material, [Si$_{7.76}$Al$_{0.24}$]^{IV}[Al$_{3.10}$Fe^{3+}$_{0.38}$Fe^{2+}$_{0.04}$Mg$_{0.52}$]^{VI}O$_{20}$(OH)$_4$Na$_{0.68}$. X-ray data by MacEwan[43] (unpublished elsewhere).

3. Black Jack Mine, Carson County, Idaho; [Si$_{6.92}$Al$_{1.08}$]^{IV}[Al$_{3.92}$Fe^{3+}$_{0.05}$]^{VI}O$_{20}$(OH)$_4$Ca$_{0.46}$Na$_{0.04}$Mg$_{0.04}$K$_{0.02}$; spacings and formula from Nagelschmidt.[44]

4. Behenjy, Madagascar; [Si$_{7.14}$Al$_{0.86}$]^{IV}[Al$_{0.16}$Fe^{3+}$_{3.68}$Fe^{2+}$_{0.08}$Mg$_{0.16}$]^{VI}O$_{20}$(OH)$_4$Ca$_{0.30}$Mg$_{0.26}$; spacings and formula from Nagelschmidt.[44]

5. Thompsons, Utah; analysis Al$_2$O$_3$ 19.70, Fe$_2$O$_3$ (incl. FeO) 3.10, Cr$_2$O$_3$ 1.67, MgO 3.72, NiO 0.02, MnO 0.21, CaO 0.23, TiO$_2$ 0.45, SiO$_2$ and H$_2$O ± not determined; spacings and analysis from McConnell.[45]

6. Spacings from Kerr; d values appear to be too low especially near top of table.[46]

7. Hector, San Bernardino County, California; [Si$_{7.78}$Al$_{0.10}$Mg$_{0.12}$]^{IV}[Mg$_{5.34}$Li$_{0.66}$]^{VI}O$_{20}$(OH)$_4$Ca$_{0.08}$Na$_{0.84}$K$_{0.02}$; spacings and formula from Nagelschmidt.[44]

8. Yankee Doodle Mine, Leadville, Colorado; [Si$_{6.60}$Al$_{1.40}$]^{IV}[Zn$_{3.70}$Mg$_{0.28}$Al$_{1.58}$Fe^{3+}$_{0.04}$]^{VI}O$_{20}$(OH)$_4$Ca$_{0.23}$Na$_{0.10}$; X-ray data after Faust[47]; formula after Ross.[48]

9. Synthetic material of small particle size; formula given as Si$_8$Ni$_6$O$_{20}$(OH)$_4$ by Voorthuijsen and Franzen[49]; X-ray spacings deduced from their microphotometer trace by MacEwan.

10. Carden Wood, Aberdeenshire; [Si$_{6.12}$Al$_{1.81}$]^{IV}[Al$_{0.88}$Fe^{3+}$_{1.36}$Fe^{2+}$_{0.30}$Mg$_{2.96}$Ca$_{0.24}$]^{VI}O$_{20}$(OH)$_4$M$^+$$_{0.46}$; spacings and formula from MacEwan.[50]

11. Moniwa, Japan; an iron-rich saponite [Si$_{7.24}$Al$_{0.76}$]^{IV}[Al$_{0.08}$Fe^{2+}$_{2.90}$Mg$_{3.04}$]^{VI}O$_{20}$(OH)$_4$Ca$_{0.46}$; spacings and formula from Sudo.[51] (Formula shows imperfect charge balance.)

Notes on Table

(i) Indices are those for a two-dimensional orthohexagonal lattice a = 5.21 Å, b = 9.02 Å; d calc. gives hk0 spacings which are approximately coincident with band heads.

(ii) Lattice parameters of the various minerals given in last two lines are calculated from the observed spacings; for specimens 1, 3, 4 and 7 from 26.40 for a and 33.06 for b, for specimens 2, 5, 6 and 10 from the higher-order reflections and adjusted to make b = √3a, for specimens 8, 9, and 11 they are obtained from 33.06 and the relation b = √3a.

(iii) When two or more spacings are given against a group of indices they probably represent subsidiary maxima or apparent edges of bands.

(iv) Reflections in parentheses may be due to impurities.

(v) I = relative intensity, S = shape of band, v = very, sh = sharp, br = broad.

44 Nagelschmidt, G., Atomic Arrangement of Montmorillonite Group, *Mineral Mag.*, **25**: 140–155 (1938).
45 McConnell, D., Volkhonskoite, *Natl. Acad. Sci., Publ.* 327, pp. 152–157 (1957).
46 Kerr, P. F., Attapulgite Clay, *Am. Mineralogist*, **22**: 534–535 (1937).
47 Faust, G. T., Sauconite, *Am. Mineralogist*, **36**: 795–822 (1951).
48 Ross, C. S., Sauconite, *Am. Mineralogist*, **31**: 411–424 (1948).
49 Voorthuijsen, J. J. B. van E., and P. Franzen, Synthetic Pimelite, *Rec. Trav. Chim.*, **70**: 793–812 (1951).
50 MacEwan, D. M. C., Cardenite, *Clay Minerals Bull.*, **2**(11): 120–126 (1954).
51 Sudo, T., Iron Saponite, *J. Geol. Soc. Japan*, **60**: 18–27 (1954).

Table 5-8   X-ray powder data for micas
(After Bradley and Grim[52])

$2M_1$ indices	Muscovite		Illite (Gilead)		$1M$ indices	Illite (Ballater)		Illite (St. Austell)		Illite (trioctahedral) (Carden Wood)		Biotite	
	d	I	d	I		d	I	d	I	d	I	d	I
002	9.99	s	9.98	s	001	9.9	s	10.1	s	10.0	10	10.1	vs
004	4.98	m	4.97	w	002	4.9	m	4.98	m	4.94	2		
110 $11\bar{1}$	4.47	vs	4.47	s	020 110	4.45	vs	4.50	s	4.47	9	4.58	w
111	4.29	w			$11\bar{1}$	4.28	w	4.35	vw				
022	4.11	w	4.11	vw	021	4.10	w	4.10	vw				
112	3.95	vw											
$11\bar{3}$	3.87	m			111	3.87	m	3.85	vw				
023	3.72	m	3.7	vw									
					$11\bar{2}$	3.64	mw	3.62	ms	3.68	2b		
113	3.55	vw											
$11\bar{4}$	3.48	m	3.4	vwd									
024 006	3.32	vs	3.31	m	022 003	3.35	vs	3.32	s	3.32	9	3.36	vs
114	3.20	ms	3.2	vw									
$11\bar{5}$	3.1	vw			112	3.09	mwd	3.08	ms	3.16	½	3.15	vw
025	2.98	s	2.98	w									
115	2.86	m	2.84	vw	$11\bar{3}$	2.85	md	2.89	mw	2.86	1	2.91	vw
$11\bar{6}$	2.78	m			023			2.67	w				
$13\bar{1}$ 200	2.585	w											
$20\bar{2}$ 131	2.56	vs	2.56	s	200; $13\bar{1}$	2.56	vs	2.57	vs	2.60	6	2.65	s
008	2.49	w			004					2.50	1		
202; $13\bar{3}$	2.46	w	2.44	w	$20\bar{2}$; 131	2.45	mw	2.47	w			2.51	w
$20\bar{4}$	2.39	mdb											
133	2.38	mdb	2.38	m	201; $13\bar{2}$	2.39	m	2.38	m	2.41	4	2.45	s
204; $13\bar{5}$	2.245	wd	2.24	m	$20\bar{3}$; 132	2.235	mw	2.25	mw			2.28	vw
$22\bar{3}$	2.185		2.18	w									
$20\bar{6}$ 135	2.14 2.13	mdb			202; $13\bar{3}$	2.14	m	2.14	m	2.158	2	2.183	s
$13\bar{6}$			2.11	w									
044	2.05	vw											
0,0,10	1.99	s	1.98	m	005	1.988	m	1.99	md	1.982	1	2.002	s
206; $13\bar{7}$	1.95	w			$20\bar{4}$; 133	1.94	w						
	1.83	vw			203; $13\bar{4}$							1.911	vw
138	1.76	w			$20\bar{5}$; 134			1.71	vw			1.752	vw
					204; $13\bar{5}$	1.647	md	1.65	md	1.689 1.639	3	1.672	s
2,0,$\bar{1}$0	1.65	w											
139	1.64	m	1.65	w									
								1.58	vvw			1.551	s
060 $33\bar{1}$	1.504	s	1.50	s	060; $33\bar{1}$	1.497	s	1.50	s	1,53	6	1.527	

Sources of data:

Muscovite and illite (Gilead); Grim, Bray, and Bradley.[53]

Illite (Ballater); Mackenzie, Walker, and Hart.[54]

Illite (St. Austell); Levinson.[55]

Illite (Carden Wood); Walker.[56]

Biotite; Nagelschmidt.[57]

Sources do not necessarily distinguish between angstrom and kX units but collective precision is less than conversion factor.

(Table footnotes continued on the following page)

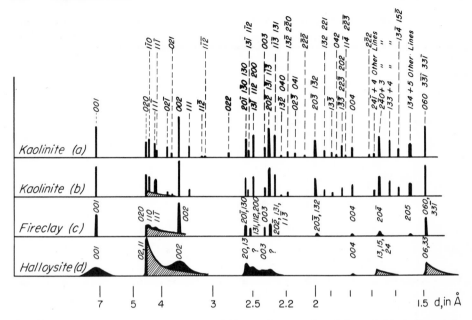

Fig. 5-1. Diagrammatic representation of X-ray photographs of (*a* and *b*) well-crystallized kaolinite, (*c*) poorly crystallized kaolinite, and (*d*) halloysite 2H₂O. (After Brindley and Robinson.⁴¹ᵃ)

with a consequent shift in the *c* dimension, depending on the size and configuration of the organic molecule. Such resulting smectite-organic complexes, as Bradley[18] showed, have a high degree of regularity, yielding sharp (00*l*) lines of many orders. For example, treatment of a sample of smectite with ethylene glycol provides a sharp (001) reflection at about 17 Å. Treatment with organic compounds permits the detection of small amounts of the mineral which would otherwise be missed in complex mixtures.

Illite minerals. Diffraction data for several illites, muscovite, and biotite are given in Table 5-8, and data for distinguishing the polymorphic varieties

Probable best fits for moderately strong diagnostic reflections:

Gilead: $b = 9.00$ Å, $\beta = 95°$
Ballater: $b = 9.00$ Å, $\beta = 101°$
St. Austell: $b = 8.98$ Å, $\beta = 101°$
Carden Wood: $b = 9.2$ Å, $\beta = 100–101°$

Indices become uncertain for spacings less than about 26 Å.

⁵³ Grim, R. E., R. H. Bray, and W. F. Bradley, Mica in Argillaceous Sediments, *Am. Mineralogist*, **22**: 813–829 (1937).

⁵⁴ Mackenzie, R. C., G. F. Walker, and R. Hart, Illite from Ballater, *Mineral. Mag.*, **28**: 704–714 (1949).

⁵⁵ Levinson, A. A., Polymorphism among Illites and Hydrous Micas, *Am. Mineralogist*, **40**: 41–49 (1955).

⁵⁶ Walker, G. F., Trioctahedral Minerals in Soil Clays, *Mineral. Mag.*, **29**: 72–84 (1950).

⁵⁷ Nagelschmidt, G., X-ray Investigations on Clays, *Z. Krist.*, **97**: 514–521 (1937).

Table 5-9   Powder diffraction data for distinguishing polymorphic varieties of mica
(After Bradley and Grim[52])

1			2			3			4				5		
hkl	d(Å)	I	hkl	d(Å)	I	hkl	d(Å)	I	hkl(1M)	hkl(3T)	d(Å)	I	hkl	d(Å)	I
			$11\bar{1}$	4.46	m	101	4.46	w	110	101	4.55	vw			
			021	4.39	vw										
$11\bar{1}$	4.35	w	111	4.30	w								021	4.52	vvw
021	4.12	vw	022	4.11	vw										
			112	3.97	vw										
			$11\bar{3}$	3.89	w	104	3.87	w	111	104	3.93	vvw	112	4.08	vvw
			023	3.74	w								023	3.81	w
$11\bar{2}$	3.66	w				105	3.60	w	$11\bar{2}$	105	3.66	vw			
			$11\bar{4}$	3.50	m								$11\bar{4}$	3.54	m
003 / 022 {3.36	vs		006 / 024 {3.35	vs		009	3.33	vs	022 / 003	106 / 009	3.39 / 3.35	m / vs	006	3.36	vs
			114	3.21	m								114	3.28	m
112	3.07	m				107	3.11	w	112	107	3.14	w	$11\bar{5}$	3.16	vvw
			025	3.00	m								025	3.04	s
$11\bar{3}$	2.93	m	115	2.87	w	108	2.88	w	$11\bar{3}$	108	2.92	w	115	2.93	vvw
			$11\bar{6}$	2.80	w								$11\bar{6}$	2.82	w
023	2.69	m							023	109	2.71	vw			

1. Synthetic 1M muscovite, $a = 5.208$ Å, $b = 8.995$ Å, $c = 10.275$ Å, $\beta = 101°\ 35'$ (Yoder and Eugster[58]).

2. Synthetic 2M₁ muscovite, $a = 5.189$ Å, $b = 8.995$ Å, $c = 20.097$ Å, $\beta = 95°\ 11'$ (Yoder and Eugster[58]).

3. Natural 3T muscovite, $a = 5.203$ Å, $c = 29.988$ Å (Yoder and Eugster[58]).

4. Natural 1M and 3T phlogopites; 1M polymorph, $a = 5.32$ Å, $b = 9.20$ Å, $c = 10.19$ Å, $\beta = 99°\ 49'$; 3T polymorph, $a = 5.32$ Å, $c = 30.15$ Å. These two polymorphs can be distinguished only by single-crystal methods (Smith and Yoder[59]).

5. Natural 2M₁ phlogopite, $a = 5.347$ Å, $b = 9.227$ Å, $c = 20.252$ Å, $\beta = 95°\ 1'$ (Yoder and Eugster[60]).

[58] Yoder, H. S., and H. P. Eugster, Synthetic and Natural Muscovites, *Geochim. Cosmochim. Acta*, **8**: 225–280 (1955).

[59] Smith, J. V., and H. S. Yoder, Studies of Mica Polymorphs, *Mineral. Mag.*, **31**: 209–235 (1956).

[60] Yoder, H. S., and H. P. Eugster, Phlogopite Synthesis, *Geochim. Cosmochim. Acta*, **6**: 157–185 (1954).

of mica are given in Table 5-9 from Bradley and Grim.[52]   The illite minerals can usually be identified on the basis of X-ray diffraction by their (00*l*) spacings, with the first order at about 10 Å.   Some difficulty may be encountered with completely collapsed smectite, but the slightly lower (001) reflection for it and the characteristics of the remainder of the pattern are generally adequate to distinguish it from illite.

Differentiation of the polymorphic forms of mica by X-ray diffraction is not always an easy matter.   Bradley and Grim[52] showed that the reflections in the region between 4.4 and 2.6 Å can usually be used to distinguish micas of one-, two-, and three-layer unit cells (see Table 5-9).   This region includes only (02*l*) and (11*l*) reflections except for one strong basal order at

[52] Bradley, W. F., and R. E. Grim, Mica Clay Minerals, "The X-ray Identification and Crystal Structures of Clay Minerals," chap. V, pp. 208–241, Mineralogical Society of Great Britain Monograph (1961).

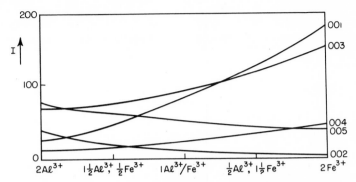

Fig. 5-2. Graph showing the change in intensities of the (00*l*) reflections with varying composition of the octahedral layer for a dioctahedral mica. (After Grim, Bradley, and Brown,[61])

3.33 Å. Grim, Bradley, and Brown,[61] Walker,[56] and others[62] have pointed out that the position of the (060) reflection and the intensity of the second-order basal reflection can usually be used to distinguish between dioctahedral and trioctahedral micas. For the dioctahedral forms, (060) is close to 1.50 Å and (002) is strong. For the trioctahedral forms, (060) lies between 1.525 and 1.535 Å and (002) is weak or absent. There are some exceptions to this rule, e.g., glauconite, and it must be used with caution. The (060) line is useful in identifying other clay minerals; for kaolinite it is at 1.485 Å and for chlorites at 1.53 to 1.56 Å. This line cannot be used to separate illites and smectites, and care must be taken to avoid confusion with quartz, which can give a reflection at 1.53 Å, if present in abundance of more than about 10 percent.

Grim, Bradley, and Brown[61] attempted a correlation of the effect of major substitutions in the mica lattice and the intensities of the (00*l*) reflections. They made no computations for substitutions of Li^+ and Ti^{4+}. Substitutions of Al^{3+} for Si^{4+} in the tetrahedral sheet make little or no difference in basal intensities, whereas substitutions in the octahedral sheet cause large intensity variations. The results of the analyses are given in Figs. 5-2 and 5-3. They pointed out that a comparison of calculated and observed values should make possible a guess at the composition of the octahedral layer, and hence a guess at whether the mica is dioctahedral or trioctahedral (if Li^+ and Ti^{4+} are absent or very scant).

[61] Grim, R. E., W. F. Bradley, and G. Brown, The Mica Clay Minerals, "X-ray Identification and Structures of Clay Minerals," chap. V, pp. 138–172, Mineralogical Society of Great Britain Monograph (1951).

[62] MacEwan, D. M. C., Les Minéraux argileux de quelques sols écossais, *Verre Silicates Ind.,* **12:** 3–7 (1947).

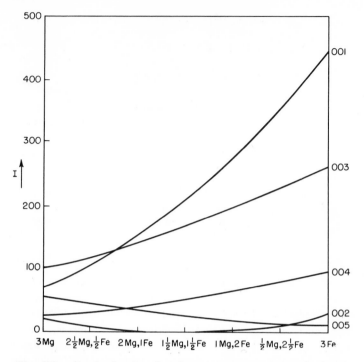

Fig. 5-3. Graph showing the change in intensities of the $(00l)$ reflections with varying composition of the octahedral layer for a trioctahedral mica. (After Grim, Bradley, and Brown.[61])

Suppose the mica is dioctahedral, then if the second-order basal reflection is of the same order of strength as the first and third order, the mica is a highly aluminous member. If the second-order basal reflection is weak or missing, then the mica is mainly ferriferous in octahedral positions. If the mineral is known to be of the trioctahedral variety, the third and fifth orders should be compared. If the third and fifth orders are of similar intensity, the mica is of the magnesium end of the series, but if the third order is stronger than the fifth order, then the mica is highly ferriferous.

The intensities shown in Figs. 5-2 and 5-3 were calculated from the formula for a single crystal

$$I_{00l} = F_{00l}^2 \, \Xi$$

where F_{00l} = structure factor for a particular plane $(00l)$

Ξ = Lorentz and polarization factor for a single crystal = $\dfrac{1 + \cos^2 2\theta}{\sin 2\theta}$

θ = Bragg reflection angle

The radiation is assumed to be Cu Kα with $\lambda = 1.54$ kX.

The intensities for the corresponding power diagrams can be calculated by multiplying each intensity value by the factor

$$\frac{\Xi_{powder}}{\Xi_{single\ crystal}} = \frac{1}{\sin\theta}$$

for each plane.

In a more recent extension of this principle, Brown[63] tabulated sets of the pertinent partial sums for the amplitudes of the first three orders of reflection through pertinent composition ranges for iron- and magnesium-bearing compositions.

Chlorite minerals. X-ray-diffraction data for several chlorite minerals are given in Tables 5-10 and 5-11 taken from Brindley.[64] The variation from one chlorite to another is to be found in slight modifications of intensities or lattice spacings due to small variations of unit-cell dimensions and population of various cation positions. The identification of the various forms of chlorite is very difficult and may be impossible unless chemical and optical data are available to supplement the X-ray analysis. The small size of the clay-mineral chlorite particles and their less regular crystallinity cause some diffuseness of reflections and the absence of some ordinary weak reflections. As a consequence, usually it is now possible to identify a clay mineral only as belonging to the chlorite group of clay minerals.

Care must be taken in the identification of the chlorite minerals by X-ray diffraction, particularly to avoid confusion with smectite, vermiculite, and kaolinite. Chlorites in which the proportion of octahedral positions occupied by Fe ions is not greater than about 30 percent have medium to strong intensities for the first five orders of the basal reflections. Chlorites rich in iron give relatively weak (001) and (003) reflections and strong (002) and (004) reflections. Iron-rich chlorites, therefore, are readily confused with kaolinite.

[63] Brown, G., The Effect of Isomorphous Substitutions on the Intensities of $(00l)$ Reflections of Mica- and Chlorite-type Structures, *Mineral. Mag.*, **30:** 657–665 (1955).

[64] Brindley, G. W., The Chlorite Minerals, "The X-ray Identification and Crystal Structures of Clay Minerals," chap. VI, pp. 242–246, Mineralogical Society of Great Britain Monograph (1961).

Table 5-10 X-ray powder data for monoclinic chlorites

(After Brindley[64])

hkl	1 d (Å)	1 I	2 d (Å)	2 I	3 d (Å)	3 I	4 d (Å)	4 I	5 d (Å)	5 I	6 d (Å)	6 I
001	14.1	7	14.2	8	14.2	7	13.6	6	14.3	6	14.7	41
002	7.07	9	7.12	10	7.10	10	7.02	7	7.18	10	7.14	>90
003	4.72	8	4.75	8	4.72	7	4.70	9	4.79	10	4.75	57
020	4.59	1							4.61	2		
004	3.54	10	3.56	10	3.54	10	3.525	10	3.592	10	3.56	>90
005	2.84	5	2.85	4	2.83	4	2.822	6	2.873	6	2.840	23
131; 20$\bar{2}$	2.58	3	2.58	3	2.60	4	2.570	1	2.590	3	2.600	5
13$\bar{2}$; 20$\bar{1}$	2.53	6	2.55	5	2.56	4	2.526	2	2.548	4	2.564	6
132; 20$\bar{3}$	2.43	5	2.44	4	2.45	4	2.430	2	2.449	4	2.462	6
13$\bar{3}$; 20$\bar{2}$	2.37	4	2.38	2	2.39	3	2.355	2	2.392	3	2.398	6
006							2.353	2			2.356	4
133; 20$\bar{4}$	2.25	3	2.27	3	2.27	3	2.244	2	2.268	2	2.238	5
134; 20$\bar{5}$	2.06	1			2.07	½					?2.151	4
007	2.026	2	2.04	2	2.02	1	2.016	4	2.053	2	2.027	10
13$\bar{5}$; 20$\bar{4}$	2.000	6	2.01	4	2.00	5	1.993	4	2.018	4	2.014	11
135; 20$\bar{6}$	1.883	3	1.891	2	1.888	2	1.872	3	1.896	1	1.895	5
13$\bar{6}$; 20$\bar{5}$	1.825	3	1.833	2	1.830	1	1.815	3	1.838	3	1.824	4
136; 20$\bar{7}$	1.73	½B	1.732	1			1.708	½	1.729	½	?1.762	3
13$\bar{7}$; 20$\bar{6}$	1.660	1	1.672	1	1.668	1	1.653	½	1.675	½	1.713	5
137; 20$\bar{8}$	1.562	4	1.577	2	1.570	4	1.559	4	1.582	4	1.578	8
060; 33$\bar{1}$	1.534	7	1.541	6	1.549	5	1.530	3	1.539	2	1.549	5
062; 331	1.500	2	1.507	2	1.513	2			1.505	1	1.514	4
063+	1.458	1										
0,0,10	1.417	2	1.429	1	1.415	2	1.412	3	1.437	1	1.425	8
064+	1.407	½										
208+	1.392	5	1.403	4	1.395	4	1.389	5	1.409	3	1.397	8

Index	d	I	d	I	d	I	Sheridanite d	I	Penninite d	I
065+	1.349	½								
262+ } 2,0,10+	1.317	2	1.323	2	1.327	2	1.315	1	1.323	½
263+	1.297	½	1.294	3	1.299	3				
066+	1.287	2	1.228	1						
2,0,$\overline{11}$+	1.220	2			1.222	1				
266+ } 2,0,10+	1.190	½			1.196	½				
0,0,12	1.182	1	1.192	1	1.180	1				

	1	2	3	Sheridanite	Penninite
Si	2.730 } 4.0	3.096 } 4.0	2.694 } 4.0		2.64 } 4.0
Al^{IV}	1.270	0.904	1.306		1.36
Al^{VI}	1.155	0.791	1.127		1.17
Fe^{3+}	0.074 } 6.019	0.113 } 6.002	0.144 } 6.018		0.52 } 5.84
Fe^{2+}		0.553	1.846		1.24
Mn	0.054	0.008	0.041		
Mg	4.736	4.537	2.860		2.82
Ca					0.09

1–3: Data from Shirozu.[65]

4–5: Data by Brindley and Ali, from Brindley and Robinson.[66]

6: Data from Sato and Sudo.[67]

B signifies broad.

+ signifies other possible indices.

[65] Shirozu, H., X-ray Patterns and Cell Dimensions of Chlorites, *Mineral. J.* (*Tokyo*), **2:** 209–223 (1958).

[66] Brindley, G. W., and K. Robinson, The Chlorite Minerals, "X-ray Identification and Crystal Structures of Clay Minerals," chap. VI, pp. 173–198, Mineralogical Society of Great Britain Monograph (1951).

[67] Sato, H., and T. Sudo, Chlorite from Hitachi Mine, *Mineral. J.* (*Tokyo*), **1:** 395–397 (1956).

Table 5-11   X-ray powder data for orthohexagonal chlorites
(After Brindley[64])

hkl	1 d (Å)	1 I	2 d (Å)	2 I	3 d (Å)	3 I	4 d (Å)	4 I
001	14.1	5	14.1	4	14.2	3	14	3
002	7.05	10	7.05	10	7.1	10	6.93	10
003 / 020	}4.69	5	}4.70	4	}4.68	2B	}4.64	5
004	3.52	9	3.54	8	3.55	8	3.50	9
005	2.83	4	2.83	1	2.85	$\frac{1}{2}$	2.784†	3
200	2.67	1	2.67	3	2.68	3	2.694	1
202	2.51	5	2.53	5*	2.52	5*	2.512	7
203			2.47	1			2.335	$\frac{1}{2}$
025	2.40	$\frac{1}{2}$	2.39	$\frac{1}{2}$	2.43	$\frac{1}{4}$		
204	2.14	4	2.15	1*	2.15	2*	2.138	3
007	2.02	2			2.03	$\frac{1}{4}$		
206	1.776	4						
240			1.77	1B	1.77	1B	1.770	2
060	1.559	5	1.558	6	1.557	5	1.559	5
062	1.523	3	1.521	4	1.521	3	1.527	2
063 / 208	}1.479	4	}1.479	1	}1.480	$\frac{1}{2}$B	}1.476	1
064	1.424	1	1.423	1	}1.422	1B	1.428	1
0,0,10	1.412	3	1.416	1				
065	1.365	$\frac{1}{2}$						
400 / 401	}1.347	$\frac{1}{2}$	}1.345	1	}1.343	$\frac{1}{2}$		
402	1.326	1	1.326	1	1.323	1		
066	1.300	1	1.298	$\frac{1}{2}$	1.302	$\frac{1}{2}$		
404			}1.26	$\frac{1}{4}$	}1.26	$\frac{1}{4}$		
2,0,10	1.253	1						
0,0,12	1.179	2	1.180	$\frac{1}{2}$	1.183	$\frac{1}{2}$		
406	1.171	$\frac{1}{2}$						

	1		2		3		4	
Si	2.537 }4.0		2.633 }4.0		2.797 }4.0		3.1 }4.0	
Al^{IV}	1.463		1.367		1.203		0.9	
Al^{VI}	1.204		1.196		0.953		1.3	
Fe^{3+}	0.405		0.212		0.436		0.6	
Fe^{2+}	3.242 }5.928		3.612 }5.980		3.300 }5.908		3.35 }6.0	
Mn	0.053		0.056		0.137			
Mg	0.012		0.904		1.045		0.75	
Ca^{2+}	0.012				0.037			
							Chamosite	

1–3: Data from Shirozu.[65]

4: Data from Von Engelhardt.[68]

* = head of a diffuse band.   B = broad line.   † Coincidence with siderite line.

[68] Von Engelhardt, Structures of Thuringite, Bavalite, and Chamosite, Z. Krist., **104**: 142–159 (1942).

Table 5-12   Powder diffraction data for two macroscopic Mg
vermiculites (Cu Kα radiation)

(After Walker[70])

Indices	Batavite		West Chester vermiculite	
	d(Å)	I est.	d(Å)	I est.
002	14.4	vvs	14.4	vvs
004	7.18	vvw	7.20	vw
006	4.79	vw	4.79	vw
02l; 11l*	4.60	s	4.60	s
008	3.602	m	3.587	m
0,0,10	2.873	m	2.869	m
130; 200; 20$\bar{2}$	2.657	mw	2.657	mw
132; 20$\bar{4}$	2.602	ms	2.597	m
13$\bar{4}$; 202	2.550	m	2.550	mw
0,0,12; 13$\bar{6}$; 204	2.392	ms	2.392	ms
136; 20$\bar{8}$	2.277	vvw	2.266	vvw
13$\bar{8}$; 206	2.209	vvw	2.214	vw
138; 2,0,$\overline{10}$	2.082	w	2.081	w
0,0,14			2.048	vw
208	2.016	w	2.011	vw
1,3,$\overline{12}$; 2,0,10	1.835	vvw	1.835	vvw
2,0,$\overline{14}$	1.744	mw	1.748	w
1,3,$\overline{14}$; 2,0,12	1.673	mw	1.677	mw
1,3,14; 2,0,$\overline{16}$	1.576	vvw	1.574	vvw
060; 1,3,16; 2,0,14; 330; 33$\bar{2}$; 33$\bar{4}$	1.537	s	1.537	ms
332; 33$\bar{6}$	1.506	vvw	1.508	vvw
0,0,20; 1,3,16; 2,0,$\overline{18}$	1.444	vw	1.449	w
338	1.356	vw	1.357	vw
1,3,18; 2,0,$\overline{20}$; 3,3,$\overline{12}$; 40$\bar{2}$	1.332	mw	1.334	mw
2,0,$\overline{20}$; 400; 40$\bar{6}$	1.319	mw	1.320	mw
1,3,20; 2,0,18; 3,3,10; 3,3,14; 402	1.296	w	1.298	w
404	1.278	w	1.275	vw

* Reflections with k = 3n are diffuse; the 02 and 11 series, therefore, show
asymmetric powder lines with a sharp cutoff on the low-angle side and a
tail extending toward high angles.

Bradley[69] pointed out that typical chlorite compositions give little or no intensity at the sixth-order position, whereas kaolinite has at about 2.38 Å a distinct third order of approximately one-eighth the intensity of the (002) reflection of this mineral.

On heating chlorites to about 600°C, the temperature for the decomposition of the brucite layer, there are a marked increase in the intensity of

[69] Bradley, W. F., X-ray Diffraction Criteria of Chloritic Material, *Natl. Acad. Sci., Publ.* 327, pp. 324–334 (1954).

Table 5-13 Basal spacings (in angstroms) of powdered macroscopic vermicu-
lites saturated with various cations

(After Walker[70])

Inter-layer cation	Macon County vermiculite (Barshad[71,72])		Kenya vermiculite at 24°C (less than 300-mesh B.S.S.)			
	Air-dry	In water	30 percent RH*	50 percent RH*	70 percent RH*	In water
H	14.3					
Li	12.6	15.1	12.3	15.0	15.0	†
Na	12.6	14.8	12.3	14.8	14.8	14.8
K	10.4	10.6	10.5 vd		10.6 vd	13 vd
NH₄	11.2	11.2	10.8 vd		10.8 vd	10.9 vd
Rb	11.2	11.2				
Cs	12.0	12.0				
Mg	14.3	14.5	14.4	14.4	14.4	14.7
Ca	15.1	15.4	15.0	15.0	15.1	15.3
Sr				14.9		15.4
Ba	12.6	15.4	12.3	15.0	15.2	15.7
Al			14.0			14.0

vd = very diffuse.

* RH = relative humidity.

† High spacings of several hundred angstroms, obtained with Li vermiculite in distilled water, tend to revert to 15 Å after immersion for several hours or days.

the (001) reflection and decreases in the intensities of the (002), (003), and (004) reflections. Caution must be followed in using this factor to distinguish kaolinite and chlorite, because the temperature for the dehydration of both kaolinite and chlorite varies with particle size.

The characteristic 14-Å spacing may be confused with that of smectite or vermiculite even though it is generally sharper in the chlorites. For chlorites the 14-Å spacing is not changed on moderate heating to about 200°C, as it is in vermiculite, nor is it changed by treatment with a polar organic molecule (like glycol), as it is in the smectite.

Vermiculite minerals. Diffraction data for vermiculites are given in Tables 5-12 to 5-14 from Walker.[70] The strong reflection at 14 Å is characteristic of the mineral, but care must be used to differentiate vermiculite from chlorite, which also has a reflection at 14 Å, and from smectite, which may show a reflection at this position.

[70] Walker, G. F., Vermiculites and Some Related Mixed-layer Minerals, "The X-ray Identification and Crystal Structures of Clay Minerals," chap. VII, pp. 297–324, Mineralogical Society of Great Britain Monograph (1961).

Table 5-14 *Basal spacings (in angstroms) of heated vermiculite powders*

(*After Walker*[70])

Temperature °C	Interlayer cation						
	Mg	Ca	Ba	H	Li	Na	K
Room	14.3	15.1	12.6	14.3	12.6	12.6	10.4
170	11.8	11.8	10.3		11.0	10.3	10.5
250	10.1	10.1	10.3	10.1	10.3	10.1	10.4
410	10.0	10.0	10.1		9.7	10.0	10.4
610	9.4	9.6	10.1		9.4		10.4

From Macon County, North Carolina, less than 140 mesh (A.S.T.M.); samples sealed in glass capillaries at indicated temperature and photographed after cooling (after Barshad[71-73]).

[71] Barshad, I., Vermiculite and Its Relation to Biotite, *Am. Mineralogist.*, **33:** 655–678 (1948).

[72] Barshad, I., Effect of Interlayer Cations on Expansions of Mica Lattice, *Am. Mineralogist*, **35:** 225–238 (1950).

[73] Barshad, I., Interlayer Expansion of Vermiculite and Montmorillonite with Organic Substances, *Soil Sci. Soc. Am. Proc.*, **16:** 176–182 (1952).

Vermiculite can usually be distinguished from chlorite on the basis of the intensity of the (00*l*) reflections: In vermiculite the 14-Å line is the strongest reflection, and the 7-Å line and subsequent basal orders are relatively weaker, whereas in chlorite several basal orders of approximately equal intensity may occur. Vermiculite can be further distinguished from chlorite by the fact that the basal spacings are more readily shifted by heating. Also, treatment with certain organic molecules causes a shift in the basal reflections for vermiculite and not for chlorite.

Possible confusion of kaolinite with minerals giving a 14-Å reflection has already been considered (see page 130). Care must be taken not to confuse mixtures of kaolinite and vermiculite with chlorite; this usually requires that diffraction data be obtained for samples that have been heated as well as for the natural sample.

Vermiculites and smectite may both show a 14-Å reflection. A distinction can usually be made because this reflection changes readily for slight variations of relative humidity in smectites, whereas the vermiculites are stable over a wide range of moisture contents. Further, it appears that vermiculite forms only a single molecular layer of organic molecules between the mica sheet, whereas smectites may form multimolecular layers with certain organic molecules.

As shown in Table 5-13, the c-axis dimension of vermiculite varies with the nature of the interlayer cation. This factor distinguishes the mineral from chlorite and may be used to resolve the confusion with kaolinite.   However, smectites have this same characteristic, but the variation is of a greater range than in vermiculites.

Sepiolite-attapulgite-palygorskite minerals. Diffraction data for sepiolite and attapulgite-palygorskite taken from Caillere and Henin[74-75] are given in tables 5-15 and 5-16, respectively.   The general difference in the diffraction patterns of these minerals as compared with those of other clay minerals makes them readily distinguishable.

Sepiolite shows a characteristic diffraction at about 12.1 Å and attapulgite at 10.48 Å.   The scant available diffraction data for material called palygorskite are generally similar to the data for attapulgite.   However, the extent of variation in the diffraction characteristics of attapulgite-palygorskite is not yet known.   These spacings do not change with variations in relative humidity or moderate heating up to about 300°C.   Also they are not changed by treatment with organic polar molecules, and therefore they should not be confused with the expanding-layer clay minerals or with halloysite.   They do exhibit a change in diffraction characteristics on heating to over 300°C (see Chap. 9), which can be used as a further means of identifying them.

Mixed-layer structures. In the case of mixtures of discrete clay-mineral particles, a diffraction pattern contains reflections typical of each component. The characteristics are no different from that resulting from a mixture of clay minerals and nonclay minerals, for example, when quartz is present in the clay material.

In the case of a mixture which is a regular interstratification of layer clay minerals, the diffraction effect is equivalent to a larger unit cell which is a multiple of the individual layers.   The regularity of the structures is maintained on a larger scale, and an integral series of (00l) reflections is obtained from the larger unit cell.   The structure of chlorite is an example of this type of mixing.

If the mixture is a random interstratification of layer clay minerals, and only a very few layers of second type are present, the reflections differ very little from those of the dominant layer.   If, however, the second type of layer is present in considerable abundance ($> 10\%$ $\pm$), new diffraction effects arise; in particular, a nonintegral series of reflections is obtained from

[74] Caillere, S., and S. Henin, Sepiolite, "The X-ray Identification and Crystal Structures of Clay Minerals," chap. VIII, pp. 325–342, Mineralogical Society of Great Britain Monograph (1961).

[75] Caillere, S., and S. Henin, Palygorskite, "The X-ray Identification and Crystal Structures of Clay Minerals," chap. IX, pp. 343–353, Mineralogical Society of Great Britain Monograph (1961).

Table 5-15   Observed and calculated X-ray powder data for sepiolite
(After Caillere and Henin[74])

hkl	d(Å) calc.	1		2		3		4	
		d(Å) obs.	I	d(Å) obs.	I	d(Å) obs.	I	d(Å) obs.	I
020	13.48								
110	12.07	12.05	100	12.1	100	12.1	100	12.3	60
130	7.482	7.47	10	7.5	7	7.7	5B	7.6	5
200	6.750	}6.73	5	6.7	4B	6.7	5B		
040	6.742								
220	6.036								
001	5.255								
150	5.008	5.01	7	5.04	3B	5.0	5B	4.9	6B
021	4.896								
111	4.818								
240	4.780								
060	4.495	4.498	25	4.49	25	4.47	18	4.5	
310	4.438								20NR
131	4.301	4.306	40	4.29	35	4.31	25	4.3	
201	4.146					4.17	5		
041	4.145								
330	4.023	4.022	7	4.02	7				
221	3.963								
260	3.741	3.750	30	3.738	25	3.738	20	3.746	20B
170	3.705								
151	3.626								
241	3.532	3.533	12	3.506	5			3.49	5
350	3.455								
061	3.416								
311	3.391								
400	3.375								
080	3.370	3.366	30	3.339	45	3.339	35	3.34	
420	3.274								
331	3.195	3.196	35	3.181	15B	3.187	12		
261	3.048	3.050	12			3.048	5		20NR,B
171	3.028								
440	3.018								
280	3.016								
370	2.928	2.932	4	2.950	5			2.98	
190	2.925								
351	2.887								
401	2.840								
081	2.837	2.825	7						
421	2.778	2.771	4			2.79?	4		
460	2.699								

Table 5-15   Observed and calculated X-ray powder data for sepiolite (Continued)

(After Caillere and Henin[74])

hkl	d(Å) calc.	1		2		3		4	
		d(Å) obs.	I	d(Å) obs.	I	d(Å) obs.	I	d(Å) obs.	I
0,10,0	2.697	}2.691	20	2.66	8NR	2.675	8NR	2.67	
510	2.687								
002	2.627								
441	2.618	}2.617	30	2.59		2.59			
281	2.617								
530	2.586	2.586	NR						2.56
022	2.580				45NR		40NR		40NR,B
112	2.567								
371	2.557	}2.560	55						
191	2.556			2.56		2.56			
2,10,0	2.505								
390	2.495								
132	2.479	2.479	5					2.49	
202	2.448	}2.449	25						
042	2.448			2.43		2.44		2.43	
550	2.414				20NR		15NR		10NR
1,11,0	2.412								
222	2.409	}2.406	15	2.395		2.39		2.36	
461	2.401								
062	2.268								
312	2.261	}2.263	30	2.256	20	2.259	18B	2.24	20B
2,10,1	2.260								
620	2.220								
570	2.211	}2.206	3						
332	2.200								
640	2.134								
2,12,0	2.130	}2.125	7	2.117	4B	2.117	5		
4,10,0	2.107								
402	2.073								
082	2.072	2.069	20	2.060	10	2.071	7B	2.08	6B
601	2.069								
571	2.038	2.033	4						
		1.957	4						
		1.921	2						
		1.881	7	1.873	4				
		1.818	2						
		1.760	6						
		1.700	10	1.716	7	1.722	5		
				1.691	10	1.692	8	1.69	5B
		1.637	3						
		1.592	10	1.598	4	1.583	9B	1.58	7
		1.550	15	1.578	7	1.548	10	1.551	10
				1.540	8				

Table 5-15   Observed and calculated X-ray powder data for sepiolite (Continued)

(*After Caillere and Henin*[74])

hkl	$d(\text{Å})$ calc.	1		2		3		4	
		$d(\text{Å})$ obs.	I	$d(\text{Å})$ obs.	I	$d(\text{Å})$ obs.	I	$d(\text{Å})$ obs.	I
		1.518	15	1.517	15	1.517	15	1.517	14
		1.502	8						
		1.468	4	1.465	3				
		1.416	9	1.406	4	1.412	4B		
		1.349	6						
		1.312	6			1.310	4B		
		1.299	15			1.296	10		

Experimental data (Brindley[76]) for sepiolites from:

1. Little Cottonwood, Utah.   2. Vallecas, Spain.   3. Kenya.

4. Eski Chehir, Asia Minor.   B signifies broad; NR not resolved.

the basal planes. Also the layers can contribute to the scattering only as individuals. Prism zone reflections and the scattering distribution from the bases are to be expected, but special sequences such as the (11l) and (02l) interference for muscovite, which depend on a specific mutual orientation of neighboring layers, cannot be given by a mixed structure. There is, therefore, likely to be no diffraction distinction between muscovite and biotite in mixed-layer structures.

In mixed-layer structures the basal reflections are composites of adjacent reflections of the same orders of the different layers, and at an intermediate position between them, or a composite of overlapping reflections of different orders of the different layers. The position and intensity of the composite reflections vary with the relative abundance of the different individual layers. For example, in a random mixed-layer structure of 10-Å and 15-Å layers, the first observed reflection will be a composite of (001) of both layers and will have a spacing intermediate between 10 and 15 Å, the exact position depending on the relative abundance of the individual layers. A strong reflection at 5 Å will appear; this will be a composite of (002) of the 10 Å layers and (003) of the 15-Å layers. Only if the relative amount of the 15-Å layer is large will a slight reflection be found at about 7.5 Å.

In the case of a random mixture of a 17.7-Å layer and a 10-Å layer, where both are present in considerable abundance, the first reflection will be a composite of (001) of both layers, and the second reflection will be a composite of (001) of the 10-Å layer and (002) of the 17.7-Å layer. The position and relative intensity of the composite peaks or, more properly, bands will vary with the relative abundance of the two layers. It should be pointed out that the position and intensity of the composite peaks do not vary with

Table 5-16   X-ray data for attapulgite-palygorskite
(After Caillere and Henin[75])

hkl	d(Å) calc.	d(Å) obs.	I calc.	I obs.	d(Å) obs.	I obs.
		Data by Bradley[77]			Data by De Lapparent[78]	
110	10.48	10.50	330	10	10.2	12
200	6.45	6.44	17	6	6.44	2
130	5.44	5.42	12	5	5.3	2
220	5.24		1			
040	4.50	4.49	66	8		
310	4.18	4.18	13	3	4.3	10
240	3.69	3.69	18	5		
330	3.49	} 3.50	12	} 3		
150	3.47		2			
400	3.23	3.23	120	10	3.25	10
420	3.04	3.03	3	1		
350	2.76		2			
440	2.62	2.61	43	8		
510	2.56	2.55	8	3	2.55	10
530	2.38	2.38	5	3		
080	2.25		1			
600	2.15	2.15	15	5	2.25	4
550	2.10		1			
480	1.845		1			
390	1.815	1.82	3	1	1.80	2
660	1.75		1			
800	1.615	1.62	3	1	1.67	2
680	1.555	1.56	17	3		
0,12,0	1.50	1.50	40	5	1.49	4

[76] Brindley, G. W., X-ray and Electron Diffraction Data for Sepiolite, *Am. Mineralogist,* **44**: 495–500 (1959).

[77] Bradley, W. F., Structure of Attapulgite, *Am. Mineralogist,* **25**: 405–410 (1940).

[78] De Lapparent, J., Structure, *Bull. Soc. Franc. Mineral.,* **61**: 253–283 (1938).

the relative abundance of the constituent layers in a necessarily straight-line relationship. However, in a preliminary analysis, no great error will result by assuming that the peaks move linearly from the position for one pure compound toward the nearest position of the other component as the proportions of the components vary. Weaver[79] published a method for binary mixtures based on a visual inspection of the diffraction data, and Jonas and Brown[80] interpreted patterns for three-component systems.

[79] Weaver, C. E., Mixed-layer Clays in Sedimentary Rocks, *Am. Mineralogist,* **41**: 202–221 (1956).

[80] Jonas, E. C., and T. E. Brown, Analysis of Interlayer Mixtures of Three Clay Mineral Types by X-ray Diffraction, *J. Sediment. Petrol.,* **29**: 77–86 (1959).

The mathematical theory of X-ray diffraction for such statistical structures has been discussed by Hendricks and Teller[81] and later by Mering.[82] Brown and MacEwan[42] and MacEwan, Amil, and Brown[83] more recently presented further discussions of this matter and an extremely useful series of graphs, showing the expected basal reflections of mixtures of various proportions of a variety of layers.   McAtee,[84] MacEwan,[85] and MacEwan and Amil[86] discussed especially the application of the Fourier-transform method to the analysis of random interstratifications.   The reader is referred to these works for further details of mixed-layer structures.

⌐ *Electron diffraction*

Electron-diffraction data can commonly be obtained in the same equipment as that used for electron microscopy.   With small apertures, single crystals may be analyzed, thereby making possible the study of individual particles.   The identification of many of the individual clay minerals is difficult because most of them consist of similar layer lattices, and because the plate-shaped particles tend to lie on their basal planes and thus yield identical or nearly identical hexagonal spot patterns in the diffractional diagrams.   As a consequence, it may be impossible to distinguish the patterns of the various clay minerals.

The electron-diffraction technique has not yet been widely used in the study of the clay minerals.   Pinsker,[87,88] Rees,[89] Honjo et al.,[90] Vajnstejn,[91]

[81] Hendricks, S. B., and E. Teller, X-ray Interference in Partially Ordered Layer Lattices, *J. Chem. Phys.*, **10**: 147–167 (1942).

[82] Mering, J., L'Interférence des rayons X dans les systèmes à stratification disordonnée, *Acta Cryst.*, **2**: 371–377 (1949).

[83] MacEwan, D. M. C., A. R. Amil, and G. Brown, Interstratified Clay Minerals, "The X-ray Identification and Crystal Structures of Clay Minerals," chap. XI, pp. 393–445, Mineralogical Society of Great Britain Monograph (1961).

[84] McAtee, J. L., Determination of Random Interstratification in Montmorillonite, *Am. Mineralogist,* **41**: 627–631 (1956).

[85] MacEwan, D. M. C., Fourier Transform Methods for Studying Scattering from Lamellar Systems, I, A Direct Method for Analyzing Interstratified Mixtures, *Kolloid-Z.,* **149**: 96–108 (1956).

[86] MacEwan, D. M. C., and A. R. Amil, Fourier Transform Methods for Studying X-ray Scattering from Lamellar Systems, III, Some Calculated Diffraction Effects of Practical Importance in Clay Mineral Studies, *Kolloid-Z.,* **162**(2): 93–100 (1959).

[87] Pinsker, Z. G., Electron Diffraction, Butterworth Scientific Publications, London (1953). Translated by J. A. Spink and E. Feigl.

[88] Pinsker, Z. G., Study of Clay Minerals by Electron Diffraction and Electron Microscope, *Tr. Biogeokhim. Lab., Akad. Nauk SSSR,* **10**: 116–141 (1954).

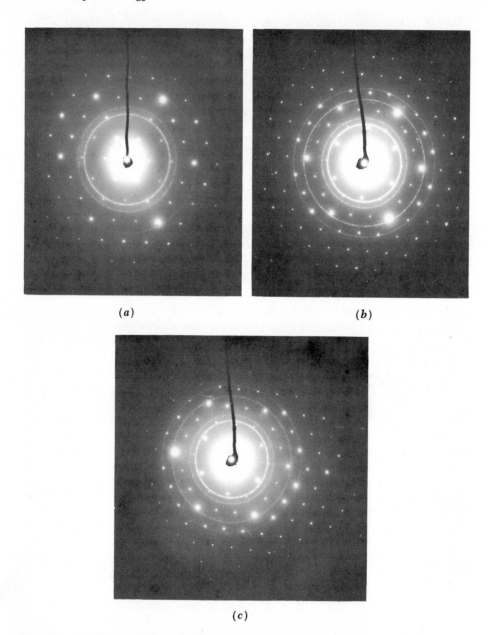

(a)

(b)

(c)

Fig. 5-4. Single-crystal electron-diffraction diagrams of (*a*) kaolinite, (*b*) dickite, (*c*) nacrite, with calibration rings from aluminum metal. (After Brindley and DeKimpe.[94])

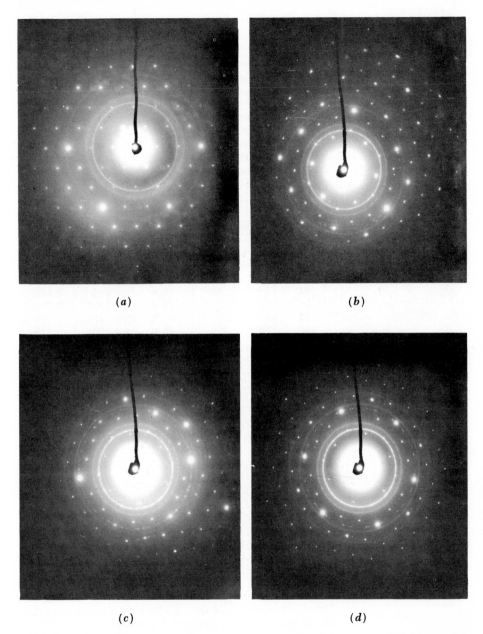

Fig. 5-5.   Single-crystal electron-diffraction diagrams of (*a*) muscovite, (*b*) biotite, (*c*) clinoclore, (*d*) daphnite, with calibration rings from aluminum metal.   (After Brindley and DeKimpe.[94])

Table 5-17   Electron diffraction data for hk0 reflections
of montmorillonite
(After MacEwan[43])

Hexagonal hk indices	Orthohexagonal hk indices	I obs.	d(Å) obs.	d (Å) calc.
10	11; 02	(357)	4.50	4.50
11	13; 20	(177)	2.60	2.61
20	22; 04	n.d.	2.26	2.27
21	31; 15; 24	(138.5)	1.706	1.709
30	33; 06	(211)	1.503	1.498
22	26; 40	(113.9)	1.295	1.305
31	35; 17; 42	(85.7)	1.246	1.247
40	08; 44	(17.2)	1.124	1.124
32	37; 28; 51	(15.1)	1.033	1.031
41	19; 53; 46	(23.2)	0.977	0.972
50	55; 0,10	10.6	0.9023	0.901
33	39; 60	25.3	0.8662	0.865
42	2,10; 48; 62	15.2	0.8507	0.851
51	1,11; 57; 64	16.6	0.8066	0.8067
60	0,12; 66	18.2	0.7505	0.7490
43	3,11; 4,10; 71	4.3	0.7419	0.739
52	73; 59; 2,12	15.0	0.7204	0.719
61	1,13; 68; 75	12.8	0.6874	0.685
44	80; 4,12	n.o.		
70	77; 0,14;	13.3	$\left\{\begin{matrix}0.650\\0.6453\end{matrix}\right\}$	0.642
53	82; 3,13; 5,11			
62	2,14; 6,10; 84	1.0	0.6172	0.624
71	86, 79; 1,15	4.3	0.5967	0.597
54	5,13; 91; 4,14	3.3	0.5772	0.5773
63	93; 6,12; 3,15	4.5	0.5681	0.568
80	0,16; 88	0.7	0.558	0.562
72	2,16; 7,11; 95	1.8	9.5506	0.550
81	1,17; 8,10; 97	2.4	0.5264	0.527
55	5,15; 10,0	0.4	0.5205	0.523
64	10,2; 6,14; 4,16	4.2	0.5170	0.517
73	10,4; 7,13; 3,17	3.9	0.5070	0.507
90	0,18; 99	0.5	0.5045	0.500
82	8,12; 10,6; 2,18	2.8	0.4910	0.495
91	1,19; 9,11; 10,8	1.5	$\left\{\begin{matrix}0.4754\\0.4719\end{matrix}\right\}$	0.472
65	11,1; 5,17; 6,16	1.5		
74	4,18; 7,15; 11,3	1.5	0.4669	0.467

Data from normal beam transmission patterns of oriented films (Finch and MacEwan, not hitherto published).

Hexagonal indices based on $a = 5.18$ Å.

Orthohexagonal indices based on $a = 5.18$ Å, $b = \sqrt{3a}$.

n.d. = not determined; n.o. = not observed.

Intensities in brackets are from X-ray photographs.

Cowley et al.,[92] and Ross and Christ[93] have discussed electron-diffraction techniques in general. Brindley and DeKimpe,[94] and Mitra and Rao[95] have discussed the application of the technique, and for details their work should be consulted.

It is not yet possible to give complete-summary electron-diffraction data for all the clay minerals. Zvyagin[96] and Oberlin and Tschoubar[97] published data for kaolinite; Brindley and DeKimpe[94] for the kaolin minerals and some of the micas (Figs. 5-4 and 5-5); Honjo et al.[90] for halloysite; Brindley[76] for sepiolite; and MacEwan[43] for montmorillonite (Table 5-17). Sufficient work has been done, especially using single crystals, to show that the method yields very important data for clay-mineral studies.

Additional references

Brindley, G. W., and J. Mering, Diffraction by Random Layers, *Nature,* **161:** 774 (1948).

Cesari, M., G. L. Morelli, and L. Favretto, The Determination of the Type of Stacking in Mixed-layer Clay Minerals, *Acta Cryst.,* **18**(2): 189 (1965).

Correns, C. W., and M. Mehmel, Ueber den optischen und röntgenographischen Nachweis von Kaolinit, Halloysit, und Montmorillonit, *Z. Krist.,* **94:** 337–348 (1936).

Favejee, J. C. L., Zur Methodik der röntgenographischen Bodenforschung, *Z. Krist.,* **100:** 425–436 (1939); **101:** 259–270 (1939).

[89] Rees, A. L. G., Electron Diffraction in the Chemistry of the Solid State, *J. and Proc. Roy. Soc. New South Wales,* **86:** 38–54 (1953).

[90] Honjo, G., N. Kitamura, and K. Mihama, Study of Clay Minerals by Single-crystal Electron Diffraction Diagrams, *Clay Minerals Bull.,* **2**(12); 133–141 (1954); see also *Acta Cryst.,* **7:** 511–513 (1954).

[91] Vajnstejn, B. K., Structure Analysis by Electron Diffraction, *Nuovo Cimento, Suppl.,* **3:** 773–797 (1956).

[92] Cowley, J. M., P. Goodman, and A. L. G. Rees, Crystal Structure Analysis from Fine Structure in Electron Diffraction Patterns, *Acta Cryst.,* **10:** 19–25 (1957).

[93] Ross, M., and C. L. Christ, Mineralogical Applications of Electron Diffraction, *Am. Mineralogist,* **43:** 1157–1178 (1958).

[94] Brindley, G. W., and C. DeKimpe, Identification of Clay Minerals by Single-crystal Electron Diffraction, *Am. Mineralogist,* **46:** 1005–1016 (1961).

[95] Mitra, R. P., and M. V. R. Rao, Electron Diffraction by Orientation Aggregates of Clays as a Method for Identifying Clay Minerals, *Soil Sci.,* **96**(5): 299 (1963).

[96] Zvyagin, B. B., Electronographic Determination of the Structure of Kaolinite, *Kristallografia,* **5:** 40–50 (1959).

[97] Oberlin, A., and C. Tschoubar, Examination of Certain Minerals of the Kaolinite Group by Electron Microdiffraction, *Compt. Rend.,* **248:** 3184–3186 (1959).

Gibbs, R. J., Error Due to Segregation in Quantitative Clay Mineral X-ray Diffraction Mounting Techniques, *Am. Mineralogist,* **50:** 741–751 (1965).

Guinier, A. J., X-ray Focusing Cameras, *Ann. Phys.,* **12:** 161–237 (1939).

Guinier, A. J., Imperfections of Crystal Lattices, *Proc. Phys. Soc. (London),* **57:** 310–324 (1945).

Hendricks, S. B., Continuous Scattering of X-rays from Layer Silicates, *Phys. Rev.,* **57:** 448–454 (1940).

Kerr, P. F., P. K. Hamilton, and R. J. Pill, X-ray Diffraction Measurements, "Analytical Data on Reference Clay Minerals," pp. 1–37, Prelim. Rept. 7, American Petroleum Institute Project 49, Columbia University, New York (1950).

Lucas, J., T. Camez, and G. Millot, Practical X-ray Determination of Simple and Interstratified Clay Minerals, *Bull. Serv. Carte Geol. Alsace Lorraine,* **12**(2)**:** 21–31 (1959).

MacEwan, D. M. C., The Identification and Estimation of the Montmorillonite Group of Minerals with Special Reference to Soil Clays, *J. Soc. Chem. Ind. (London),* **65:** 298–304 (1946).

MacEwan, D. M. C., Effect of Structural Irregularities on the Quantitative Determination of Clay Minerals by X-rays, *Acta Univ. Carolinae, Geol. Suppl.,* **1:** 83–90 (1962).

MacEwan, D. M. C., and H. H. Sutherland, The Use of Periodograms for Studying Two- and Three-dimensional Irregularities in Crystals, *Proc. Intern. Clay Conf., Stockholm,* **2:** 9–18 (1965).

Mehmel, M., Datensammlung zum Mineralbestimmung mit Röntgenstrahlen, *Fortschr. Mineral. Krist. Petrog.,* **23:** 91–118 (1939).

Nagelschmidt, G., Mineralogy of Soil Colloids, *Imp. Bur. Soil Sci. Tech. Comm.* **42,** Harpenden, England (1944).

Nagelschmidt, G., and D. Hicks, The Mica of Certain Coal Measure Shales in South Wales, *Mineral. Mag.,* **26:** 297–303 (1943).

Nishkanen, E., Reduction of Orientation Effects in the Quantitative X-ray Diffraction Analysis of Kaolin Minerals, *Am. Mineralogist,* **49:** 705–714 (1964).

Shearer, J., An X-ray Camera for Clay Studies, *J. Sci. Instr.,* **21:** 198–200 (1944).

Sutherland, H. H., and D. M. C. MacEwan, Use of a Weissenberg Technique in the Quantitative Determination of Clay Minerals, *Acta Univ. Carolinae, Geol. Suppl.,* **1:** 91–96 (1962).

Van der Marel, H. W., Identification of Minerals in Soil Clays by X-ray Diffraction, *Soil Sci.,* **70:** 109–136 (1950).

Von Engelhardt, W., Ueber silikatische Tonminerale, *Fortschr. Mineral. Krist. Petrog.,* **21:** 276–340 (1937).

Winkler, H., Zur Kristallstruktur von Montmorillonit, *Z. Krist.,* **107:** 291–303 (1943).

Zvyagin, B. B., Structural Modifications of the Layer Silicates and Possibilities of Their Determination by Means of Electron and X-ray Diffraction, *Proc. Intern. Clay Conf., Stockholm,* **2:** 29–42 (1965).

Zvyagin, B. B., and Z. G. Pinsker, Electronographic Determinations of the Structure of Montmorillonite, *Dokl. Akad. Nauk. SSSR,* **68:** 65–67 (1949).

Six
Shape and Size—
Electron Microscopy

General statement

The development of the electron microscope has permitted the precise determination of the shape of the particles of the various clay minerals, and it use has shed light on the range of particle size of the components of clays and on the degree to which the particle size can be reduced when clays are worked mechanically with water. With modern techniques and present-day microscopes resolutions of better than 5 Å and instrumental magnifications up to 250,000× can be obtained.

Most current equipment provides for electron diffraction as well as micrography so that phases can often be determined at the same time that morphology is studied. The addition of specimen heating devices permits electron microscopy and diffraction at elevated temperatures. With this equipment morphological changes on heating can be observed continuously. The most recent advance in equipment combines electron microscopy, electron diffraction, and electron probe microanalysis. Application of this instrument allows chemical, structural, and morphological data to be obtained from individual particles as small as 0.5 μ.

Numerous investigators have published electron micrographs of the clay minerals and discussed electron-microscopic techniques as applied to the clay-mineral researches. Worthy of mention are the early contributions of Ardenne,[1] Eitel,[2] Middel,[3] Humbert,[4] Shaw,[5] Marshall,[6] Alexander,[7] Moore,[8] Bates,[9] Kerr,[10] Comer,[11] and their various collaborators.

The techniques of electron microscopy are numerous and varied, and the texts of Kay,[12] Hirsch et al.,[13] Siegel,[14] Thomas,[15] and Zvyagin[16] should

[1] Ardenne, M. von, K. Endell, and U. Hofmann, Investigation of the Finest Fraction of Bentonite and Clay Soil with the Universal Electron Microscope, *Ber. Deut. Keram. Ges.*, **21**: 209–227 (1940).

[2] Eitel, W., and O. E. Radczewski, On Recognition of Montmorillonite Clay Minerals in Supermicroscope Pictures, *Naturwissenshaften*, **28**: 397–398 (1940).

[3] Middel, V., R. Reichmann, and G. S. Kausche, Supermicroscopic Investigation of the Structure of Bentonites, *Wiss. Veröffentl. Siemens-Werken*, pp. 334–341 (1940).

[4] Humbert, R. P., and B. Shaw, Studies of Clay Particles with the Electron Microscope, I, Shape of Clay Particles, *Soil Sci.*, **52**: 481–487 (1941).

[5] Shaw, B. T., The Nature of Colloidal Clay as Revealed by the Electron Microscope, *J. Phys. Chem.*, **46**: 1032–1043 (1942).

[6] Marshall, C. E., R. P. Humbert, B. T. Shaw, and O. G. Caldwell, Studies of Clay Particles with the Electron Microscope, II. Beidellite, Nontronite, Magnesium Bentonite, and Attapulgite, *Soil Sci.*, **54**: 149–158 (1942).

[7] Alexander, L. T., G. T. Faust, S. B. Hendricks, H. Insley, and H. F. McMurdie, Relationship of the Clay Minerals Halloysite and Endellite, *Am. Mineralogist*, **28**: 1–18 (1943).

[8] Moore, C. A., Some Geological Applications of the Electron Microscope, *Proc. Oklahoma Acad. Sci.*, **27**: 86–90 (1947).

[9] Bates, T. F., F. A. Hildebrand, and A. Swineford, Morphology and Structure of Endellite and Halloysite, *Am. Mineralogist*, **35**: 463–484 (1950).

[10] Kerr, P. F., P. K. Hamilton, D. W. Davis, T. G. Rochow, F. G. Rowe, and M. L. Fuller, "Electron Micrographs of Reference Clay Minerals," Prelim. Rept. 6, American Petroleum Institute Project 49, Columbia University, New York (1950).

[11] Comer, J. J., and J. W. Turley, Replica Studies of Bulk Clays, *J. Appl. Phys.*, **26**: 346–350 (1955).

[12] Kay, D. H., "Techniques for Electron Microscopy," 2d ed., Davis, Philadelphia (1965).

[13] Hirsch, P. B., A. Howie, R. B. Nicholson, D. W. Pashley, and M. J. Whelan, "Electron Microscopy of Thin Crystals," Butterworth, Inc., Washington, D.C. (1965).

[14] Siegel, B. M., ed., "Modern Developments in Electron Microscopy," Academic, New York (1964).

[15] Thomas, G., "Transmission Electron Microscopy of Metals," Wiley, New York (1962).

[16] Zvyagin, B. B., "Electron-Diffraction Analysis of Clay Mineral Structure," Plenum Press, New York (1966).

be consulted for details. The earlier works of Zworykin et al.,[17] Cosslett,[18] Pinsker,[19] Hall,[20] and Fischer[21] are also useful.

In transmission electron microscopy, image formation is due to the scattering of electrons as the focused electron beam passes through the sample. Clay-mineral particles can be dispersed on thin plastic films and observed directly in the electron beam, or the dispersions can be "shadowed" in order to increase contrast and provide better three-dimensional information. In the shadow method, mounted specimens are placed in a high-vacuum evaporator, and a heavy metal, frequently chromium or platinum, is evaporated *in vacuo* obliquely onto the specimens. The angle of deposition of the metal varies according to the thickness of the particles in the sample. Angles of near 10° are used for thin particles while angles up to 30° are used for thicker ones.

In the electron microscope, the specimens, particularly thicker ones, can become heated by the absorption of energy or altered by the high vacuum of the microscope column. Mounting clay minerals directly in the beam, therefore, has the disadvantage that they may dehydrate, with accompanying changes in character. This situation is particularly troublesome with allophane, halloysite, smectite, and vermiculite. Current instrumentation can minimize the heating effects through the use of double condenser illumination and specimen cooling devices. The development of the replica technique provided early workers with a method of circumventing the heating problem in the study of clay morphology. Because it is an indirect approach, however, it cannot be used where electron-diffraction data are also desired. The replica technique has a further advantage in that it makes possible the study of natural or freshly fractured clay surfaces which show something of the textural arrangements of the clay-mineral particles in the bulk clay itself.

Of the replica techniques available at the present time, the most useful in the study of the clay minerals is the single-stage, preshadowed or self-shadowed carbon replica method. The technique as described by Bradley[22]

[17] Zworykin, V. K., C. A. Morton, E. G. Remberg, J. Hillier, and A. W. Vorce, "Electron Optics and the Electron Microscope," Wiley, New York (1945).

[18] Cosslett, V. E., "Introduction to Electron Optics," Oxford University Press, Fair Lawn, N.J. (1946).

[19] Pinsker, Z. G., "Electron Diffraction," Butterworth, London (1953).

[20] Hall, C. E., "Introduction to Electron Microscopy," 2d ed., McGraw-Hill, New York (1966).

[21] Fischer, R. B., "Applied Electron Microscopy," Indiana University Press, Bloomington, Ind. (1954).

[22] Bradley, D. E., Evaporated Carbon Films for Use in Electron Microscopy, *J. Appl. Phys.*, **5:** 65–66 (1954).

was initially adapted to the study of clay surfaces by Bates and Comer[23] and has since been widely used by many others. The method has numerous variations but basically it involves the direct shadowing *in vacuo* of the bulk clay with a heavy metal (for contrast) and carbon (for strength). The entire specimen is then placed in appropriate solvents to dissolve the clay, leaving the thin replica of the surface for study in the electron microscope. Details of the order of 20 Å can often be resolved.

All the electron micrographs presented herein were made by Dr. Kenneth Towe of the Electron Microscope Laboratory of the Smithsonian Institution, Washington, D.C. It is desired to acknowledge the great kindness and courtesy of Doctor Towe and the Smithsonian Institution for supplying the micrographs and permitting their publication. Doctor Towe is not only a superb technician with the electron microscope but has extensive training and experience in clay mineralogy.

Data for the clay minerals

Allophane. As would be expected, this material with little or no definite structural organization is found in particles without any definite and regular shape. Electron micrographs generally show fluffy aggregates with a rounded nodular appearance (Fig. 6-1). Electron-diffraction data indicate that this sample is truly amorphous. Sometimes the aggregates appear to be composed of flake-shaped units. At other times, as pointed out by Aomine and Wada,[24] the rounded forms appear to be composed of fibrous units.

Halloysite. Numerous investigators have shown that the morphology of halloysite is quite different from that of kaolinite, the former being elongate instead of flake-shaped (Fig. 6-2). Primarily because of the work of Bates[25] et al., halloysite is now known to appear in electron micrographs as elongate tubular particles. Convincing evidence of the tubular development is furnished by the doughnut forms that are the ends of tubular particles seen in some electron micrographs (Fig. 6-2a,c).

Bates[25] et al. determined the dimensions of halloysite tubes from the electron micrographs of many samples. The outside diameters were found to range from 0.04 to 0.19 μ, with a median value of 0.07 μ. The inside diameters were found to have the median value of 0.04 μ, with a range of from 0.02 to 0.1 μ. The average wall thickness was found to be 0.02 μ.

[23] Bates, T. F., and J. J. Comer, Electron Microscopy of Clay Surfaces, *Natl. Acad. Sci.*, Publ. 395, pp. 1–25 (1955).

[24] Aomine, S., and K. Wada, Differential Weathering of Volcanic Ash and Pumice Resulting in Formation of Hydrated Halloysite, *Am. Mineralogist,* **47:** 1024–1048 (1962).

[25] Bates, T. F., F. A. Hildebrand, and A. Swineford, Morphology and Structure of Endellite and Halloysite, *Am. Mineralogist,* **35:** 463–484 (1950).

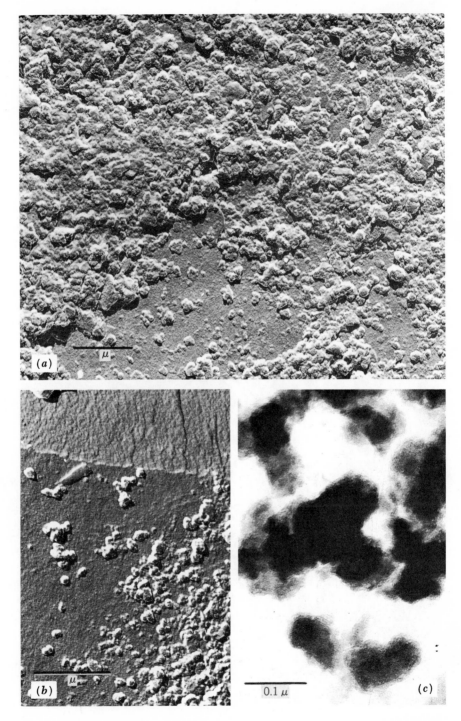

Fig. 6-1. Allophane, Georgia: (a) and (b) replicas; (c) transmission. (Courtesy Dr. Kenneth Towe, Smithsonian Institution, Electron Microscope Laboratory, Washington, D.C.)

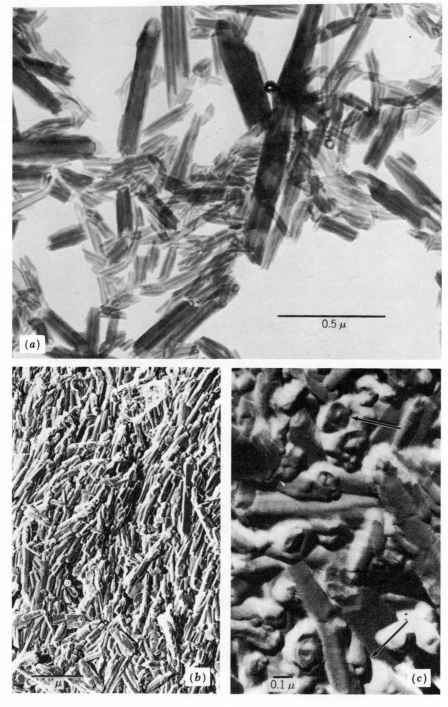

Fig. 6-2. Halloysite, Utah; (a) transmission; (b) and (c) replicas. (Courtesy Dr. Kenneth Towe, Smithsonian Institution, Electron Microscope Laboratory, Washington, D.C.)

Figure 6-2*a* and *b* shows the higher hydrated form of halloysite. Figure 6-2*c* presents replicas of freeze-dried specimens. The micrographs show that the tubular form is preserved on drying. The replicas of some halloysites, particularly those in the lower hydration state, commonly show flat sides meeting at fairly definite angles, so that in cross section the tubular particles are angular rather than rounded. In some cases, micrographs of dried samples show split ends and partially unrolled tubes. Probably this phenomenon develops as a consequence of the change from the $4H_2O$ to the $2H_2O$ form of the mineral. Very high magnification sometimes shows characteristics of the tubes which suggest that twinning has played a role in their development.

Sudo and Takahashi[26] showed halloysite composed of extremely fine hair-like twisted fibers and elongate crystals having sharp edges occurring in clays which also contain allophane. They considered that this halloysite is a transition form from allophane. The occurrence of allophane and halloysite together in clays is not unusual and, as Sudo and Takahashi[26] pointed out, it is probable that all transition forms exist between these minerals.

Kaolinite. Electron micrographs of the well-crystallized mineral (Fig. 6-3) show well-formed six-sided flakes, frequently with a prominent elongation in one direction. The elongation is parallel to either (010) or (110). Commonly the edges of the particles are beveled instead of being at right angles to the flake surface. Occasionally the particles appear to be twinned.

Electron micrographs of various kaolinite samples have shown particles with maximum dimensions of flake surfaces from 0.3 to about 4 μ and thicknesses from 0.05 to about 2 μ. This does not mean that larger particles of kaolinite are not present in some clays, since such larger particles may have been split or otherwise reduced in size in the preparation of the sample for electron microscopy. It does mean, however, that kaolinite particles are easily reduced to this size but are not reduced to smaller sizes except with difficulty, e.g., after much mechanical work.

Commonly, disordered kaolinite shows flakes with poorly developed hexagonal outlines. The edges of the flakes are somewhat ragged and irregular and the hexagonal outline is only crudely shown (Fig. 6-4). There are, however, exceptions, and Robertson et al.[27] and Von Engelhardt and Goldschmidt[28] described disordered kaolinites showing small but excellent hexagonal crystals.

In general, poorly crystallized kaolinite occurs in smaller and thinner particles than the well-crystallized mineral. However, it does not follow that

[26] Sudo, T., and H. Takahashi, Shape of Halloysite Particles in Japanese Clays, *Natl. Acad. Sci., Publ.* 456, pp. 67–79 (1956).

[27] Robertson, R. H. S., G. W. Brindley, and R. C. Mackenzie, Mineralogy of Kaolin Clays from Pugu, Tanganyika, *Am. Mineralogist,* 39: 118–138 (1954).

[28] Von Engelhardt, W., and H. Goldschmidt, A Clay Mineral from Provins (France), *Heidelberger Beitr. Mineral. Petrog.,* 4: 319–324 (1954).

Fig. 6-3. Well-crystallized kaolinite, Georgia; (a) replica; (b) transmission. (Courtesy Dr. Kenneth Towe, Smithsonian Institution, Electron Miscroscope Laboratory, Washington, D.C.)

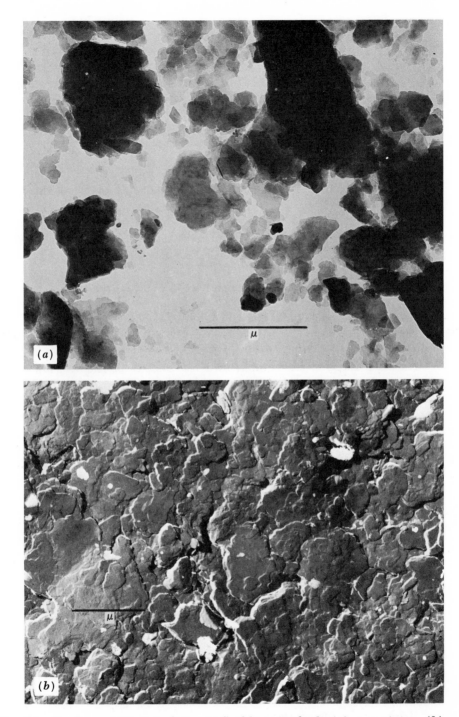

Fig. 6-4. Poorly crystallized kaolinite, Staffordshire, England; (a) transmission; (b) replica. (Courtesy Dr. Kenneth Towe, Smithsonian Institution, Electron Microscope Laboratory, Washington, D.C.)

the kaolinite with very small particle size always has a low degree of crystallinity. For example, the kaolinite component of some so-called flint clays is extremely fine grained and extremely well crystallized.

Kaolin clays have been described (Visconti[29]) for which X-ray-diffraction data indicate the presence of kaolinite as the sole constituent but electron micrographs indicate the presence of elongate tubular particles as well as definite hexagonal flake-shaped units.

De Souza Santos et al.[30] concluded that the tubular particles were halloyite rather than tubular kaolinite but were present in too small quantities to be revealed by X-ray diffraction in the presence of abundant kaolinite. Later work by these same authors has thrown some doubt on this conclusion, and there remains the possibility of the existence of tubular particles which are well enough organized to be classed structurally as kaolinite rather than halloysite.

Dickite. Dickite occurs in well-formed six-sided flake-shaped particles frequently showing a definite elongation in one direction. Samples that have been examined show flake-surface dimensions ranging from about 2.5 to 8 μ and thickness dimensions of 0.07 to 0.25 μ. Dickite particles are often large enough to be studied with the light microscope.

Nacrite. Electron micrographs of a few samples of nacrite show somewhat irregular rounded flake-shaped units. In some particles a crude hexagonal outline can be seen. In these samples, the flakes are generally less than about 1 μ in diameter, and the thickness is about 0.025 to 0.15 μ.

Smectite. Frequently electron micrographs of montmorillonite (Figs. 6-5 and 6-6) show broad undulating mosaic sheets that when disturbed and dispersed break easily into irregular fluffy masses of extremely small particles. Commonly the individual particles can barely be discerned and are too small to reveal any characteristic outlines. Micrographs of some montmorillonites reveal irregular flake-shaped aggregates which appear to be stackings of units without regular outlines (Fig. 6-6b). In some cases, the flake-shaped units are discernible but frequently they are too small to be seen individually. The aggregates sometimes appear to have curled edges (Fig. 6-6a) and, as this characteristic is revealed in replicas, it must be inherent in the natural mineral. Accurate estimations of the areal dimensions of the flakes are difficult to obtain because of the irregularity but are probably of the order of 10 to 100 times the thickness.

[29] Visconti, Y. S., and B. N. F. Nicot, Further Comment on Tubular Kaolin Crystals, *Ceramica (Brazil)*, **5:** 2–10 (1959).

[30] De Souza Santos, P., H. de Souza Santos, and G. W. Brindley, Mineralogical Studies of Kaolinite-Halloysite Clays, Part I, *Am. Mineralogist*, **49:** 1543–1548 (1964); Part II, *Am. Mineralogist,* **50:** 619–628 (1965).

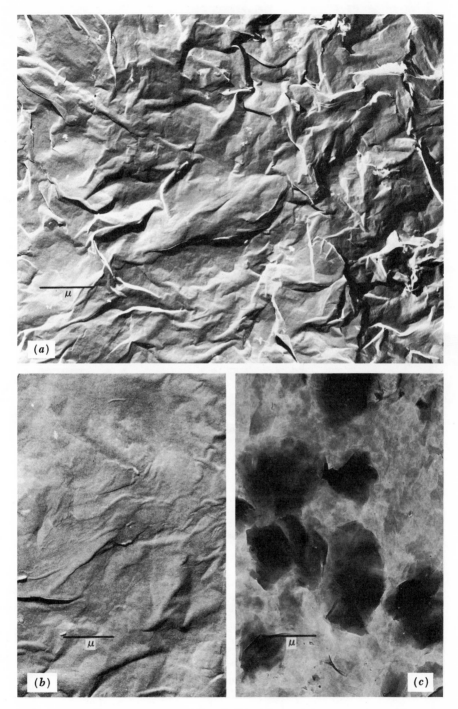

Fig. 6-5. Montmorillonite, Wyoming; (a) and (b) replicas; (c) transmission. (Courtesy Dr. Kenneth Towe, Smithsonian Institution, Electron Microscope Laboratory, Washington, D.C.)

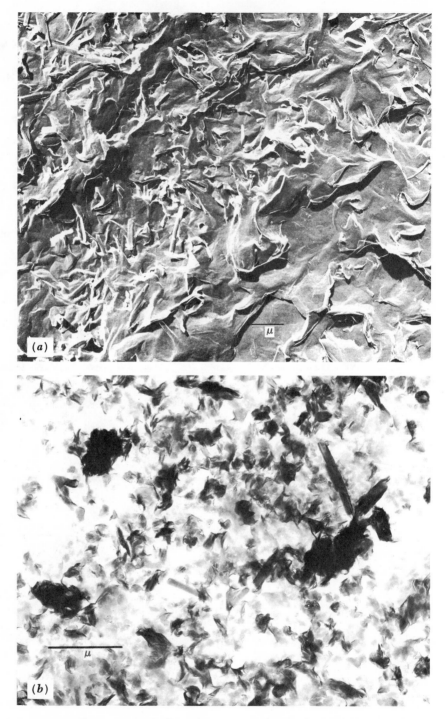

Fig. 6-6.   Montmorillonite, Montmorillon, France; (a) replica; (b) transmission.   (Courtesy Dr. Kenneth Towe, Smithsonian Institution, Electron Microscope Laboratory, Washington, D.C.)

Mathieu-Sicaud, Mering, and Perrin-Bonnet[31] showed that variations in the exchangeable cation carried by montmorillonites may be reflected in their appearance in electron micrographs. According to them, Na montmorillonites appear as more or less continuous nebulous film-like assemblages frequently showing cracks intersecting at 120°. H montmorillonites appear as a mosaic of aggregate masses with distinct hexagonal outlines which are often 300 Å in diameter and 50 to 80 Å thick. Ca montmorillonite appears as irregular aggregates which increase in size as the preparatory suspension is aged. At concentrations of calcium at least equal to the cation-exchange capacity, the aggregates grow in thickness as well as laterally. According to Mathieu-Sicaud et al.,[31] these differences are due to variations in the attraction between the montmorillonite particles and in the relative strength of the attractive forces at the edge of the particles and on their basal planes.

Electron micrographs of some nontronites (Fig. 6-7) show no indication of morphology except that the particles are flake-shaped, whereas in other samples there is a distinct tendency toward elongate lath-shaped units, with flake-, needle-, and rod-like particles of varying size also evident. The length of the laths may reach several microns and is frequently about five times the width dimension.

Some of the magnesium-rich members of the smectite group appear to be composed of equidimensional flake-shaped units about like those of the aluminum-rich members. For example, this is commonly true of saponite. Hectorite, the fluorine-bearing magnesium-rich smectite, is found in thin laths which tend to lose their identity when gathered into aggregates (Fig. 6-8). The laths attain the length of several microns and a width of about 0.1 μ. The laths appear to be extremely thin, some probably only 12 to 18 Å thick.

Illite. Electron micrographs of illite show small, poorly defined flakes commonly grouped together in irregular aggregates (Fig. 6-9a,c). Occasionally the flakes show crude hexagonal outlines (Fig. 6-9b). The thinnest flakes are approximately 30 Å thick. Many of the flakes have a diameter of 0.1 to 0.3 μ. In general, the electron micrographs of illite resemble those of montmorillonite, but the particles are larger and thicker and have better-defined edges.

Weaver[32] described an illite composed of lath-shaped particles which had all the attributes of other illites. It occurred in the -1-μ fraction of a sandstone. Jasmund and Riddle[33] described a ribbon-shaped illite having a maxi-

[31] Mathieu-Sicaud, A., J. Mering, and I. Perrin-Bonnet, Etude su microscope électronique de la montmorillonite et de l'hectorite saturées par différent cations, *Bull. Soc. Franc. Mineral. Crist.*, **74**: 439–455 (1951).

[32] Weaver, C. E., A Lath Shaped Non-expanded Dioctahedral 2:1 Clay Mineral, *Am. Mineralogist*, **38**: 279–289 (1953).

[33] Jasmund, K., and D. Riddle, Untersuchungen des tonigen zwischemmittels in Hauptbuntsandstein der Norfels, *Bull. Geol. Inst. Univ. Upsala*, **40**: 247–257 (1961).

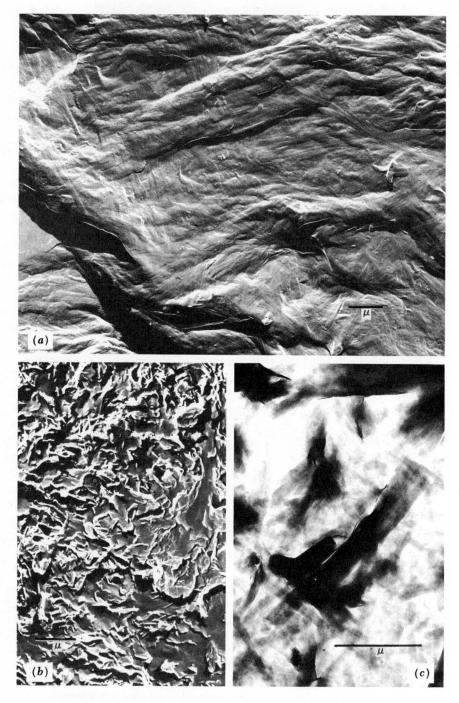

Fig. 6-7. Nontronite, Washington; (a) and (b) replicas; (c) transmission. (Courtesy Dr. Kenneth Towe, Smithsonian Institution, Electron Microscope Laboratory, Washington, D.C.)

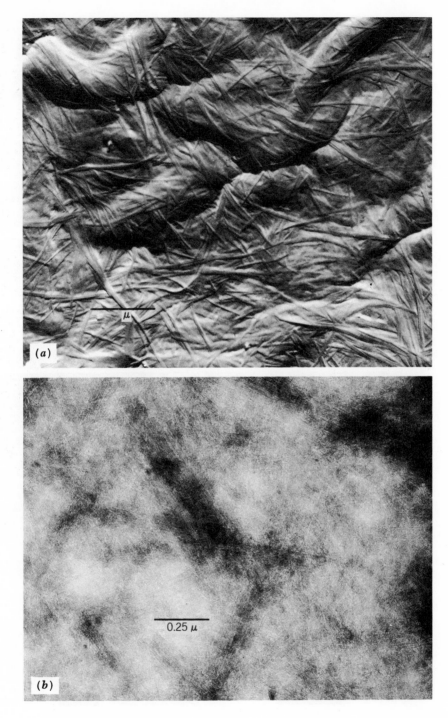

Fig. 6-8. Hectorite, Hector, California; (a) replica; (b) transmission. (Courtesy Dr. Kenneth Towe, Smithsonian Institution, Electron Microscope Laboratory, Washington, D.C.)

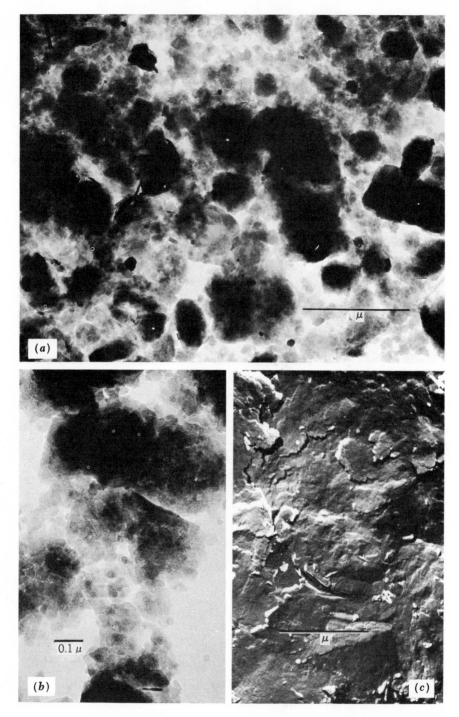

Fig. 6-9. Illite, Fithian, Illinois; (*a*) and (*b*) transmission; (*c*) replica. (Courtesy Dr. Kenneth Towe, Smithsonian Institution, Electron Microscope Laboratory, Washington, D.C.)

Fig. 6-10. Attapulgite, Attapulgus, Georgia; (*a*) and (*b*) replicas; (*c*) transmission. (Courtesy Dr. Kenneth Towe, Smithsonian Institution, Electron Microscope Laboratory, Washington, D.C.)

Fig. 6-11. Sepiolite, Eski-sher, Turkey; (a) transmission; (b) replica. (Courtesy Dr. Kenneth Towe, Smithsonian Institution, Electron Microscope Laboratory, Washington, D.C.)

mum length of $10\,\mu$, a width of 0.1 to $0.3\,\mu$, and a thickness of less than 100 Å, again from a sandstone. The significance of the elongate form of illite has not been determined.

Vermiculite and Chlorite. Little information is available regarding the electron micrography of the vermiculite and chlorite clay minerals. These clay minerals invariably occur in association with other clay minerals so that the individual components cannot be isolated to permit the determination of their specific characteristics. It appears probable from structural consideration that electron micrographs of these minerals would be similar to those of illites except that vermiculite might occur in thinner flakes.

Attapulgite, Sepiolite, Palygorskite. Attapulgite is shown by electron micrographs (Fig. 6-10) to occur in elongate lath-shaped units and bundles of laths. Individual laths frequently are many microns in length and have a width of 50 to 100 Å. There is no evidence of a tubular form like that found for halloysite. Frequently in micrographs of high magnification, the gutter-and-channel characteristic exterior of the laths is revealed. Sometimes the individual laths are bent, and the aggregate mass appears to be interwoven. Tangling of the fibers produces a woven texture like that found in paper and is undoubtedly responsible for the leather- or paper-like form in which the mineral sometimes is found.

Sepiolite (Fig. 6-11) is found in elongate lath-shaped units like attapulgite, with the difference that the individual laths appear to be thicker and shorter. The gutter-and-channel surface of the laths is sometimes easily discernible. The individual laths are frequently bent and show the interwoven characteristic of attapulgite. However, the aggregate masses appear to be denser than those of attapulgite.

Additional references

Eitel, W., The Electron Microscope and Its Use in Ceramic Problems, *Ber. Deut. Keram. Ges.,* **24:** 37–53 (1943).

Endell, J., The Study of the Fine Structure of Clays by Means of the Supermicroscope, *Keram. Rundschau,* **49:** 23–26 (1941).

Hope, E. W., and J. A. Kittrick, Surface Tension and the Morphology of Halloysite, *Am. Mineralogist,* **49:** 859–866 (1964).

Jackson, M. L., W. J. Mackie, and R. P. Pennington, Electron Microscope Applications in Soils Research, *Soil Sci. Soc. Am. Proc.,* **22:** 57–63 (1947).

Kelley, O. J., and B. T. Shaw, Studies of Clay Particles with the Electron Microscope, III, Hydrodynamic Considerations in Relation to Shape of Particles, *Soil Sci. Soc. Am. Proc.,* **7:** 58–62 (1942).

Kotel'nikov, D. D., Morphological Characteristics of Montmorillonites in Sedimentary Rocks, *Mineralog. Sb., L'vovsk. Geol. Obshchestvo pri L'vovsk. Gos. Univ.,* no. 17, pp. 60–68 (1963).

Noll, W., The Electron Microscope in the Study of Hydrothermal Silicate Reactions, *Kolloid-Z.*, **107**: 181–190 (1944).

Popov, N. M., and B. B. Zvyagin, Investigation of Minerals by the Microdiffraction Method with an Electron Microscope of 400 kv. Acceleration Voltage, *Izv. Akad. Nauk SSSR*, **23**: 670–672 (1959).

Prebus, A. F., The Electron Microscope, *Ohio State Univ. Studies, Eng. Expt. Sta. News*, **14**: 7–32 (1942).

Raman, K. V., and M. L. Jackson, Vermiculite Surface Morphology, Proceedings of the 12th National Clay Conference, pp. 423–429, Pergamon Press, New York (1964).

Sudo, T., and H. Takahashi, Shapes of Halloysite Particles in Japanese Clays, *Natl. Acad. Sci., Publ.* 456, pp. 67–79 (1956).

Waterman, A. T., The Electron Microscope, *Am. J. Sci.*, **239**: 386–388 (1941).

Seven
Ion Exchange and Sorption

The clay minerals have the property of sorbing certain anions and cations and retaining them in an exchangeable state; i.e., these ions are exchangeable for other anions or cations by treatment with such ions in a water solution. [The exchange reaction also takes place sometimes in a nonaqueous environment (see page 207).] The exchange reaction is stoichiometric and thereby differs from simple sorption. This distinction, however, is sometimes difficult to apply since nearly every ion-exchange process is accompanied by sorption or desorption. The heat evolved in the course of an ion-exchange reaction is usually rather small, $2\pm$ kcal/mole (Helfferick[1]), unless ion exchange is followed by a reaction such as neutralization, etc.

The exchangeable ions are held around the outside of the silica-alumina clay-mineral structural unit, and the exchange reaction generally does not affect the structure of the silica-alumina packet. A simple and well-known example of the ion-exchange reaction is the softening of water by the use of zeolites, permutites, or carbon exchangers.

The property of exchange capacity is measured in terms of milliequivalents per gram or more frequently per 100 g. One equivalent of Na^+ expressed as Na_2O would be a combining weight of 31, and 1 meq/100 g would be

[1] Helfferick, F., "Ion Exchange," McGraw-Hill, New York (1962).

equal to 0.031 percent Na_2O. Exchange capacity is determined at neutrality, that is, pH 7.

Vastly more information is available regarding cation exchange than anion exchange, and although they will be considered separately, an elaborate discussion of all the aspects of anion exchange is not possible at this time. In clay materials the commonest exchangeable cations are Ca^{++}, Mg^{++}, H^+, K^+, NH_4^+, Na^+, frequently in about that order of general relative abundance. The common anions in clay materials are SO_4^{--}, Cl^-, PO_4^{3-}, NO_3^-. The general relative abundance of the anions is not yet known.

In recent years, the reaction of organic ions and the clay minerals has been of much interest (see Chap. 10). Weiss[2] pointed out that even the interlayer cations in nonswelling layer silicates, e.g., micas, can be exchanged slowly for voluminous alkylammonium ions. Weiss and Thamerus[3] concluded that the alkylammonium ions can be replaced in turn by other ions, for example, aluminum and silicon ions. With such ions the products are very brittle and show a basal spacing smaller than that of the original micas.

Importance of ion exchange

The property of ion exchange and the exchange reaction are of very great fundamental and practical importance in all the fields in which clay materials are studied and used. The significance of exchange reaction in many fields has not always been appreciated even by those actively working in those fields. Therefore, some significant applications of ion exchange will be given so that its importance will be apparent.

In the field of soils, plant foods are frequently held in the soils as exchangeable ions, and consequently their persistence in the soil and their availability for plant growth depend on exchange reactions. For example, the retention and availability of potash added in fertilizers depend on cation exchange between the potash salt and the clay mineral in the soil. Brown[4] showed that the availability of calcium, magnesium, and potassium to certain plants varies depending on the identity of the clay minerals in the soil. Fur-

[2] Weiss, A., Der Kationenaustausch bei den Mineralen der Glimmer-Vermikulit—und Montmorillonitgruppe, *Z. Anorg. Allgem. Chem.*, **297**: 257–286 (1958).

[3] Weiss, A., and G. Thamerus, Über die Einführung von Al^{3+} und Si^{4+} in glimmerartige Schichtsilikate and Stelle der austauschfähigen Kationen, *Z. Anorg. Allgem. Chem.*, **317** (1–2): 142–148 (1962).

[4] Brown, D. A., Ion Exchange in Soil Plant Environments, II, The Effect of Type of Clay Mineral upon Nutrient Uptake by Plants, *Soil Sci. Soc. Am. Proc.*, **19**: 296–300 (1955).

ther, the tilth of the soil is frequently determined by the character of the exchangeable ion, and it may be controlled by an exchange reaction. Thus the presence of appreciable Na^+ in a soil makes it unsuitable for agriculture. The replacement of the Na^+ by another ion, usually Ca^{++}, will generally make the soil suitable for agriculture. According to Toth,[5] in an "ideal soil" the cation population consists of 65 percent calcium, 10 percent magnesium, 5 percent potassium, and 20 percent hydrogen.

In the field of geology many examples could be given, but two will suffice. Weathering processes involve the liberation of alkalies and alkaline earths, which may or may not be retained in the secondary material, depending on exchange reactions. The nature of the weathering product depends to a very great extent on whether or not the alkalies and alkaline earths are retained and on which of them are preferentially retained. Weathering is not simply the breakdown of the primary minerals followed by leaching.

Variations in the amount and kind of certain ions in the environment of accumulation of sedimentary rocks must be reflected in the exchangeable-ion composition of argillaceous sediments. Therefore, exchangeable-ion data should be of significance in determining the environment of accumulation of ancient sediments.

In oceanography the concentration of sodium in sea water is to a considerable extent a consequence of the cation-exchange properties of clay materials which have accumulated in the sea. The relative exchangeability of the common cations brought to the sea and the property of some clay minerals to fix K^+ would lead to a concentration of sodium (see pages 220–223).

Ion exchange is of very great importance in all the applied arts where clay materials are used, or where the properties of clays are important, because the physical properties of clay materials are frequently dependent to a large extent on the exchangeable ions carried by a clay. Again many examples of the great importance of this fact could be given, but a few will suffice. In general the plastic properties of a clay or soil are very different, depending on whether Na^+ or Ca^{++} is the exchangeable cation. The ceramist, therefore, can change the plastic characteristics of many clays to meet his needs by carrying out a base-exchange reaction. Thus, it is common practice in the brick industry to add soda ash to the plastic clay to improve its properties. The construction engineer, also, can sometimes vary to suit his needs the property of a soil material on which, through which, or with which he proposes to work.

Sometimes the construction engineer inadvertently causes an ion-exchange reaction, by a shift of water table, emplacement of a mass of concrete, etc., with an unexpected change in the properties of the soil. If the changes in

---

[5] Toth, S. J., The Physical Chemistry of Soils, "Chemistry of the Soil," chap. III, pp. 142–162, American Chemical Society Monograph (1964).

the plastic, compaction, and shrinkage properties resulting from such exchange reactions are not foreseen, the consequences may be disastrous.

Bentonite clay is widely used for many purposes. For certain uses it must form thixotropic suspensions in water. Only bentonites composed of the clay mineral montmorillonite carrying Na^+ as the exchangeable ion mix with water to give suspensions of pronounced thixotropic character.

Cation exchange

History. According to Kelley,[6] who reviewed the history of cation exchange in detail, the discovery that soils have the power of exchanging cations with solutions containing other cations was the outgrowth of observations dating back into the remote past. For example, it has been known for centuries that liquid manures become decolorized and deodorized when filtered through soils. Thompson[7] is generally credited with being the first person who systematically studied cation exchange. The term base exchange was used to describe the reaction for many years, even long after it was established that the hydrogen ion may take part in the exchange reaction. In experiments begun in 1845 and published 5 years later, Thompson showed that, when soils were mixed with ammonia and then leached with water, the greater part of the ammonia was held back.

Following Thompson, Way[8,9] began a detailed study of the phenomenon and began to publish his results in 1850. Way showed that cation exchange in soils was restricted to the clay fraction and that it was connected with the silicate compounds in the soil. Also in 1850 Forschamer[10] showed that calcium and magnesium are released from soil by leaching with sea water. Following the pioneering labors of Thompson, Way, and Forschamer, a large number of investigators, particularly in the field of soil chemistry, have studied all aspects of the exchange reaction.

Cation-exchange capacity. The range of the cation-exchange capacity of the clay minerals is given in Table 7-1.

[6] Kelley, W. P., "Cation Exchange in Soils," Reinhold, New York (1948).

[7] Thompson, H. S., On the Absorbent Power of Soils, *J. Roy. Agr. Soc. Engl.,* **11:** 68–74 (1850).

[8] Way, J. T., On the Power of Soils to Absorb Manure, *J. Roy. Agr. Soc. Engl.,* **11:** 313–379 (1850).

[9] Way, J. T., On the Power of Soils to Absorb Manure, *J. Roy. Agr. Soc. Engl.,* **13:** 123–143 (1852).

[10] Forschamer, G., reference in Wiklander, Studies on Ionic Exchange, *Ann. Roy. Agr. Coll. Sweden,* **14:** 1–171 (1946).

It follows from a consideration of the factors influencing cation-exchange capacity (see pages 193–194) that there is no single capacity value that is characteristic of a given group of clay minerals. A range of capacities must be shown for each group. Since the cation-exchange capacity of a given mineral type may vary with so many factors, capacity values are rigorously comparable only if they have been obtained by the same standard procedure on material of comparable textural and structural attributes.

The large range shown for the cation-exchange capacity for illite and chlorite is due to the large variation in particle size, crystallinity, and time of treatment (see pages 193–204). The potassium and magnesium in interlayer positions become increasingly replaceable with decreasing particle size, decreasing crystallinity, and increasing time of treatment.

Table 7-1 Cation-exchange capacity of clay minerals, in milliequivalents per 100 g

Kaolinite	3–15
Halloysite $2H_2O$	5–10
Halloysite $4H_2O$	40–50
Smectite	80–150
Illite	10–40
Vermiculite	100–150
Chlorite	10–40
Sepiolite-attapulgite-palygorskite	3–15

The values for halloysite $4H_2O$ are somewhat questionable as they probably vary substantially with the procedure used for determination.

Precise values for attapulgite are difficult to obtain because of the almost universal association of montmorillonite with attapulgite. The high cation-exchange capacities reported in the past for attapulgite are probably due to montmorillonite contamination.

Birrell and Gradwell[11] reported cation-exchange capacities for allophane ranging from about 25 to 50 meq/100 g and showed that the values varied greatly, depending on the procedure used in making the determination. They also suggested that generally the reactions are not really those of cation exchange as considered for other clay minerals but are chemical reactions with the constituents of the allophane. Birrell[12] found that the values for cation exchange increased with decreasing pH and for anion exchange increased with decreasing pH.

Chemical pretreatment can have substantial influence on the cation-exchange capacity of the clay minerals generally. This is especially true for

[11] Birrell, K. S., and M. Gradwell, Ion-exchange Phenomena in Some Soils Containing Amorphous Mineral Constituents, *J. Soil Sci.*, **7**: 130–147 (1956).

[12] Birrell, K. S., Ion Fixation by Allophane, *New Zealand J. Sci.*, **4**: 393–414 (1961).

allophane, as Jackson[13] has shown that pretreatment of his samples with a mildly alkaline sodium carbonate solution caused a relatively high capacity whereas pretreatment with sodium acetate buffered at pH 3.5 caused a relatively low capacity. The difference in capacity is reported to be about 100 meq/100 g when determined with potassium acetate at pH 7.

The cation-exchange capacities in Table 7-1 are taken at pH 7. Titration curves showing the relation between pH and milliequivalents of added NaOH are given in Figs. 7-1 to 7-6. Marshall,[14,15] Mukherjee,[16] Mitra,[17] and others have considered in detail the significance of such titration curves. At relatively low and high pH values, attack of acids and alkalies on the clay-mineral structure is the factor largely controlling the shape of the curves. It can be seen from Figs. 7-1 to 7-6 that the character of the clay mineral, its concentration, and the base used[18,19] also affect the nature of the curves. Other attributes to be discussed later (see pages 193 and 202), such as variations in degree of crystallinity and particle size, also affect the shape of the curves.

Marshall[14] and others[16] have shown that a single cation may be sorbed by a clay mineral with a wide range of bonding energies and that this is fundamentally related to the position on the silica-alumina packet at which the cation is sorbed, e.g., whether it is held between the sheets of the layered minerals or around their edges. According to these workers, this is an important cause of the variation in titration curves. The possible positions for sorbed cations vary with the lattice structure of the minerals, and consequently the variation of bonding energy for a given cation would not be expected to be the same for all types of clay minerals (see page 195).

It can be seen from Figs. 7-1 to 7-4 that the cation-exchange capacity

[13] Jackson, M. L., Chemical Composition of Soils, "Chemistry of the Soil," chap. II, pp. 71–141, American Chemical Society Monograph (1964).

[14] Marshall, C. E., "The Colloid Chemistry of the Silicate Minerals," Academic, New York (1949).

[15] Marshall, C. E., and C. A. Krinbill, The Clays as Colloidal Electrolytes, *J. Phys. Chem.*, **46**: 1077–1090 (1942).

[16] Mukherjee, J. N., and R. P. Mitra, Some Aspects of the Electrochemistry of the Clays, *J. Colloid Sci.*, **1**: 141–159 (1946).

[17] Mitra, R. P., Electrochemical Aspects of Ion Exchanges in Clays, Bentonites, and Clay Minerals, *Indian Soc. Soil Sci. Bull.*, **4**: 41–61 (1942).

[18] Mukherjee, J. N., R. P. Mitra, and D. K. Mitra, Electrochemical Properties of Clay Minerals and the Differentiation of Hydrogen Clays and Bentonites by Electro-chemical Methods, I, *J. Phys. Chem.*, **47**: 543–549 (1943).

[19] Mitra, R. P., S. N. Bagchi, and S. P. Ray, Electrochemical Properties of Clay Minerals and the Differentiation of Hydrogen Clays and Bentonites by Electrochemical Methods, II, *J. Phys. Chem.*, **47**: 549–553 (1943).

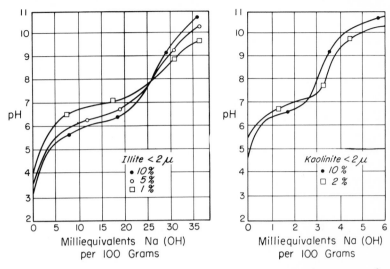

Fig. 7-1. Titration curves for hydrogen illite. (After Marshall and Krinbill.[15])

Fig. 7-2. Titration curves for hydrogen kaolinite. (After Marshall and Krinbill.[15])

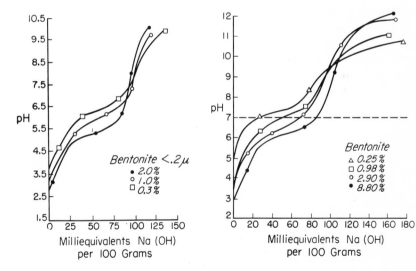

Fig. 7-3. Titration curves for hydrogen montmorillonite. (After Marshall and Krinbill.[15])

Fig. 7-4. Titration curves for hydrogen montmorillonite (Indian bentonite). (After Mukherjee and Mitra.[16])

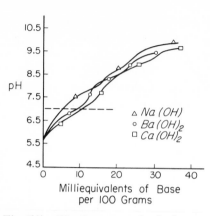

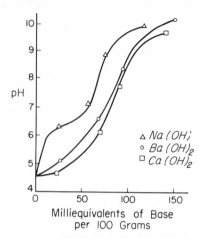

Fig. 7-5.   Titration curves for hydrogen kaolinite. (After Mukherjee, Mitra, and Mitra.[18])

Fig. 7-6.   Titration curves for hydrogen bentonite (montmorillonite). (After Mitra, Bagchi, and Ray.[19])

of montmorillonite may vary with the concentration of the clay, particularly at relatively high concentrations.   Thus the cation-exchange capacity based on NaOH for an Indian bentonite (montmorillonite) calculated at the inflection point of its potentiometric-titration curve is shown to vary from 81 to 103 meq/100 g as the clay concentration increased from 0.25 to 8.80 percent. The variation with clay concentration is considerably less for kaolinites and illites than for smectites.

The exchange capacity may vary also with the nature of the cation.   Figures 7-5 and 7-6, showing titration curves for kaolinite and montmorillonite with cations of different valence, illustrate the relatively great possible differences for monovalent and divalent cations.   In both cases the exchange capacity is considerably larger when determined with Ca^{++} than with Na^+.   As will be shown (pages 217–220), other factors such as particle size, lattice distortion, clogging of exchange positions, etc., may also affect the cation-exchange capacity.

Other minerals with cation-exchange capacity. The clay minerals are not the only components of clay materials that have cation-exchange capacity.   All inorganic minerals of extreme fineness have a small cation-exchange capacity as a result of broken bonds around their edges.   This capacity increases as the particle size decreases, but even in the small size in which nonclay minerals occur in clays, the exchange capacity is generally insignificant.   Zeolite minerals, which are occasionally found in some clays, have cation-exchange capacities of the order of 100 to 300 meq/100 g.

Some organic materials have cation-exchange capacity, and values ranging from 150 to 500 meq/100 g are reported[20,21] for the organic fraction of some soils. In general, organic material with high exchange capacity is restricted to Recent sediments and to soils. The organic material in ancient sediments which has undergone even a small amount of metamorphism is not likely to have significant cation-exchange capacity.

Causes of cation exchange. The causes of the cation-exchange capacity of the clay minerals can be considered under three headings:

1. Broken bonds around the edges of the silica-alumina units would give rise to unsatisfied charges, which would be balanced by adsorbed cations. The broken bonds would tend to be on noncleavage surfaces and hence on the vertical planes, parallel to the *c* axis, of the layer clay minerals and on horizontal planes, perpendicular to the *c* axis, of the sepiolite-palygorskite-attapulgite minerals.

The number of broken bonds and hence the exchange capacity due to this cause would increase as the particle size decreased. Also lattice distortions would tend to increase broken bonds, and the exchange capacity would be expected to increase as the degree of crystallinity decreased.

In the kaolinite and halloysite minerals, broken bonds are probably the major cause of exchange capacity. In the illite, chlorite, and sepiolite-palygorskite-attapulgite minerals, broken bonds are an important cause of exchange capacity, and when these minerals are well crystallized and have relatively low capacity, it may be the major cause. Substitutions within the lattice, particularly in poorly crystalline examples of illite, chlorite, and sepiolite-palygorskite-attapulgite, may partially explain their exchange capacity.

In smectites and vermiculites, broken bonds are responsible for a relatively small portion ($20\% \pm$) of cation-exchange capacity, the remainder probably resulting from substitutions within the lattice. Johnson[22] argued that the total exchange capacity of montmorillonites results from broken bonds, but his evidence is not convincing.

Numerous investigators have studied in detail the character of the broken bonds on clay-mineral units. Wiklander[23] reviewed much of this work, with

[20] Malquori, A., Behavior of Humus in Clay-bearing Soils, II, Base Exchange Capacity of Organic Substances, *Ann. Chim. Appl.,* **23:** 111–126 (1944).

[21] Francis, M., Sur la matière organique dans les argiles, *Verre Silicates Ind.,* **14:** 155–158 (1949).

[22] Johnson, A. L., Surface Area and Its Effect on Exchange Capacity of Montmorillonite, *J. Am. Ceram. Soc.,* **32:** 210–213 (1949).

[23] Wiklander, L., Cation and Anion Exchange Phenomena, "Chemistry of Soils," chap. IV, pp. 163–205, American Chemical Society Monograph (1964).

the general conclusion that hydroxyls would be attached to the silicons of broken tetrahedral units and that the hydroxyls would be ionized similarly to ordinary silicic acid, that is, $Si\text{-}OH + H_2O = SiO^- + H_3O^+$, causing a negative charge on the lattice. The positive charges may originate from exposed octahedral groups which react as bases by accepting protons, thus acquiring a positive charge. The negative charge would grow and the positive charge decrease with rising pH as a result of increasing ionization of the acid groups and decreasing proton addition to the basic groups. The charge goes in the opposite direction with decreasing pH.

2. *Substitutions within the lattice structure* of trivalent aluminum for quadrivalent silicon in the tetrahedral sheet and of ions of lower valence, particularly magnesium, for trivalent aluminum in the octahedral sheet result in unbalanced charges in the structural units of some clay minerals. Sometimes such substitutions are balanced by other lattice changes, for example, OH for O, or by filling more than two-thirds of the possible octahedral positions, but frequently they are balanced by adsorbed cations.

Schofield and Samson[24] pointed out that only one silicon out of 400 needs to be replaced by an aluminum to get a cation-exchange capacity of 2 meq/100 g for kaolinite. Such small replacements could hardly be detected by chemical analyses so that the possibility of replacement as a cause for the exchange property in minerals with low capacity cannot be eliminated.

White[25] found, in the case of dioctahedral minerals, that lithium from molten lithium nitrate could penetrate into vacant octahedral positions, with the consequent reduction in cation-exchange capacity.

Exchangeable cations resulting from lattice substitutions are to be found mostly on cleavage surfaces, e.g., the basal cleavage surfaces of the layer clay minerals. Since the charges resulting from substitutions in the octahedral sheet would act through a greater distance than the charges resulting from substitutions in the tetrahedral sheet, it would be expected that cations held because of the latter substitutions would be bonded by a stronger force than those held by forces resulting from substitutions in the octahedral sheet. In some cases, cations held by forces due to substitutions of aluminum for silicon seem to be substantially nonexchangeable, e.g., the potassium in the micas. In the clay minerals, replacements in the octahedral layer are probably the major substitutions causing cation-exchange capacity.

In smectite and vermiculite, substitutions within the lattice cause about 80 percent of the total cation-exchange capacity. In a smectite with a ca-

[24] Schofield, R. K., and H. R. Samson, The Deflocculation of Kaolinite Suspensions and the Accompanying Change Over from Positive to Negative Chloride Adsorption, *Clay Minerals Bull.,* **2**(9): 45–51 (1953).

[25] White, J. L., Reactions of Molten Salts with Layer-Lattice Silicates, *Natl. Acad. Sci., Publ.* **456**, pp. 133–146 (1956).

tion-exchange capacity of about 100 meq/100 g (1 equivalent per 1,000 g) and a molecular weight of about 720, substitutions of less than one-sixth of the aluminum by magnesium or one-twelfth of the silicon by aluminum would account for the capacity if there were no internal balancing of charges.

3. *The hydrogen of exposed hydroxyls* (which are an integral part of the structure rather than due to broken bonds) may be replaced by a cation which might be exchangeable. However, it seems probable that such hydrogen would be relatively tightly held as compared with those associated with broken bonds and hence, in the main, not replaceable. This cause of exchange capacity would be important for kaolinite and halloysite because of the presence of the sheet of hydroxyls on one side of the basal cleavage plane.

Edelman and Favejee[26] suggested structures for smectite and halloysite, and McConnell[27] suggested for smectite an alternative structure which would have hydroxyls in basal plane cleavage surfaces. On the basis of these structural concepts a considerable amount of the cation-exchange capacity of these minerals would be due to this cause. However, neither of these suggested structures has been generally accepted.

Position of exchangeable cations. In the clay minerals in which the cation exchange results from broken bonds, the exchangeable cations would be expected around the edges of the flakes and elongate units. In the clay minerals where the exchange is due to lattice substitutions, the cations are mostly on the basal plane surfaces. In smectite and vermiculite about 80 percent are on basal plane surfaces, with the remainder on the edges. In the case of illite, chlorite, and sepiolite-palygorskite-attapulgite minerals, most of the cations are at the edges with a relatively few on cleavage surfaces. Weiss[28] and later Weiss and Russow[29] concluded that for kaolinite the exchangeable cations occur only on the basal oxygen surface of the silica tetrahedral units, and therefore cation-exchange capacity depends on the thickness of kaolinite particles.

In a smectite with a cation-exchange capacity of 100 meq/100 g and a unit-cell weight of 720, and assuming that 80 percent of the exchange positions are on basal plane surfaces, there would be 1 equivalent for $1\frac{3}{4}$ unit

[26] Edelman, C. H., and J. C. L. Favejee, On the Crystal Structure of Montmorillonite and Halloysite, *Z. Krist.,* **102:** 417–431 (1940).

[27] McConnell, D., The Crystal Chemistry of Montmorillonite, *Am. Mineralogist,* **35:** 166–172 (1950).

[28] Weiss, A., Der Kationenaustausch bei Kaolinit, *Z. Anorg. Allgem. Chem.,* **299:** 92–120 (1959).

[29] Weiss, A., and J. Russow, Über die Lage der austauschbaren Kationen bei Kaolinit, *Proc. Intern. Clay Conf., Stockholm,* **1:** 203–214 (1963).

cells. With a basal surface area of $92.26 \, A^2$ per unit cell, there would be a total surface area of about $160 \, A^2$ per exchange position, or an area of about $80 \, A^2$ per exchange position on each basal plane surface.

In masses of clay with relatively small amounts of adsorbed water, i.e., with no more water than is required to develop plasticity, it is likely that the adsorbed cations around the edges of the flakes are held directly in contact or at least very close to the clay-mineral surface. Brown[30] presented X-ray data based on Fourier syntheses indicating that, for montmorillonite under such conditions of clay-water concentration, the adsorbed cations between the basal plane surfaces are held midway between the clay-mineral surfaces. Pezerat and Mering[31] found that, for montmorillonite, sodium is in the hexagonal cavity in the tetrahedral layer in the anhydrous material when there is a single layer of water present, and calcium is also present in the cavity in the anhydrous form but is in the water layer when a single layer of water is present.

In clay-water systems in which the amount of water is at least greater than that required for the plastic state, the exchangeable cations may be at greater distances from the clay-mineral surfaces and separated from them by water molecules. In any given system of this kind, the position of all the exchangeable cations with respect to the clay-mineral surface is not the same, and even the relative positions of all the same sort of cations probably are not the same; some cations of a given type are closer to the clay-mineral surface than others.

Mering and Glaeser[32] pointed out that the negative charges on the clay mineral can be neutralized locally only by monovalent cations. The saturation of the mineral by divalent or polyvalent cations inevitably creates deficiencies of local neutralization. The cations can solvate or hydrate only when they are detached from the silicate units, and the cations can detach themselves only where they do not participate in local neutralization. Monovalent cations cannot detach themselves and, therefore, cannot solvate. As a consequence, clay saturated with polyvalent cations has relatively greater stability in the solvated states. These authors showed that, when there is less than about 30 percent calcium in a montmorillonite containing calcium and sodium, the calcium ion is not effective; in other words, the clay acts as if saturated completely with sodium. Glaeser and Mering[33] determined

[30] Brown, G., A. Fourier Investigation of Montmorillonite, *Clay Minerals Bull.*, no. 4, pp., 109–111 (1950).

[31] Pezerat, H., and J. Mering, Détection des cation échangeables de la montmorillonite par l'emploi des séries différences, *Bull. Groupe Franc. Argiles*, 10(5): 26–30 (1958).

[32] Mering, J., and R. Glaeser, On the Role of the Valency of Exchangeable Cations in Montmorillonite, *Bull. Soc. Franc. Mineral*, 77: 519–530 (1954).

[33] Glaeser, R., and J. Mering, Isotherms d'hydration des montmorillonites bi-ioniques (Na,Ca), *Clay Minerals Bull.*, 2(12): 188–193 (1954).

adsorption isotherms for mixed calcium-sodium montmorillonites containing various proportions of calcium and sodium. They found that a "demixing" occurs when more than 30 percent of the charges on the lattice are satisfied by calcium. This is interpreted to mean that a system is formed in which some interlayer spaces are occupied mainly by calcium and others mainly by sodium. They stated that this observation can be explained by the occurrence of a limited number of doubly charged sites (by alumina-for-silicon exchange spots being accidentally contiguous). In a later paper these same authors[34] reported the same behavior in hectorite, a trioctahedral smectite. Fripiat and his colleagues[35] determined, on the basis of X-ray fluorescence, that the electropositive character of calcium in montmorillonites saturated with calcium and sodium increased as the proportion of the calcium ion to sodium increased and as the hydration of the mineral increased. They concluded that the first observation can be explained by the conclusions of Mering and Glaeser[32] and that the second observation can in some way be explained by the fact that dehydration brings the silicate sheets closer together and the cations closer to the negative sites, thereby contributing to the neutralization of the charge.

Marshall[14] showed that only a portion of the adsorbed cations are likely to be ionized, and the percentage ionized depends on the particular clay mineral, the amount of water, i.e., the concentration of the clay-water system, the nature of the cations, the relative concentration of the cations, and the nature of the adsorbed anions.

Many investigators have shown that there may be a significant difference in the character and intensity of the charge on different parts of clay-mineral units. Schofield and Samson[36] concluded that the charge on the edge and basal surfaces of kaolinite particles can behave differently. Garrels and Christ[37] found that a sample labeled beidellite had two different types of exchange sites on the basis of titration curves prepared by Marshall and Bergman.[38] Free-energy calculations indicated that hydrogen was bonded to the edges more strongly than to the center surface site, whereas potassium had no preference.

[34] Glaeser, R., and J. Mering, Diionic Hectorite, *Bull. Groupe Franc. Argiles,* **10:** 71–75 (1958).

[35] Fripiat, L., P. Cloos, and A. Jornoy, Note préliminaire sur l'état de valence du calcium absorbé par la montmorillonite, *Bull. Groupe Franc. Argiles,* **13:** 65–68 (1962).

[36] Schofield, R. K., and H. R. Samson, Flocculation of Kaolinite Due to the Attraction of Oppositely Charged Crystal Faces, *Discussions Faraday Soc.,* **18:** 135–145 (1954).

[37] Garrels, R. M., and C. L. Christ, Application of Cation-exchange Reactions to the Beidellite of the Putnam Silt Loam Soil, *Am. J . Sci.,* **254:** 372–379 (1956).

[38] Marshall, C. E., and W. E. Bergman, The Electronic Properties of Mineral Membranes II, Measurement of Potassium-ion Activities in Colloidal Clays, *J. Phys. Chem.,* **46:** 52–61 (1942).

Wey et al.[39] interpreted breaks in absorption curves of ammonia on kaolinite to mean acid groups of varying strength on the surface of the kaolinite. In agreement with earlier work by Fripiat et al.,[40] they concluded that this is due to Si-OH and Al-OH at broken bonds from which the hydrogen dissociates at different pH values.

Pommer and Carroll[41] and Pommer[42] concluded, on the basis of a titration curve for a hydrogen montmorillonite, that it behaved as a mixture of two acids resulting from two types of exchange sites. Interlayer sites held hydrogen ions less tightly, causing a stronger acid function than edge sites, which form stronger hydrogen bonds and are responsible for the weaker acid functions. Correlation of these data with information from the literature indicated to Pommer[42] that in the case of beidellite the pattern was reversed. The beidellites considered were probably mixed-layer structures. Jonas[43] earlier made the opposite assignment regarding the strength of the edge and interlayer sites of montmorillonite.

Volk and Jackson[44] found that their titration curves of acid montmorillonite revealed four buffer ranges which they correlated with hydronium ions, monomeric trivalent aluminum ions, and two pH-dependent charge ranges. Earlier Mehlich[45] presented an elaborate technique for measuring various pertinent charge characteristics of soils. It is not certain that in every case such correlations have the fundamental significance ascribed to them.

Fieldes and Schofield[46] concluded that in allophane and probably also in smectite and similar clays the characteristic variable charge on the lattice with varying pH is due to the behavior of specific sites originating from

[39] Wey, R., O. Sieskind, and V. P. Leiberguth, L'Influence du pH sur l'adsorption des cations par les argiles, *Bull. Assoc. Franc. Etude Sol* (1959).

[40] Fripiat, J. J., M. C. Gastuche, and G. Vancompernolle, Les Groupes hydroxyles de surface de la kaolinite et sa capacité d'échange ionique, *Compt. Rend. Congr. Intern. Sci. Sol,* **2**: 401–422 (1954).

[41] Pommer, A. M., and D. E. Carroll, Interpretation of Potentiometric Titration of H-Montmorillonite, *Nature,* **185**: 595–596 (1960).

[42] Pommer, A. M., Relation Between Dual Acidity and Structure of H-Montmorillonite *U.S. Geol. Surv., Profess. Paper* 386-C, 23 pp. (1962).

[43] Jonas, E. C., Mineralogy of the Micaceous Clay Minerals, *Intern. Geol. Congr.,* 21st, XXIV, pp. 7–16 (1961).

[44] Volk, V. V., and M. L. Jackson, Inorganic pH Dependent Cation Exchange Charge of Soils, Proceedings of the 12th National Clay Conference, pp. 281–296, Pergamon Press, New York (1964).

[45] Mehlich, A., Charge Characterization of Soils, *Trans. 7th Intern. Congr. Soil Sci.,* **2**: 292–302 (1961).

[46] Fieldes, M., and R. K. Schofield, Mechanisms of Ion Adsorption by Inorganic Soil Colloids, *New Zealand J. Sci.,* **3**: 563 (1960).

broken bonds of aluminum in tetrahedral coordination. They concluded further that the concept of tetrahedral aluminum explains the increase of cation retention with increasing pH, the mechanism of anion retention, and the order of affinity of soil clays for different anions.

Denny and Roy[47] concluded that, in general, bonding energies are greater for materials with greater charge density.

Fripiat and his colleagues[48,49] investigated in great detail the surface characteristics of kaolinite, especially the character of the hydroxyls that develop at the broken edges of tetrahedral and octahedral units in relation to the cation-exchange capacity. They found that the hydroxyls attached to the silicons are thermally more stable than those bound to the aluminum. Gastuche et al.[50] pointed out that mild acid treatment removes external aluminum and that there is a good correlation between amount extractable and cation-exchange capacity; following this work, Fripiat[48] suggested that tetrahedral alumina, perhaps in a gel, serves as a coating of clay-mineral particles and accounts for the exchange capacity of kaolinites.

Marshall[51,52] discussed in detail the electrochemistry of clays, and the following is taken from his work. Kaolinite was found to have two inflections on the pH titration curve as compared with one (with certain exceptions) for smectite clays. More complex relationships were found in the titration curves of illites, which appeared to give three inflections. Considering the cationic bonding energy of various clay minerals, it was found that with monovalent cations there was the following order for dissociation: montmorillonite (Arizona bentonite) > illite-montmorillonite mixed-layer (Putnam soil) > illite > montmorillonite (Wyoming bentonite) > kaolinite > attapulgite. For divalent cations the order varied somewhat with the cation used and with the degree of saturation chosen. Certain special features were noticed: the very high bonding energy of illite for barium as contrasted to the relatively low bonding energy for calcium and magnesium, and the change in order

[47] Denny, P. J., and R. Roy, Cation Exchange Between Mixtures of Clay Minerals and Between a Zeolite and a Clay Mineral, Proceedings of the 12th National Clay Conference, pp. 567–580, Pergamon Press, New York (1964).

[48] Fripiat, J. J., and P. Dondeyne, Hydration de la kaolinite, *J. Chim. Phys.*, pp. 543–552 (1960).

[49] Fripiat, J. J., Surface Properties of Alumino-Silicates, Proceedings of the 12th National Clay Conference, pp. 327–358, Pergamon Press, New York (1964).

[50] Gastuche, M. C., B. Delmon, and L. Vielvoye, La Cinetique des réactions hétérogènes. Attaque du réseau silico-alumique des kaolinites par l'acide chlorhydrique, *Chim. de France*, pp. 60–70 (1960).

[51] Marshall, C. E., "The Physical Chemistry and Mineralogy of Soils," Wiley, New York (1964).

[52] Marshall, C. E., The Electrochemistry of the Clay Minerals in Relation to Pedology, *Trans. 4th Intern. Congr. Soil Sci.*, 1: 71–82 (1950).

of dissociation from barium > calcium > magnesium > for kaolinite to magnesium > calcium > barium for illite, beidellite, and montmorillonite at degrees of saturation below 70 percent. Above this point the curves begin to cross over one another, tending toward barium > calcium > magnesium with increasing alkalinity. It is pointed out that minerals of the smectite group vary widely in their dissociation. Within this group a high charge in the silica layer seems to parallel a high cationic bonding energy, whereas when the charge is on the alumina layer a much lower bonding energy results. Attapulgite, in general, is the most highly dissociated of the clays studied. Jackson[13] stated that allophane is spectacular in the degree to which exchangeable cations hydrolize in aqueous alcohol solutions. Considerable differences between different clay minerals are revealed by the shapes of bonding-energy curves, especially those of divalent cations, which relate the percent saturation to the bonding energy. The bonding-energy curve of kaolinite gives a fairly sharp maximum in bonding energy at about 70 percent saturation, whereas smectites, illites, and attapulgites show a broad plateau at this concentration.

Rate of exchange reaction. The rate of cation exchange varies with the clay mineral, the nature and concentration of the cations, and the nature and concentration of the anions. In general the reaction for kaolinite is most rapid, being almost instantaneous. It is slower for smectite and for attapulgite and requires an even longer time, perhaps hours or days, to reach completion for illites. Apparently exchange on the edge of the particles, as in kaolinite, can take place quickly, but penetration between the sheets of smectite or in the channels of attapulgite requires more time. In the case of illite a part of the exchange is between basal flake surfaces firmly held together, and this is likely to be slow, resulting in a long time to complete the reaction.

Chloritic clay minerals are likely to have a rate of exchange similar to the illites. Vermiculite is likely to be similar to smectite, except that it may be somewhat slower because of the larger areal size of the flakes of vermiculite, so that more time is required for penetration between them.

Exchange in some of the zeolite minerals is relatively slow, and many hours are required for completion. Apparently much time is required to penetrate the channel-like openings of some of these minerals.

Ion exchange is a diffusion process, and its rate depends on the mobility of the ions. Thus ion-exchange kinetics has no resemblance to chemical-reaction kinetics in the usual sense. However, the simple well-known rate loss of diffusion holds only in exceptional cases (Helfferick[1]).

When the exchange reaction is accompanied by swelling, e.g., in montmorillonite, and the change from one stable interlayer distance to another requires a certain activation energy, a hysteresis phenomenon may be observed.

Walker[53] pointed out that cation exchange in smectites and vermiculites is often accompanied by a change in the c dimension of the unit cell and that this influences the rate of exchange. Replacing the interlayer exchangeable magnesium by strontium ions in vermiculite, for example, involves an increase from 14.4 to 15 Å, because of the larger size (0.6 Å) of the strontium ion as compared with magnesium, since double sheets of interlayer water molecules are present in both cases.

The water-molecule sites are arranged in two sheets and are related in an equivalent manner to the adjacent silicate layer surfaces, whereas the cation layer sites are located in a plane midway between the two water sheets. The cation sites are regularly spaced 5.3 Å apart, but with divalent cations only about one-third of the available sites are occupied, and the mean distance between cations is about 9 Å (Mathieson and Walker[54]). When an increase in unit-cell height occurs during an exchange, as in the magnesium- to strontium-vermiculite system, the reaction proceeds in a highly regular fashion from the edges toward the interior of the flake and can be readily followed in single crystals by means of low-power microscopy.

Walker[55] showed that the distance traveled by the boundary is proportional to the square root of the time of immersion at a given temperature, indicating a diffusion-controlled process. If vermiculites of differing surface charge density are examined under comparable conditions, the rate is found to be slower, the higher the charge. Since smectites are essentially finely divided vermiculites with a lower charge, although they cannot be examined directly by the optical method, it is reasonable to infer that their exchange rates will be faster than those of vermiculites and also will vary according to their charge. The activation energy of the reaction was calculated to be 10.1 kcal/gram ion and the diffusion coefficients at 30 and 60°C to be 5.5×10^{-8} and 2.8×10^{-7} cm²/sec, respectively.

The replacement of strontium by magnesium was studied by immersing flakes previously saturated with strontium in magnesium chloride solutions. In this instance, exchange involved a contraction of the lattice from 15 to 14.4 Å, and no optical boundaries were observed. X-ray measurements suggested a rate initially similar to that of magnesium for strontium but with progressive slowing up as exchange proceeded. This appears to be due to the fact that the strontium ions must expand the magnesium lattice locally in order to move through it. As the area of magnesium replacement spreads

[53] Walker, G. F., The Cation Exchange Reaction in Vermiculite, *Proc. Intern. Clay Conf., Stockholm,* **1**: 177–181 (1963).

[54] Mathieson, A. McL., and G. F. Walker, Crystal Structure of Magnesium-Vermiculite, *Am. Mineralogist,* **39**: 231–255 (1954).

[55] Walker, G. F., Diffusion of Exchangeable Cations in Vermiculite, *Nature,* **184**: 1392–1393 (1959).

Table 7-2   Variations in the cation-exchange capacity of kaolinite with particle size
(After Harmon and Fraulini[57])

Particle size, μ	10–20	5–10	2–4	1–0.5	0.5–0.25	0.25–0.1	0.1–0.05
Cation-exchange capacity, meq/ 100 g	2.4	2.6	3.6	3.8	3.9	5.4	9.5

Table 7-3   Variations in the cation-exchange capacity of illite with particle size
(After Grim and Bray[58])

Particle size, μ	. . . .	1–0.1	0.1–0.06	−0.06
Cation-exchange capacity, meq/100 g	Sample *A*	18.5	21.6	33
	Sample *B*	13.0	20.0	27.5
	Sample *C*	20.0	30.0	41.7

[57] Harmon, C. G., and F. Fraulini, Properties of Kaolinite as a Function of Its Particle Size, *J. Am. Ceram. Soc.*, **23:** 252–258 (1940).

[58] Grim, R. E., and R. H. Bray, The Mineral Constitution of Various Ceramic Clays, *J. Am. Ceram. Soc.*, **19:** 307–315 (1936).

inward from the edge, strontium ions find it increasingly difficult to penetrate and tend to be trapped in the interior of the crystal. Inhibition of the exchange reaction was also observed in other systems where contraction of the lattice accompanied the exchange. The generalization appears to be valid that exchange reactions involving lattice contraction are more difficult to push to completion than those reactions involving lattice expansion.

Low[56] studied the influence of adsorbed water on exchangeable-ion movement in montmorillonite clays and concluded that electrical interaction between ions and clays appears to have little effect on the activation energy and that the most important factor governing exchangeable-ion movement in the pores of clay-water systems is the structure of the adsorbed water.

Variation due to particle size. As shown in Tables 7-2 and 7-3 the cation-exchange capacities of kaolinite and illite increase as the particle size decreases. It is difficult to separate the influence of particle size from that of varying perfection of crystallinity. Ormsby et al.[59] found a linear relation between surface area and exchange capacity for kaolinites from Georgia. They also investigated the effect of variations in crystallinity and concluded that surface area is more important than crystallinity.

[56] Low, P. F., Influence of Adsorbed Water on Exchangeable-ion Movement, Proceedings of the 9th National Clay Conference, pp. 219–228, Pergamon Press, New York (1962).

[59] Ormsby, W. C., J. M. Shartsis, and K. H. Woodside, Exchange Behavior of Kaolins of Varying Degrees of Crystallinity, *J. Am. Ceram. Soc.*, **45:** 361–366 (1962).

It has generally been considered that the cation-exchange capacity of smectite does not change substantially with particle size, and Hauser and Reed[60] showed that there is no variation in the 87- to 14-mμ particle range. Caldwell and Marshall[61] (see Table 7-4) showed no variation in the cation-exchange capacity of nontronite in the particle range from 2 to 0.05 μ, but they showed a slight increase in this same range for saponite. The same authors showed a slight increase for attapulgite with decreasing particle size.

Table 7-4   *Variation in the cation-exchange capacity of nontronite, attapulgite, and saponite with particle size*
(After Caldwell and Marshall[61])

Particle size, μ	. . . .	2–1	1–0.5	0.5–0.2	0.2–0.05	−0.05
Cation-exchange capacity, meq/100 g	Nontronite	60.8	61.0	64.3	57.0	
	Attapulgite	18.0	19.0	22.2		
	Saponite	69.3	76.0	81.5	86.3	81.4

In minerals such as kaolinite and illite, in which the exchange capacity is due primarily to broken bonds, an increase is to be expected with decreasing particle size. In the case of expanding-lattice minerals, where most of the exchange is on accessible basal plane surfaces, it seems that particle size should make little difference. However, it would be expected that, in some montmorillonites, perhaps because of the location of lattice substitutions, nature of exchangeable ion, areal size of the flakes, etc., the accessibility of the basal plane surfaces would increase with decreasing particle size, and therefore some types of smectite would be expected to show moderate variation of cation-exchange capacity with particle size. Johnson[22] presented data showing a variation of cation capacity with particle size for several montmorillonites and concluded that the capacity is entirely derived from broken bonds. This conclusion, however, does not necessarily follow from Johnson's data, and Osthaus[62] presented analytical data indicating no variation of cation-exchange capacity with particle size for the smectites he studied.

[60] Hauser, E., and C. E. Reed, Studies in Thixotropy. II. The Thixotropic Behavior and Structure of Bentonite, *J. Phys. Chem.*, **41**: 911–934 (1937).

[61] Caldwell, O. G., and C. E. Marshall, A Study of Some Chemical and Physical Properties of the Clay Minerals Nontronite, Attapulgite, and Saponite, *Missouri, Univ., Agr. Expt. Sta., Res., Bull.* 354 (1942).

[62] Osthaus, B. B., Interpretation of Chemical Analysis of Montmorillonite, *Calif., Dept. Nat. Resources, Div. Mines, Bull.* 169, pp. 95–100 (1955).

Effect of grinding. Kelley and Jenny[63] showed that grinding the clay minerals, as well as many nonclay minerals, caused an increase in cation-exchange capacity, as is shown in Table 7-5. The experiments of these investigators were carried out in a rubber-lined ball mill, using polished agate balls. The grinding caused a variation in particle size, an increase in surface, and an increase in the number of broken bonds. X-ray examination of the ground material showed a progressive broadening and diffuseness of the diffraction bands, with their final disappearance after very long grinding, indicating a gradual breaking down of the structure.

Table 7-5   Cation-exchange capacity, in milliequivalents per 100 g, in relation to grinding

(After Kelley and Jenny[63])

Minerals	100 mesh	Ground 48 hr	Ground 72 hr	Ground 7 days
Muscovite	10.5	....	76.0	
Biotite	3.0	62.0	72.5	
Kaolinite	8.0	57.5	70.4	100.5
Montmorillonite	126	....	238.0	

Laws and Page[64] found that there is little further increase in the cation-exchange capacity of kaolinite after 96 hr of grinding, and they presented evidence to indicate that the kaolinite structure has been destroyed at this point and that a new permutite-like structure has developed. Perkins[65] showed that only 48 hr of grinding is enough to seriously disrupt the structure of both kaolinite and muscovite. Takahashi[66] obtained similar results and emphasized that wet versus dry grinding and crystallinity of starting material influence the rate of structural breakdown.

Relation to temperature. According to Kelley[6] the temperature effect of cation exchange is generally small. Wiegner[67] reported a small negative tem-

[63] Kelley, W. P., and H. Jenny, The Relation of Crystal Structure to Base-Exchange and Its Bearing on Base-Exchange in Soils, *Soil Sci.*, **41**: 367–382 (1936).

[64] Laws, W. D., and J. B. Page, Changes Produced in Kaolinite by Dry Grinding, *Soil Sci.*, **62**: 319–336 (1946).

[65] Perkins, A. T., Kaolin and Treated Kaolins and Their Reactions, *Soil Sci.*, **65**: 185–192 (1948).

[66] Takahashi, H., Effects of Dry Grinding on Kaolin Minerals, I–III, Proceedings of the 6th National Clay Conference, pp. 279–291, Pergamon Press, New York (1959).

[67] Wiegner, G., Zum Basenaustausch in der Ackererde, *J. Landw.*, **60**: 11–150 (1912).

perature coefficient, and various other workers[68] have found that the exchange reaction is accelerated somewhat by raising the temperature. Chapman and Kelley[69] pointed out that the disadvantages of heating may outweigh the advantage because of the increased solubility of certain constituents at the higher temperatures. Wiklander[23] stated that temperature increases slightly the rate of rapid exchange. For slower exchange, temperature is likely to have a greater effect since ion diffusion and, for anions, processes of a chemical nature may be involved.

The change in cation-exchange capacity of montmorillonite saturated with Ca^{++}, Na^+, and Li^+ on heating to various temperatures is given in Table 7-6. The exchange capacity is reduced on heating, but the reduction is not uniform and varies with the cation present. Thus Ca montmorillonite shows a gradual loss of cation-exchange capacity on heating to 300°C (93 to 41 meq/100 g) and an abrupt drop from 41 to 12 meq between 300 and 390°C, which is the temperature interval during which inner crystalline swelling is lost. Na montmorillonite, unlike the calcium variety, shows only a slight drop in exchange capacity up to 300°C (95 to 90 meq), and a moderate drop from 300 to 390°C. Between 390 and 490°C the reduction is only to 39 meq, even though the property of inner crystalline swelling is lost in this temperature interval. At 700°C, following the loss of lattice OH water from the Na montmorillonite, the exchange capacity drops to 3.4 meq. For Li montmorillonite, the exchange capacity is reduced to 56 meq on heating to only 105°C, which is below the point of loss of the swelling characteristic. Swelling of this montmorillonite is lost at 125°C, with a drop in exchange capacity to 31 meq/100 g. Ferrandis et al.[70] and Tibikh, Barshad, and Overstreet[71] provided additional data on the effect on cation-exchange capacity of heating montmorillonite to various temperatures.

The data of Hofmann and Endell[72] and Hofmann and Klemen[73] showed that the exchange capacity of montmorillonite is reduced to a considerable

[68] Kelley, W. P., and S. M. Brown, Replaceable Bases in Soils, *Calif., Univ., Agr. Expt. Sta., Tech. Paper* **15** (1924).

[69] Chapman, H. D., and W. P. Kelley, The Determination of the Replaceable Bases and the Base-Exchange Capacity of Soils, *Soil Sci.,* **30:** 391–406 (1930).

[70] Ferrandis, V. Aleixandre-, J. C. Vicente, and M. C. Rodriguez-Pascual, Modification of the Adsorptive Properties of Clay Minerals Produced by Thermal Treatment and Base-Exchange Cations, II, *Anales Edafol. Fisiol. Vegetal (Madrid),* **17:** 133–161 (1958).

[71] Tibkh, A. A., I. Barshad, and R. Overstreet, Cation-exchange Hysteresis in Clay Minerals, *Soil Sci.,* **90:** 219–226 (1960).

[72] Hofmann, U., and J. Endell, Die Abhängigkeit des Kationenaustausches und der Quellung bei Montmorillonit von der Vorerhitzung, *Ver. Deut. Chemiker Beihefte,* **35:** 10 (1939).

[73] Hofmann, U., and R. Klemen, Verlust der Austauschfähigkeit von Lithiumionen an Bentonit durch Erhitzung, *Z. Anorg. Chem.,* **262:** 95–99 (1950).

Table 7-6   Effect of heating on the cation-exchange capacity of montmorillonite
and illite

(After Hofmann and Klemen[73])

| Mineral | Heating temperature, °C | Drying time, days | d(001), Å | | Water content* over 35% H₂SO₄ | Exchangeable cations, meq/100 g |
			Dry clay	Wetted clay		
Ca montmorillonite	105	2	10.2	20	33	93
	300	2	9.8	20	26	41
	390	14	9.6	9.6	16	12
	490	14	9.6	9.6	5.7	6.1
	700	2	9.6	9.6	4.7	2.6
Na montmorillonite	105	2	9.8	30	24	95
	300	2	9.8	30	22	90
	390	14	9.6	21	16	68
	490	14	9.6	9.6	10	39
	700	2	9.6	9.6	1.5	3.4
Li montmorillonite	20	...	10.2	30	14.7	98
	105	2	10.5	30	12.5	56
	125	2	10	10	10	31
	200	2	10	10	10	20
Illite, Sárospatak, Hungary	105	2	....	....	....	17
	300	2	....	....	....	14
	500	2	....	....	....	11
	700	2	....	....	....	9

* Computed on dry basis on heating to 900°C.

extent by heating before the swelling property is lost. The amount of reduction is large for montmorillonite saturated with Li^+ and Ca^{++} and small when it is saturated with Na^+. Complete loss of swelling follows the reduction in exchange capacity. Hofmann and Endell[72] interpreted their data to mean that, when the clay is heated, the exchangeable cations tend to move inside the montmorillonite lattice. Since Li^+ is a small ion, it can fit easily into the structure, possibly in vacancies in the octahedral sheet, and consequently only a low temperature is required for the shift into the structure (Greene-Kelly[74]). The move into the lattice is followed by a loss of swelling. Because the Na^+ ion is a large one, it would fit with great difficulty into the structure, and a high temperature would be required for the move. The exchange capacity and swelling of Na montmorillonite would persist, therefore, to a relatively high temperature.

In the case of nonexpanding clay minerals the available data indicate a gradual reduction in cation-exchange capacity with increasing temperature

[74] Greene-Kelly, R., A Test for Montmorillonite, *Nature*, **170**: 1130–1131 (1952).

of heating. Greene-Kelly[74] found that lithium ions introduced into exchange positions on kaolinite were not exchangeable after heating to 200°C, which suggests that the ions moved into vacant octahedral positions.

Scott et al.[75] studied the release of NH_4 ions from various clay minerals on heating. They found that, in the case of kaolinite and illite, these ions were largely exchangeable and started to decompose at temperatures below 100°C and were completely decomposed by heating for 24 hr at 400°C. On the other hand, in the case of vermiculite, the NH_4 ions were fixed and were stable until nearly 400°C, requiring 24 hr of heating at 600°C to decompose entirely. In the case of montmorillonite, some of the NH_4 ions which were initially exchangeable were fixed when the clay was heated at 300 to 350°C, whereas some of the other NH_4 ions were exchangeable and decomposed at the lower temperature and still other ions were fixed and did not decompose until the temperature exceeded 400°C. However, it was impossible to make a sharp distinction between the exchangeable and nonexchangeable NH_4 ions on the basis of temperature of decomposition.

Sen and Guha[76] found that on heating a kaolinite clay to 1000°C, its cation-exchange capacity showed no change until about 400°C. Above 400°C there was a sharp increase to about 550°C, followed by a gradual decrease broken by a relatively more moderate decrease between 700 and 800°C (Fig. 7-7). The increase in cation-exchange capacity is correlated with the disruption of the kaolinite structure on dehydroxylation. The subsequent decrease in capacity suggests reorganization in the resulting product.

Environment of the exchange reaction. Cation exchange usually takes place in an aqueous environment, and the ions generally have considerable solubility. However, it has been shown that clays may take ions from water suspensions of very insoluble substances and resistant minerals by means of ionic sorption reactions, and it is probable that the reaction can take place in suspensions of high concentration, i.e., in the presence of relatively little water. Indeed this is a means of the natural disintegration of minerals that is important and is not generally appreciated. Thus, Bradfield[77] found that sodium-saturated clays were able to take enough barium from barium sulfate to fill about one-fifth of the exchange positions of the clay. Graham[78]

[75] Scott, A. D., J. J. Hanway, and G. Stanford, Thermal Studies of Ammonium Fixations and Release in Certain Clay Minerals, *Am. Mineralogist*, **41:** 701–721 (1956).

[76] Sen, S., and S. K. Guha, The Effect of Heat on the Base Exchange Capacity of a Kaolinite Clay and Its Structural Implications, *Proc. Intern. Clay Conf., Stockholm*, **1:** 215–229 (1963).

[77] Bradfield, R., The Concentration of Cations in Clay Soils, *J. Phys. Chem.*, **36:** 340–347 (1932).

[78] Graham, E. R., Calcium Transfer from Mineral to Plant through Colloidal Clay, *Soil Sci.*, **51:** 65–71 (1941).

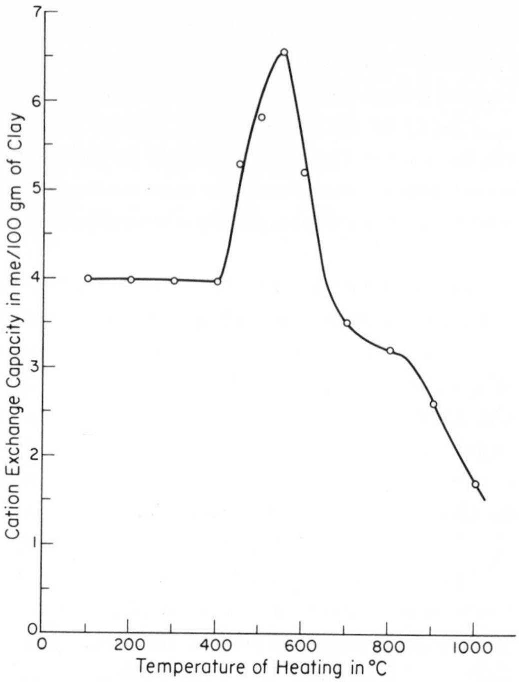

Fig. 7-7. Variation of cation-exchange capacity with temperature by the ammonium acetate method. (After Sen and Guha.[76])

showed that H clay will extract calcium from anorthite by cation exchange. This exchange has been explained[79] as follows: The resistant minerals in water suspensions are in equilibrium with traces of ions which dissolve from their surfaces. The clays destroy this equilibrium by sorbing the ions, and if the equilibrium is to be maintained, ions must move from the resistant mineral into the solution.

It was suggested, by Kelley[80] and by Jenny and Overstreet,[81] and shown later by Jenny and his coworkers,[82] using tracer elements, that ion exchange can take place directly between plant roots and clays without the intermediate solution of the ions. The cation moves directly from the clay to the plant in return for another ion, which moves directly from the plant to the clay. This probably requires that ions be able to migrate on clay-mineral surfaces

[79] Gieseking, J., The Clay Minerals in Soils, *Advan. Agron.*, 1: 159–204 (1949).

[80] Kelley, W. P., A General Discussion of the Chemical and Physical Properties of Alkali Soils, *Proc. 1st Intern. Congr. Soil Sci.*, 4: 483–489 (1927).

[81] Jenny, H., and R. Overstreet, Cation Interchange between Plant Roots and Soil Colloids, *Soil Sci.*, 47: 257–272 (1939).

[82] Jenny, H., R. Overstreet, and A. D. Ayers, Contact Depletion of Barley Roots as Revealed by Radioactive Indicators, *Soil Sci.*, 48: 9–24 (1939).

from exchange spot to exchange spot. Jenny[83] postulated that exchangeable ions are in a continuous state of thermal agitation, and when neighboring zones of agitation overlap, there should be an opportunity for a given cation to jump from one spot to another, provided that there is another ion of like charge simultaneously jumping in the opposite direction. It may be that direct exchange can take place between clay minerals and inorganic materials as well as between clay minerals and plants.

Buswell and Dudenbostel[84] showed that H montmorillonite will react with dry NH_3 gas to form NH_4 montmorillonite. Cornet[85] showed the same thing and went further to indicate that some of the NH_3 is taken up between the basal plane surfaces of the montmorillonite. Mortland et al.[86] found that hydrogen montmorillonite and base-saturated montmorillonite which have reacted with ammonia show mainly the same infrared-absorption features as NH_4 montmorillonite, thereby indicating that chemisorbed ammonia is present as NH_4^+ ions rather than in molecular form. This is explained for the base-saturated montmorillonite on the basis that it contains some residual interlayer water, thereby providing the mechanism necessary for NH_4^+ ion formation. The present use of dry gaseous ammonia as a fertilizer makes use of this sorption of NH_3 by the clay minerals.

Magistad and Burgess[87] showed that the cation-exchange reaction can take place in alcohol.

Hydrogen clays. It was shown by Paver and Marshall,[88] Chatterjee and Paul,[89] and Mukherjee and others[90] that hydrogen montmorillonites and hydrogen kaolinites are in reality hydrogen-aluminum systems. It is substan-

[83] Jenny, H., Simple Kinetic Theory of Ionic Exchange: Ions of Equal Valency, *J. Phys. Chem.*, **40**: 501–517 (1936).

[84] Buswell, A. M., and B. F. Dudenbostel, Spectroscopic Studies of Base-Exchange Materials, *J. Am. Chem. Soc.*, **63**: 2554–2559 (1941).

[85] Cornet, I., Sorption of NH_3 on Montmorillonite Clay, *J. Chem. Phys.*, **11**: 217–226 (1943).

[86] Mortland, M. M., J. J. Fripiat, J. Chaussidon, and J. Uytterhoeven, Interaction Between Ammonia and the Expanding Lattices of Montmorillonite and Vermiculite, *J. Phys. Chem.*, **67**: 248–258 (1963).

[87] Magistad, O. C., and P. S. Burgess, The Use of Alcoholic Salt Solutions for the Determination of Replaceable Bases in Calcareous Soils, *Ariz., Univ., Agr. Expt. Sta., Tech. Bull.* 20 (1928).

[88] Paver, H., and C. E. Marshall, The Role of Alumina in the Reactions of Clays, *J. Soc. Chem. Ind. (London)*, **53**: 750–760 (1934).

[89] Chatterjee, B., and M. Paul, Interaction between Hydrogen Clays and Neutral Salts, *Indian J. Agr. Sci.*, **12**: 113–120 (1942).

[90] Mukherjee, J. N., B. Chatterjee, and P. C. Goswami, Limiting Exchange of Aluminum Ions from Hydrogen Clays on the Addition of Neutral Salts, *J. Indian Chem. Soc.*, **19**: 40–407 (1942).

tially impossible to prepare a clay in which all the exchange positions are occupied by H^+, since Al^{3+} moves from the lattice to exchange positions before saturation with H^+ becomes complete. Coleman and Craig[91] pointed out that reversion of hydrogen clays to aluminum-hydrogen or magnesium-aluminum-hydrogen systems is appreciable even at room temperatures. These conclusions are probably applicable to the other clay minerals as well and apply in the greatest degree to the expanding-lattice minerals and sepiolite-attapulgite-palygorskite. Kaolinite is affected to the least degree, with the illite and chlorite minerals showing an intermediate effect. Coleman and Craig[91] found that the rate of change from hydrogen to magnesium or aluminum saturation increased with temperature, with the reaction rate about doubling for a 10° increase in temperature. These authors also found that various smectites altered at different rates, but always faster than kaolinite.

Paver and Marshall[88] showed that the electrolysis of clays removes not only the mobile cations, although this predominates in the earlier stages, but also basic constituents, such as iron and magnesium, which are frequently found at the anode, while silicic acid moves in considerable quantity with the bulk of the bases to the cathode. Electrodialyzed clays, on treatment with neutral salts, liberate aluminum, and the amount increases with the concentration of the salt to a maximum which is approximately equivalent to the exchange capacity of the clay.

Numerous workers have tried many procedures to prepare hydrogen clays without the replacement of the hydrogen by aluminum. Ojea and Taboadela[92] claimed to have succeeded for kaolinite, using an exchange resin. Fruhauf and Hofmann[93] attempted to determine the relative amounts of hydrogen and aluminum ions developed on kaolinite by various procedures.

Mukherjee and his colleagues[94,95] studied in detail the changes that occur when the clay minerals undergo repeated desaturation and treatment with barium chloride. With montmorillonite, repeated cycles of desaturation and neutral salt treatment caused a steady decrease in the aluminum and a smaller decrease in the hydrogen released by exchange. After four cycles the ratio

[91] Coleman, N. T., and D. Craig, The Spontaneous Alteration of Hydrogen Clay, *Soil Sci.*, **91:** 14–18 (1961).

[92] Ojea, F. G., and M. M. Taboadela, Preparation of Hydrogen-kaolinite with Amberlite IR-120, *Anales edafol. Fisiol. Vegetal (Madrid)*, **18:** 49–57 (1959).

[93] Fruhauf, K., and U. Hofmann, Kaolinite and Montmorillonite with Exchangeable Hydrogen Ions, *Z. Anorg. Chem.*, **307:** 187–191 (1961).

[94] Mukherjee, J. N., and B. Chatterjee, Liberation of H^+, Al^{+++} and Fe^{+++} Ions from Hydrogen Clays by Neutral Salts, *Nature,* **155:** 200–269 (1945).

[95] Mukherjee, J. N., B. Chatterjee, and A. Ray, Liberation of H^+, Al^{+++} and Fe^{+++} Ions from Pure Clay Minerals on Repeated Soil Treatment and Desaturation, *J. Colloid Sci.*, **3:** 437–446 (1946).

of hydrogen to aluminum released became very large. Successive treatments caused a large reduction in the cation-exchange capacity of the montmorillonite, indicating considerable decomposition, which was confirmed by the appearance of soluble silica in the earlier desaturations. With kaolinite, aluminum and hydrogen were similarly brought into solution by the neutral salt treatment, but there was no reduction in cation-exchange capacity, and no soluble silica was found. Probably the attack on kaolinite is restricted to the aluminum exposed on external surfaces. It is likely that the sepiolite-palygorskite and vermiculite minerals would act like the montmorillonite and that illites and chlorites would be intermediate, but more like the kaolinite.

Unpublished work by Michelson[96] suggests that the movement of aluminum from positions in the lattice to exchange positions is facilitated by drying. So long as the sample is not dried, the amount of movement is relatively small. Michelson's work was done on montmorillonite, but it probably applies also to the other clay minerals.

The fact that hydrogen clays are in reality hydrogen-aluminum systems is of great importance in clay and soil investigations. The failure to recognize this fact has caused many erroneous conclusions to be reported and much confusion in interpreting results. For example, numerous studies have been made of the physical properties of supposedly monoionic clays prepared by treating hydrogen clays with various cations. Actually in many cases monoionic clays were not attained, and the reported results are to a considerable extent a consequence of the presence of Al^{3+} and not, as supposed, of the cation with which the clay was treated. Further it makes extremely difficult the study of the properties of calcareous clays, since the carbonate cannot readily be removed without damaging the clay-mineral component and altering its properties.

Clogging of cation-exchange positions. It has just been pointed out that the development of aluminum on exchange positions reduces the exchange capacity of montmorillonite clays. This is in part due to damaging of the montmorillonite lattice, but also in part to clogging of the exchange positions by aluminum. Dion[97] pointed out that Fe_2O_3 alone or in the hydrated form may serve to reduce the cation-exchange capacity of clay minerals by a clogging action.

A somewhat similar effect may be produced by organic ions. Hendricks[98]

[96] Michelson, G., personal communication.

[97] Dion, H. G., Iron Oxide Removal from Clays and Its Influence on Base-Exchange Properties and X-ray Diffraction Patterns of the Clays, *Soil Sci.,* **58:** 411–424 (1944).

[98] Hendricks, S. B., Base Exchange of the Clay Mineral Montmorillonite for Organic Cations and Its Dependence on Adsorption Due to van der Waals Forces, *J. Phys. Chem.,* **45:** 65–81 (1944).

pointed out that large, flat organic ions adsorbed on the basal surface of montmorillonite may be of sufficient size to blanket more than one exchange position and thereby seem to reduce the exchange capacity of the montmorillonite. Organic molecules with an area greater than about 80 A^2 and flat-lying could spread over more than one exchange position.

Other effects may serve to reduce the exchange reaction. Sulfur compounds may form on the adsorbing surface[99] of the clay minerals, which would serve to reduce exchange. This matter is not well understood but studies of clay-mineral catalysts reveal that they can be poisoned by sulfur, probably because of sulfur compounds which form on the clay-mineral surface and reduce the sorption activity.

Replaceability of exchangeable cations. Very early studies of cation exchange showed that, under a given set of conditions, various cations were not equally replaceable and did not have the same replacing power. Way[9] concluded that the replacing power of the common ions was Na < K < Ca < Mg < NH₄, which means, for example, that in general Ca^{++} will more easily replace Na^+ than Na^+ will replace Ca^{++}.

As the cation-exchange reaction was studied, it became obvious that there was no single universal replaceability series. The series varied depending on the experimental conditions, on the cations involved, and on the kind of clay material. Gedroiz[100] in 1922, for example, gave the following order of replacing power based on the replacement of Ca^{++} by 0.01 N solution of chlorides: Li < Na < K < Mg < Rb < NH₄ < Co < Al. In the replacement of Ba by 0.1 N solution using the same soil material, Gedroiz found Li < Na < NH₄ < K < Mg < Rb < Ca < Co < Al.

The matter of cation replaceability is of very great importance, and a very large amount of research on the problem has been reported. Replaceability is not yet completely understood, but it is known that it is controlled by a considerable number of factors, and much has been learned about the nature and the influence of each of the factors. In the following discussion of the factors involved, no attempt is made to place them in an order of importance.

Effect of Concentration. Kelley and Cummins[101] found that the replacement of Ca^{++} and Mg^{++} by Na^+ in the Yolo soil of California increased as the concentration of Na^+ in the solution increased. Gedroiz[100] pointed

[99] Davidson, R. C., Cracking Sulphur Stocks with Natural Catalysts, *Petrol. Refiner,* September, 1947.

[100] Gedroiz, K., On the Absorptive Power of Soils. Translated by S. Waksman and distributed by U.S. Department of Agriculture (1922).

[101] Kelley, W. P., and A. B. Cummins, Chemical Effects of Salts on Soils, *Soil Sci.* **11:** 139–159 (1921).

out that the replacement of Ca^{++} and Mg^{++} by NH_4^+ in a Chernozem soil increased as the concentration of NH_4^+ increased. This is to be expected since cation exchange is a stoichiometric reaction and the laws of mass action would hold. In general, therefore, increased concentration of the replacing cation causes greater exchange by that cation.

Gedroiz[100] in his experiments found that the increased replacement of Ca^{++} and Mg^{++} by NH_4^+ was not in direct proportion to the concentration and that the ratio of Ca^{++} to Mg^{++} was also not in direct proportion. That is to say, the concentration of replacing cation is important, but it is not the sole factor. Schachtschabel[102] demonstrated conclusively that the effects of concentration depend on the kind of cation that is being replaced and also on the valence of the cation, as well as on other factors.

The complexity of this factor of concentration is shown by the following data brought out by Kelley:[6] "With cation pairs of about similar replacing power and of the same valence such as K^+ vs. NH_4^+ or Ca^{++} vs. Ba^{++}, dilution has relatively little effect on exchange, while with cations of different replacing power and different valence, for example Na^+ vs. Ca^{++} or NH_4^+ vs. Ca^{++}, dilution produces marked effect on exchange."

Population of Exchange Positions. Jenny and Ayers[103] and later Wiklander,[104] in considerable detail, showed that the ease of release of an ion depends not only on the nature of the ion itself but also upon the nature of the complementary ions filling the remainder of the exchange positions, and on the degree to which the replaced ion saturates the exchange spots. Thus, Wiklander showed that, as the amount of exchangeable calcium on the clay mineral becomes less, the calcium becomes more and more difficult to release. Sodium, on the other hand, tends to become easier to release as the degree of saturation with sodium ions becomes less. Magnesium and potassium are not affected by the degree of saturation to the same extent as calcium and sodium. Toth[5] stated that NH_4 and Ca^{++} are more difficult to replace if present initially upon the clay rather than introduced later.

Nature of Anion in Replacing Solution. Numerous investigators have found that the replaceability of a given ion varies with the nature of the anion which may be present. Thus Neznayko[105] found considerable variation in the replaceability of Na^+ from montmorillonite by Ca^{++}, depending on

[102] Schachtschabel, P., Untersuchungen über die Sorption der Tonmineralien und organischen Bodenkolloide und die Bestimmung des Anteils dieser Kolloide an der Sorption in Boden, *Kolloid-Beihefte,* **51**: 199–276 (1940).

[103] Jenny, H., and A. D. Ayers, The Influence of the Degree of Saturation of Soil Colloids on the Nutrient Intake by Roots, *Soil Sci.,* **48**: 443–459 (1939).

[104] Wiklander, L., Studies of Ionic Exchange with Special Reference to the Conditions in Soils, *Ann. Roy. Agr. Coll. Sweden,* **14**: (1946).

[105] Neznayko, M., personal communication.

whether calcium hydroxide or calcium sulfate was used. Marshall[14] reported that considerable variation in the cation-exchange capacity is obtained by leaching with different neutral salts of a given ion. The whole matter of the effect of anions is complicated by the possibility of the formation of "basic" salts with the clay and a soluble anion, e.g., clay-$(ZnOH)^+$,[106] and, at relatively high concentration of anions, by the probability of reactions with the clay mineral, altering its structure and forming new compounds. Toth[5] pointed out that it is to be expected from the Paneth-Fajans-Hahn rule of adsorption by crystal lattices that, when the anion is strongly adsorbed, cation adsorption will increase.

Nature of the Ion. Other things being equal, the higher the valence of the ion, the greater is its replacing power and the more difficult it is to displace when already present on the clay. Hydrogen is an exception since for the most part it behaves like a divalent or trivalent ion.

Kelley[6] pointed out that the replacing power increases qualitatively with atomic number in ions of the same valence.

Also, in ions of the same valence, replacing power tends to increase as the size of the ion increases, i.e., the smaller ions are less tightly held than the larger ions. An exception to the effect of ion size occurs in those ions which have almost the correct size and coordination properties to fit on the basal oxygen sheet of the layer clay minerals. Potassium, as pointed out by Page and Baver,[107] has these characteristics, and as a consequence the potassium ion is relatively very difficult to replace.

Wiegner and Jenny[108] suggested that the size of the hydrated ion, rather than the size of the nonhydrated ion, controls replaceability. According to Wiegner,[109] for ions of equal valence, those which are least hydrated have the greatest energy of replacement and are the most difficult to displace when already present upon the clay. Thus, lithium, although a very small ion, is considered to be highly hydrated and, therefore, to have a very large hydrated size. The low replacing power of Li^+ and its ready replaceability are said to be a consequence of this large hydrated size.

The magnitude of the computed hydration of the various ions varies with the base chosen for comparison. Hydration values suggested by several investigators and obtained in different ways are given in Table 7-7 to show

[106] Jenny, H., and M. M. Elgabaly, Cation and Anion Interchange with Zinc Montmorillonite Clays, *J. Phys. Chem.*, 47: 399–410 (1943).

[107] Page, J. B., and L. D. Baver, Ionic Size in Relation to Fixation of Cations by Colloidal Clay, *Soil Sci. Soc. Am. Proc.*, 4: 150–155 (1939).

[108] Wiegner, G., and H. Jenny, Ueber Basenaustausch an Permutiten, *Kolloid-Z.*, 43: 268–272 (1927).

[109] Wiegner, G., Ionenumtausch und Struktur, *Trans. 3rd, Intern. Congr. Soil Sci., Oxford*, 3: 5–28 (1935).

the order of magnitude of the hydration generally given for the common ions.

The matter of the hydration of the adsorbed ions in clay-water systems is a matter of much controversy at the present time. Wiegner and Jenny[108] and later Alten and Kurmies[110] presented strong evidence that all the common

Table 7-7 Ion sizes and ionic hydration

Ion	Ionic radii, Å				Hydration, moles H_2O				
	Not hydrated		Hydrated						
	A	B	C	D	E	F	G	H	I
Li	0.68	0.78	10.03	7.3	12.6	10	15	11–13	13–14
Na	0.98	0.98	7.90	5.6	8.4	5	8	9–11	8–9
K	1.33	1.33	5.32	3.8	4.0	1	4	5–6	5
NH_4	....	1.43	5.37	....	4.4	....	....	2–3	
Rb	....	1.49	5.09	3.6	....	0.5			
Cs	....	1.65	5.05	3.6	....	0.2			
Mg	0.89	0.78	....	10.8	13.3	33	21	20–23	
Ca	1.17	1.06	.....	9.6	10.0	22	22	19–22	
Sr	1.34	1.27	.....	9.6	8.2	21			
Ba	1.49	1.43	.....	8.8	4.1	17	14	18–20	
Al	0.79	0.57	.....	....	....	....	57		
La	1.30	1.22	.....	....	....	....	30.5		

A. Zachariassen, W. H., *Z. Krist.*, **80:** 137 (1931).

B. Goldschmidt, V., *Norske. Videnskaps- Akad. Oslo, Skrifter. Mat.-Naturv. Klasse*, **7, 8** (1926, 1927).

C. Jenny, H., *J. Phys. Chem.*, **36:** 2217–2258 (1935).

D and *F*. Pallman, H., *Bodenk. und Forsch.*, **6:** 21 (1938).

E. Remy, H., *Z. Phys. Chem. A.*, **39:** 467 (1915).

G. Brintziner, H., and C. Ratanarat, *Z. Anorg. Allgem. Chem.*, **222:** 119 (1935).

H. Bourion, F., E. Ronyer, and O. Hun, *Compt. Rend.*, **204:** 1420 (1937).

I. Baborovski, J., J. Velisch, and A. Wagner, *J. Chem. Phys.*, **25:** 452–482 (1928).

cations are hydrated and that hydration is important in exchange reactions. The evidence is based on investigations carried out in alcohol rather than water. It is stated that alcohol, in which the ions would not be hydrated, reverses the usual lyotropic replacement series, so that ions are held in the order of their true ionic size and not in the order of their hydrated size.

[110] Alten, F., and B. Kurmies, "Handbuch der Bodenlehre," vol. 8, Springer, Berlin (1931).

Bernal and Fowler[111] some years ago presented data that threw doubt on the hydration of some of the cations. Bar and Tenderloo[112] enumerated difficulties encountered by explanations of clay properties based on cation hydration, and Baver[113] summarized experimental data on the swelling and heat of wetting of soil colloids that do not fit well with cation hydration. Hendricks and his colleagues[114] presented compelling evidence, based on careful dehydration studies, that Na^+, H^+, and K^+ and the trivalent ions are not hydrated when adsorbed by the clay minerals. According to them, Ca^{++} and Mg^{++} are hydrated to $6H_2O$, and Li^+ is hydrated to $3H_2O$. This is directly contrary to conclusions of those favoring the hydration concept, who consider Li^+ and Na^+ to be among the most highly hydrated ions.

According to Bar and Tenderloo,[112] replaceability is related to the polarization of the ion, with increasing polarization being accompanied by increasing difficulty of exchange. Also, highly polar ions are thought to be held closer to the adsorbing surface. Polarization increases as the valence increases and as the size of the ion decreases.

Effect of Heating. The work of Hofmann and J. Endell[72] indicated that heating to moderate temperatures not only reduces the cation-exchange capacity but changes the relative replaceability of the cations (see page 205). Thus, for montmorillonite, heating to 125°C tends to fix Li^+ in an unreplaceable form and not to affect the replaceability of Na^+, whereas at ordinary temperatures Li^+ is more easily replaced than Na^+. The extensive and excellent work of Marshall and his colleagues[115] in developing and using clay membranes as electrodes to measure ion activity also showed the great variation of ion replaceability with increasing temperatures. Andrews and Maldonado[116] showed that the relative amounts of the cations that are replaceable change when the clay fraction of the Crowley silt loam soil is heated up to 100°C. They found that the relative amounts of replaceable K^+, Ca^{++}, and H^+ decreased on prolonged heating, whereas Na^+ and Mg^{++} increased.

It seems likely that the change in replaceability of cations when heated would be relatively greater for the expanding-lattice minerals than for those

[111] Bernal, J. D., and R. H. Fowler, A Theory of Water and Ionic Solution with Particular Reference to Hydrogen and Hydroxyl Ions, *J. Chem. Phys.,* **1**: 515–548 (1933).

[112] Bar, A. L., and H. J. Tenderloo, Ueber die Doppelschicht der Tonkolloide, *Kolloid-Beihefte,* **44**: 97–124 (1936).

[113] Baver, L. D., "Soil Physics," Wiley, New York (1940).

[114] Hendricks, S. B., R. A. Nelson, and L. T. Alexander, Hydration Mechanism of the Clay Mineral Montmorillonite Saturated with Various Ions, *J. Am. Chem. Soc.,* **62**: 1457–1464 (1940).

[115] Marshall, C. E., et al., The Electrical Properties of Mineral Membranes, series of papers beginning in 1941 in *J. Am. Chem. Soc.* and *J. Phys. Chem.*

[116] Andrews, J. S., and J. F. Maldonado, Effect of temperature on the Base-Exchange Capacity of Clays, *J. Agr. Univ. Puerto Rico,* **24**: 133–142 (1940).

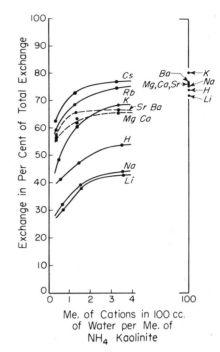

Fig. 7-8. NH₄ kaolinite; ionic exchange with alkali and alkaline-earth chlorides. (After Schachtschabel.[102])

minerals in which the capacity is due largely to broken bonds. In the former case, most of the adsorbed cations are held between the basal layers, and the change in replaceability is probably largely in these ions. At elevated temperatures, when there is little or no water present between basal layers in addition to the sorbed cations, the size of the ion and its geometrical fit into the structure of the oxygen layers are probably major factors in determining replaceability.

Nature of the Clay Mineral. Early investigations of many soils suggested that the replaceability of various cations varied with the nature of the adsorption complex, all other factors being equal. The work of Jarusov,[117] Bar and Tenderloo,[112] and Gieseking and Jenny[118] particularly showed this to be true, and the careful work on pure kaolinite, muscovite, and montmorillonite by Schachtschabel[102] clearly demonstrated that there was not a single replaceability series characteristic of all clay materials, but separate replaceability series for the various clay minerals.

The replaceability of NH₄⁺ from kaolinite, montmorillonite, and muscovite by various cations at various concentrations is shown in Figs. 7-8 to

[117] Jarusov, S. S., Mobility of Exchangeable Cations in the Soil, *Soil Sci.*, **43**: 285–303 (1937).

[118] Gieseking, J. E., and H. Jenny, Behavior of Polyvalent Cations in Base Exchange, *Soil Sci.*, **42**: 273–280 (1936).

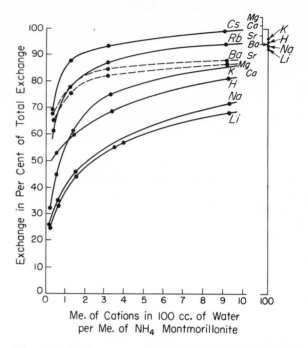

Fig. 7-9. NH₄ montmorillonite; ionic exchange with alkali and alkaline-earth chlorides. (After Schachtschabel.¹⁰²)

7-10, taken from Schachtschabel. These data show, for example, that, for NH₄ montmorillonite, H⁺ and K⁺ are about equally exchangeable, that all the univalent cations except Rb⁺ and Cs⁺ are more exchangeable than the divalent ions, and that all the divalent ions have about the same replaceability. For NH₄ kaolinite the exchangeability of the cations is about the same as for NH₄ montmorillonite. NH₄⁺, however, is more lightly held by kaolinite. For NH₄ muscovite, H⁺ and K⁺ are more lightly held than the divalent ions, and NH₄⁺ is even more lightly held than in kaolinite. When a mixture of montmorillonite and muscovite is treated with mixtures of calcium and ammonium acetate, the mica adsorbs relatively much more of the NH₄⁺, and the montmorillonite adsorbs relatively more of the Ca⁺⁺. Also Ca⁺⁺ in competition with equivalent concentrations of K⁺ would be taken up preferentially by montmorillonite, while the opposite is the case for muscovite. Schachtschabel¹⁰² used this latter behavior as the basis for an analytical method for the determination of the relative abundance of montmorillonite and the mica-like clay minerals when mixed together in soil. Merriam and Thomas¹¹⁹ found that the order of replaceability of alkali ions on attapulgite

¹¹⁹ Merriam, C. N., and H. C. Thomas, Adsorption Studies on Clay Minerals, VI, Alkali Ions on Attapulgite, *J. Phys. Chem.,* **24:** 993–995 (1956).

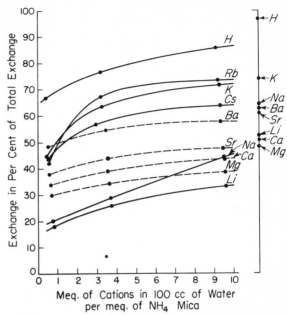

Fig. 7-10.   NH₄ muscovite; ionic exchange with alkali and alkaline-earth chlorides.   (After Schachtschabel.[102])

was no different from that found by Schachtschabel for montmorillonite. They also determined the standard free energies for the successive replacement reactions of the alkali ions on this clay mineral.

Hendricks and Alexander[120] also studied the relative acceptance of Ce^{3+} and H^+ by montmorillonite and illite and found that the illites preferentially take up H^+, while the montmorillonites take up much more Ce^{3+}. They used this difference as the basis for a method of estimating the amount of montmorillonite present in mixtures of these two minerals.

Barshad[121] investigated the cation-exchange characteristics of vermiculite and showed that the exchange process is reversible between Na^+, Ca^{++}, Mg^{++}, and K^+ but is not completely reversible between K^+, NH_4^+, Rb^+, and Cs^+. The latter ions tend to become fixed and relatively nonreplaceable. Walker and Milne[122] also investigated the replaceability of the exchangeable cations of

[120] Hendricks, S. B., and L. T. Alexander, Semiqualitative Estimation of Montmorillonite in Clays, *Soil Sci. Soc. Am. Proc.*, **5:** 95–99 (1940).

[121] Barshad, I., Vermiculite and Its Relation to Biotite as Revealed by Base Exchange Reactions, Differential Thermal Curves, and Water Content, *Am. Mineralogist*, **33:** 655–678 (1948).

[122] Walker, G. F., and A. Milne, Hydration of Vermiculite Saturated with Various cations, *Trans. 4th Intern. Congr. Soil Sci.*, **2:** 62–67 (1950).

vermiculite with much the same results. They showed that K⁺ and NH₄⁺ replace Mg⁺⁺ with difficulty, whereas Mg⁺⁺ is readily replaced by Li⁺ and Na⁺. However, a vermiculite prepared with Li⁺ or Na⁺ readily loses these ions by replacement with K⁺ or NH₄⁺. Any difference in replaceability between vermiculite and montmorillonite is of interest, since the only major difference between these two minerals seems to be in the size of the layers in the *a* and *b* crystallographic directions.

Allaway[123] investigated the replaceability of Ca⁺⁺ in various clay minerals by determining its availability to plants. He found that for Ca⁺⁺ the availability series is peat > kaolinite > illite > montmorillonite and that the availability increases as the saturation of the Ca⁺⁺ increases.

Marshall's[14] work on the ionization of the cations sorbed by the clay minerals also showed a difference in relative cation replaceability for the various clay minerals (see pages 199–200). In general the higher the degree of ionization, the greater is the ease of replaceability.

Fixation of cations. It is well known by soil investigators that certain cations may be sorbed by the clay minerals in a nonexchangeable or difficultly exchangeable state. Potassium is the commonest ion that is "fixed" to a considerable degree and the ion that has been most studied because of its importance in soil fertilization. Wood and DeTurk,[124] Bray and DeTurk,[125] Stanford,[126] and others have shown that illite will fix potassium, and most of the K⁺ fixed in soils is probably due to the action of this clay mineral. Volk[127] showed that potash fixation is often accompanied by the formation of mica in the soil. The potassium fixation in illite occurs by the emplacement of K⁺ between the basal surfaces of the mineral in the positions normally occupied by K⁺ in this mineral. Much illite, particularly in soils that are undergoing leaching, is somewhat degraded; i.e., it is potash-deficient in that leaching has removed some K⁺ from positions between the layers. On the addition of potassium to such material, K⁺ goes back into these normal positions, and the illite is rebuilt. Under natural soil conditions, fixation appears to be completed in a matter of months following the addition of potassium as fertilizer.

[123] Allaway, W. H., Availability of Replaceable Calcium from Different Types of Colloids as Affected by Degree of Calcium Saturation, *Soil Sci.,* **59:** 207–217 (1945).

[124] Wood, L. K., and E. E. DeTurk, The Chemical Effect of Soluble Potassium Salts in Some Illinois Soils, *Soil Sci. Soc. Am. Proc.,* **5:** 152–161 (1941).

[125] Bray, R. H., and E. E. DeTurk, The Release of Potassium from Non Replaceable Forms in Illinois Soils, *Soil Sci. Soc. Am. Proc.,* **3:** 101–106 (1939).

[126] Stanford, G., Fixation of Potassium in Soils under Moist Conditions and on Drying in Relation to the Type of Clay Mineral, *Soil Sci. Soc. Am. Proc.,* **12:** 167–171 (1947).

[127] Volk, G. W., Nature of Potash Fixation in Soils, *Soil Sci.,* **45:** 263–276 (1938).

Drake[128] discussed the variation in the availability of various plant nutrients for different clay minerals, for variations in the cation population of the soil, and for various types of plants using the nutrients. Wiklander[129] showed that the amount of fixation in illite-type material varies with the cations already adsorbed. Thus fixation is relatively high for materials saturated with Ca^{++} and Na^+ and low for those saturated with H^+ and NH_4^+. Stanford[126] also showed that the presence of H^+, Fe^{3+}, and Al^{3+} tends to block the fixation of K^+ by illite. Drying tends to increase the fixation by illite. Kardos[130] recently discussed all aspects of ion fixation in soils.

According to Raney and Hoover[131] and others, smectite has some power to fix potassium, but less than illite. Other investigators, notably Stanford,[126] have stated that K^+ is fixed in smectite only if the material is dried. Barshad[121] showed that vermiculite can fix some potassium. In the case of smectite and vermiculite, as with illite, any fixation of K^+ would take place on the basal planes between the unit layers.

Barshad, [132] studying cation exchange in micaceous minerals, concluded that the replaceability of interlayer NH_4^+ and K^+ is strongly affected by the magnitude of the interlayer crystal lattice charge, particle size, presence of difficultly replaceable H^+, and the nature of the replacing cations. He concluded that the magnitude of interlayer charge rather than the origin of the charge determines the fixation of NH_4^+ and K^+. Earlier, Wear and White[133] had concluded that, for smectite, potassium fixation was distinctly more marked when the charge on the lattice had a tetrahedral rather than an octahedral origin, and Weaver[134] later reached the same conclusion.

Mackenzie[135] showed that certain ions such as lead, potassium, rubidium,

[128] Drake, M., Soil Chemistry and Plant Nutrition, "Chemistry of the Soil," chap. 10, pp. 395–444, American Chemical Society Monograph (1964).

[129] Wiklander, L., Fixation of Potassium by Clays Saturated with Different Cations, *Soil Sci.*, **69**: 261–271 (1950).

[130] Kardos, L. T., Soil Fixation of Plant Nutrients, "Chemistry of the Soil," chap. 9, pp. 369–394, American Chemical Society Monograph (1964).

[131] Raney, W. A., and C. D. Hoover, The Release of Artificially Fixed Potassium from Kaolinitic and Montmorillonitic Soils, *Soil Sci. Soc. Am. Proc.*, **11**: 231–237 (1946).

[132] Barshad, I., Cation Exchange in Micaceous Minerals, I. Replaceability of the Interlayer Cations of Vermiculite with Ammonium and Potassium Ions, *Soil Sci.*, **77**: 463 (1954).

[133] Wear, J. I., and J. L. White, Potassium Fixation in Clay Minerals as Related to Crystal Structures, *Soil Sci.*, **71**: 1–14 (1951).

[134] Weaver, C. E., The Effects and Geologic Significance of Potassium "Fixation" by Expandable Clay Minerals Derived from Muscovite, Biotite, Chlorite, and Volcanic Material, *Am. Mineralogist*, **43**: 839–861 (1958).

[135] Mackenzie, R. C., Retention of Exchangeable Ions by Montmorillonite, *Proc. Intern. Clay Conf., Stockholm*, **1**: 183–193 (1963).

and strontium may be rendered nonexchangeable on smectite whenever the clay containing them is dried, even without the influence of appreciable heat, and that the tendency toward retention is greater as the charge on the lattice is increasingly due to substitutions within the tetrahedral sheet. The same author also found that ions with an ionic radius of less than about 0.85 Å are significantly fixed in the smectite lattice on alternately wetting and heating to 300°C, irrespective of the charge on the ion.

Chaussidon[136] studied the fixation of potassium in smectites on heating to 80°C with increasing amounts of calcium ions associated with potassium ions in exchange positions. He found that when calcium ions occupied more than about 30 percent of the exchange positions little potassium was fixed.

Kaolinite has substantially no power to fix potassium, either in a moist condition or after drying. Joffe and Levine[137] reported K+ fixation by very finely ground kaolinite, but the grinding may well have destroyed the kaolinite structure.

Bray and DeTurk[125] and Wood and DeTurk,[124] working with illitic Prairie soils, showed that the K+ is held in various degrees of replaceability and availability to plants and that some substantially nonreplaceable K+ may be available to plants. These authors showed that an equilibrium between the exchangeable and nonexchangeable potassium exists in such soils.

NH4+ is very similar to K+ in its ion-exchange properties, and Chaminade,[138] Page and Baver,[107] and Stanford and Pierre[139] demonstrated the fixation of NH4+ by clays in difficultly exchangeable form. The latter workers concluded that the ammonium ion and the potassium ion are fixed in soils by the same mechanism.

Allison et al.[140] pointed out that considerable fixation of ammonium occurs in moist soils if the predominant clay mineral is illite or vermiculite and that the amount is increased by drying. Where montmorillonite was the dominant clay mineral, the fixation values were about half those for illitic soils. Stevenson and Dhariwal[141] reported similar findings and further that values for kaolinite are lower than those for smectite.

[136] Chaussidon, J., Evolution des caractéristiques chimiques et cristallographique de montmorillonites biioniques K-Ca au cours d'alternances répétées d'humectation-dessication, *Proc. Intern. Clay Conf., Stockholm,* **1**: 195–204 (1963).

[137] Joffe, J. S., and A. K. Levine, Fixation of Potassium in Relation to the Exchange Capacity of Soils, *Soil Sci.,* **63**: 241–247 (1947).

[138] Chaminade, R., Fixation d'l'ion NH4 par les colloides argileux des sols sous forme non-échangeable, *Compt. Rend.,* **210**: 264–266 (1940).

[139] Stanford, G., and W. H. Pierre, The Relation of Potassium Fixation to Ammonium Fixation, *Soil Sci. Soc. Am. Proc.,* **11**: 155–160 (1946).

[140] Allison, F. E., M. Kefauver, and E. M. Roller, Ammonium Fixation in Soils, *Soil Sci. Soc. Am. Proc.,* **17**: 107–110 (1953).

[141] Stevenson, F. J., and A. P. S. Dhariwal, Distribution of Fixed Ammonium in Soils, *Soil Sci. Soc. Am. Proc.,* **23**: 121–125 (1959).

Barshad[121] investigated vermiculite and found that this clay mineral can fix at least a part of any adsorbed Rb^+ or Cs^+. Barshad[132] pointed out that various amounts of interlayer Mg^{++}, Ca^{++}, Ba^{++}, or Na^+ can become "trapped" in the interior of the particles of vermiculite and be rendered difficultly exchangeable upon replacement of the most accessible portion of these cations by NH_4^+ or K^+ which causes collapse of the interlayer space of the crystal lattice adjacent to the edges of the particles. This trapping phenomenon would increase with increasing size of the vermiculite flakes (see also pages 200–202).

Although the matter has not been established experimentally, it seems likely that Mg^{++} could in some conditions be fixed by "degraded chlorite" by a mechanism similar to that operating for K^+ and "degraded illite." Also the synthesis experiments of Caillere, Henin, and Mering[142] indicate that Mg^{++} can be fixed by smectite with the development of a chlorite type of structure under certain conditions of concentration of the Mg^{++}.

The fixation of magnesium and potassium is of considerable importance in diagenetic processes in sediments accumulating at the present time (see pages 535–544).

Theory of cation exchange. Numerous attempts have been made to develop a theory of cation exchange that would permit the quantitative expression of exchange data by an equation. Kelley,[6] Marshall,[14] and Du Rietz[143] reviewed this matter in detail and commented on the deficiencies of the theories that have been proposed.

Equations based on several considerations have been suggested. Wiegner and Jenny[108] and Vageler and Woltersdorf[144] derived empirical equations from, or connected with, adsorption isotherms. The general basis of these theories is Freundlich's[145] adsorption equation or modifications of it. Thermodynamic arguments involving more or less crude approximations have led to various mass-action equations by Kerr,[146] Vanselow,[147] and Rothmund and Kornfeld.[148] Davis[149] criticized these formulations, and Kelley[6] in turn objected

[142] Caillere, S., S. Henin, and J. Mering, Experimental Transformation of Montmorillonite to a Phyllite of Stable *c*-Distance of 14Å, *Compt. Rend.*, **224**: 842–843 (1947).

[143] Du Rietz, C., "Das Ionenbindungsvermögen fester Stoffe," Stockholm (1938).

[144] Vageler, P., and J. Woltersdorf, Beiträge zur Frage des Basen-austausches und der Aziditären, *Z. Pflanzenernähr. Düng. Bodenk.*, **15**: 329–342 (1930).

[145] Freundlich, H., "Kappillarcheme," 2d ed., Akademische Verlagsgesellschaft, Leipzig (1922).

[146] Kerr, H. W., The Nature of Base-Exchange and Soil Activity, *J. Am. Soc. Agron.*, **20**: 309–355 (1928).

[147] Vanselow, A. P., Equilibria of the Base-Exchange Reactions of Bentonites, Permutite, Soil Colloids, and Zeolites, *Soil Sci.*, **33**: 95–113 (1932).

[148] Rothmund, V., and G. Kornfeld, Der Basenaustausch im Permutit, *Z. Anorg. Allgem. Chem.*, **103**: 129–162 (1918).

[149] Davis, L. E., Theories of Base-Exchange Equilibrium, *Soil Sci.*, **59**: 379–395 (1945).

to some of Davis' remarks.   A third approach to the problem based on kinetic considerations was made by Gapon[150] and Jenny.[83]   Davis[151] attempted to improve Jenny's equation.

Erickson[152] derived an equation based on the Gouy theory of the electric double layer for the monodivalent cation-exchange equilibrium.   Erickson himself and others (Bolt,[153] Babcock,[154] and Helmy[155,156]) pointed out shortcomings of the equation.

As Kelley[6] showed, none of these equations is completely satisfactory, and indeed they cannot be universally applied because of the large number of variations dependent on the nature of the clay mineral, nature of ion, concentration of ion, concentration of clay, etc.   The application of any such equations would probably need to be restricted to similar kinds of exchange material, to a given range of concentrations, and perhaps to other factors.

Determination of cation-exchange capacity and exchangeable cations.   It should be obvious from the foregoing discussion of the factors controlling cation exchange in clay minerals that accurate determinations of cation-exchange capacity and exchangeable cations are very difficult to accomplish. Literally dozens of methods have been suggested, and a tremendous amount of time has been spent on this problem.   Kelley[6] and Peech et al.[157,158] considered in detail the various methods and pointed out various pitfalls.

The determination of cation-exchange capacity is at best a more or less arbitrary matter, and no high degree of accuracy can be claimed.   The measurement is generally made by saturating the clay with NH_4^+ or Ba^{++} and determining the amount held at pH 7.   In the absence of water-soluble or slightly soluble salts, and for a clay mineral which itself is not moderately soluble, the determination is not very difficult.   However, even for such materials, if the clay mineral is degraded illite or degraded chlorite, it is difficult

[150] Gapon, E. N., Theory of Exchange Adsorption in Soils, *J. Gen. Chem. USSR (Eng. Transl.)*, 3: 144–152, 153–158 (1933).

[151] Davis, L. E., Simple Kinetic Theory of Ionic Exchange for Ions of Unequal Charge, *J. Phys. Chem.*, **49**: 473–479 (1945).

[152] Erickson, E., Cation-Exchange Equilibria on Clay Minerals, *Soil Sci.*, **74**: 103–113 (1952).

[153] Bolt, G. H., Ion Adsorption by Clays, *Soil Sci.*, **79**: 267–276 (1955).

[154] Babcock, K. L., Theory of the Chemical Properties of Soil Colloid Systems at Equilibrium, *Hilgardia*, **34**: 417–542 (1963).

[155] Helmy, A. K., On Cation Exchange Stoichiometry, *Soil Sci.*, **95**: 204–205 (1963).

[156] Helmy, A. K., An Exchange Equation Based on Positive Adsorption, *J. Soil Sci.*, **13**: 79–83 (1964).

[157] Peech, M., L. T. Alexander, and L. A. Dean, Methods of Soil Analysis for Soil-fertility Investigations, *U.S. Dept. Agr.*, *Circ.* 757 (1947).

[158] Peech, M., Determinations of Exchangeable Cations and Exchange Capacity of Soils, *Soil Sci.*, **59**: 25–38 (1945).

frequently to separate exchangeable from nonexchangeable ions. Also in acid clays the presence of Al^{3+}, or in ferruginous clays the presence of Fe^{3+}, tends partially to clog some of the exchange positions and make the determination difficult. The determination is particularly difficult for relatively soluble clay minerals, such as the high-magnesium smectites and the palygorskites. Trustworthy determinations of cation-exchange capacity can be accomplished only by a skilled analyst who is aware of the fundamental causes of the difficulties that beset the problem. Weiss[159] discussed in detail the problem of accurate determinations of cation-exchange capacity and showed the wide variations in values obtained in using different methods. He pointed out that quantitative determinations of all cations and anions both in the exchange solution and in the solid substance are necessary to obtain accurate values and that simpler methods, for example, using ammonium acetate, give only approximate values.

Determination of the exchangeable-cation composition is also a very difficult problem. It involves the complete replacement of all exchangeable cations by some cation which is not present in the sample, accurate analysis of the solution obtained, and determination of and suitable correction for the cations that pass into solution from any soluble substances that may be present or from the decomposition of some insoluble material in the sample. Again the problem is particularly difficult for clay minerals of relatively high solubilities, for acid clays, and for materials with a fairly high content of soluble or moderately soluble salts. The use of the flame photometer for the analysis of the ions in the replacing solution has in recent years provided satisfactory data in a very short time, particularly for some of the alkali ions and calcium.

Anion exchange

It has been shown by Mattson,[160] Ravikovitch,[161] Scarseth,[162] Toth,[163] and many others[164,65] that the constituent minerals of many soil clays exhibit anion-

[159] Weiss, A. Über das Kationenaustauchvermögen der Tonminerale, 1, Verleich der Untersuchungmethoden, *Z. Anorg. Allgem. Chem.*, **297**: 232–256 (1958).

[160] Mattson, S., The Laws of Soil Colloidal Behavior, VI, Amphoteric Behavior, *Soil Sci.*, **32**: 343–365 (1931).

[161] Ravikovitch, S., Anion Exchange, I, Adsorption of Phosphoric Acid Ions by Soils, *Soil Sci.*, **38**: 219–239, 279–290 (1934).

[162] Scarseth, G. D., The Mechanism of Phosphate Retention by Natural Alumina-Silicate Colloids, *J. Am. Soc. Agron.*, **27**: 596–616 (1935).

[163] Toth, S. J., Anion Adsorption of Soil Colloids in Relation to Changes in Free Iron Oxides, *Soil Sci.*, **44**: 299–314 (1937).

[164] Dean, L. A., and E. J. Rubins, Anion Exchange in Soils, I, *Soil Sci.*, **63**: 377–406 (1947).

exchange reactions. The investigations of anion exchange in clay materials have been to a considerable extent associated with studies of the adsorption of phosphate by soils.

The investigation of anion-exchange reactions in soils is very difficult, primarily because of the possibility of the decomposition of the clay minerals in the course of the reaction. Thus, in studies of the adsorption of phosphate by kaolinite, there has been considerable argument as to whether many of the results observed are due to adsorption, to replacement of OH ions in the kaolinite lattice by phosphate ions, or to a reaction between the phosphate and alumina produced by the destruction of the kaolinite lattice.

Table 7-8 gives the anion-exchange capacity for some of the clay minerals.

Table 7-8 Anion-exchange capacities (in milliequivalents per 100 g) (After Hofmann et al.[165])

Montmorillonite, Geisenheim	31
Montmorillonite, Wyoming	23
Beidellite, Unterrupsroth	21
Nontronite, Untergrieshach	20
Nontronite, Pfreimdtal	12
Saponite, Groschlattengrun	21
Vermiculite, South Africa	4
Kaolinite (colloidal)	20.2
Kaolinite, Melos	13.3
Kaolinite, Schnaittenbach	6.6

[165] Hofmann, U., A. Weiss, G. Koch, A. Mehler, and A. Scholz, Intracrystalline Swelling, Cation Exchange, and Anion Exchange of Minerals of the Montmorillonite Group and of Kaolinite, *Natl. Acad. Sci., Publ.* 456, pp. 273–287 (1956).

Schoen[166] presented similar data indicating that an average ratio for cation- to anion-exchange capacity is about 0.5 for kaolinite, 2.3 for illite, and 6.7 for montmorillonite. Grinding in a ball mill reduces the ratio for kaolinite. For a smectite clay, the ratio decreased to about 0.5 owing to a decrease in cation- and an increase in anion-exchange capacity.

There seem to be two, and possibly three, types of anion exchange in the clay minerals, as follows:

1. Replacement of OH ions, as has been suggested by many authors for the phosphate adsorption by kaolinite. Buswell and Dudenbostel[84] pre-

[166] Schoen, U., Identification of Clay by Phosphate Fixation and Cation Exchange, *Z. Pflanzenernähr. Bodenk.*, **63**: 1–17 (1953).

sented strong evidence, based on infrared absorption, that this can take place, and McAuliffe and coworkers,[167] using deuterium-tagged hydroxyls, showed conclusively that the OH ions of clay-mineral surfaces can enter into exchange reactions.

Dickman and Bray[168] presented very clear evidence for the replacement of hydroxyls by fluorine in kaolinite. They showed that the liberation of hydroxyls caused a well-marked increase in the alkalinity of the suspensions. The OH and fluorine ions are of about the same size, and the exchange would involve no lattice rearrangement. In the case of exchange due to replacement of OH ions, the extent of the reaction depends on the accessibility of the OH ions, and in general the only factor preventing complete substitution is the fact that many OH ions are within the lattice and, therefore, not accessible.

2. Hendricks[169] suggested that another factor in anion exchange is the geometry of the anion in relation to the geometry of the clay-mineral structure units. Anions such as phosphate, arsenate, borate, etc., which have about the same size and geometry as the silica tetrahedron, may be adsorbed by fitting on to the edges of the silica tetrahedral sheets and growing as extensions of these sheets. Other anions such as sulfate, chloride, nitrate, etc., because their geometry does not fit that of the silica tetrahedral sheets, cannot be so adsorbed.

In the case of both types 1 and 2 listed above, anion exchange would take place around the edges of the clay minerals. In contrast to cation exchange, as in smectite, there would be substantially no anion exchange on the basal plane surface of any of the clay minerals. For these types of anion exchange, and in the case of the clay minerals like kaolinite in which cation exchange is due to broken bonds, the cation- and anion-exchange capacities should be substantially equal, and Dean and Rubins[164] and Hofmann et al.[165] showed this to be the case. In the case of smectite and vermiculite, in which cation exchange is due mostly to lattice substitutions, anion capacity should be only a small fraction of the cation-exchange capacity. In the case of illitie, chlorite, and the sepiolite-palygorskite minerals, anion-exchange capacity should be slightly less than cation-exchange capacity.

Dean and Rubins[164] showed that anion-exchange capacity of the clay mineral in the Sassafrass soil is proportional to the surface area. In addition, anion-exchange capacity would be expected to vary with the degree of crystal-

---

[167] McAuliffe, C. D., M. S. Hall, L. A. Dean, and S. B. Hendricks, Exchange Reactions between Phosphates and Soils, *Soil Sci. Soc. Am. Proc.*, **12:** 119–123 (1947).

[168] Dickman, S. R., and R. H. Bray, Replacement of Adsorbed Phosphate from Kaolinite by Fluoride, *Soil Sci.*, **52:** 263–275 (1941).

[169] Hendricks, S. B., personal communication.

linity of the clay minerals. Thus, there should be a difference in the anion-exchange capacity between well-crystallized and poorly crystallized kaolinite. In the latter, because of imperfections in the stacking of the alumina-silica layers, there would be considerably more exposed OH ions available for anion exchange.

Schofield[170,171] suggested a third manner of anion exchange. According to him, the clay minerals may have anion-exchange spots, as well as cation-exchange spots, on basal plane surfaces. Such active anion-exchange positions would be due to unbalanced charges within the lattice, e.g., an excess of aluminum in the octahedral positions. It is difficult to see how this could be, since positive and negative deficiencies would tend to balance each other, unless they occurred at considerable distances from each other. That is to say, there would be either positive or negative deficiencies, but not both. Schofield indicated that in strongly acid solutions clays can take up both potassium and chloride ions, the former greatly predominating, and that both the K^+ and Cl^- are exchangeable for other cations and anions. Mattson,[172] on the other hand, interpreted experiments on the Donnan equilibrium in such a way as to show an apparent negative adsorption of chloride and sulfate. Further experimentation is necessary to resolve this matter, and indeed much remains to be learned about the whole subject of anion exchange in the clay minerals.

A further factor complicating anion-exchange studies is that any free or exchangeable iron, aluminum, or alkaline earths present in the clay may form insoluble salts with the anions, and it is very difficult to separate the effects due to such reactions from those which may be due to reaction with the clay minerals.

Wey[173] pointed out that adsorption of anions is not comparable to the absorption of cations because of greater dependence on the nature of the anion than cation and on pH. In cation-exchange experiments with kaolinite, Weiss[28] found that polyvalent cations may act like monovalent ions and may bind an equivalent amount of anions on the surface of the particle; that is, one of the valences is tied to the kaolinite lattice and the others are free for union with anions.

Phosphate fixation. Students of soils have known for a long time that some of the phosphate added to soils in fertilizers is frequently converted to an

[170] Schofield, R. K., Clay Mineral Structures and Their Physical Significance, *Trans. Brit. Ceram. Soc.*, 39: 147–158 (1940).

[171] Schofield, R. K., Calculation of Surface Area of Clays from Measurements of Negative Adsorption, *Trans. Brit. Ceram. Soc.*, 48: 207–213 (1949).

[172] Mattson, S., The Laws of Soil Colloidal Behavior, I, *Soil Sci.*, 28: 179–201 (1929).

[173] Wey, R., The Mechanism of the Adsorption of Anions by Montmorillonite, *Silicates Ind.*, 24: 376 (1959).

insoluble form and fixed in the soil. Such fixation can be due to the formation of insoluble salts of iron, aluminum, or alkaline earths. The formation of phosphates of aluminum and iron probably explains to a considerable extent the results of soil-phosphate fixation experiments carried out at low pH, and the formation of phosphates of alkaline earths probably accounts for many of the results obtained at a high pH.

Black,[174] Murphy,[175] Scarseth,[162] and Stout[176] showed that phosphate fixation also is carried out by the clay minerals, and this probably accounts for the phosphate that is fixed at about neutrality. The fixation power of the various clay minerals is relatively low, and it would seem from the work of Murphy that kaolinite, or possibly halloysite, has the highest power of all the clay minerals for phosphate fixation. Data presented by Steele[177] indicate that the phosphate retained at pH 7 for a smectite soil is about 0.03 mmole/g. For a kaolinitic soil it is about 0.07 mmole/g. The other clay minerals would probably retain about as much phosphate as the smectites.

Mitra and Prakash[178] found that, in general, the order in which minerals fix phosphate under the experimental conditions they used is smectite > vermiculite > biotite > kaolinite > halloysite > muscovite. They further concluded that smectite and vermiculite possess a high phosphate-fixing ability because they have a high content of exchangeable calcium. Hemwall[179] concluded that phosphate is fixed by clay minerals by reacting with soluble alumina which originates from exchange sites or from dissociation from the clay minerals. The maximal amount of aluminum phosphate was formed in 15 min for smectite, where aluminum was primarily on the exchange complex, but for kaolinite fixation continued for weeks, indicating dissociation from the lattice. De,[180,181,182] in a series of papers, reported in detail the

[174] Black, C. A., Phosphate Fixation of Kaolinite and Other Clays as Affected by pH, Phosphate Concentration, and Time of Contact, *Soil Sci. Soc. Am. Proc.*, **7**: 132–137 (1943).

[175] Murphy, H. F., Clay Minerals and Phosphate Availability, *Proc. Oklahoma Acad. Sci.*, **20**: 79–81 (1940).

[176] Stout, P. P., Alterations in the Crystal Structure of the Clay Minerals as a Result of Phosphate Fixation, *Soil Sci. Soc. Am. Proc.*, **4**: 177–182 (1939).

[177] Steele, J. G., The Effect of Other Anions on the Retention of Phosphate by Colloidal Clays, Ph.D. thesis, Ohio State University (1934).

[178] Mitra, S. P., and D. Prakash, Phosphate Fixation by Minerals in Alkaline Medium, *Proc. Natl. Acad. Sci., India*, **24A**: 169–175 (1955).

[179] Hemwall, J. B., The Role of Soil Clay Minerals in Phosphorus Fixation, *Soil Sci.*, **83**: 101–108 (1957).

[180] De, S. K., and A. L. Misra, Intake of Phosphate by Indian Montmorillonite (Kashmir Bentonite) in Presence of Ammonium Salts, *Proc. Natl. Acad. Sci., India*, XXIX, pt. III, pp. 250–254 (1960).

[181] De, S. K., Adsorption of Phosphate Ion by Homoionic Indian Montmorillonites, *J. Indian Soc. Soil Sci.*, **9**: 169–177 (1961).

adsorption of phosphate ions by montmorillonite and kaolinite and concluded that the adsorption might involve the release of hydroxyl ions from the minerals; this may be a simple substitution or may be due to reactions with hydrous iron oxides or hydrous aluminum oxides on the surface, forming iron or aluminum phosphates in an acidic medium.

In the case of kaolinite, phosphate fixation increases with decreasing particle size, and Stout[176] showed that fixation greatly increases in finely ground kaolinite. Schoen[166] showed the same thing for kaolinite and also for montmorillonite. It may well be that the fine grinding of the kaolinite so disrupts its structure that the results obtained are not a simple matter of clay-mineral fixation. Schoen also showed that, on heating kaolinite and halloysite, phosphate fixation increases up to 500°C and then decreases on heating to 1000°C; on heating montmorillonite, it increases at 500°C and then decreases sharply at 1000°C; on heating illite, it increases at both 500 and 1000°C.

The manner in which phosphate fixation takes place in clay minerals is probably a matter of sorption of the phosphate ion in the way developed in discussing anion exchange, with the portion that is fixed being adsorbed in relatively inaccessible positions in the clay-mineral structure.

Selective sorption

In recent years there has been much investigation of the selective-sorption properties of the clay minerals and other minerals, especially the zeolites. In some cases the sorption is definitely related to ion exchange, in other cases it is not, and in still other cases the mechanism is not clearly understood. It is not feasible to review all this work, and the following citations are given simply to illustrate the range of such investigations.

Barrer and his colleagues have been particularly active in this field. For example, Barrer and MacLeod[183] studied the sorption of nonpolar and polar gases and vapors on a sodium-rich montmorillonite and showed that nonpolar species are not intercalated but that polar molecules may be. They pointed out that the shape of a polar molecule may be so unfavorable that the influence of polarity, which normally brings about interlamellar penetration, is neutralized. Von Engelhardt and Smolinski[184] found that kaolinite and montmoril-

---

[182] De, S. K., Intake of Phosphate by Indian and Georgia Kaolinites, *J. Japan. Assoc. Mineral Petrog. Econ. Geol.*, pp. 31–38 (1962).

[183] Barrer, R. M., and D. M. MacLeod, Intercalation and Sorption by Montmorillonite, *Trans. Faraday Soc.*, **50**: 980–989 (1954).

[184] Von Englehardt, W., and A. v. Smolinski, Reaction of Polyphosphates with Clay Minerals, *Kolloid-Z.*, **151**(1): 47–52 (1957).

lonite adsorb sodium tripolyphosphate from aqueous solutions according to a Freundlich isotherm. The adsorbed amount increases with temperature, indicating a reaction with an activation barrier. After adsorption, a slow decomposition of the clay minerals takes place with the formation of SiO_2 and aluminum polyphosphate in solution, indicating decomposition of the clay mineral. Heydemann[185] showed that copper is adsorbed by clay minerals according to the Freundlich adsorption isotherm and that the adsorption capacity in clay minerals is increased by raising the pH as well as the copper concentration. The influence of concentration and pH values is most pronounced with kaolinite; there is only a small influence with the smectites. The adsorption capacity related to 100 g of clay increases in the following order: kaolinite, fireclay-type kaolinite, illite, smectite. However, the adsorption capacities related to a 100-m^2 surface of kaolinite and to fireclay-type kaolinite are of the same order of magnitude. The adsorption capacity per surface unit for illite and smectite can be considerably smaller under certain conditions. Basu and others[186] studied the order of selectivity of various cations by montmorillonite. They showed that the variations of selectivity coefficients are more marked at low concentration than at higher concentrations.

Problems associated with the disposal of waste products from atomic reactors have led to many studies of the selective-sorption properties of the clay minerals. Thus, Nishita and others[187] made a general study of fixation of various fission products on smectite and kaolinite.

Frysinger[188] found that vermiculite-biotite minerals had high selective-adsorption capacity for the cesium ion. Tamura and Jacobs[189,190] showed that, at low cesium-ion concentrations, extremely high selectivities for cesium ions were exhibited by layer lattice silicates with an unexpanded 10-Å c-axis spacing and that the adsorption was not directly related to the exchange capacities of the mineral. Their investigation showed higher cesium removal

[185] Heydemann, A., Adsorption aus sehr verdunnten Kupferlosungen an reinen Tonmineralen, *Geochim. Cosmochim. Acta,* 15: 305–329 (1959).

[186] Basu, A. N., B. K. Seal, and S. K. Mukherjee, Selectivity Coefficients of Trace Elements on a Montmorillonite Clay and a Humic Acid System, *J. Indian Chem. Soc.,* 39(2): 71–78 (1962).

[187] Nishita, H., B. W. Kowalewsky, A. J. Steen, and K. H. Larson, Fixation and Extractability of Fission Products Contaminating Various Soils and Clays, I, Sr^{90}, Y^{91}, Ru^{106}, Cs^{137}, and Ce^{144}, *Soil Sci.,* 81: 317–326 ((1956).

[188] Frysinger, G. R., Cation Exchange Behavior of Vermiculite-Biotite Mixtures, *Clays, Clay Minerals,* 9: 116–121 (1960).

[189] Tamura, T., and D. G. Jacobs, Structural Implications in Cesium Sorption, *Health Phys.,* 2: 391–398 (1960).

[190] Tamura, T., and D. G. Jacobs, Improving Cesium Selectivity of Bentonite by Heat Treatment, *Health Phys.,* 5: 149–154 (1961).

by illite, as compared with smectites and kaolinite, and that improved cesium removal was attained by the treatment of hydrobiotites with potassium chloride, and by bentonites following their heating to temperatures of 500 to 700°C.

Darab and Schonfeld[191] pointed out that the sorption of cesium 137 on kaolinite, illite, and smectite is much higher than would be anticipated by the cation-exchange capacity and that increasing amounts of inactive cesium caused a marked decrease in the sorption of cesium 137. Gaudette, Metzger, and Grim[192] studied the sorption of cesium by illite in great detail and concluded that the cation was fixed in an interlayer position in the "frayed edges" of illite particles.

Yammamoto[193] showed that subjecting kaolinite to a radiation with intense gamma rays tended to increase surface acidity.

Nahin[194] presented experimental data to show that polyethylene can be bonded directly to clay surfaces by means of ionizing radiation and that at least in one case an organic clay complex is more effective than a wholly inorganic clay for the purpose of radiatively linking polyethylene to clay surfaces. Experiments were run on both kaolinite and montmorillonite.

Perhaps the most interesting work on selective sorption is that directed toward the preparation of atomic sieves. Some zeolite materials contain small pores of a very uniform size so that large ions or molecules cannot be accommodated where a smaller species can be sorbed. Barrer and Mackenzie[195] have shown that some of the clay minerals have potential value as atomic sieves.

Additional references

Bower, C. A., and E. Truog, Base Exchange Capacity Determinations as Influenced by Nature of Cation Employed and Formation of Base-Exchange Salts, *Soil Sci. Soc. Am. Proc.,* **5:** 86–90 (1940).

Chaminade, R., and G. Drouineau, Recherches sur la mécanique chimique des cations échangeables, *Ann. Agron.,* **6:** 677–690 (1936).

[191] Darab, K., and T. A. Schonfeld, A Cs⁺ Ion Adzorpciójának Vizsgalata agyagásványon (Investigation of the Adsorption of Cs⁺ of Clay Minerals), *Agrokem. Talajtan,* **10:** 539–546 (1961).

[192] Gaudette, H. E., R. E. Grim, and C. F. Metzger, Illite: A Model Based on Sorption Behavior of Cesium, *Am. Mineralogist,* **51:** 1649–1656 (1966).

[193] Yammamoto, D., Increase of Surface Acidity of Kaolinite by Gamma-ray Irradiation of Large Dosage, *Nippon Kagaku Zasshi,* **83:** 115–116 (1962).

[194] Nahin, P., Organoclays Bonded to Polyethylene by Ionizing Radiations, Proceedings of the 13th National Clay Conference, pp. 317–330, Pergamon Press, New York (1966).

[195] Barrer, R. M., and N. Mackenzie, Sorption by Attapulgite, *J. Chem. Phys.,* **58:** 560–571 (1955).

Friedrich, Heinz, and U. Hofmann, Activities of the Exchangeable Cations of Mineral Clays, *Z. Anorg. Allgem. Chem.*, 342(1–2): 10–19 (1966).

Gedroiz, K., Die Lehre von Adsorptionsvermögen der Boden, *Kolloid-Chem. Beih.*, 33: 317–448 (1931).

Halevy, E., The Exchangeability of Hydroxyl Groups in Kaolinite, *Geochim. Cosmochim. Acta*, 28: 1139–1145 (1964).

Hendricks, S. B., Base Exchange of Crystalline Silicates, *Ind. Eng. Chem.*, 37: 625–630 (1945).

Hissink, D. J., Beitrag zur Kenntnis der Adsorptionvorgang im Bodem, *Intern. Mitt. Bodenk.*, 12: 81–172 (1922).

Hofmann, U., and W. Bilke, Ueber die innerkristalline Quellung und das Basenaustauschvermögens der Montmorillonits, *Kolloid-Z.*, 77: 238–251 (1936).

Hofmann, U., and A. Hausdorf, Kristallstruktur und innerkristalline Quellung der Montmorillonits, *Z. Krist.*, 104: 205–293 (1942).

Jorgensen, P., and I. T. Rosenqvist, Replacement and Bonding Conditions for Alkali Ions and Hydrogen in Dioctahedral and Trioctahedral Micas, *Norsk Geol. Tidsskr.*, 43: pt. 4: 497–536 (1963).

Kielland, J., Thermodynamics of Base Exchange Equilibria of Some Different Clays, *J. Soc. Chem. Ind. (London)*, 54: 232–247 (1935).

Lai, T. M., and M. M. Mortland, Self-diffusion of Exchangeable Cations in Bentonite, Proceedings of the 9th National Clay Conference, pp. 229–247, Pergamon Press, New York (1962).

Lemberg, J., Ueber Silikatumverwandlungen, *Z. Deut. Geol. Ges.*, 28: 519–621 (1876).

Marshall, C. E., and R. S. Gupta, Base Exchange Equilibria in Clays, *J. Soc. Chem. Ind. (London)*, 52: 433–443 (1933).

Marshall, C. E., and L. L. McDowell, Surface Reactivity of Micas, *Soil Sci.*, 99(2): 115–131 (1965).

Mattson, S., The Laws of Soil Colloidal Behavior, V, Ion Adsorption and Exchange, *Soil Sci.*, 31: 311–331 (1931).

Mattson, S., and V. Gustafson, The Electrochemistry of Soil Formation, I, Gel and Soil Complex, *Ann. Roy. Agr. Coll. Sweden*, 4: 1–54 (1937).

Mehlich, A., Soil Properties Affecting the Proportionate Amounts of Calcium, Magnesium, and Potassium in Plants and HCl Extracts, *Soil Sci.*, 62: 373–405 (1946).

Murphy, H. F., The Role of Kaolinite in Phosphate Fixation, *Hilgardia*, 12: 341–382 (1939).

Sieling, D. H., Role of Kaolin in Anion Sorption and Exchange, *Soil Sci. Soc. Am. Proc.*, 11: 161–170 (1946).

Vageler, P., "Der Kationen und Wasserhaushalt des Mineralbodens," Springer, Berlin (1932).

Weyl, W. A., Surface Structure and Surface Properties of Crystals and Glasses, *J. Am. Ceram. Soc.*, 32: 367–374 (1949).

Wiegner, G., Some Physico-Chemical Properties of Clays, *J. Soc. Chem. Ind. (London)*, 59: 65–71, 105–112 (1931).

Eight
Clay-Water System

This chapter is concerned with water which can be held by clay materials only at relatively low temperatures and is driven off by heating to about 100 to 150°C. The OH lattice water which is lost at temperatures above about 300°C is considered elsewhere (Chap. 9). The nature of the low-temperature water and the factors that control its characteristics are of great importance, since they largely determine the plastic, bonding, compaction, suspension, and other properties of clay materials, which in turn frequently control their commercial utilization. Thus, an understanding of this low-temperature water in relation to the clay minerals must precede an understanding of the plastic and many other properties of clay materials.

The water lost at low temperatures may be classed in three categories, depending on its relation to the mineral components and to the texture of the clay materials, as follows: (1) the water in pores, on the surfaces, and around the edges of the discrete particles of the minerals composing the clay material; (2) in the case of vermiculite, smectite, and the hydrated form of halloysite, the interlayer water between the unit-cell layers of these minerals (this is the water that causes the swelling of smectite); and (3) in the case of the sepiolite-attapulgite-palygorskite minerals, the water that occurs within the tubular opening between the elongate structural units. Type

1 water requires generally very little energy for its removal, and drying at only a little above ordinary room temperatures is adequate for its substantially complete elimination.   Water of types 2 and 3 requires definite energy for its complete removal.   In the case of the hydrated form of halloysite, drying at room temperature is adequate to remove most of the interlayer water, but higher temperatures (see pages 71–73) are required for total removal.   In the case of the vermiculite and smectite minerals, temperatures at least approaching 100°C are necessary for substantially complete elimination of interlayer water.   At this temperature some little time is required for complete removal of this water, whereas at somewhat higher temperatures it is driven off rapidly.   In the case of halloysite, the reaction is not reversible and the hydrated mineral ordinarily cannot be formed again.   Vermiculites and smectites rehydrate with difficulty if the dehydration is absolutely complete, but easily if only a trace of water is allowed to remain between the unit layers.   The water in the channels of the sepiolite-attapulgite-palygorskite minerals is lost at about the same temperature as the interlayer water of the layer clay minerals.   The channel water is regained readily if removal is by drying only at low temperatures.   The energy necessary for dehydration and the precise temperatures at which it occurs are shown for the various clay-mineral groups by differential thermal and dehydration analyses in Chap. 9.

Nature of adsorbed water

Langmuir,[1] Terzaghi,[2] Hardy,[3] Baver,[4] Winterkorn,[5] Low,[6] and many others[7] have presented evidence to show that the water held directly on the surfaces of the clay particles is in a physical state different from that of liquid water. The specific characteristics of this water which delimit it from ordinary water would probably be restricted ordinarily to relatively short distances from

[1] Langmuir, I., The Constitution and Fundamental Properties of Solids and Liquids, *J. Am. Chem. Soc.*, **39**: 1848–1906 (1917).

[2] Terzaghi, K., The Physical Properties of Clays, *M.I.T., Tech. Eng. News,* **9**: 10, 11, 36 (1928).

[3] Hardy, W. B., Friction, Surface Energy, and Lubrication, "Colloid Chemistry," J. Alexander, ed., vol. 1, pp. 301–330, Reinhold, New York (1926).

[4] Baver, L. D., and H. W. Winterkorn, Sorption of Liquids by Soil Colloids, II, Surface Behavior in the Hydration of Clays, *Soil Sci.*, **40**: 403–418 (1936).

[5] Winterkorn, H. W., The Condition of Water in Porous Systems, *Soil Sci.*, **55**: 109–115 (1943).

[6] Low, P. F., Physical Chemistry of Clay-Water Interaction, *Advan. Agron.*, **13**: 269–327 (1961).

[7] Grim, R. E., and F. L. Cuthbert, Some Clay-Water Properties of Certain Clay Minerals, *J. Am. Ceram. Soc.*, **28**: 90–95 (1945).

the clay-particle surfaces.   The possible thickness of the nonordinary water can vary a good deal even for a given clay mineral, and the transition from nonordinary water to ordinary water can be abrupt or gradual, depending on factors that will be considered presently.

It appears certain that the possible thickness of the nonordinary, or so-called nonliquid, water is relatively small on irregular surfaces, such as those around the edges of clay-mineral particles, and relatively large on the flat surfaces of the clay minerals.   The film of nonordinary water is best developed and appears to reach its greatest thickness on the basal plane surfaces of the expanding-lattice minerals of the smectite group.   It follows that the water in pores would be substantially liquid water, with nonliquid water forming only a thin film on the surface of the pores and where adjacent clay-mineral particles come together.

Many early workers, particularly Terzaghi,[2] explained the initially adsorbed water on the basis of the dipole character of the water molecule; the latter possesses positive and negative charges, the centers of which do not coincide.   Since the surface of the clay particle is normally negatively charged, the positive ends of the water molecules are considered to lie toward the clay surface with the negative ends extending outward.   The initial layer of water is believed to consist of water molecules all oriented in the same direction.   According to this concept, the first layer of oriented-dipole water molecules forms another surface of negative charges on which can be built another layer of completely oriented water molecules.   This process of building up layers could be continued indefinitely were it not for the fact that the water molecules possess thermal energy and tend to be in a state of continuous motion.   In accordance with classical concepts of colloidal theory, the motion due to thermal energy will oppose the regular orientation.   Therefore, at any given instant and at a certain distance from the surface some of the molecules will be oriented at right angles to the surface.   An instant later these molecules will have moved, but others will have become oriented in their place.   At the actual clay-mineral surface the molecules will be highly oriented, and the degree of orientation will decrease going outward, as the relative effect of thermal movement becomes greater.   Macey[8] pointed out difficulties encountered by this concept, particularly in view of the facts that the clay-mineral surface is not a uniformly charged plane and that the water molecules strictly do not act as little rods with positive and negative ends.

Investigations by Bernal and Fowler[9] and Bernal and Megaw[10] of the

[8] Macey, H. H., Clay-Water Relationships and the Internal Mechanism of Drying, *Trans. Brit. Ceram. Soc.*, **41**: 73–121 (1942).

[9] Bernal, J. D., and A. H. Fowler, A Theory of Water and Ionic Solution with Particular Reference to Hydroxyl Ions, *J. Chem. Phys.*, **1**: 515–548 (1933).

[10] Bernal, J. D., and H. D. Megaw, The Function of Hydrogen in Intermolecular Forces, *Proc. Roy. Soc.* (*London*), *A*, **151**: 384–420 (1935).

hydrogen bond and of the distribution of charges about a water molecule provided the basis for a more satisfactory concept of the nature of the adsorbed water on clay-mineral surfaces. Low[6] discussed this matter in detail, and the following is taken largely from the work of this investigator and his colleagues.

The water molecule consists of a V-shaped arrangement of the atomic nuclei, the internuclear O-H distance being 0.96 Å and the internuclear angle being 103 to 106°, which is very close to the tetrahedral angle of 109°. In the molecule, there are four regions where the density of the outer electron is maximal. Two of these regions are associated with the OH bonds and coincide with the positions of the protons; the other two are associated with lone pairs of electrons and are located above and below the plane of the atomic nuclei on the opposite side of the oxygen nucleus from the proton. Therefore, the net charge distribution of the water molecule resembles a tetrahedron with two positive and two negative corners. The resultant center of positive electricity, midway between the protons, is separated from the resultant center of negative electricity near the oxygen nucleus on the side next to the protons. Hence, the water molecule has a dipole moment, which is equal to 1.83×10^{-18} esu.

When two water molecules approach each other, there is electrostatic attraction between the positive tetrahedral corner of one molecule and the negative tetrahedral corner of the other; that is, there is an electrical interaction between the proton of the former and the lone electron pair of the latter. Since the proton of the hydrogen is involved, the bond is called the hydrogen bond. Each water molecule can form four hydrogen bonds, one in each tetrahedral corner. Therefore, in an assembly of water molecules, there is a tendency for each molecule to be hydrogen-bonded to four neighboring water molecules which surround it tetrahedrally.

At temperatures below 0°C in the ice lattice, the water molecules exist in fixed positions, with each molecule tetrahedrally coordinated to four others. When ice melts, there is an increase in density from 0.917 for ice to nearly 1 for water. The consideration of the magnitude of this density increase has led to the conclusion that the radius of the water molecule is nearly the same as in ice. It appears that liquid water retains a high degree of hydrogen bonding and that this bonding decreases with increasing temperature. It has been concluded that the structure of water must be similar to that of ice at least for short distances or perhaps it may be stated that water has a loose ice-like arrangement or, in the terminology of Morgan and Warren,[11] water has a "broken down ice structure."

The surface of the clay mineral is made up of either oxygen atoms or hydroxyl groups arranged in a hexagonal pattern which, according to Hen-

[11] Morgan, J., and B. E. Warren, X-ray Analysis of the Structure of Water, *J. Chem. Phys.*, **6**: 666–673 (1938).

dricks and Jefferson,[12] Macey,[8] and Forslind,[13] can coincide at points with a similar pattern in a hydrogen-bonded water structure. Further, the crystal lattice of many clay minerals contains excess electrons which arise from the isomorphic substitution of cations in the lattice. From the work of Lennard-Jones and Pople[14] and Frank,[15] there is reason to believe that covalency may occur in hydrogen-bond formation if one of the systems involved is capable of having its lone pair of electrons distorted by the proton of positive element of the other. Such distortion is conducive to the formation of additional hydrogen bonds in a cooperative manner (Frank[15]). The lone-pair electrons of the oxygen atoms in the surface of a clay mineral should be easily distorted because of the excess electrons in the lattice, and therefore it is reasonable to believe that water molecules adjacent to a clay-mineral surface are bonded to the oxygen atoms of the surface by covalent hydrogen bonds. The existence of the covalent bonds should alter the electron distribution in these molecules and make it easier for them to form additional covalent bonds with other molecules in the same and next layer. Those in the next layer in turn should be expected to form hydrogen bonds of partially covalent character with their neighbors, and so on. The bonded water molecules should be arranged in a tetrahedral fashion because of the directional properties of the bonds. However, the degree of covalency in the bonds should decrease with distance from the surface and for this reason the tetrahedral arrangement should become less rigid in the same direction. Thus, it is possible for a tetrahedral structure of water molecules to be attached to and propagated with decreasing rigidity away from the oxygen surface of a clay mineral.

It is not unlikely that a hydrogen-bonded water structure builds up also on the hydroxylic surface of a clay mineral. Here the excess electrons in the mineral lattice should help to screen the protons of the hydroxyl groups and render them less electropositive. Consequently, the lone-pair electrons in the oxygen atoms in the bonded water molecules should experience little distortion and the degree of covalency in the hydrogen bonds should be slight. For this reason the water structure on the hydroxylic surface may be expected to be less stable than on an oxygen surface but the balance between order and disorder is delicate; therefore, even the hydroxylic surface, by fixing

[12] Hendricks, S. B., and M. E. Jefferson, Structure of Kaolin and Talc-Pyrophyllite Hydrates and Their Bearing on Water Sorption of Clays, *Am. Mineralogist,* **23**: 863–875 (1938).

[13] Forslind, E., The Crystal Structure and Water Adsorption of the Clay Minerals, *Trans. 1st Intern. Ceram. Congr.,* pp. 98–110 (1948).

[14] Lennard-Jones, J., and J. A. Pople, Molecular Association in Liquids, I, Molecular Association Due to Lone-Pair Electrons, *Proc. Roy. Soc. (London),* A, **205**: 155–162 (1951).

[15] Frank, H. S., Covalency in the Hydrogen Bond and the Properties of Water and Ice, *Proc. Roy. Soc. (London),* A, **247**: 481–492 (1958).

the positions of the layer of bonded molecules, should tip the balance in favor of order for considerable distances.

On the basis of the foregoing structural factors, Hendricks and Jefferson[12] suggested a concept of the nature of adsorbed water which, in general terms and with some modification, has found widespread acceptance.

In the language of Hendricks and Jefferson,[12] a water layer is composed of water molecules joined into hexagonal groups of an extended hexagonal net, as shown in projection in Fig. 8-1. Each side of the hexagon (Fig. 8-1) must correspond to a hydroxyl bond, the hydrogen bond of one water molecule being directed toward the negative charge of a neighboring molecule. One-fourth of the hydrogen atoms, or a hydrogen atom of half of the water molecules, are not involved in bonding within the net (K, M, and O of Fig.

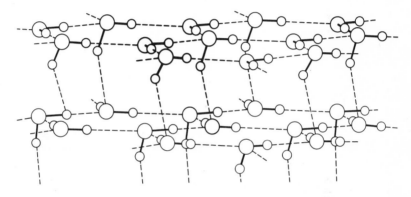

Fig. 8-1.   Configuration of water adsorbed directly adjacent to the basal plane surfaces of the clay minerals.   (After Hendricks and Jefferson.[12])

8-2). The net is tied to the surface of the clay minerals by the attraction between those hydrogen atoms not involved in bonding within the net and the surface oxygen layer of the clay-mineral units (Figs. 8-2 and 8-3). The suggested[12] superposition of the water net and basal layers of oxygens and hydroxyls, together with the type of bonding between them, is shown in Figs. 8-4 and 8-5, respectively.

The net has just the a and b dimensions of the silicate layer minerals if the separation of the oxygen atoms of the water molecules is about 3.0 Å in projection. It is assumed that the oxygen atoms are in one plane. In this configuration, there is relatively loose packing of the water molecule, there being four water molecules per molecular layer per unit cell of the clay mineral instead of six, as would be the case in close packing. The structure is essentially that of ice except that the oxygen atoms are planar.

The stability of the layer of water molecules arises from the geometrical relationship to the oxygen atoms or hydroxyl groups of the silicate framework. The presence of the first layer would favor the formation of a second

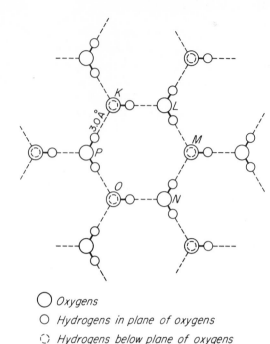

○ Oxygens
○ Hydrogens in plane of oxygens
◌ Hydrogens below plane of oxygens

Fig. 8-2.   Arrangement of oxygens and hydrogens in a water net.   (After Hendricks and Jefferson.[12])

layer, and the water structure would thus be propagated away from the clay-mineral surface.

In the Hendricks and Jefferson[12] configuration every other water molecule lies about over an oxygen of the surface layer of the three-layer clay minerals (Fig. 8-4).   Since half of the water molecules have hydrogen available for vertical bonding, the directly superimposed oxygens and water molecules may be assumed to be tied together through these hydrogen bonds.   As successive molecule layers develop, the exterior layer can be tied to the layer next below by a hydrogen from every other water molecule.   According to this scheme no hydrogens would be available for tying together series of water layers growing outward from two neighboring clay-mineral surfaces, unless the directly adjacent water layers of two water envelopes were tied together by relatively fewer bonds.   A plane or planes of relatively weakly bonded water molecules would exist, therefore, at the junction of two water envelopes.

A second mechanism by which water may be attracted to a clay surface is hydration of exchangeable cations.   Since the cations cannot escape from the negatively charged surface, neither can the water of hydration.   This mechanism of attracting water should be most important at low water contents.   At higher water contents, exchangeable cations should still play a

Silica layer
of clay mineral

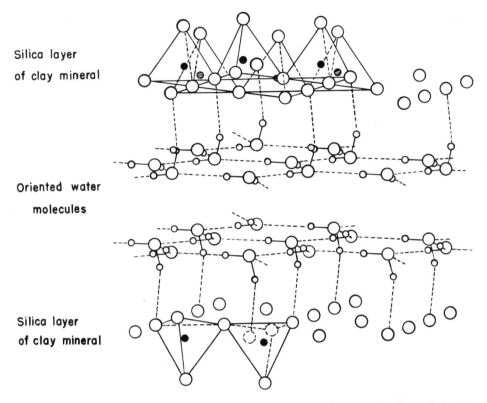

Oriented water
molecules

Silica layer
of clay mineral

Fig. 8-3. Configuration of the water net proposed by Hendricks and Jefferson,[12] showing the binding through hydrogens to the adjacent clay-mineral surfaces.

role in clay-water interaction. Those exchangeable ions that are dissociated from the surface may be regarded as being in solution. Undoubtedly they lower the activity of the water in the vicinity of the clay surface in the same manner as ions lower the activity of water in solution. Consequently, water should tend to move into the surface region. In short, clays may be expected to attract water by osmosis.

Numerous investigators, for example, Hofmann and Hausdorf,[16] Mackenzie,[17] Walker,[18] and Low[6] have pointed out that the presence of adsorbed cations and their hydration envelope would influence the arrangement of water molecules directly adjacent to them.

[16] Hofmann, U., and A. Hausdorf, Kristallstruktur und innerkristalline Quellung der Montmorillonits, *Z. Krist.*, **104**: 265–293 (1942).

[17] Mackenzie, R. C., Some Notes on the Hydration of Montmorillonite, *Mineral. Soc. Gr. Britain Clay Mineral Bull.* 4, pp. 115–120 (1950).

[18] Walker, G. F., Water Layers in Vermiculite, *Nature,* **163**: 726 (1949).

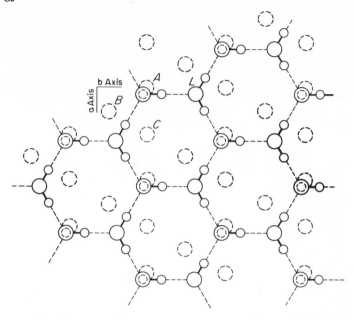

Fig. 8-4. Superposition of a water net on the basal oxygen layer of vermiculite, suggested by Hendricks and Jefferson.[12] Large dashed circles (A, B, C, etc.) are oxygen atoms of the clay-mineral surface, 2.73 Å below the plane of the water molecules.

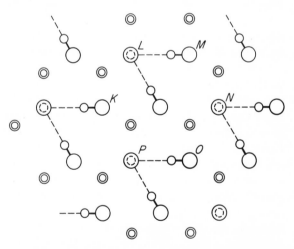

Fig. 8-5. Superposition of a water net on the basal hydroxyl layers of two-layer clay minerals and the type of bonding between hydroxyl groups and water layers. At L, N, and P, hydrogens of hydroxyl groups are free for binding to oxygens of water molecules. (After Hendricks and Jefferson.[12])

Mackenzie[17] particularly emphasized this point and indicated that the matter of cation hydration would be particularly important in the part of the adsorbed water net closest to the clay-mineral surfaces. In the case of smectite, there would be about one monovalent cation for each two hexagonal configurations of water in the first molecular water layer. According to Mackenzie, there is a space problem in fitting such a number of ions into the water net. Mackenzie presents computations of the energy of hydration of the sheet surface of smectite carrying various ions; these data are in accord with his suggestion that, at low water contents, water sorption depends primarily on the exchangeable ion present, the sheet surface being of subsidiary importance.

Macey, arguing from the similarity between the structure of ice and of the oxygen atoms exposed at a sheet surface of the layer clay minerals, postulated that the initially adsorbed water has the structure of ice. He considered that it fits on top of the oxygen net of the basal plane of the three-layer clay minerals, as shown in Fig. 8-6. The fit of the water molecules with the oxygen net as suggested by Macey is different from that suggested by Hendricks and Jefferson. According to the Macey concept, the distance between oxygens in the water layer would be 4.52 Å, and the packing would be even looser than that suggested by Hendricks and Jefferson.[12] There would be three molecules of water per unit cell per molecular layer. Such loose packing of water in a given layer would be predicated on hydrogen binding to additional water not in the layer under consideration. As illustrated in Fig. 8-6*b*, showing the structure of ice, other water molecules would be just out of the plane in contact with the silicate structure, and there is no epitaxial arrangement between them and the silicate surface. In the Hendricks and Jefferson concept, the ice structure is stretched so that the offset water molecules come into the same plane and there is no change in the hydrogen binding. Also the stretch permits a complete epitaxial arrangement of all the water molecules and the silicate surface. Forslind[19] reported electron-diffraction data which seem to indicate a structure at least similar to ice in the initially adsorbed water of montmorillonite. DeWit and Arens[20] published some density measurements that seem to be in agreement with the Macey concept.

Barshad[21] suggested another concept of the nature of the adsorbed water

[19] Forslind, E., The Clay Water System, I, Crystal Structure and Water Adsorption of Clay Minerals, *Swed. Cement Concrete Res. Inst., Bull.* 11 (1948).

[20] DeWit, C. T., and P. L. Arens, Moisture Content and Density of Some Clay Minerals and Some Remarks on the Hydration Patterns of Clay, *Trans. 4th Intern. Congr. Soil Sci.,* **2:** 59–62 (1950).

[21] Barshad, I., The Nature of Lattice Expansion and Its Relation to Hydration in Montmorillonite and Vermiculite, *Am. Mineralogist,* **34:** 675–684 (1949).

on the basis of careful dehydration determinations. According to him, at very low states of hydration for montmorillonite, the water molecules tend to form tetrahedrons with the oxygens of the top layer of the linked silica tetrahedrons of the lattice. This type of packing would give rise to hexagonal

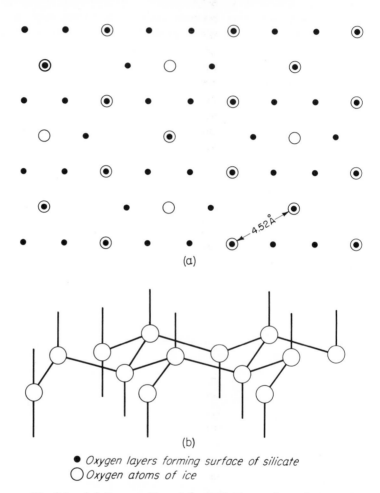

● *Oxygen layers forming surface of silicate*
○ *Oxygen atoms of ice*

Fig. 8-6. (*a*) **Superposition of the ice lattice on the surface oxygen layer of the clay minerals (after Macey**[8]**).** (*b*) **Structure of ice.**

rings of water molecules which are similar to the hexagonal rings of oxygens of the vertices of the linked silica tetrahedrons of the individual silicate sheets. In Fig. 8-7, *a* to *f* represent such water molecules forming tetrahedral units with the oxygens of the underlying silica tetrahedral network. The packing in this configuration would be loose, as there would be only four molecules of water per unit cell per molecular layer, and the height added for a single layer of water molecules would be 1.78 Å, according to Barshad. It appears that this value should be 2.1 Å, using Barshad's value of 2.55 Å for the water molecules, which reduces the agreement of Barshad's computed and analytical data for c dimensions at certain stages of hydration. Barshad postulated that at higher states of hydration the water adsorbed by montmorillonite tends to form hexagonal rings of water molecules; these are similar to the hexagonal rings of the oxygens of the montmorillonite basal plane which forms the bases of the linked silica tetrahedrons. In Fig. 8-7, 1 to 6 represent such water molecules. In this configuration the packing is more dense, and there are six molecules of water per unit cell per layer of water molecules. The height added for a single water layer would be about 2.55 Å, since the water molecules would be directly superimposed on the oxygens. At still higher states of hydration, it is believed that water molecules fill even the centers of these hexagonal water rings and the centers of the hexagonal oxygen

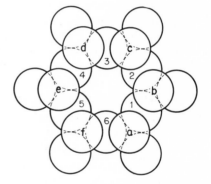

Fig. 8-7. Arrangement of the water molecules in the interlayer space of montmorillonite and vermiculite. (After Barshad.[21])

rings of the linked silica tetrahedrons not occupied by exchangeable ions. Such water layers would consist of closely packed water molecules, and there would be eight water molecules per unit cell per molecular layer. Barshad[21] stated that adsorption does not progress beyond the state of two molecular layers until dense packing is accomplished. It must be pointed out that Barshad's interpretations were based on dehydration data, which require extremely pure mono-mineral material. They would be subject to another interpretation if the montmorillonite and vermiculite investigated consisted of mixed layers with hydration layers of different thicknesses. Investigations by Hendricks et al.[22] and Mering[23] suggest the possibility of this alternative explanation.

[22] Hendricks, S. B., R. A. Nelson, and L. T. Alexander, Hydration Mechanism of the Clay Mineral Montmorillonite Saturated with Various Ions, *J. Am. Chem. Soc.,* **62:** 1457–1464 (1940).

[23] Mering, J., The Hydration of Montmorillonite, *Trans. Faraday Soc.,* **42B:** 205–219 (1946).

Thus, Mering suggested that, in Ca montmorillonite, the initially adsorbed water is packed octahedrally about the Ca^{++} and that, with increasing available water, a double water layer of superimposed water molecules first develops, followed by successive layers of superimposed water molecules. In Na montmorillonite, increasing amounts of water develop successive layers of superimposed water molecules. It appears that Barshad's concept is not substantiated geometrically as well as those of Mering and Hendricks et al.

Various investigators, notably Williamson,[24] have objected to the foregoing concepts, because of their feeling that the sorption forces are strong enough to cause dense packing of at least the initially adsorbed water layers.

It is the opinion of the author that the evidence to be presented in the following pages is overwhelmingly in favor of the concept that the water adsorbed directly on clay-mineral surfaces and for some distance outward is composed of water molecules in an organized arrangement, with consequent properties different from those of liquid water. It seems most likely that the water structure is generally similar to that of ice. Present concepts of clay-mineral structures indicate that the surface layers are not exactly planar and that the surface atoms of the minerals would not coincide with protons or oxygens of the ice lattice as neatly as has been postulated. Thus, for structural reasons as well as because of the presence of adsorbed cations, distortions would be expected in the adsorbed water lattice so that the water might well be described as quasi-crystalline.

Not all investigators accept the quasi-crystalline nature of the adsorbed water. For example, Martin[25] reviewed the data that have been presented on the character of adsorbed water on clay-mineral surfaces and concluded that "unfortunately one may reasonably question virtually all the data cited for various reasons." He further concluded that the data are inadequate to prove whether the adsorbed phase is a solid in which there is more order than normal water or a two-dimensional liquid. Martin favored the two-dimensional-liquid model.

Density of initially adsorbed water

As indicated in discussing the configuration of the adsorbed water molecules on the clay-mineral surfaces, rather large differences in the denseness of packing are possible. The density of the adsorbed water would vary, of course, depending on the nature of the packing. Accurate determinations of the density of adsorbed water would provide evidence for or against a particular

[24] Williamson, W. O., The Physical Relationship Between Clay and Water, *Trans. Brit. Ceram. Soc.*, **50**: 10–34 (1951).

[25] Martin, R. T., Adsorbed Water on Clay: A Review, Proceedings of the 9th National Clay Conference, pp. 28–70, Pergamon Press, New York (1962).

configuration. Unfortunately, the determination of the density of the adsorbed water is a very difficult problem experimentally. Attempts using a pycnometer technique have not been very successful because of the difficulty of assigning a correct specific volume to the clay mineral. A wide variety of results have been obtained, emphasizing their unsatisfactory character.

Tscapek[26] arrived at a value of 1.7, and Hauser and Le Beau,[27] studying bentonite suspensions, presented values greater than 1, suggesting a dense packing. On the other hand, Nitzsche[28] presented determinations giving values less than 1, and DeWit and Arens[20] obtained values as low as 0.73 for the density of the first few molecular layers of water adsorbed by smectite. The configurations having the greatest similarity in geometry to that of the oxygen layer of the surface of the clay minerals, namely, those suggested by Hendricks and Jefferson[12] and Macey[8] and the low-hydration types postulated by Barshad,[21] require a specific gravity less than 1.

Low and Anderson[29] used a unique method, involving no assumptions, that gave results which seem entirely satisfactory. They compressed a homoionic Wyoming smectite paste between a mercury piston and a stainless-steel filter and observed the corresponding changes in volume of the paste and expressed water. From the latter quantities, the initial water content of the paste, and the grams of water in the paste, the different paste volumes were calculated, and a plot was made of paste volumes against grams of water in the paste. The slope of the resulting line in any water content equals the partial specific volume of the water at that water content. For the lithium, sodium, and potassium smectite with which they worked, the following conclusions were drawn:

1. The partial specific volume of water is different from normal water out to distances in excess of 60 Å from the mineral surface.

2. The partial specific volume increases continuously as the surface is approached.

3. Within 10 Å of the clay surface a partial specific volume is as much as 3 percent greater than that of normal water (ice has a specific volume of only about 8 percent greater).

4. As the temperature is lowered, the partial specific volume of the water increases.

[26] Tscapek, W., The Density of Adsorbed Water in Soils, *Z. Pflanzenernähr. Düng. Bodenk.,* **34**: 265–271 (1934).

[27] Hauser, E., and D. S. Le Beau, Studies of Colloidal Clay, I, *J. Phys. Chem.,* **42**: 1031–1050 (1938).

[28] Nitzsche, W., On the Structure of the Hydration Hull of Inorganic Soil Colloids, *Kolloid-Z.,* **93**: 110–115 (1940).

[29] Low, P. F., and D. M. Anderson, The Partial Specific Volume of Water in Bentonite Suspensions, *Soil Sci. Soc. Am. Proc.,* **22**: 22–24 (1958).

5. Exchangeable ions influence the partial specific volume of the water.

Confirmation of the partial-specific-volume results of Anderson and Low[30] has been provided by the calculations of Bradley.[31]   Anderson and Low found that, at distances of about 10 Å, the density of water on lithium, sodium, and potassium smectite surfaces was 0.97, 0.972, and 0.981, respectively.

Evidence for the crystalline state of the initially adsorbed water

There is abundant evidence for some sort of definite configuration of the water molecules initially adsorbed on the surfaces of the clay minerals.   The characteristics of water molecules themselves and the nature of hydrogen bonding afford evidence for a grouping into a definite network, as has been discussed previously.   The oxygen atoms in the clay-mineral surface have a definite pattern, and consequently there would be a pattern of charges on this surface that could be carried over into the adsorbed water molecules. Therefore, the natures of the clay-mineral surface and of the water molecules both favor the development of a definite configuration in the initially adsorbed water.

Strong evidence that the character of initially adsorbed water is different from that of liquid water comes from a study of the properties of clay-water systems.   Thus Grim and Cuthbert[32] suggested a theory of the bonding action of clay and water in molding sands, based on the nonliquid nature of initially adsorbed water, which accounts for the strength and other properties of such materials and explains many of their attributes which had not been understood.   Grim[33] and many others[8,24] have shown that the plastic properties of clays are difficult or impossible to account for without postulating some change in the physical state of the adsorbed water.   Thus, in the case of sand-clay-water mixtures, the maximum bonding force is developed when the sand and clay are wet with an extremely restricted amount of water.   With a very small amount less or more than the optimum amount of water, the bond is greatly reduced.   This can readily be explained by considering that the maximum bond is developed when the mixture contains the maximum

[30] Anderson, D. M., and P. F. Low, The Density of Water Adsorbed by Lithium-, Sodium-, and Potassium-Bentonite, *Soil Sci. Soc. Am. Proc.*, **22**(2) : 99–103 (1958).

[31] Bradley, W. F., Density of Water Sorbed on Montmorillonite, *Nature*, **183**: 1614–1615 (1959).

[32] Grim, R. E., and F. L. Cuthbert, The Bonding Action of Clays, I, Clays in Green Molding Sands, *Illinois State Geol. Surv., Rept. Invest.* 102 (1945).

[33] Grim, R. E., Some Fundamental Factors Influencing the Properties of Soil Materials, *Proc. 2d Intern. Congr. Soil Mech.*, **3**: 8–12(1948).

amount of water that can develop a definite configuration. Additional water will, at least in part, partake of the nature of liquid water and greatly weaken the bond between the sand grains. Similarly, clays develop optimum plasticity when a definite amount of water is added to a dry clay. If water is added in small increments to a dry powdered clay, almost no experience is necessary to determine just when enough water has been added to the clay to develop plasticity. Again this phenomenon is readily explained by considering that plasticity develops when just enough water is added to satisfy the requirements of all available surfaces for water with a definite configuration *plus* a little more water which would develop little or no definite configuration of water molecules. For a further discussion of this point, see Grim.[33]

Russell[34] pointed out that the structural characteristics of soils require that the initially adsorbed water be in a nonliquid state. Richards[35] stated "when the thickness of adsorbed water film is reduced to 6 or 8 monomolecular layers of water the soil water is so tightly bound that crop growth ceases. All agriculture is conducted in the soil-water film thickness range from this value up to two or three times this thickness."

The investigators of the phenomenon of heat of wetting in clay materials (see page 271) have generally attributed it, at least in part, to a change in the nature of the water adsorbed on the surfaces of the clay-mineral particles. Many soil investigators have shown that the freezing point of the initially adsorbed water in soils is depressed. In other words, the soil contains a certain amount of water in a physical state that makes it difficult to freeze. Bodman and Day[36] showed that smectite depresses the freezing point to a greater degree than kaolinite, which is in accordance with the high water-sorption powers of smectite.

On the basis of an analysis of the phase relations of water, Winterkorn[5] concluded that the water held directly on the clay-mineral surfaces must be solid and not liquid and that the change in state of the water with distance from the clay-mineral surface is an exponential one.

From a study of the viscosity of water in clay-water systems, Low[37] concluded that a water structure, which varies in extent with particle arrangements and the adsorbed cationic species, exists at the surface of clay particles and that this structure bestows a high viscosity to the adsorbed water.

[34] Russell, E. W., Soil Structure, *Imp. Bur. Soil Sci. Tech. Commun.* 37 (1938).

[35] Richards, L. A., Advances in Soil Physics, *Trans. 7th Intern. Congr. Soil Sci.,* I, pp. 67–79 (1960).

[36] Bodman, G. B., and P. R. Day, Freezing Points of a Group of California Soils and Their Extracted Clays, *Soil Sci.,* **55**: 225–246 (1943).

[37] Low, P. F., The Viscosity of Water in Clay Systems, Proceedings of the 8th National Clay Conference, pp. 170–182, Pergamon Press, New York (1960).

Numerous investigators (Macey,[38] Winterkorn,[39] Schmid,[40] and Von Engelhardt and Tunn[41]) have concluded that the inapplicability of conventional flow equations to water movement through clay materials is an indication of an immobile or highly viscous water layer on the particle surfaces, which makes the effective porosity much less than the measured porosity.

Ionic activation energies (Low[37]) for movement in clays are greater than those for movement in pure water, which is in keeping with the concept that a coherent hydrogen-bonded water structure exists in the neighborhood of clay surfaces. The case of ionic movement through this structure should depend on the viscosity resistance that it offers.

If water in the vicinity of surfaces has a quasi-crystalline structure, its flow properties should be as follows: (1) It should have a yield value leading to a threshold hydraulic gradient below which flow will not occur; (2) after flow commences, there should be a range of hydraulic gradients over which non-newtonian flow occurs, that is, the viscosity should be dependent on the shearing force; (3) the viscosity should increase with proximity to the clay surface; and (4) near the clay surface the viscosity of the water should be greater than the viscosity of free water. Low[6] summarized experimental observations which are in complete harmony with these requirements.

Leonard and Low[42] measured the water tension at the upper plastic limit of several montmorillonite-water systems at several temperatures. From the resulting data, the relative partial molar free energy, entropy, and heat content of the water were calculated. In all cases the values were negative, suggesting that the water in the suspension had more order than pure bulk water. When the water tension was measured at different concentrations of two montmorillonites, it was found that the tension remained zero until the clay concentration was sufficient to allow gelation of the suspension. Thereafter the tension increased rapidly with clay concentration and, consequently, it was concluded that the sol-gel transformation was accompanied by a change in the energy status of the included water. The authors concluded that at the sol-gel transformation where the fields of influence of adjacent surfaces overlap there is a coalescence of the quasi-crystalline water structures extending from these

[38] Macey, H. H., Clay-Water Relationships and the Internal Mechanisms of Drying, *Trans. Brit. Ceram. Soc.,* **41**: 73–121 (1943).

[39] Winterkorn, H. F., Water Movement Through Porous Hydrophilic Systems Under Capillary, Electric, and Thermal Potentials, *Am. Soc. Testing Mater., Spec. Tech. Publ.* 163, pp. 27–35 (1955).

[40] Schmid, W. E., The Permeability of Soils and the Concept of a Stationary Boundary-Layer, *Am. Soc. Testing Mater., Proc.,* **57**: 1195–1218 (1957).

[41] Von Engelhardt, W., and W. L. M. Tunn, The Flow of Fluids Through Sandstones, *Illinois State Geol. Surv., Circ.* 194 (1955). Translated by P. A. Witherspoon.

[42] Leonard, R. A., and P. F. Low, Effect of Gelation on the Properties of Water in Clay Systems, Proceedings of the 12th National Clay Conference, pp. 311–325, Pergamon Press, New York (1964).

surfaces and, hence, there is no free water present in the system and tension develops. Both the coalescence of the water structures and the formation of interparticle bonds impart rigidity to the system.

If the water structure in the clay-water suspension changes with gelation, it should be possible to detect a change in the volume of the suspension as this process occurs. Anderson, Leming, and Spozido[43] detected such a change. They liquefied a suspension of montmorillonite by repeatedly dropping a steel ball through it; then they kept the ball stationary and found that the suspension expanded as gelation developed.

Yamaguchi[44] found that the electrical resistance of a clay-water system increased as the gel strength increased. Presumably the increase in electrical resistance was due to the altered nature of the water. Yamaguchi's work provides additional evidence for the concept that the properties of water in the clay suspension change with gelation.

Thickness of adsorbed nonliquid water

Because of the kinetic nature of the water molecules, adsorbed water with a definite configuration of water molecules could be expected to extend for only a limited distance from the clay-mineral surface. Numerous attempts have been made to estimate the thickness of adsorbed water with definite nonliquid characteristics, without very good agreement. However, the values probably do indicate the correct order of magnitude.

Houwink[45] arrived at a value of 25 Å, which is equivalent to about 10 molecular layers of water, and Mattson[46] gave the figure of 40 Å, equal to about 16 water layers. Grim and Cuthbert,[32] from their study of the relation of bonding strength to water content, concluded that, for montmorillonite with Na+ as the exchangeable cation, the nonliquid water has a thickness of three molecular layers and, for Ca montmorillonite, the thickness is four molecular layers, or about 10 Å. These same authors also suggested that for Na montmorillonite there is a gradual transition to liquid water beyond the 7.5-Å thickness, whereas in Ca montmorillonite the transition to liquid water is abrupt. In Na montmorillonite some orientation of water molecules may

[43] Anderson, D. M., G. F. Leming, and G. Spozido, Volume Changes of a Thixotropic Sodium-Bentonite Suspension During the Sol-Gel-Sol Transition, *Science,* **41:** 1040–1141 (1963).

[44] Yamaguchi, S., On the Sensitivity of Clay, *Disaster Prevent. Res. Inst., Kyoto Univ., Bull.* 28, (1959).

[45] Houwink, R., "Elasticity, Plasticity and Structure of Matter," Cambridge, New York (1937).

[46] Mattson, S., Laws of Soil Colloidal Behavior, VII, Form and Function of Water, *Soil Sci.,* **33:** 301–322 (1932).

persist to a distance of a hundred or more angstroms from the montmorillonite surface, whereas in Ca montmorillonite little or none persists beyond about 15 Å. An explanation of the variation in physical properties due to variations in the exchangeable cations seems to require that the cations influence the thickness of the oriented layers of water molecules as well as the perfection of the arrangement, and perhaps also the actual nature of the configuration of the water molecules. It is a well-known fact that, other things being equal, Na clays require less water to develop plasticity than do Ca clays. This may be explained by the reduced thickness requirement for oriented water when Na⁺ is the dominant adsorbed cation.

In contrast to the relatively small thickness of nonliquid water suggested by the previous authors, Jaeger[47] and later Spiel[48] concluded that the thickness of adsorbed water at optimum plasticity for kaolinite clays is of the order of several hundred angstroms. Opposed to these very large values, DeWit and Arens[20] concluded that the thickness of adsorbed water held by kaolinite is to be measured in a very few molecular layers.

Low and Anderson[29] presented data indicating that the specific volume of water adsorbed by montmorillonite is different from normal water out to distances in excess of 60 Å.

On structural grounds and from the properties of clay-water systems, it would seem unlikely that kaolinite would adsorb water with a definite configuration to a greater thickness than would smectite. The low values for the thickness of the nonliquid water for kaolinite therefore seem most reasonable. It seems likely that the illite, chlorite, and possibly also the vermiculite minerals would yield values similar to kaolinite. No information is available on this characteristic for the palygorskite-attapulgite-sepiolite minerals.

Rate of water adsorption

White[49] and White and Pichler[50] measured the rate of water adsorption of some of the clay minerals with various exchangeable cations. They found (Figs. 8-8 to 8-10) that for air-dried clay there is an initial rapid pickup of water up to, or slightly in excess of, the liquid limit and that there is usually little

[47] Jaeger, F. M., Viscosity of Liquids in Connection with Their Chemical and Physical Constitution, *Verhandel. Koninkl. Ned. Akad. Wetenschap., Afdel. Natuurk.*, Sec. II, **16** (1938).

[48] Spiel, S., Effect of Adsorbed Electrolytes on the Properties of Monodisperse Clay-Water Systems, *J. Am. Ceram. Soc.*, **23**: 33–38 (1940).

[49] White, W. A., Water Sorption Properties of Homoionic Clay Minerals, Ph.D. thesis, University of Illinois (1955).

[50] White, W. A., and E. Pichler, Water Sorption Characteristics of Clay Minerals, *Illinois State Geol. Surv.*, Circ. **266** (1959).

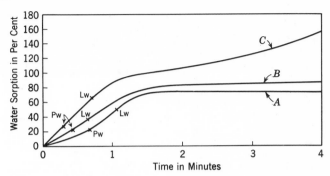

Fig. 8-8. Water-sorption curves for (A) illite, (B) kaolinite with calcium, and (C) sodium. (After White.[49])

further water adsorption except when sodium and lithium are the adsorbed cations.   When sodium and lithium are present there is generally no sharp break in the adsorption rate in the range of the liquid limit; i.e., water adsorption continues far beyond the liquid limit.   The break is particularly pronounced for calcium clays, and since many, perhaps most, natural clays carry this cation, the liquid limit often approximately marks the upper limit of water-holding capacity.

Grim and Cuthbert[51] showed that, when rammed samples of sand, water,

[51] Grim, R. E., and F. L. Cuthbert, The Bonding Action of Clays, II, Clays in Dry Molding Sands, *Illinois State Geol. Surv., Rept. Invest.* 110 (1946).

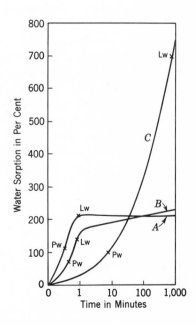

Fig. 8-9. Water-sorption curves for (A) attapulgite, (B) calcium montmorillonite, and (C) sodium montmorillonite.   (After White and Pichler.[50])

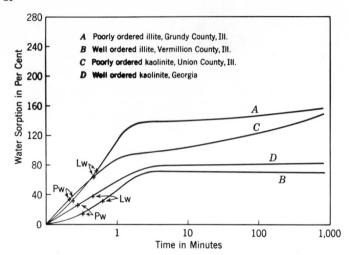

Fig. 8-10.   Water-sorption curves for kaolinites and illites.   (After White and Pichler.[50])

and certain clays are allowed to stand under certain conditions, there is an increase in the compression strength without any loss of water.   This "air-set" strength is explained by them on the basis that a certain amount of time is required for the water to penetrate to some of the surfaces of the clay-mineral particles and for the water molecules to assume their orientation.   According to them, the compressive strength is dependent on the development of nonliquid water.

These authors[7] emphasized the general fact that the properties of clay-water systems may change with time without a change in water content because of a change in the nature of the water.   This is a matter of very great importance in certain fields of utilization of clays.   For example, in engineering practice, it makes it difficult to translate laboratory findings on strength, sensitivity, etc., to actual field application because of the difficulty of duplicating field conditions and of evaluating the time factor.   Engineers are aware of this factor in their constant use of undisturbed samples with carefully retained natural moisture content for their laboratory tests.   Its significance is illustrated by the very different values for such properties as compressive strength and plasticity that are obtained from undisturbed and remolded samples.

Influence of cations and anions

The ions adsorbed on the surfaces of the clay minerals may affect the adsorbed water in several ways:

1. A cation may serve as a bond to hold the clay-mineral particles together or to limit the distance to which they can be separated.

In general, multivalent cations have a greater tendency to tie clay-mineral flakes together than univalent cations. The potassium ion is an exception, because its size and coordination number permit it to fit with the oxygen net of the surface of the three-layer clay minerals. The ammonium ion acts like the potassium ion. In the case of other common monovalent cations, there would probably be little tendency for the ion to tie the particles together, and under some conditions it might enhance the repulsive force between the particles.

2. The size of adsorbed cations and their tendency to hydrate would influence the overall nature of the arrangement of the water molecules and the thickness to which orientation could develop. Sullivan,[52] Forslind,[53] Low,[6] and others[17] have particularly studied this matter. The following statements are mainly from Low. In view of the charge distribution in the water molecule, one would expect charged ions in ionic solutions to attract water molecules electrostatically. In other words, one would expect the ion to hydrate by the formation of ion-dipole bonds. Hydration will occur if the potential energy of the water molecule is less in the hydration shell of the ion than it is in the hydrogen-bonded water structure. Bernal and Fowler[9] made calculations to show that this is the case, especially for small or multiple charged ions. However, the water molecules around the ions can exchange with other water molecules in the medium, the frequency of exchange depending on the intensity of the ion-dipole bonds (Samoilov[54]).

In normal water each molecule is surrounded by four others in tetrahedral fashion. Two of the neighboring water molecules are oriented with their protonic corners toward the central molecule; the other two are oriented with their lone-pair-electron corners toward it and their protonic corner away from it. When an ion is introduced into the water structure, the situation is different. All the water molecules around the cation have their resultant electronic centers directed inward. Around an anion all the water molecules have their protonic corners directed inward. Therefore, even if the ion is of the right size to fit into the space normally occupied by a water molecule the water of hydration cannot "match" or coordinate with the surrounding water. The result is a disruption of the quasi-crystalline water structure. The disruption

[52] Sullivan, J. D., Physical Chemical Control of Properties of Clays, *Trans. Electro-chem. Soc.,* **75:** 71–98 (1939).

[53] Forslind, E., Some Remarks on the Interaction Between the Exchangeable Ions and the Adsorbed Water Layers in Montmorillonite, *Trans. 4th Intern. Congr. Soil Sci.,* **1:** 110–113 (1950).

[54] Samoilov, O. Ya., A New Approach to the Study of Hydration of Ions in Aqueous Solutions, *Discussions Faraday Soc.,* **24:** 141–146 (1957).

will be enhanced if the ion differs in size from the water molecule. In general, the larger the ion, the greater is the disruptive effect. Bernal and Fowler[9] concluded that the presence of ions in solutions either contracts the water structure or breaks it down to cause a closer packing of water molecules. Gurney[55] arrived at similar conclusions. Other investigators (Brady[56] and Wang[57]) presented evidence indicating that some ions reduce the fluidity of water molecules whereas other ions disrupt the quasi-crystalline structure of water to make the water more fluid.

Low[6] presented the following working hypothesis of clay-water interaction. When the clay is exposed to water vapor the exchangeable cations hydrate first if they are small enough to be capable of hydration, and then the remainder of the surface hydrates by the formation of hydrogen bonds between the surface hydroxyls, or oxygens, and the water molecules. The water molecules in the initial layer are not arrayed in perfect order as the competition between adsorbed ions and surface atoms for these molecules is too great. Nevertheless, the initially adsorbed molecules have very low free energy, heat content, and entropy, owing to the intensity with which they are held. As other molecules come within the force fields of those in the first layer, they are captured by them. The first-layer molecules, having their electron distributions affected somewhat by covalent bonding to surface oxygens, form partially covalent bonds with the captured molecules. These in turn are induced to form partially covalent bonds with their neighbors, including third-layer molecules. As additional layers accrue for this type of cooperative action the degree of bond covalency, and hence rigidity, decreases. Therefore, the degree of order decreases gradually with distances from the surface. The adsorbed ions promote disorder. If they are small and monovalent, their contribution in this direction is relatively small, but if they are large and multivalent, their disordering effect is large.

Further, the more they dissociate from the critical surface region where the structure is "anchored," the less they disturb the structure. Those ions that create the least disturbance in the quasi-crystalline water "dissolve" in it most readily. When these ions are present, the water structure may extend with considerable regularity for distances of the order of 75 to 100 Å. An attenuated structure may persist as far as 200 to 300 Å. The water structure may be connected from mineral surface to mineral surface without any intervening region of disorder. However, in the presence of large or multivalent cations, there may be little or no order in the adsorbed water. It should be noted that nowhere is the water structure so rigid that ions cannot

[55] Gurney, R. W., "Ionic Processes in Solution," McGraw-Hill, New York (1953).

[56] Brady, G. W., Structure of Ionic Solutions, IV, *J. Chem. Phys.*, 33: 1079–1082 (1960).

[57] Wang, J. H., Effect of Ions on the Self-diffusion and Structure of Water in Aqueous Electrolytic Solutions, *J. Phys. Chem.*, 58: 686–692 (1954).

diffuse through it, nor is it so rigid that it will not shear under stress. Nevertheless it has a yield point which depends on the proximity of adjacent mineral surfaces.

3. The geometry of the adsorbed ions is probably also of importance, in relation to the possible fit of the adsorbed ions into the water net and the resulting disruption or weakening of the water configuration. It seems that, in the case of some anions, such as the phosphate ion, this matter of geometry is of particular importance. For example, it is well known in the drilling-mud art that an extremely small amount of the phosphate ion may greatly influence some of the thixotropic characteristics of montmorillonite-water suspensions. The probable explanation resides in the similarity between the tetrahedral nature of the phosphate ion and the configuration of the water molecules; this similarity would permit the phosphate tetrahedrons almost to fit into the water structure. They would fit without completely disrupting the water structure but would cause some strain and weakening of the water configuration, which would account for the alteration of the thixotropic characteristics of the system.

Hendricks, Nelson, and Alexander[22] made a very careful study of the water adsorbed by montmorillonite saturated with various cations. They determined the amount of water adsorbed at various relative humidities, the resulting variation in the c-axis spacing of the mineral, and the corresponding differential thermal curves. The differential thermal curves show an endothermic peak due to the energy necessary to drive off the adsorbed waters from the montmorillonite. Some of the results of the work of Hendricks and his colleagues are shown in Figs. 8-11 and 8-12. The endothermic peak due to loss of adsorbed water is single, double, or triple for certain of the cations prepared at certain relative humidities. These workers interpreted the composite nature of the multiple peaks—and their interpretation is in accord with their other data—as follows: In the case of dual peaks of samples prepared at comparatively low relative humidities, the larger of the maxima is attributed to the dehydration of the cation, and the lesser of the maxima to the loss of water from the surface of the clay mineral away from the hydrating cation. At higher relative humidities the dual peaks may become triple, and the third maximum is attributed to the development of an additional layer of oriented water molecules. Thus Mg montmorillonite shows a dual peak after adsorbing water at 5 percent relative humidity and a triple peak after being treated at only 10 percent relative humidity. Ba and Li montmorillonites show no evidence of the development of a second water layer unless the relative humidity is at least 40 percent.

Certain of the univalent ions, for example, Na^+, K^+, and Cs^+, show a single peak after treatment at low relative humidities, and this is interpreted to mean that these ions are not hydrated. As has been mentioned elsewhere,

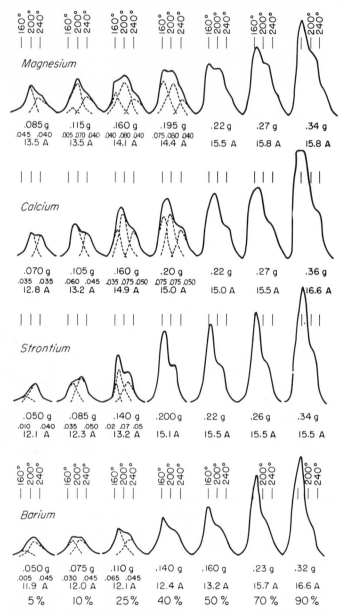

Fig. 8-11. Thermal-analysis curves, amounts of adsorbed water, and values of the apparent cleavage space of magnesium, calcium, strontium, and barium salts of Mississippi montmorillonite. (After Hendricks, Nelson, and Alexander.[22])

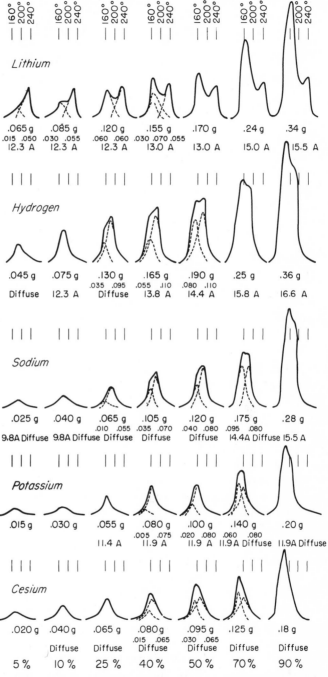

Fig. 8-12. Thermal-analysis curves, amounts of adsorbed water, and values of the apparent cleavage space of lithium, hydrogen, potassium, and cesium salts of Mississippi montmorillonite. (After Hendricks, Nelson, and Alexander.[22])

there is other evidence for the nonhydration characteristics of these cations.

Hendricks, Nelson, and Alexander[22] concluded from their investigation that, for montmorillonite carrying magnesium and alkaline-earth ions, the first step of water sorption is the hydration of the cation with six molecules of water; this is followed by completion of a water layer having a hexagonal type of structure. An additional water layer of similar structure is taken up at higher relative humidities. Similar results were obtained for Li montmorillonite, except that only three molecules of water were required for the hydration of the lithium ion. In montmorillonites carrying sodium, potassium, and cesium, the cation apparently was not hydrated. Mackenzie[17] questioned the interpretation of the single endothermic peak of Na montmorillonite as evidence for the nonhydration of the sodium ion. According to him, all the water molecules in a single molecular water layer of Na montmorillonite could be tied to the sodium ion and, therefore, could be expected to yield a single endothermic peak.

Barshad's[21] concept (see page 243) of a change in the configuration of the water molecules as the amount of water sorbed increases provides another possible explanation for the multiple nature of the differential thermal endothermic peaks resulting from the loss of adsorbed water. Mering[23] presented evidence suggesting that, at low relative humidities (below about 30 percent relative humidity), a Ca montmorillonite does not form a single water layer but instead a skeletal double layer, corresponding to the octahedral coordination of hydration water about the Ca^{++}, and that the thickness of such a water layer is about 4.4 Å. At relative humidities between 30 and 80 percent, two complete water layers develop, and the thickness of the water layer increases to about 5.9 Å. Mering showed that saturation of about 30 percent of the base-exchange capacity with Ca^{++} is adequate to produce the hydration characteristics of a Ca montmorillonite. (See pages 196–197.)

Numerous investigations have brought out the interesting fact that Na montmorillonite, on drying at room temperature, tends to develop a single water layer between the silicate layers and that Ca montmorillonite under the same conditions tends to develop two water layers, whereas, at high relative humidities and in the presence of an abundance of water, Na montmorillonite sorbs by far the larger quantity of water. Further, in the case of Ca montmorillonite, unlike Na montmorillonite, the boundary between oriented and nonoriented water molecules is abrupt (Grim[58]).

Barshad[59] studied the sorption of water by vermiculite. According to

[58] Grim, R. E., "Applied Clay Mineralogy," McGraw-Hill, New York (1962).

[59] Barshad, I., Vermiculite and Its Relation to Biotite as Revealed by Base-Exchange Reactions, X-ray Analyses, Differential Thermal Curves, and Water Content, *Am. Mineralogist*, **33**: 655–678 (1948).

him, vermiculite saturated with Mg^{++}, Ca^{++}, Ba^{++}, H^+, Li^+, or Na^+ shows, when immersed in water, total sorption between the silicate layer equal to only two molecular water layers. When vermiculite saturated with K^+, NH_4^+, Rb^+, or Cs^+ is immersed in water, there is no expansion of the lattice, indicating no water sorption between the silicate layers. Walker[60] suggested a detailed hydration sequence for Mg vermiculite.

Siefert[61] presented the results, shown in Table 8-1, for the sorption of water by kaolinite. These data show the relative influence of the various cations on water sorption by this clay mineral.

Table 8-1 *Influence of cations on water sorption*
of kaolinite
(After Siefert[61])

Relative humidity, %	Order of decreasing sorption
10	Ca > H = Na = K
81.5	H = Ca > Na ≫ K
99.9	H = Ca = Na ≫ K

Mackenzie[17] suggested that a given ion like Na^+ might hydrate to a different degree on different clay minerals and even on the same type of clay mineral if the bonding force holding the ion varied. Thus, in montmorillonite the degree of hydration of Na^+ might vary depending on whether the charge holding it derived from a substitution within the tetrahedral or the octahedral sheet.

Stepwise hydration of smectite

Careful study by Bradley, Clark, and Grim[62] of a hydrogen montmorillonite produced by electrodialysis of a Wyoming bentonite indicated that the swelling of the lattice during hydration took place in a stepwise fashion. A series of four apparently definite and discrete hydrates were found, having cell heights of 12.4, 15.4, 18.4 and 21.4 Å. Near the range where any given hydrate is stable, successive orders of $(00l)$ reflections appear for it alone. In intermediate ranges two suites of $(00l)$ reflections appear simultaneously, one

[60] Walker, G. F., Vermiculite Minerals, "The X-ray Identification and Crystal Structures of Clay Minerals," chap. VII, pp. 297–324, Mineralogical Society of Great Britain Monograph (1961).

[61] Siefert, A. C., Studies on the Hydration of Clays, Ph.D. thesis, Pennsylvania State College (1942).

[62] Bradley, W. F., G. F. Clark, and R. E. Grim, A Study of the Behavior of Montmorillonite on Wetting, *Z. Krist.*, **97:** 216–222 (1937).

to be identified with the hydrate next higher in the sequence and one with the hydrate next lower.

In their study, Bradley et al.[62] used oriented aggregates in which the basal cleavages were substantially parallel. Hofmann,[63] working with powders, failed to check the conclusions of Bradley et al., but later Hofmann and Hausdorf,[16] using oriented aggregates, also concluded that the hydration of montmorillonite took place by the formation of successive monomolecular layers of water. The latter investigators, and later Hendricks and Jefferson,[12] showed theoretically that powder diffraction should indicate that the c-axis-dimension spacing varies continuously but not uniformly with water content, and that this apparently would be a result of an averaging effect from a lattice that contains various numbers of water layers in different parts. That is, the apparently continuous change in the c dimension results from a random alternation of successive discrete hydrates.

Table 8-2 *Interlayer spacings, d_{001} Å, formed during water uptake by montmorillonite*
(*From Norrish*[65])

				Interlayer cation					
H$_3$O$^+$	Li$^+$	Na$^+$	K$^+$	NH$_4^+$	Cs$^+$	Mg^{2+}	Ca^{2+}	Sr^{2+}	Ba^{2+}
~10.0	9.5*	9.5*	10.0	10	~12	9.5*	9.5*	9.5*	9.8*
12.4	12.4†	12.4† ‡	12.4	?				~12.0†	~12.0‡
15.4	15.4†	15.4	15.0	15		15.4‡	15.4	~15.5†	~15.5‡
19.0	19.0	19.0				19.2	18.9	?	18.9
22.4	22.5								

* Greene-Kelly, R., Studies of the Sorption of Polar molecules by Layer Lattice Silicates, Ph.D. Thesis, University of London (1953).

† Hendricks, Nelson, and Alexander.[22]

‡ Mooney, Keenan, and Wood.[69]

Mering[23] and others have shown that the development of discrete monomolecular water layers probably does not hold precisely at low moisture contents when the clay carries an adsorbed ion that hydrates, for example, Ca^{++} or Mg^{++}. In these cases, the hydration net around the cation prevents the orderly development of the initial hydration layers.

Norrish,[64] using low-angle X-ray-diffraction methods, made a very careful study of the interlayer spacings of montmorillonite with various adsorbed cations during water uptake. In general, he found that there was a stepwise hydration up to a c-axis dimension of approximately 19 Å, beyond which the regularity decreased. Norrish[65] tabulated his own and pertinent data from the literature (Table 8-2) on these successive interlayer steps and the

[63] Hofmann, U., Neues aus der Chemie des Tons, *Chemie*, **55**: 283–294 (1942).

[64] Norrish, K., Crystalline Swelling of Montmorillonite—Manner of Swelling of Montmorillonite, *Nature*, **173**: 256–257 (1954).

maximum basal spacing for what he calls crystalline swelling. It is interesting that Cs montmorillonite shows no regularity of spacing above about 12 Å; that is, the spacing is independent of the water content, which is probably due to the size of the ion which in turn keeps the silicate sheets apart so that water can enter readily the interlayer space. The initial swelling is believed to be crystalline and dependent on hydration energy of the interlayer cation.

Norrish[65] studied the swelling of montmorillonites in electrolytes and concluded that 21 Å may be considered the effective thickness of the silicate sheet for infinite electrolyte concentration as this was about the maximum spacing for which the diffraction effects were relatively sharp. Norrish believed that this suggests that, even for large separations, each clay surface retains two layers of water molecules which are effectively solid.

Stability of smectite hydration

A considerable body of evidence, largely unpublished, shows that smectite clays, when subjected for long periods of time to substantially uniform moisture conditions, develop hydration characteristics of considerable stability. Such materials resist change in the degree of hydration, but when the hydration is changed even to a very slight degree, the stability may be abruptly and completely lost. Thus a high-swelling bentonite sample carefully collected to preserve its natural hydration can be placed in water without any slaking, even when moderately stirred. However, if such a bentonite is dried only a very small amount, it will slake and swell immediately when placed in water. Apparently, if hydration layers develop with a high degree of uniformity of thickness and distribution, they have considerable stability, which is not present if there is even a very minor variation in thickness from layer to layer.

This matter of stability of hydration is of great practical importance. In soil mechanics, for example, tests run on samples with the natural moisture carefully retained (as is the usual procedure) are likely to give results completely misleading if applied to the materials as used in the field, where some loss of water may take place.

A further point in regard to the stability of smectite hydration is worthy of note. When the dried mineral is placed in water, it may rapidly adsorb water up to a certain point and then become a thick pasty mass. No more water is sorbed unless the mass is stirred vigorously. Adsorption tends to stop at a given point, and an equilibrium is developed at a very high moisture

[65] Norrish, K., The Swelling of Montmorillonite, *Discussions Faraday Soc.,* **18:** 120–134 (1954).

content. This characteristic is shown particularly by the Na montmorillonite in the bentonite from the Wyoming area. It is of considerable commercial utility in that it permits the mineral to be used for water impedance.

Influence of adsorbed organic molecules

Gieseking[66] reported that smectite clays lost their tendency to swell by water sorption when saturated with a variety of organic cations. These cations are adsorbed on the basal plane surfaces of the smectite (see Chap. 10). Hendricks[67] showed that the amount of water adsorbed by smectite carrying certain adsorbed amine ions approaches closely the quantity expected from the difference between the total extent of the surface and the probable part of the surface covered by the amine ions. Hendricks also pointed out that the reduction in water sorption should not be exactly correlative with the size of the organic ion, since the shape of the organic ion may be such as to destroy the configuration of water molecules in the adsorbed water layer.

Grim, Allaway, and Cuthbert[68] showed that water adsorption of kaolinite clays is reduced after their treatment with organic ions and that the decrease in water adsorption is relatively less for kaolinite than for smectite. It appears certain that similar results would be obtained for the other clay minerals and that the values would be intermediate between those for smectite and kaolinite.

Sorption and surface-area data

Numerous investigators, Mooney et al.,[69] Barrer and McLeod,[70] Barshad,[71] Orchiston,[72,73] Slabaugh,[74] Martin,[75] Johansen and Dunning,[76] Brooks,[77]

[66] Gieseking, J. E., Mechanism of Cation Exchange in the Montmorillonite-Beidellite-Nontronite Type of Clay Mineral, *Soil Sci.*, **47**: 1–14 (1939).

[67] Hendricks, S. B., Base Exchange of the Clay Mineral Montmorillonite for Organic Cations and Its Dependence upon Adsorption Due to van der Waals Forces, *J. Phys. Chem.*, **45**: 65–81 (1941).

[68] Grim, R. E., W. H. Allaway, and F. L. Cuthbert, Reaction of Different Clay Minerals with Some Organic Cations, *J. Am. Ceram. Soc.*, **30**: 137–142 (1947).

[69] Mooney, R. W., A. G. Keenan, and L. A. Wood, Adsorption of Water Vapor by Montmorillonite, I, Heat of Desorption and Application of BET Theory, *J. Am. Chem. Soc.*, **74**: 1367–1371 (1952).

[70] Barrer, R. M., and D. M. McLeod, Intercalation and Sorption by Montmorillonite, *Trans. Faraday Soc.*, **50**: 980–989 (1954).

[71] Barshad, I., Adsorptive and Swelling Properties of Clay-Water Systems, *Calif., Dept. Nat. Resources, Div. Mines, Bull.* 169, pp. 70–77 (1955).

Jurinak and Volman,[78] and many others have presented sorption curves and surface-area data for the clay minerals carrying various ions under varying conditions and prepared in various ways. These data have formed the basis of thermodynamic investigations of free energy, etc., for the reaction between the clay minerals and water and other liquids. For details of this work the reports of these workers should be consulted. The following statements are merely illustrative of the character of the data available and the conclusions that have been presented.

Table 8-3   Surface area of clay samples determined by nitrogen and water-vapor sorption
(After Johansen and Dunning[76])

Sample	Geographical location	Nitrogen	Water vapor	Ratio W/N
Bentonite I (montmorillonite)	Clay Spur, Wyo.	30 ± 1.0	164*	5.5
			206†	6.9
Bentonite II (montmorillonite)	Wyoming	38 ± 3.0	138*	3.6
			195†	5.1
Bentonite II (montmorillonite) Na treated	Wyoming	38 ± 1.0	203*	5.3
			250†	6.6
Illite	Fithian, Ill.	56 ± 1.0	52*	0.9
			82†	1.5
Kaolinite	Bath, S.C.	16 ± 1.0	12*	0.8
			12†	0.8

* Adsorption.
† Desorption.

Table 8-3, after Johansen and Dunning,[76] gives surface-area data for samples of montmorillonite, illite, and kaolinite determined by nitrogen ad-

[72] Orchiston, H. D., Adsorption of Water Vapor, III, Homoionic Montmorillonites at 25°C, *Soil Sci.*, **79:** 71 (1955).

[73] Orchiston, H. D., Adsorption of Water Vapor, VII, Homoionic Montmorillonites at 25°C, *Soil Sci.*, **87:** 276–282 (1959).

[74] Slabaugh, W. H., Adsorption Characteristics of Homoionic Bentonites, *J. Phys. Chem.*, **63:** 436–438 (1959).

[75] Martin, R. T., Water Vapor Sorption on Kaolinite: Entropy of Adsorption, Proceedings of the 8th National Clay Conference, pp. 102–114, Pergamon Press, New York (1960).

[76] Johansen, R. T., and H. N. Dunning, Water Vapor Adsorption on Clays, Proceedings of the 6th National Clay Conference, pp. 249–258, Pergamon Press, New York (1959).

[77] Brooks, C. S., Free Energies of Immersion for Clay Minerals in Water, Ethanol, and *n*-Heptane, *J. Phys. Chem.*, **64**(5): 532 (1960).

[78] Jurinak, J. J., and D. H. Volman, Cation Hydration Effects on the Thermodynamics of Water Adsorption by Kaolinite, *J. Phys. Chem.*, **65:** 1853 (1961).

sorption and water-vapor adsorption. As expected, the surface-area values
for montmorillonites by water-vapor adsorption are much higher than those
obtained by nitrogen adsorption, owing to the penetration of water between
the basal planes of the montmorillonite units. It is of interest that the values
obtained by desorption of water vapor are substantially higher than those
obtained by adsorption for both montmorillonite and illite. These authors
pointed out that sorption isotherms of montmorillonite show a large hysteresis
effect which seems to be characteristic of this clay mineral (Fig. 8-13). The

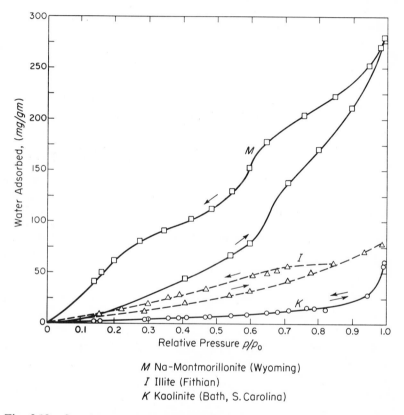

M Na-Montmorillonite (Wyoming)
I Illite (Fithian)
K Kaolinite (Bath, S. Carolina)

Fig. 8-13. Sorption curves of montmorillonite, illite, and kaolinite. (After Johansen and Dunning.[76])

isotherm for kaolinite, on the other hand, is readily reproducible and reversible. The isotherm for illite is intermediate between that for montmorillonite
and kaolinite. The steep section in the middle of the isotherm is characteristic of montmorillonite and is believed to be associated with the degree
of hydration of the clay platelets. The desorption isotherm shows an even
more pronounced change at about the same relative pressure (0.5 to 0.6).

Johansen and Dunning[76] concurred with the observations of Mooney et al.[69] that the desorption isotherms were reproducible while the adsorption isotherms were reproducible only after the first adsorption and then with difficulty. They interpreted this to indicate an apparent change in the clay structure which is not returned to the original condition upon desorption. It can easily be visualized that, with successive swelling and contraction of the montmorillonite layers in adsorption and desorption, the montmorillonite units would not in every case return to the same relative position.

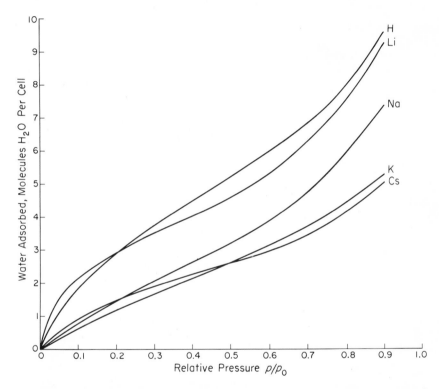

Fig. 8-14.   Adsorption isotherms at 30°C of a Mississippi montmorillonite saturated with various monovalent cations.   (After Barshad.[79])

Barrer and McLeod[70] found hysteresis loops for all substances adsorbed on montmorillonite. The loops closed at fairly high relative pressures (0.5) with nonpolar liquids and extended to very low relative pressures with polar molecules. The very unusual hysteresis with polar molecules was attributed to a barrier caused by the surface energy required for penetration of the unswollen lattice by the adsorbate or for the nucleation of the adsorbate-rich phase. According to this theory, metastable phases exist until a threshold value of relative pressure is reached.

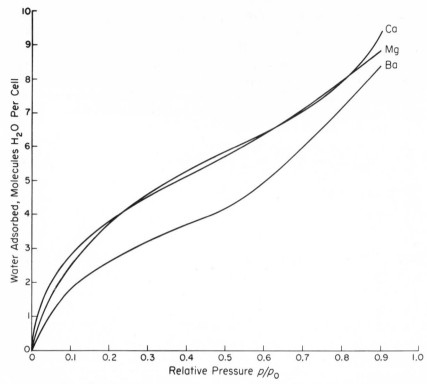

Fig. 8-15. Adsorption isotherms at 30°C of a Mississippi montmorillonite saturated with various divalent cations. (After Barshad.[79])

Barshad[79] showed that there was a difference in the shape of adsorption isotherms for montmorillonite, depending on the nature of the adsorbed cation (Figs. 8-14 and 8-15). He also showed that there was a variation depending on the cation-exchange capacity of the montmorillonite, with the sample having the higher capacity showing the larger amount of adsorbed water. Barshad considered that the first step in the hydration of montmorillonite was on the exterior surface and that only after adsorption of 50 to 75 mg of water per gram of clay did the water penetrate to interlayer positions. Barshad also studied the adsorption of water vapor by kaolinite and montmorillonite at various relative humidities. He concluded that, for montmorillonite samples saturated with cations of equal charge but varying in size, the larger the ionic radius, the higher the relative humidity at which expansion occurs. However, the degree of hydration at the beginning of the expansion is about equal for all sizes of ions. In montmorillonite saturated with potas-

[79] Barshad, I., Adsorptive and Swelling Properties of Clay Water Systems, *Calif., Dept. Nat. Resources, Div. Mines, Bull.* **169**, pp. 70–77 (1955).

sium, rubidium, and cesium ions, expansion beyond a unimolecular water layer does not take place. In samples saturated with cations of equal radius but varying charge, the larger the charge, the lower is the relative humidity at which expansion occurs but, again, the degree of hydration at the beginning of expansion is about equal for all the ions. In samples saturated with the same cation but varying in amount, the larger the number of cations, the lower is the relative humidity at which expansion occurs. But here too, when expansion is taking place, the degree of hydration is about equal for the different cations.

Barshad found that the degree of hydration of kaolinite at any given vapor pressure is much less than that of montmorillonite when the degree of hydration is expressed on a unit-weight basis, but it is considerably higher when expressed on a unit-area basis, particularly in the range where montmorillonite is in an expanded state. He attributed the greater reactivity of the kaolinite surfaces of the water to two causes: (1) All the adsorbing surface of kaolinite is on the exterior of the particle and as a result the water molecules in the water vapor impinge directly on it, whereas in montmorillonite the larger portion of the absorbing surface is an interior one, with the result that the molecules in the water vapor must first be adsorbed on the edges of the particle and then migrate to the interior, a process which tends to retard adsorption; (2) it may be attributed in part to the larger cation charge per unit area if it be true that the ability of a surface to adsorb water is proportional to the cation-exchange capacity per unit area.

In the case of kaolinite, Martin[75] found that when desorption is started from P/P_0 near saturation all ionic modifications show reversible hysteresis. The size and P/P_0 range in the hysteresis loop vary with the nature of the exchangeable ion and the sample history. No hysteresis occurs on lithium and potassium kaolinite when desorption is started at P/P_0 less than 0.8, which along with other data indicates that lithium and potassium exchangeable ions have not truly hydrated below this value. The hysteresis for all ionic modifications is traced to hydration of the exchangeable ions, and the apparent anomalous position of lithium and potassium is attributed to the fact that control of exchangeable ions is a matter of the hydration energy of the ion minus the specific adsorption energy of the ion for the clay. The difference must be positive before true ionic hydration occurs. Martin presented a molecular model for sorption in which it is postulated that sorption takes place at specific sorption sites on the surface. The energy of different sorption sites is believed to be different and also to be a function of the hydration state of the exchangeable ion.

Slabaugh[74] presented excellent adsorption curves for montmorillonite with various exchangeable cations and water vapor at 14, 27, and 37.5°C. The data show that the Na, K, and Mg clays adsorbed water vapor in distinct steps and that the calcium montmorillonite was unique in that the adsorption isotherm exhibited a discontinuity at 14 and 27° with no discontinuity at 37.5°C.

Heat of wetting

A dry clay material evolves heat when placed in water, and this phenomenon is known as the heat of wetting. Heat is also evolved when clay materials are placed in some liquids other than water, as, for example, alcohols and various organic liquids. Janert[80] showed that heat of wetting may be higher for organic liquids and that values generally are higher for polar than for nonpolar liquids. Janert also showed that there is no direct correlation between polarity and heat of wetting, so that polarity is not the sole determining factor. Figure 8-16, giving data for the heat of wetting for water and for carbon tetrachloride as developed by a "brick clay," illustrates the kind of difference in values for water and some other liquids. The heat of wetting decreases as the moisture content of the clay increases at the time of making the determination (Fig. 8-16). Miller et al.[81] presented heat-of-wetting values for attapulgite, obtained with water and several organic liquids (Table 8-4). The organic liquids used by them also gave lower values than water,

[80] Janert, H., The Application of Heat of Wetting Measurements to Soil Research Problems, *J. Agr. Sci.,* **24**: 136–150 (1934).

[81] Miller, J. G., H. Heinemann, and W. S. W. McCarter, Heat of Wetting of Activated Bauxite and Attapulgus Clay, *Ind. Eng. Chem.,* **42**: 151–153 (1950).

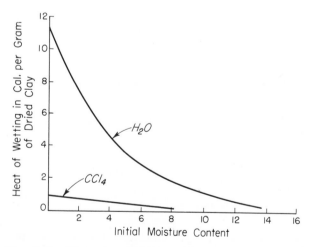

Fig. 8-16. Heat of wetting for water and carbon tetrachloride in relation to initial moisture content. (After data from Janert.[80])

and for a series of normal alcohols the values decreased markedly as the chain length of the alcohol increased. On the other hand, Slabaugh[82] found that immersion of montmorillonite in a high-molecular-weight amine salt evolved a large quantity of heat. Large variations in heat of wetting are to be expected, depending on the character of the organic liquid.

Heat of wetting is measured in calories per gram of dried clay, with the drying being done usually at 110°C.

Table 8-4 *Heat-of-wetting values for attapulgite, in calories per gram*
(After Miller, Heinemann, and McCarter[81])

Liquid	Dried raw clay	Clay dried at			
		400°C	450°C	550°C	900°C
Water	10.6	17.7	22.0	25.6	10.9
Benzene	4.2	....	4.2	4.3	2.4
Ethyl alcohol	....	11.6	....	17.1	

Table 8-5 *Heat-of-wetting values for kaolinite, montmorillonite, and illite, in calories per gram*
(After Siefert,[61] Endell, Loos, et al.,[83] and Parmelee and Frechette[84])

Reference	Kaolinite					Montmorillonite					Illite
	Natural	Ca++	H+	Na+	K+	Natural	Ca++	H+	Na+	K+	Natural
Siefert[61]	...	1.45	1.40	1.30	1.22	....	22.1	20.1	12.1	9.1	
Endell, Loos, et al.[83]	2.2	....	....	....	....	....	22	....	16.1		
Parmelee and Frechette[84]	1.9	....	....	....	....	11.8	....	....	....	...	4.0

Values for heat of wetting. Heat-of-wetting values for kaolinite and montmorillonite saturated with various cations and for an illite sample are given in Table 8-5. Values for attapulgite, given in Table 8-4, are only slightly lower than those of montmorillonite.

[82] Slabaugh, W. H., Heats of Immersion of Some Clay Systems in Aqueous Media, *J. Phys. Chem.*, **59**: 1022–1024 (1955).

[53] Endell, K., W. Loos, H. Meischeider, and V. Berg, Ueber Zusammenhange zwischen Wasseraushalt der Tonminerale und Boden physikalischen Eigenschalten, *Inst. Deut. Forsh. Bodenmechanik*, **5**: (1938).

[84] Parmelee, C. W., and D. Frechette, Heat of Wetting Values of Fired and Unfired Clays, *J. Am. Ceram. Soc.*, **25**: 108–112 (1942).

Values for the other clay minerals are not yet available, but they will undoubtedly range between those for montmorillonite and kaolinite. Chlorite will probably yield values closer to those for illite, and vermiculite closer to those for montmorillonite.

Pate,[85] Anderson,[86] Janert,[87] and others[61] have shown that the heat of wetting varies with the adsorbed cation and that it is generally higher for divalent than for univalent cations. Janert gave the order $Ca > Mg > H > Na > K$. According to Rosenqvist[88] and Slabaugh,[74] generally speaking, heat of wetting increases with decreasing ion size.

Harmon and Fraulini[89] presented data showing an increase in heat of wetting for kaolinite as the particle size of the kaolinite decreases (Table 8-6). This is an expected relationship, since heat of wetting should increase

Table 8-6 *Heat of wetting of kaolinite in relation to particle size and cation-exchange capacity*

(After Harmon and Fraulini[89])

Particle size, μ	10–20	.5–10	.2–4	.1–0.5	0.5–0.25	0.25–0.10	0.10–0.05
Heat of wetting, cal/g	0.95	0.99	1.15	1.38	1.42	1.87	
Exchange capacity, meq/100 g	2.40	2.60	3.58	3.76	3.88	5.43	9.50

as the surface area increases with decreasing particle size. It should apply equally well to the other nonexpanding clay minerals. In the case of expanding-lattice minerals, such as smectite and vermiculite, there should be no strict correlation with particle size, since the total surface is theoretically available to water sorption regardless of the size of the particles. However, in the case of large units, as in vermiculite, there would probably be some variation in the ease of wetting of the particle surfaces, so that some variation of heat of wetting with particle size probably would be found. Also, the surfaces of smectites carrying certain adsorbed ions, for example, K^+, are not completely available for water adsorption, and some correlation with particle size would be expected for such smectites.

[85] Pate, W. N., The Influences of the Amount and Nature of the Replaceable Bases upon the Heat of Wetting of Soils and Soil Colloids, *Soil Sci.*, **20**: 329–375 (1925).

[86] Anderson, M. S., The Influence of Substituted Cations on the Properties of Soil Colloids, *J. Agr. Res.*, **38**: 565–584 (1929).

[87] Janert, H., Cation Exchange and Water Absorption in soils, *Z. Pflanzenernähr. Düng. Bodenk. A*, **34**: 100–108 (1934).

[88] Rosenqvist, I. Th., Investigations in the Clay-Electrolyte-Water System, *Norweg. Geotech. Inst., Publ.* 9 (1955).

[89] Harmon, C. G., and F. Fraulini, Properties of Kaolinite as a Function of Its Particle Size, *J. Am. Ceram. Soc.*, **23**: 252–258 (1940).

Anderson and Mattson,[90] Baver,[91] and others have pointed out a relation between cation-exchange capacity and heat of wetting for a series of natural soils. Smectites with a very high cation-exchange capacity yield high values for heat of wetting; kaolinites with low cation-exchange capacity yield low heat-of-wetting values; and illites give intermediate values. Soils composed of one or a mixture of these common clay minerals would show some correlation between cation-exchange capacity and heat of wetting. It is doubtful if a close correlation would exist between these two characteristics in a series of natural soils—certainly it would not be a straight-line relationship—since other factors, such as particle size and nature of adsorbed cation, would influence the heat evolved on wetting.

Cause of heat of wetting. Behrends[92] narrated early ideas offered to explain heat of wetting, such as the compression of water at the adsorbing surface, capillary condensation, etc. It is now generally agreed that the phenomenon is due to two factors: (1) a change in state of the water directly adjacent to the adsorbing surface and (2) the possible hydration of adsorbed ions. Baver and Winterkorn[4] emphasized the importance of the development of orientation of the water molecules in the adsorbed water as a cause of heat of wetting.

Janert[80] showed that the heat of wetting, calculated per equivalent, is only a fraction of the values usually assigned to the total heat of hydration of various cations in dilute solutions. This is explained on the basis that the ion is not completely free and is, therefore, capable of only partial hydration. According to Janert, the ratios of the heat of wetting to the heat of hydration for various ions are as follows: for H clays, 1:11.5; for Mg clays, 1:9; for Ca clay, 1:7; for K clays, 1:5.1; and for Na clays, 1:4.9. Janert was of the opinion that ion hydration is a major cause of the heat of wetting.

Siefert[61] computed the heat evolved per milliequivalent of cation and found that a given cation adsorbed by kaolinite evolves more heat than when adsorbed by montmorillonite. Since, in montmorillonite, there is little surface aside from that occupied by the cation, Siefert concluded that the heat of wetting due to the surface is greater than that due to cation hydration.

It seems likely that the relative importance of surface and cation hydration would vary for different clay minerals and for various adsorbed cations.

[90] Anderson, M. S., and S. Mattson, Properties of Soil Colloidal Material, *U.S. Dept. Agr. Bull.* 1452 (1926).

[91] Baver, L. D., The Effect of the Amount and Nature of Exchangeable Cations on the Structure of a Colloidal Clay, *Missouri, Univ., Agr. Expt. Sta., Res. Bull.* 129 (1929).

[92] Behrends, W. U., The Relation between the Surface, Hygroscopicity, and Heat of Wetting of Soils, *Z. Pflanzenernähr. Düng. Bodenk.*, **40:** 255–309 (1935).

It is probably incorrect to assume that either factor is always the more important cause.

Heat of wetting in solutions of electrolytes. Siefert[61] showed that when hydrogen kaolinite is wetted in 0.1 N NaOH or 0.015 N Ca(OH)$_2$ solution there is an increase in heat evolved of 0.2 cal/g over that of pure water; this is probably due in part to the neutralization of the clay acid by the base. However, Na kaolinite wetted in NaOH and K kaolinite wetted in KOH showed a similar increase in heat evolved in comparison to that developed in water. Somewhat similar results were obtained by Siefert for montmorillonite clays. The explanation for this phenomenon is not clear, but it may be merely the result of the better dispersion of the clay in the alkaline solution.

Effect of preheating. Slabaugh[74] presented excellent data on the heat of immersion of montmorillonites which had been heated to various temperatures up to 300°C prior to immersion. Calcium and hydrogen montmorillonites undergo rapid increases in heat of wetting as the result of preheating to temperatures up to 100 to 150°C. Higher temperatures reduce the heats of immersion of these two clays. The sodium montmorillonite and the raw clay (essentially a sodium montmorillonite) are affected only slightly by preheating. The calcium clays appeared to show a transition temperature at approximately 29°C while the hydrogen clay showed only a gradual increase, with no distinguishable transition temperatures. Adsorption isotherms are in agreement (Slabaugh[74]), suggesting that below this temperaure a two-step adsorption process occurs, whereas above this temperature a single adsorption isotherm is produced. In most of the observations, particularly with sodium clays and with calcium clays at lower pretreatment temperatures, oven-dried clays gave off less heat upon immersion than freeze-dried clays. This is attributed to the higher degree of dispersion of the freeze-dried clays. Slabaugh considered that the great difference between calcium and sodium montmorillonite clays may result from the probably complete dehydration of the sodium ion at room temperature, whereas the hydrates of calcium, on the other hand, decompose at higher temperatures, and only at about 150°C can it be presumed that the exchangeable calcium ion is completely dehydrated.

The data given in Fig. 8-17, after Parmelee and Frechette,[84] show the relation of heat of wetting to firing temperature for montmorillonite, kaolinite, and illite. In the case of the montmorillonite, there is a sharp drop in heat of wetting after firing at the temperature where hydroxyl water is lost from the lattice. The illite shows a slight increase in heat of wetting on firing up to about 450°C and a reduction in heat of wetting after temperatures somewhat above that required for the loss of hydroxyl water. The kaolinite shows a slight reduction in the amount of heat evolved when heated to temperatures somewhat in excess of that required for the loss of hydroxyl water.

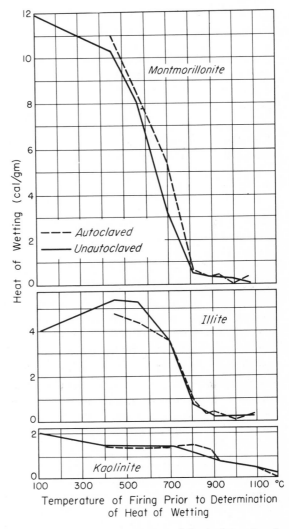

Fig. 8-17. Heat of wetting of fired clay minerals. (From Parmelee and Frechette.[84])

Data given by Miller et al.[81] (Table 8-4) show the heat of wetting of attapulgite to increase greatly on firing up to about 550°C and then decrease, so that after firing at 900°C the same value as that for the unfired mineral is obtained. The reason for the changes in heat of wetting of the various clay minerals when heated to these particular temperatures is not clear. Additional experimental data are needed before heat-of-wetting values for fired clays can be understood.

Additional references

Alexander, L. T., and T. M. Shaw, Determination of Ice-Water Relationships by Measurement of Dielectric Constant Changes, *J. Phys. Chem.,* **41**: 955–960 (1937).

Bangham, D. H., and S. Mosallam, The Adsorption of Vapours at the Plane Surfaces of Mica, II, Heats of Adsorption and the Structure of Multimolecular Films, *Proc. Roy. Soc. (London), A,* **156**: 558–571 (1938).

Baver, L. D., "Soil Physics," Wiley, New York (1940).

Bouyoucos, G. J., State in Which the Hygroscopic Moisture Exists in Soils as Indicated by Its Determination with Alcohol, *Soil Sci.,* **41**: 443–447 (1936).

Brasseur, H., Water of Hydration in Crystals, *Assoc. Franc. Avance. Sci. Liége,* pp. 339–342 (1939).

Buergers, J. M., Introductory Remarks on Recent Investigations Concerning the Structure of Liquids, "Second Report on Viscosity and Plasticity," North Holland Publishing Company, Amsterdam (1938).

Chessick, J. J., and A. C. Zettlemoyer, Surface Chemistry of Silicate Minerals, IV, Adsorption and Heat-of-wetting Measurements of Attapulgite, *J. Phys. Chem.,* **60**: 1181–1184 (1956).

Denisov, P. I., Nonfreezing Clay Solutions, *Azerb. Neft. Khoz.,* pp. 31–32 (1941); *Khim. Referat. Zhur.,* **4**: 110 (1941).

East, W. H., Fundamental Study of Clays, X, Water Films in Monodisperse Kaolinite Fractions, *J. Am. Ceram. Soc.,* **33**: 211–218 (1950).

Endell, K., and P. Vageler, Der Kationen- und Wasseraushalt keramischer Tone im rohen Zustand, *Ber. Deut. Keram. Ges.,* **13**: 377–411 (1932).

Freundlich, H., U. Schmidt, and G. Lindau, Ueber die Thixotropie von Bentonit-Suspensions, *Kolloid-Beihefte,* **36**: 43–81 (1932).

Goates, J. R., and S. J. Bennett, Thermodynamic Properties of Water Adsorbed on Soil Minerals, II, Kaolinite, *Soil Sci.,* **83**: 325–330 (1957).

Graham, J., G. F. Walker, and G. W. West, Nuclear Magnetic Resonance Study of Interlayer Water in Hydrated Layer Silicates, *J. Chem. Phys.,* **40**(2): 540 (1964).

Hofmann, U., Ueber die Grundlagen der Plastizitat der Kaoline und Tone, *Ber. Deut. Keram. Ges.,* **29**: 21–32 (1949).

Kapp, L. C., The Approximate Size of Soil Particles at Which the Heat of Wetting Is Manifest, *Soil Sci.,* **29**: 401–412 (1930).

Kiefe, C., Sur les possibilités de liaisons et d'orientation des molécules d'un liquide sur un solide. Epitaxie entre la kaolinite et l'eau, Paris (1947). Mimeographed.

Kul'chitskii, L. I., Study of Surface Crystallochemistry of Highly Dispersed Alumino-Silicates, *Tr. Vses. Nauchn.-Issled. Inst. Gidrogeol. i Inzh. Geol. N.S.,* no. 9, pp. 253–266 (1964).

Longuet-Escard, J., The Effect of Progressive Dehydration on the Area of the Surface of Montmorillonites, *J. Chim. Phys.,* **47**: 113–117 (1950).

Makower, B., T. Shaw, and L. T. Alexander, The Specific Surface and Density of Some Soils and Their Colloids, *Soil Sci. Soc. Am. Proc.,* **2**: 101–108 (1937).

Norton, F. H., and A. L. Johnson, Fundamental Study of Clays, V, Nature of Water Films in Plastic Clays, *J. Am. Ceram. Soc.,* **27**: 77–80 (1944).

Posner, A. M., and J. P. Quirk, Adsorption of Water from Concentrated Electrolyte Solutions by Montmorillonite and Illite, *Proc. Royal Soc. (London),* **278**(1372): 35–56 (1964).

Puri, A. N., and R. C. Hoon, Physical Characteristics of Soils, III, Heat of Wetting, *Soil Sci.,* **47**: 415–423 (1939).

Rios, E. G., and J. L. Vivaldi, Silicates of Laminar Structure, I, Hydration, *Anales Fis. y Quim.* (*Madrid*), *B*, **45**: 291–342 (1949).

Sonders, L. R., D. P. Enright, and W. A. Weyl, Wettability, a Function of the Polarizability of the Surface Ions, *Penn. State Univ., Mineral Ind. Expt. Sta., Tech. Rept.* 12, N. R. 032-265 (1949).

Van Olphen, H., Unit-Layer Interaction in Hydrous Montmorillonite Systems, *J. Colloid Sci.*, **17**: 660–667 (1962).

Wu, T. H., Nuclear Magnetic Resonance Study of Water in Clay, *J. Geophys. Res.*, **69**(6): 1083–1091 (1964).

Nine

Dehydration, Rehydration, and the Changes Taking Place on Heating

Dehydration involves the loss of any water, adsorbed, interlayer, or lattice OH water, held by the clay minerals. A study of dehydration is concerned with the amount of water lost, rate of water loss, temperature of dehydration, and energy involved. Dehydration frequently involves significant changes in the structure of the clay minerals, and consequently changes taking place during the heating of the clay minerals cannot be considered apart from dehydration. Changes taking place in the clay minerals on heating to relatively high temperatures are not necessarily related to dehydration reactions, but for the sake of uniformity and continuity, such changes will also be considered in this chapter.

Certain of the clay minerals, after being heated to moderately high temperatures, take up water again on cooling. Such rehydration characteristics will also be considered here.

Methods of study

Various methods are available for studying the hydration properties of the clay minerals. Since some of the methods supplement each other, it is frequently desirable to use more than one procedure. This is particularly true where phase changes are important. For such studies, a combination of X-ray diffraction and some other method is necessary.

Vapor-pressure–water-content determinations. This method has been used by Thomas,[1] Puri,[2] Kuron,[3] Alexander and Haring,[4] and many other investigators of soils. Air-dried samples are allowed to absorb moisture at various relative humidities, and curves of water content versus vapor pressure are plotted. It is likely, as Nagelschmidt[5] pointed out, that the exchangeable ions are of greater importance than the clay minerals themselves in causing the variations in moisture content as determined. Only very limited application of the method has been made to monomineral samples, and its contribution to the knowledge of clay minerals remains for future studies.

Kulbicki and Grim[6] described a method wherein the water vapor released on heating is driven through a water absorber. The intensity of the exothermic reaction produced by absorption of the water is recorded against the temperature of the specimen. The method provides a simple means of determining whether or not loss of water accompanies a thermal reaction.

Dehydration curves. In this method the loss in weight of the material upon heating to higher and higher temperatures is recorded and plotted against the temperature. There are several variations of the method as follows:

The sample may be heated at a given temperature until no loss in weight occurs; it is then heated to a higher temperature and held at that temperature until no loss in weight takes place. This procedure is repeated until a temperature is reached at which there is no further loss in weight. This procedure has been used by Nutting,[7] Ross and Kerr,[8,9] Kelley,[10] and many others.[11,12]

[1] Thomas, M. D., Aqueous Vapor Pressure of Soils, *Soil Sci.,* **25:** 409–418, 485–493 (1928).

[2] Puri, A. N., E. M. Crowther, and B. A. Keen, The Relation between the Vapour Pressure and Water Content of Soil, *J. Agr. Sci.,* **15:** 68–76 (1925).

[3] Kuron, H., Adsorption von Dampfen und Gasen an Boden uund Tonen und ihre Verwendung, *Kolloid-Beihefte,* **36:** 178–256 (1932).

[4] Alexander, L. T., and M. M. Haring, Vapor Pressure–Water Relations for Certain Typical Soil Colloids, *J. Phys. Chem.,* **40:** 195–205 (1936).

[5] Nagelschmidt, G., The Identification of Minerals in Soil Colloids, *J. Agr. Sci.,* **29:** 477–501 (1939).

[6] Kulbicki, G., and R. E. Grim, A New Method for the Thermal Dehydration Study of Clay Minerals, *Mineral. Mag.,* **32:** 53–62 (1959).

[7] Nutting, P. G., Some Standard Thermal Dehydration Curves of Minerals, *U.S. Geol. Surv., Profess. Paper* 197E, pp. 197–216 (1943).

[8] Ross, C. S., and P. F. Kerr, The Kaolin Minerals, *U.S. Geol. Surv., Profess. Paper* 165E, pp. 151–175 (1931).

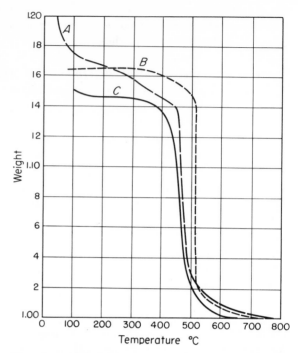

Fig. 9-1. Dehydration curves, from Nutting.[7] The weight at 800°C is taken as the base weight in plotting the curves. (A) Halloysite, Liége, Belgium; (B) kaolinite, Ione, California; (C) anauxite, Mokelumne River, California.

Results for the various clay minerals are given in Figs. 9-1 to 9-5. Various investigators have plotted their data somewhat differently, as illustrated in these figures.

The sample may be heated continuously at a constant rate, which has varied in the experiments of different investigators from 5°C/hr to 10°C/min, and the loss in weight is recorded at successively higher temperatures.

[9] Ross, C. S., and P. F. Kerr, Halloysite and Allophane, *U.S. Geol. Surv., Profess. Paper* 185G, pp. 135–148 (1934).

[10] Kelley, W. P., H. Jenny, and S. M. Brown, Hydration of Minerals and Soil Colloids in Relation to Crystal Structure, *Soil Sci.,* **41**: 259–274 (1936).

[11] Ross, C. S., and S. B. Hendricks, Minerals of the Montmorillonite Group, *U.S. Geol. Surv., Profess. Paper* 205B, pp. 23–80 (1945).

[12] Grim, R. E., R. H. Bray, and W. F. Bradley, The Mica in Argillaceous Sediments, *Am. Mineralogist,* **22**: 813–829 (1937).

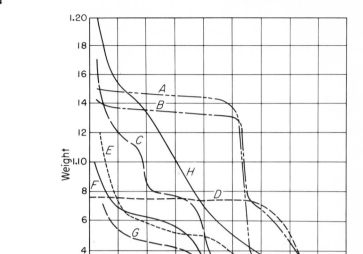

Fig. 9-2. Dehydration curves, after Nutting.[7] The weight at 700 to 900°C is taken as the base weight in plotting the curves. (A) Penninite, Paradise Range, Nevada; (B) chlorite, Danville, Virginia; (C) palygorskite (mountain leather), Montana; (D) sericite, Prince Rupert, British Columbia; (E) vermiculite, North Carolina; (F) illite, Fithian, Illinois; (G) glauconite, Lyons Wharf, Maryland; (H) sepiolite, Asia Minor.

The samples are weighed while they are hot. Migeon[13] and Longchambon[14] particularly have used this procedure, and results for some of the clay minerals are shown in Fig. 9-6. In this method equilibrium conditions are not reached at any given temperature.

A variation of this method was developed by De Keyser[15] and Erdey, Paulik, and Paulik[16,17] wherein the sample is heated in a furnace at a con-

[13] Migeon, G., Contribution à l'étude de la définition des sepiolites, *Bull. Soc. Franc. Mineral.,* **59:** 6–133 (1936).

[14] Longchambon, H., Sur les caractéristiques des palygorskites, *Compt. Rend.,* **204:** 55–58 (1937).

[15] De Keyser, W. L., Thermal Behavior of Kaolin and of Clay Materials, *Silicates Ind.,* **24:** 117–123 (1959).

[16] Erdey, L., F. Paulik, and J. Paulik, Differential Thermogravimetry, *Nature,* **174:** 885–886 (1954).

[17] Erdey, L., F. Paulik, and J. Paulik, Differential Thermogravimetry, *Közlemen. Magyar Tud. Akad. Kem. Tudomanyok Osztalyanak,* **7:** 55–90 (1955).

stantly increasing temperature while suspended from a balance to the beam of which is attached a permanent magnet free to move inside a solenoid. When a change in weight occurs the magnet moves, and the current set up in the solenoid is recorded. The starting point, the point of maximum rate, and the end point of every reaction involving weight changes may be recorded.

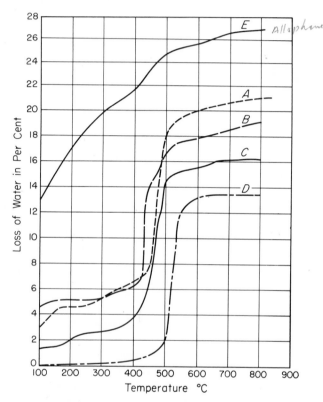

Fig. 9-3. Dehydration curves, from Ross and Kerr.[8,9] (A) Halloysite, Liége, Belgium; (B) halloysite, Adams County, Ohio; (C) halloysite, Hickory, North Carolina; (D) kaolinite, Ione, California; (E) allophane, Moorefield, Kentucky.

In the above procedures, the actual humidity of the air in contact with the sample should be kept constant, since variations will affect the results. Nagelschmidt[5] pointed out that it is generally assumed that the loss in weight of one mineral upon heating is not affected by the presence of a second mineral and that the dehydration curve of a mixture would be equal to the superposition of curves of the separate minerals in the mixture. This assumption has not been proved, and, indeed, data from differential thermal analyses (see page 347) show that it is likely to be true only in some mixtures and not in others.

In the case of natural clays that are not pure clay minerals, the loss in weight which is measured is not necessarily due entirely to a loss of water.

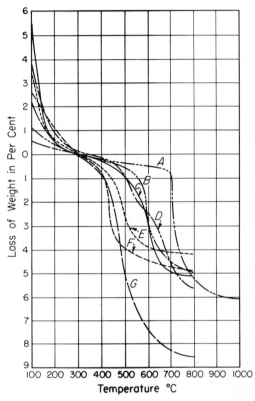

Fig. 9-4. Dehydration curves, from Ross and Hendricks.[11] (A) Hectorite, Hector, California; (B) montmorillonite, Belle Fourche, South Dakota; (C) montmorillonite, Tatatilla, Mexico; (D) montmorillonite, Montmorillon, France; (E) nontronite, Spokane, Washington; (F) nontronite, Sandy Ridge, South Carolina; (G) montmorillonite, Pontotoc, Mississippi.

It may be due to the loss of CO_2 from carbonates and to the loss of volatile constituents which may be present in nonclay-mineral components. Also, if divalent iron and/or manganese are present, their oxidation would tend to cause an increase in weight and thereby to reduce the apparent water loss. Variations in the specific characteristics of the minerals being studied, such as grain size, crystallinity, nature of adsorbed ions, etc., affect dehydra-

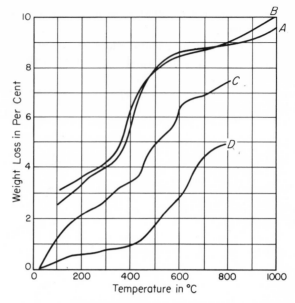

Fig. 9-5. Dehydration curves, from Grim, Bray, and Bradley.[12] (A) Illite, Gilead, Calhoun County, Illinois; (B) illite, Fithian, Vermilion County, Illinois; (C) muscovite, very finely ground; (D) muscovite, 100 mesh (coarser than C).

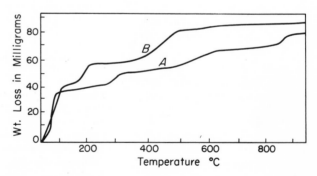

Fig. 9-6. Dehydration-weight-loss curves for (A) sepiolite (after Migeon[13]), and (B) palygorskite (after Longchambon[14]).

tion results. These factors also influence differential thermal results, and they will be considered there, as well as in the discussion of the dehydration properties of the individual minerals.

Differential thermal analyses. The method of differential thermal analysis determines, by suitable apparatus, the temperature at which thermal reactions take place in a material when it is heated continuously to an elevated temperature, and also the intensity and general character of such reactions. In the case of the clay minerals, differential thermal analyses show characteristic endothermic reactions due to dehydration and to loss of crystal structure, and exothermic reactions due to the formation of new phases at elevated temperatures. The method is, therefore, useful for clay-mineral researches as a means of studying high-temperature reactions, in addition to its value in the investigation of hydration phenomena.

The method is not restricted to clays and, in fact, has been applied to a wide variety of materials, such as carbonates, hydrates, sulfides, organics, etc. It is applicable to any material which experiences thermal reactions on heating which begin abruptly and are completed in a relatively short temperature interval.

Differential thermal results are plotted in the form of a continuous curve in which the thermal reactions are plotted against furnace temperatures, with endothermic reactions conventionally shown as downward deflections and exothermic reactions as upward deflections from a horizontal base line. The amount of divergence of the difference curve from the base line reflects the difference in temperature between the sample and the furnace at any given temperature and is, therefore, a measure of the intensity of the thermal reaction. Scales for determining the temperature differences shown by the peaks of the curves are given in Fig. 9-7. Differential curves for the various clay minerals are given in Figs. 9-8 to 9-13.

Mackenzie[18] published a history of the development of the differential thermal technique.

In the method as generally used today, the sample to be studied is placed in one hole of a specimen holder, and an inert material that experiences no thermal reaction when heated to the temperature of the experiment, usually calcined aluminum oxide (α-Al_2O_3), is placed in another hole of the specimen

[18] Mackenzie, R. C., Thermal Methods, "The Differential Thermal Investigations of Clays," chap. I, pp. 1–22, Mineralogical Society, London (1957).

[19] Barshad, I., Vermiculite and Its Relation to Biotite as Revealed by Base-Exchange Reaction, X-ray Analyses, Differential Thermal Curves, and Water Content, *Am. Mineralogist,* **33**: 655–678 (1948).

[20] Caillere, S., Étude des quelques silicates magnésiens à facies asbestiforme ou papyrace n'appartenant pas du groupe de l'antigorite, *Bull. Soc. Franc. Mineral.,* **59**: 353–386 (1936).

holder. One junction of the difference thermocouple (Fig. 9-14) is placed in the center of the sample and the other junction in the center of the inert material. The holder and thermocouples are placed in a furnace so controlled as to produce a uniform rate of temperature increase. The temperature of the inert material increases regularly as the temperature of the furnace increases. When a thermal reaction takes place in the sample, the temperature of the sample is greater or less than that of the inert material, depending

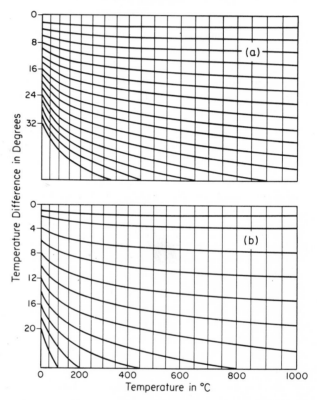

Fig. 9-7. **Scales for determining the temperature differences shown by the peaks of the thermal curves in Figs. 9-8 to 9-13.**

on whether the reaction is exothermic or endothermic, for an interval of time until the reaction is completed and the temperature of the sample again comes to the temperature of the furnace. Consequently, for an interval of time the temperature of one junction of the difference couple is different from that of the other, and an emf is set up in the differential-thermocouple circuit which is recorded as a function of time or of the temperature of the furnace. The record may be made manually with a potentiometer or galvanometer, photographically with a galvanometer, or automatically with some electronic device. When no thermal reaction is taking place in the specimen, the temperatures at both junctions of the difference couple are the same, and no emf is set up. The direction of the emf in the circuit depends on whether the temperature of the sample is above or below that of the inert material,

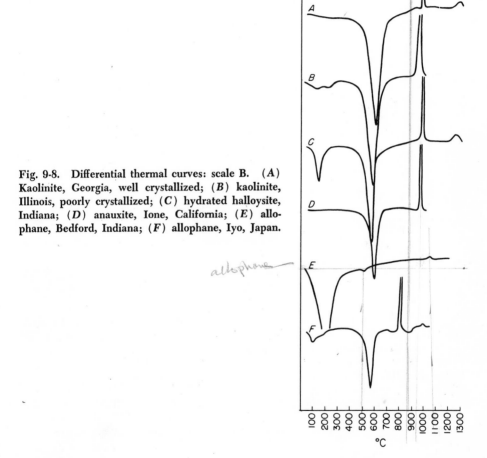

Fig. 9-8. Differential thermal curves: scale B. (*A*) Kaolinite, Georgia, well crystallized; (*B*) kaolinite, Illinois, poorly crystallized; (*C*) hydrated halloysite, Indiana; (*D*) anauxite, Ione, California; (*E*) allophane, Bedford, Indiana; (*F*) allophane, Iyo, Japan.

and consequently the recording mechanism moves in opposite directions for endothermic and exothermic reactions.

Figure 9-15 shows an idealized equilibrium dehydration curve and differential thermal curve for kaolinite. The endothermic reaction between about 500 and 700°C corresponds obviously to the dehydration of the mineral. A comparison of the curves illustrates that the differential method is a dynamic rather than a static one. The thermal reactions are not instantaneous, and

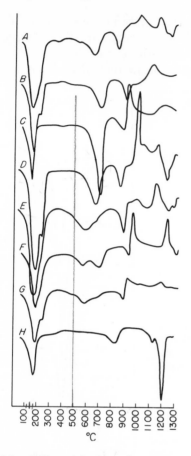

Fig. 9-9. Differential thermal curves: scale A. (A) Montmorillonite, Otay, California; (B) montmorillonite, Tatatilla, Mexico; (C) montmorillonite, Upton, Wyoming; (D) montmorillonite, Cheto, Arizona; (E) montmorillonite, Pontotoc, Mississippi; (F) montmorillonite, Palmer, Arkansas; (G) nontronite, Howard County, Arkansas, (H) hectorite, Hector, California.

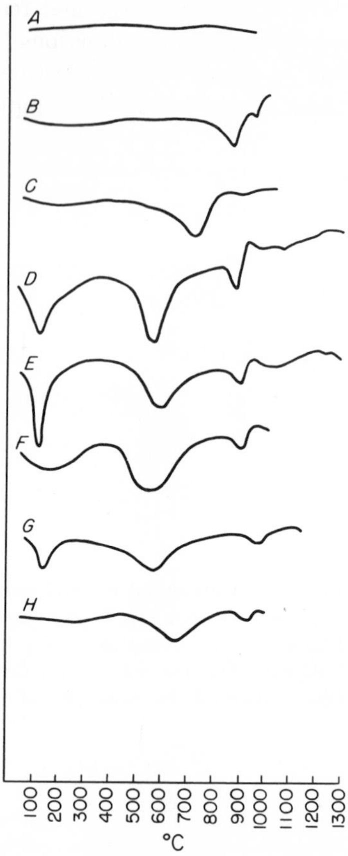

Fig. 9-10. Differential thermal curves: scale A. (A) Biotite, University of Illinois collections; (B) muscovite, University of Illinois collections; (C) muscovite, Bryman, California, (−1 μ fraction); (D) illite, Fithian, Illionis; (E) illite, Grundy County, Illinois; (F) illite, Thebes, Illinois; (G) glauconite, New Jersey; (H) glauconite, Washington.

they are recorded as functions of time or as functions of the furnace temperature, which is continuously increasing as the reaction takes place. The temperature at which the dehydration begins corresponds to the start of the endothermic reaction. The temperature of the peak of the endothermic deflection varies, depending on the details of the procedure followed, the character of the reaction involved, and the material being studied.

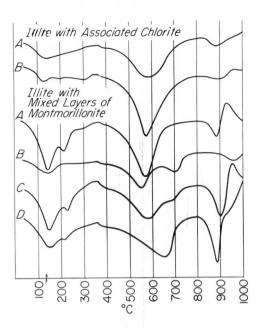

Fig. 9-11. Differential thermal curves: scale A. Illite associated with chlorite: (*A*) maquoketa shale, Illinois; (*B*) Sárospatak, Hungary (sarospatite); (*C*) Noyant Allier, France (bravaisite); (*D*) Decorah shale, Wisconsin.

Numerous attempts have been made, (Spiel et al.,[21] Kerr and Kulp,[22] Arens,[23] Boersma,[24] Deeg,[25] Kissinger,[26] and Sewell and Honeyborne[27]) to analyze mathematically the differential thermal method. It is obvious from a consideration of the equipment used and the factors affecting the results that the method has certain quantitative limitations which restrict rigorous mathematical treatment.

[21] Spiel, S., L. H. Berkelheimer, J. A. Pask, and B. Davies, Differential Thermal Analysis—Its Application to Clays and Other Aluminous Minerals, *U.S. Bur. Mines, Tech. Paper* 664 (1945).

[22] Kerr, P. F., and J. L. Kulp, Multiple Differential Thermal Analyses, *Am. Mineralogist,* 33: 387–419 (1948).

[23] Arens, P. L., "A Study of the Differential Thermal Analyses of Clays and Clay Minerals," Excelsiors Foto-offset, Wageningen, Netherlands (1951).

[24] Boersma, S. L., A Theory of Differential Thermal Analysis and New Methods of Measurement and Interpretation, *J. Am. Ceram. Soc.,* 38: 281–284 (1955).

[25] Deeg, E., Theoretical Consideration of Differential Thermal Analysis—Practical Conclusions, *Ber. Deut. Keram. Ges.,* 33(10): 321–329 (1956).

[26] Kissinger, H. E., Reaction Kinetics in Differential Thermal Analysis, *Anal. Chem.,* 29: 1702–1906 (1957).

[27] Sewell, E. C., and D. B. Honeyborne, Theory and Quantitative Use, "The Differential Thermal Investigation of Clays," chap. III, pp. 65–97, Mineralogical Society, London (1957).

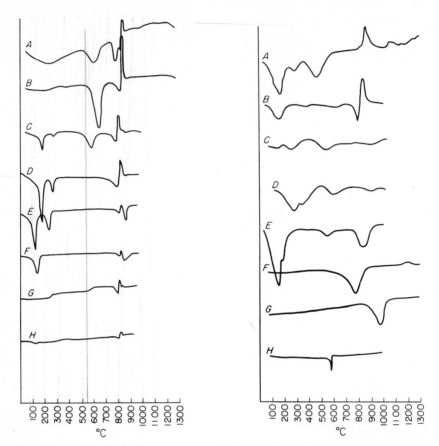

Fig. 9-12. Differential thermal curves: (A) prochlorite, Chester, Vermont; (B) clinochlore, Brewster, New York; (C) vermiculite and chlorite, Lenni, Pennsylvania; (D) vermiculite, North Carolina, natural; (E) vermiculite, North Carolina, Ca^{++}; (F) vermiculite, North Carolina, Na^+; (G) vermiculite, North Carolina, NH_4^+; (H) vermiculite, North Carolina, K^+. Curves C to H (from Barshad[19]), vertical scale slightly less than for other curves.   Scale B for curves A and B.

Fig. 9-13. Differential thermal curves: scale A. (A) Attapulgite, Attapulgus, Georgia; (B) sepiolite, Salinelles, France; (C) palygorskite, North Africa; (D) palygorskite (from Caillere[20]); (E) pyrophyllite, North Carolina; (F) vermiculite, Arizona; (G) talc, Vermont; (H) quartz, University of Illinois collections.

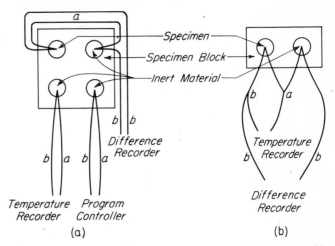

Fig. 9-14. (*a*) Thermocouple setup as used today; (*b*) thermocouple setup as used in early work and in some current work. (*a*) Platinum–10 percent rhodium wire; (*b*) platinum wire.

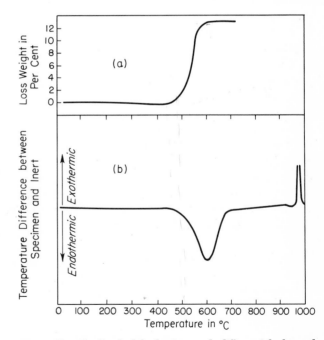

Fig. 9-15. Idealized dehydration and differential thermal curves for kaolinite. (After Spiel et al.[21])

Various investigators have discussed the different types of apparatus that may be used in making differential thermal analyses and the influence of equipment variables on the precision of the method (see Grim[28] and Mackenzie and Mitchell[29]). Variations in the material being analyzed and the manner of preparation for analysis also influence the results. For example, the temperatures at the junctions of the difference thermocouple depend to some extent on the coefficient of thermal diffusivity of the material in which they are embedded. The diffusivity of the specimen may be different from that of the inert material, and, furthermore, the diffusivity of the specimen may change as it is heated because of the formation of new phases at high temperatures or because of shrinkage of the sample.

Spiel et al.[21] showed that for some materials the thermal curve varies with the particle size of the component minerals, particularly when the maximum size is about 2 μ. In general (Fig. 9-16), the size of the thermal reaction and the temperature of the peak decrease as the particle size decreases. In some materials (see page 5) the decrease in particle size may be accompanied by a decrease in crystallinity, which is reflected in the differential thermal curve by lower intensity of reactions and a decrease in the temperature of the reaction peaks. According to Arens,[23] if the size of particle is larger than about 20 μ, the surface area is too small for dehydration reactions to occur rapidly enough to yield pronounced effects on differential thermal records. Reactions due to phase changes would be largely independent of particle size.

Norton[30] and Spiel et al.[21] showed that, in general, the slower the heating rate, the broader the peak and the lower the temperature of the peak (Fig. 9-17). Increasing the heating rate delays the attainment of the temperatures of both endothermic and exothermic peaks and increases the height of the peak and the temperature interval during which the reaction takes place. According to Spiel, the area under the curve for a given reaction and the temperature at the start of the reaction seem to be independent of the rate of heating. Arens[23] did not check this conclusion, having found some variation in the area under the curve with the heating rate. Evans and White[31] pointed out that, while the thermal decomposition of loose powdered clay is chemically

[28] Grim, R. E., Method and Application of Differential Thermal Analysis, *Ann. N.Y. Acad. Sci.,* **53**: 1031–1053 (1951).

[29] Mackenzie, R. C., and B. D. Mitchell, Apparatus and Technique for Differential Thermal Analysis, "The Differential Thermal Investigation of Clays," chap. II, pp. 23–64, Mineralogical Society, London (1957).

[30] Norton, F. H., Critical Study of the Differential Thermal Method for the Identification of the Clay Minerals, *J. Am. Ceram. Soc.,* **22**: 54–63 (1939).

[31] Evans, J. L., and J. White, The Thermal Decomposition (Dehydroxlation) of Clays, Kinetic High-Temperature Processes Conference, Dedham, Mass., pp. 301–308 (1958).

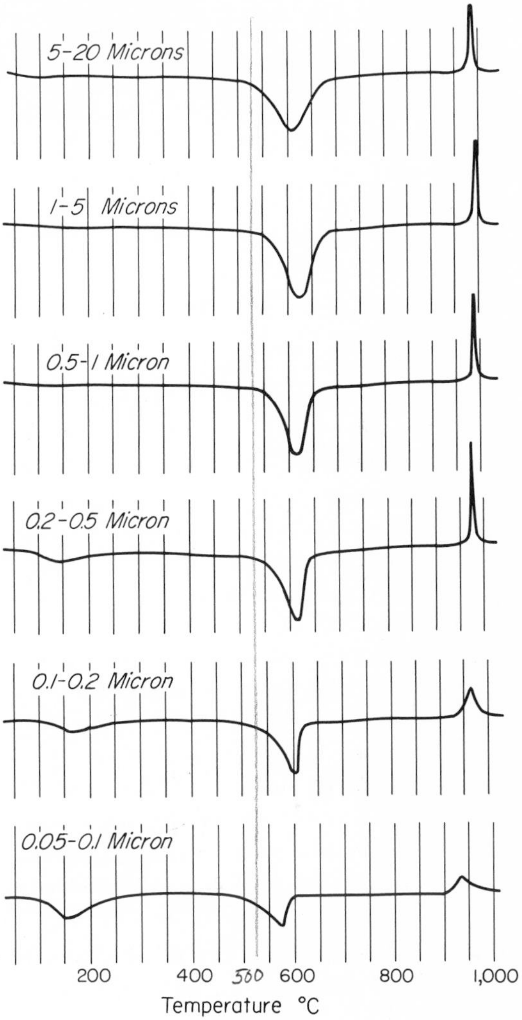

Fig. 9-16. Variation of differential thermal curves for kaolinite with particle size. (After Spiel et al.[21])

a first-order reaction, the decomposition rate decreases appreciably as the depth of the clay bed or packing density is increased. The same authors studied the influence of other factors on the rate of decomposition.

In order to obtain reproducible results for many materials, the furnace atmosphere must be kept constant, and for materials subject to oxidation it is important that the samples be prepared and placed in the furnace in such a way that they always have the same ease of oxidation. Equipment is now available permitting variation and control of the gaseous atmosphere

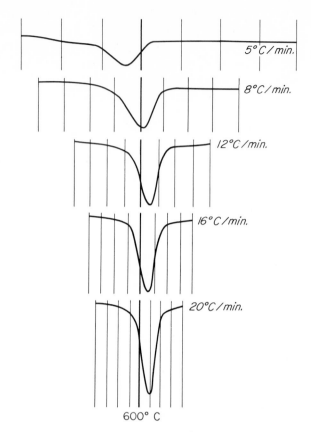

5°C/min.

8°C/min.

12°C/min.

16°C/min.

20°C/min.

600° C

Fig. 9-17. Effect of variation in heating rate on the endothermic reaction for the dehydration of kaolinite. (After Spiel et al.[21])

in which analysis is conducted (see Stone[32] and Jonas and Grim[33]).   Schramli and Becker[34] described an apparatus for carrying out the analysis in a vacuum, and Kato[35] developed equipment for studying reactions on freezing to —195°C.

[32] Stone, R. L., Apparatus for D.T.A. in Controlled Atmosphere, *J. Am. Ceram. Soc.,* **35:** 75–82 (1952).

[33] Jonas, E. C., and R. E. Grim, Differential Thermal Analysis Using Controlled Atmospheres, "The Differential Thermal Investigation of Clays," chap. XV, pp. 389–403, Mineralogical Society, London (1957).

[34] Schramli, W., and F. Becker, Comparison of the Differential Thermal Analysis of a Few Minerals and Oxides in Vacuum and in Air, *Ber. Deut. Keram. Ges.,* **37**(5)**:** 227–236 (1960).

[35] Kato, C., Differential Thermal Analysis of Clay Minerals between 0 and —195°, *Yogyo Kyokai Shi,* **67:** 243–246 (1959).

Spiel et al.[21] presented a series of curves for kaolinite diluted by various quantities of an inert material, showing that the size of the reaction and the temperature of the peak decrease as the amount of material decreases. The temperature of the peak, therefore, is not an absolute value but depends, among other things, on the amount of material used or present in a mixture.

Knofel[36] described a method of obtaining a "derivative" differential thermal analysis which requires two identical furnaces operating at a constant temperature difference.

Organic materials give exothermic reactions when heated. Several investigators[37,38] have suggested that the exothermic reactions yielded by the organic component of organic–clay-mineral complexes may vary with particular clay minerals and therefore be useful in identifying the components of clay-mineral mixtures.

Grim[39] has shown that the characteristics of curves for mixtures of some minerals depend somewhat on the intimacy of mixing of the components and that in clay minerals there is considerable difference between the curves for mixtures of discrete particles several microns or more in diameter and those for mixtures consisting of an intimate interlayering of considerably smaller units, such as are frequently found in nature. Arbitrary curves set up on the basis of dry mixtures of clay minerals, therefore, are frequently of no value in evaluating curves for natural clays. In general, the thermal reactions for the individual components decrease in intensity and sharpness of reaction as the intimacy of mixing increases.

Attempts have been made to determine heats of hydrations by the differential method. In some cases the results are in reasonable quantitative agreement with rational analyses (Murray and White[40] and Allison[41]).

Differential thermogravimetric analyses. The method of determining the differential weight loss as a sample is heated, sometimes referred to as dynamic

[36] Knofel, G. F., A Contribution on the Problem of Differential Thermal Analysis: An Apparatus for "Derivative" Differential Thermal Analysis, *Sprechsaal,* (8), a58 (1963).

[37] Oades, J. M., and W. N. Townsend, The Use of Piperidine as an Aid to Clay-mineral Identification, *Clay Minerals Bull.,* 5(29): 177–182 (1963).

[38] Kacker, K. P., and V. S. Ramchandran, Identification of Clay Minerals in Binary and Ternary Mixtures by Differential Thermal Analysis of Dye-Clay Complexes, *Trans. 9th Congr. Ceram.,* pp. 483–496 (1964).

[39] Grim, R. E., Differential Thermal Curves of Prepared Mixtures of Clay Minerals, *Am. Mineralogist,* 32: 493–501 (1947).

[40] Murray, P., and J. White, Kinetics of Clay Decomposition, *Clay Minerals Bull.,* no. 3, pp. 84–87 (1949).

[41] Allison, E. B., Quantitative Thermal Analysis of Clay Minerals, *Clay Minerals Bull.,* 2(13): 237–241, 242–254 (1955).

differential calorimetry (DDC), makes it possible to determine the weight loss associated with energy changes. The method is less sensitive than differential thermal analysis (DTA), i.e., in detecting very small reactions and the precise temperature at which they start. However, the results are said to be more reproducible. Keler[42] and later Torkar et al.[43] and Hodgson[44] described apparatus for the simultaneous determination of differential thermal and gravimetric data.

Identification of high-temperature phases. The identification of changes in phase on heating the clay minerals has been made almost entirely by means of X-ray-diffraction analyses. In early investigations the samples were heated to various temperatures, quenched or allowed to cool, and then studied. Currently the use of high-temperature X-ray-diffraction cameras has permitted the obtaining of diffraction patterns while the minerals are at high temperatures. In the author's laboratory and recently in many other laboratories, a small furnace has been mounted on the diffraction unit, so that spectrometer traces are obtained while the minerals are at the elevated temperature. The latter techniques not only save a tremendous amount of time but remove the possible difficulties resulting from phase inversions on cooling. The high-temperature X-ray-diffraction technique is providing information of great value in studies of the formation of minerals at elevated temperatures and of transformations at high temperatures. Rowland et al.[45] described an oscillating-heating X-ray-diffraction apparatus which makes possible very detailed observations of the changes which take place during heating.

Comefero, Fischer, and Bradley[46] used the electron microscope to reveal the development of mullite in heated kaolinite. Heating units that fit into electron microscopes are now available, so that studies of changes in morphology as well as identifications by electron diffraction can be made while samples are at elevated temperatures. Significant results should come from such studies in the future.

The petrographic microscope is adequate for identifying a new phase,

[42] Keler, E. K., Complex Thermal Analysis of Silicates, *Trans. 1st Congr. Thermogr., Kazan*, pp. 239–249 (1955).

[43] Torkar, K., K. Lasser, and H. P. Fritzer, Dynamic Differential Calorimetry (d.d.c.) Combined with Simultaneous Calibration and Thermogravimetry, *Sprechsaal*, **95**(10): 239 (1962).

[44] Hodgson, A. A., A Comprehensive Thermal Analysis Apparatus, *J. Sci. Instr.*, **40**: 2, 61 (1963).

[45] Rowland, R. A., E. J. Weiss, and D. R. Lewis, Apparatus for the Oscillating-Heating Method of X-ray Powder Diffraction, *J. Am. Ceram. Soc.*, **42**: 133–138 (1959).

[46] Comefero, J. E., R. B. Fischer, and W. F. Bradley, Mullitization of Kaolinite, *J. Am. Ceram. Soc.*, **31**: 254–259 (1948).

if it is well developed and relatively abundant. If the phase forms slowly, the initially formed material is frequently not sufficiently abundant nor in large enough units for identification optically; hence, the point of the beginning of the transformation cannot be fixed by microscope study.

Allophane

Dehydration curves published by Ross and Kerr[9] (Fig. 9-3) show a continuous loss of water, with no breaks or very prominent flexures but a gradual decrease in amount as the temperature is increased. Water of the order of 10 to 15 percent is lost below about 110°C, and a roughly similar amount is lost at higher temperatures up to about 1000°C. These authors concluded that the dehydration curves of allophane are characteristic of a material in which the water is held in solution and not in chemical combination. Nutting[7] published curves for two allophane samples, one of which shows a gradual water loss, whereas the other shows a distinct flexure between about 550 and 600°C.

Grim and Rowland[47] published differential thermal curves for two allophane samples (Fig. 9-8), both of which show distinct endothermic reactions at low temperatures, corresponding to the water loss below 110°C shown by the dehydration data. One of the differential thermal curves also shows an endothermic reaction at a temperature suggestive of loss of OH lattice water. The significance of this reaction and the distinct flexure shown by Nutting's[7] data might reflect some structural organization in this allophane, or possibly only the presence of a crystalline impurity. The curve for the other allophane sample shows only very slight reactions above 300°C. Birrell and Fieldes[48] presented a differential thermal curve for an allophane showing an exothermic peak at about 960°C, which also suggests some structural organization. Levitskii and Vlasov[49] presented a similar curve and identified a cristobalite structure forming above 900°C. In general, the high-temperature phases formed on heating allophane have not been studied. Since the composition of allophane can vary widely, great latitude in the high-temperature phases developed in different samples would be expected.

No information is available regarding the possible rehydration of allo-

[47] Grim, R. E., and R. A. Rowland, Differential Thermal Analyses of Clay Minerals and Other Hydrous Materials, *Am. Mineralogist,* 27: 746–761, 801–818 (1942).

[48] Birrell, K. S., and M. Fieldes, Allophane in Volcanic-ash Soils, *J. Soil Sci.,* 3: 156–166 (1952).

[49] Levitskii, V. E., and V. V. Vlasov, Allophanoid Containing Iron and Phosphorus in the Upper Jurassic Formations of the Ul'yanovsk Region, *Zap. Vses. Mineralog. Obshchestva,* 94(4): 465–468 (1965).

phane. However, Gradwell and Birrell[50] showed that, after drying an allophane sample from New Zealand, it could not be rewetted to its original character.

Kaolinite

Dehydration and phase changes on heating. Dehydration curves for kaolinite are given in Figs. 9-1 and 9-3, and differential thermal curves are given in Fig. 9-8. Ross and Kerr[8] pointed out that the dehydration curves are almost flat up to about 400°C, showing little loss of water at low temperatures, which is in accord with the absence of any thermal reactions in the low-temperature region of the differential thermal curves of many kaolinites. Differential curves for some poorly crystalline kaolinites (Fig. 9-8) show a small initial endothermic reaction, indicating that, when there is irregularity in the arrangement of the kaolinite units, a small amount of water may be present between the layers. This is in accord with the slightly greater c-axis spacing of poorly crystalline as compared with well-crystallized kaolinites.

A series of dehydration curves presented by Ross and Kerr[8] show that most of the dehydration takes place between about 400 and 525°C. The precise temperature for the loss of this OH lattice water varies from kaolinite to kaolinite, and this variation may be explained by variations in particle size, since the dehydration temperature is known to decrease with decreasing particle size. It may also be explained by variations in crystallinity, since the poorly crystalline material loses its hydroxyl water somewhat more readily than well-crystallized kaolinite. A small amount (2 to 3 percent) of water is retained at about 525°C, and this moisture is lost gradually up to about 750 to 800°C, where dehydration is essentially complete.

Ross and Kerr[8] found no difference in the dehydration characteristics of kaolinite and anauxite, but they did find that the dehydration temperature for nacrite and dickite was slightly higher than that required for kaolinite. Nutting[7] published a few curves for kaolinite, showing a slightly higher dehydration temperature than those published by Ross and Kerr. Nutting also indicated that the dehydration temperature of anauxite is slightly lower than that of kaolinite. The apparent difference in the findings of these investigators may be explained by differences in the particle size of their samples and possibly also in degree of crystallinity.

De Keyser[51] stated that kaolinite can be dehydrated completely at tem-

[50] Gradwell, M., and K. S. Birrell, Physical Properties of Certain Volcanic Clays (from New Zealand), *New Zealand J. Sci. Technol.,* 36B(2): 108–122 (1954).

[51] De Keyser, W., A Study of Kaolin and Some Belgian Clays, *Ann. Mines Belg.,* 40: 711–806 (1939).

peratures as low as 350°C with long heating times (200 hr), and indeed this is about the temperature of the very beginning of the break in the dehydration curves for the mineral. In general, very long time intervals are necessary to reach equilibrium for the clay minerals at temperatures where the loss of OH water is beginning.

Differential thermal curves of kaolinite show an intense, sharp endothermic reaction corresponding to the loss of OH water (Fig. 9-8). The reaction begins at about 400°C, and for well-crystallized kaolinite the peak is at about ??°C. The intensity of the reaction and hence the size of the peak, as well as the peak temperature, are reduced slightly as the particle size decreases and as the crystallinity decreases. The variation seems to be greater for the crystallinity factor than for the particle-size factor. According to Grimshaw et al.,[52] the peak temperature is 20 to 30°C lower for poorly crystalline kaolinite than for the well-crystallized variety.

There seems to be no difference in the differential curves for kaolinite and anauxite. The temperature of the peaks for the reaction for loss of OH water is higher (about 100°C) for nacrite and dickite than for kaolinite.

The rate of dehydration is a function of the size and shape of the specimen on which the determination is made, which suggests that entrapped water vapor exerts a controlling influence (Brindley and Nakahira[53]). At relatively low temperatures (for example, 490°C±), a stationary state may be attained. Vapor pressures as low as a few atmospheres (Pieters,[54] Stone,[55] Stone and Rowland[56]) suffice to retard considerably the endothermic reaction accompanying the loss of hydroxyls.

Weber and Roy[57] investigated the effect of water pressure on dehydration temperature and found that from 15 to about 400 psi the temperature of dehydration increases with the increasing water pressure, whereas above about 400 psi the temperature-pressure relation is reversed.

Murray and White[40,58] concluded that the dehydroxylation of kaolinite

[52] Grimshaw, R. W., E. Heaton, and A. L. Roberts, Constitution of Refractory Clays, II, Thermal Analysis Methods, *Trans. Brit. Ceram. Soc.,* **44**: 76–92 (1945).

[53] Brindley, G. W., and M. Nakahira, The Role of Water Vapour in the Dehydroxylation of Clay Minerals, *Clay Minerals Bull.,* **3**(17): 114–119 (1957).

[54] Pieters, H. A., Thesis, University of Delft, Netherlands (1928).

[55] Stone, R. L., Differential Thermal Analysis of Kaolin Group Minerals Under Controlled Partial Pressures of H_2O, *J. Am. Ceram. Soc.,* **35**: 90–99 (1952).

[56] Stone, R. L., and R. A. Rowland, DTA of Kaolinite and Montmorillonite Under Water Vapor Pressures up to 6 Atmospheres, *Natl. Acad. Sci., Publ.* 395, pp. 103–116 (1955).

[57] Weber, J. N., and R. Roy, Dehydroxylation of Kaolinite, Dickite, and Halloysite: DTA Curves Under P_{H_2O} = 15 to 10,000 psi, *J. Am. Ceram. Soc.* **48**: 309–311 (1965).

[58] Murray, P., and J. White, Kinetics of the Thermal Dehydration of Clays, *Trans. Brit. Ceram. Soc.,* **48**: 187–206 (1949).

takes place according to first-order kinetics. Brindley and Nakahira[59] found that this is true only for infinitely thin specimens. These investigators determined the activation energy as 65 kcal/mole for various kaolins. Toussaint, Fripiat, and Gastuche[60] also studied the dehydroxylation mechanism and also concluded that the rate process is limited by the outward diffusion of water molecules. They gave 25 kcal/mole as the activation energy and concluded that the reaction proceeds by successive destruction of complete octahedral layers. Holt et al.[61] studied dehydration in a vacuum and reported an activation energy of 43.5 kcal/mole. They suggested that the diffusing component may be hydroxyl ions rather than water molecules.

Nakahira[62] concluded that the "exterior" hydroxyls reacted more readily than the "interior" hydroxyls (i.e., those in the middle layer of the silicate sheet), but Brindley and Nakahira[63] were not able to confirm this conclusion.

The rate of dehydroxylation also is reported (Okuda et al.[64]) to vary with the identity of the adsorbed cations.

The sections of the differential curves for poorly crystallized kaolinite lying between the peak for loss of hydroxyl water and the intense exothermic reaction at about 950°C are relatively flat, whereas those for well-crystallized kaolinite in this temperature range tend to have an upward slope, with a slight endothermic reaction just before the exothermic reaction. Grim and Bradley[65] concluded that the loss of OH water in poorly crystallized kaolinite is accompanied by a fairly complete loss of structure but that in well-crystallized kaolinites some structural remnants persist along with some water and that this remnant is lost at the time of the higher-temperature endothermic reaction.

Numerous investigators have debated the nature of the material immediately following the loss of OH lattice water from kaolinite. Insley and

[59] Brindley, G. W., and M. Nakahira, Kinetics of Dehydroxylation of Kaolinite and Halloysite, *J. Am. Ceram. Soc.*, **40**: 346–350 (1957).

[60] Toussaint, F., J. J. Fripiat, and M. C. Gastuche, Dehydroxylation of Kaolinite, I, Kinetics, *J. Phys. Chem.*, **67**: 26–30 (1963).

[61] Holt, J. B., I. B. Cutler, and M. E. Wadsworth, Kinetics of the Thermal Dehydration of Hydrous Silicates, Proceedings of the 12th National Clay Conference, pp. 55–67, Pergamon Press, New York (1964).

[62] Nakahira, M., On the Thermal Transformations of Kaolinite and Halloysite, *Mineral. J. (Tokyo)*, **1**: 129–139 (1954).

[63] Brindley, G. W., and M. Nakahira, A Kinetic Study of the Dehydroxylation of Kaolinite, *Natl. Acad. Sci., Publ.* 556, pp. 266–278 (1958).

[64] Okuda, S., N. Tanaka, and K. Inoue, Fixation Phenomenon of the Cations Adsorbed on Kaolinite Caused by Drying and Its Influences on Some Properties of Kaolinite-Water Systems, *Mem., Fac. Ind. Arts, Kyoto Tech. Univ., Sci. Technol.*, no. 7, pp. 65–89 (1958).

[65] Grim, R. E., and W. F. Bradley, Rehydration and Dehydration of the Clay Minerals, *Am. Mineralogist*, **33**: 50–59 (1948).

Ewell[66] showed that the dehydrated kaolinite did not have the same chemical attributes, solubility in acid, etc., as mixtures of amorphous silica and alumina heated to similar temperatures. Comefero, Fischer, and Bradley[46] showed that the diffraction bands of dehydrated kaolinite were more distinct than such bands from strictly amorphous solids; also, on the basis of electron micrographs, they showed that the hexagonal shape of the kaolinite was preserved above 600°C.

On the basis of the foregoing considerations and later work, it is now generally agreed that kaolinite heated just above the temperature of complete dehydration retains some degree of order; such material has come to be called metakaolin.

Tscheischwili, Büssem, and Weyl[67] first proposed a structure for metakaolin in which the Si-O network for kaolinite remains largely intact and the Al-O network reorganizes itself in the form of edge-shared Al-O octadedral chains.

Planz and Mueller-Hesse[68] stated that infared-absorption data of poorly crystalline kaolinite confirmed the existence of a distinct "residual" (metakaolin) lattice. Miller[69] earlier studied the dehydroxylation of a well-crystallized kaolinite by infrared absorption and reported the continuous formation of a metakaolin phase as the dehydration proceeds. The infrared data are difficult to interpret in terms of specific structure. Brindley and Nakahira[70] and De Keyser[15] studied the matter in detail. Diffraction data are extremely poor for metakaolin, but, using single crystals, Brindley and Nakahira were able to obtain superior data permitting them to propose the structure shown in Figs. 9-18 and 9-19, which is in general agreement with De Keyser's data. In this structure, the Si-O network remains essentially unchanged with the possibility that the distortions in the kaolinite lattice (see pages 60–66) may be increased. With the diminished number of oxygens in metakaolin, it seems certain that the aluminum is in tetrahedral coordination and that the tetrahedrons share corners or edges rather than being face-shared. According to Brindley and Nakahira,[70] the structure is disordered in the *c*-axis direction although the structure proposed shows no obvious reason for this.

[66] Insley, H., and R. H. Ewell, Thermal Behavior of Kaolin Minerals, *J. Res. Natl. Bur. Std.,* **14:** 615–627 (1935).

[67] Tscheischwili, L., W. Büssem, and W. Weyl, Metakaolin, *Ber. Deut. Keram. Ges.,* **20:** 249–276 (1939).

[68] Planz, J. E., and H. Mueller-Hesse, The Existence of Metakaolin, *Ber. Deut. Keram. Ges.,* **40**(3): 186–190 (1963).

[69] Miller, John G., An Infrared Spectroscopic Study of the Isothermal Dehydroxylation of Kaolinite at 470°, *J. Phys. Chem.,* **65:** 800–804 (1961).

[70] Brindley, G. W., and M. Nakahira, The Kaolinite-Mullite Reaction Series, *J. Am. Ceram. Soc.,* **42:** 311–324 (1959).

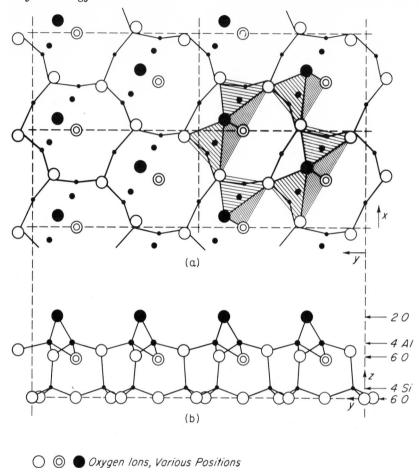

○ ◎ ● *Oxygen Ions, Various Positions*

• *Aluminum*   • *Silicon*

Fig. 9-18.   Proposed structure of metakaolin.   (a) Projection on the (001) plane; (b) elevation seen along the a axis.   (After Brindley and Nakahira.[70])

Radczewski and Schadel[71] concluded, on the basis of electron diffraction, that there was a definite periodicity in the *c*-axis direction.

Comer[72] showed that poorly crystallized kaolinite loses distinct outlines of platelet units at a lower temperature than well-crystallized kaolinite. It may well be that the order is relatively less or even somewhat different in the poorly crystallized material.

[71] Radczewski, Von O. E., and J. Schadel, Ultramikrotomschnitte von Kaolin, Ein Betrag zum Metakaolinit-Problem, *Ber. Deut. Keram. Ges.,* **39**: 48–51 (1962).

[72] Comer, J. J., Electron Microscope Studies of Mullite Development in Fired Kaolinites, *J. Am. Ceram. Soc.,* **43**(7): 378–384 (1960).

Taylor[73] suggested an inhomogeneous dehydroxylation mechanism for kaolinite (as well as for other hydrates) and arrived at a somewhat different conclusion regarding metakaolin from that of Brindley and Nakahira.[70] In this mechanism, water is not lost more or less uniformly from all over the crystal. By the inhomogeneous mechanism no oxygen is lost from those parts of the crystal where oriented conversion to a crystalline product occurs; the oxygens for the expelled water come entirely from other regions of the crystal which are wholly or partly converted into pores, while various migrations of cations occur. According to this concept, "the essential feature of the

Fig. 9-19. **Photograph of a model of the proposed metakaolin structure.** (By Brindley and Nakahira.[70])

crystalline order in metakaolin, such as it is, resides chiefly in the oxygen packing." No oxygen is lost from the acceptor regions, and the type of oxygen packing in these regions tends to remain the same as in the original kaolinite. Additional cations are incorporated into the structure to make up for the protons that have been lost. All the cations, both Si^{4+} and Al^{3+}, could occupy tetrahedral sites, but these are distributed in a largely random manner among the available tetrahedral interstices in the oxygen framework. The endothermic reaction at 600°C is attributed to cation migration and loss of water

[73] Taylor, H. W. F., Homogeneous and Inhomogeneous Mechanisms in the Dehydroxylation of Minerals, *Clay Minerals Bull.,* **5**(28) : 45–55 (1962).

from the donor regions, while the exothermic peak at 950°C is attributed to the change in oxygen packing.   At 500 to 600°C when metakaolin is formed, there is sufficient energy for cation migration but not for any significant change in oxygen packing.   At 950°C, the oxygen packing changes sufficiently to give spinel, but the change is limited in that the structure is still based on nearly close-packed layers.   Above 1000°C more profound changes in oxygen packing can occur and mullite is formed (see below).

The exothermic reaction shown by the differential thermal curves between about 900 and 1000°C is somewhat less intense and takes place over a slightly wider temperature interval for poorly crystallized kaolinite than for the well-crystallized mineral.   The explanation for this exothermic reaction has been a matter of much dispute.   Insley and Ewell[66] and many others have attributed it to the formation of γ-Al_2O_3.   When pure amorphous alumina is heated, γ-Al_2O_3 crystallizes over a long temperature interval and at a lower temperature.   Furthermore, γ-Al_2O_3 has a structure unlike one that would be anticipated from the sharpness and high intensity of this exothermic reaction.   Insley and Ewell[66] attempted to explain these difficulties on the basis of some structural order in the dehydrated kaolinite which delays the formation of γ-Al_2O_3 until a relatively high temperature, with a consequent sudden reaction releasing much energy.   Comefero, Fischer, and Bradley[46] presented cogent reasons for attributing the exothermic reaction to the formation of mullite in the sample of well-crystallized kaolinite that they studied.   These investigators, using a combination of electron microscopy and X-ray diffraction, found mullite needles developing at first while the hexagonal outline of the kaolinite was preserved, and they concluded that, after the loss of OH water, there exists a uniquely organized compound which collapses into mullite nuclei.

Several investigators[74,75] reported both mullite and γ-Al_2O_3 in kaolinite heated to 1000°C, so that the exothermic reaction could be attributed to the formation of either or both of these compounds in some samples.

Richardson[76] published the results of X-ray-diffraction analyses of a series of kaolinites heated for 20 hr at each 50°C temperature interval from 800 to 1350°C.   His results indicated the presence of γ-alumina from 950 to above 1350°C.   Richardson also investigated halloysite and found no mullite forming below 1100°C, and since halloysite shows the same exothermic

[74] McVay, T. M., and C. L. Thompson, X-ray Investigation of the Effect of Heat on China Clay, *J. Am. Ceram. Soc.*, **11:** 829–841 (1928).

[75] Wilm, D., U. Hofmann, and K. Endell, Ueber die Bedeutung von röntgeninterferenz Untersuchungen bei hohen Temperatures fur keramisch Forschung, *Sprechsaal,* **38** (1934).

[76] Richardson, H. M., Phase Changes Which Occur on Heating Kaolin Clays, "X-ray Identification and Structures of Clay Minerals," chap. III, pp. 76–85, Mineralogical Society of Great Britain Monograph (1951).

peak as kaolinite, the author concluded that the exothermic reaction in ka-olinite is probably due to the formation of γ-alumina.

Work by Glass[77] suggests that the apparent discrepancy in results may be due to variations in the crystallinity of the material investigated and to the experimental conditions. According to Glass, well-crystallized kaolinite at 1000°C yields strong diffraction lines for γ-Al_2O_3 and weak lines for mull-ite. On further heating to higher temperatures, or on long-continued heating at 1000°C, mullite continues to develop. If the heating is rapid, as it is in a differential thermal analysis, little further mullite develops until about 1250°C, when it abruptly appears in abundance and is accompanied by an exothermic reaction. Cristobalite appears in abundance at about 1300°C. In the case of poorly crystallized kaolinite, at least in the samples studied by Glass, only γ-Al_2O_3 appears at 1000°C. Mullite appears abruptly at about 1200°C, accompanied by an exothermic reaction, and cristobalite appears in abundance at about 1300°C. Johns[78] showed that the high-temperature ex-othermic reaction for kaolinite can be explained by a mullite nucleation which may be revealed by thermal data before it is clearly shown by diffraction data. Johns also showed that the nucleation would vary somewhat with the crystallinity of the kaolinite.

Bertorelli and Williams[79] claimed that mullite formation is enhanced by heating in the presence of inert gases. Caillere, Henin, and Ture[80] showed that there is a great variation in the intensity of the exothermic reaction at about 950°C as a consequence of variations in the presence of a very small amount of impurities. Parmelee and Rodriguez[81] showed that zinc, lithium, magnesium, iron, manganese, cerium, and molybdenum markedly en-hance the formation of mullite from kaolinite; boron and calcium enhance it slightly; sodium, potassium, titanium, and tin retard its formation; and the relative influence of these elements varies with temperature. Crookston[82] found that the presence of potassium reduces markedly the formation of cristobalite from kaolinite; sodium reduces it slightly; and calcium, magne-sium, and hydrogen have little effect.

[77] Glass, H. D., High-temperature Phases from Kaolinite and Halloysite, *Am. Mineralogist,* **39:** 193–207 (1954).

[73] Johns, W. D., High-temperature Phase Changes in Kaolinite, *Mineral. Mag.,* **30:** 186–198 (1953).

[79] Bertorelli, O. L., and I. Williams, Preparation of Mullite, U.S. Patent 2,536,122 (1949).

[80] Caillere, S., S. Henin, and S. Ture, Investigation of the Differential Thermal Analysis of Clays, *Compt. Rend.,* **223:** 383–384 (1946).

[81] Parmelee, C. W., and A. R. Rodriguez, Catalytic Mullitization of Kaolinite by Metallic Oxides, *J. Am. Ceram. Soc.,* **25:** 1–10 (1942).

[82] Crookston, J. A., The Effect of Exchangeable Bases on the Fired Properties of Fireclays, Ph.D. thesis, University of Illinois (1949).

Wahl[83] investigated this matter in some detail by the use of continuous X-rays. He concluded that some mullite, which he called primary mullite, is formed at the temperature of the 975°C exothermic reaction and that additional mullite, which he called secondary mullite, is formed at a temperature of about 1200°C. He found that considerable primary mullite is formed on the heating of well-organized kaolinite, whereas very little develops from poorly organized kaolinite or halloysite. He also found that the secondary mullite develops abruptly from poorly organized minerals but more gradually and beginning at a lower temperature from well-organized kaolinite. His data show that magnesium, iron, lead, boron, and calcium, when present in trace amounts, enhance the formation of mullite and that its development is retarded by alkali ions. Of particular interest is his conclusion that these elements act particularly on the development of secondary mullite and have little influence on the material that is formed at the temperature of the exothermic reaction.

Wahl[83] also recorded the formation of cristobalite at elevated temperatures. His data indicate that magnesium, fluorine, lead, calcium, and phosphorus in trace amounts enhance the development of cristobalite, whereas alkalies retard or prevent its development. These data make it clear that the effect of small amounts of impurities is undoubtedly a significant factor in explaining the apparent variation in results obtained by different investigators on heating kaolinite.

Diffraction data show unequivocally that mullite and cristobalite develop when kaolinite is heated to about 1250°C. The problem is what is the nature of material heated between the temperature of the exothermic peak and about 1250°C. The poor quality of the diffraction data amply account for the variety of the conclusions that have been reached.

Brindley and Nakahira[70] obtained superior diffraction data from single-crystal investigations. According to them, at about 950°C the metakaolin alters to γ-alumina, and the sharp exothermic reaction is a consequence of the formation of the γ-alumina. They further concluded that the γ-alumina has a spinel-type structure, and this phase carries some silicon. They pointed out a rather simple structural transformation from their proposed metakaolin structure to the cubic, silica-carrying, spinel-type structure. According to these authors, at about 1050°C the spinel-type structure transforms to a mullite phase. Silica is eliminated progressively as metakaolin transforms to a γ-alumina phase and thence to mullite. Above about 1100°C, the discarded silica appears in diffraction data as cristobalite. From 1200 to 1400°C there is continued development of cristobalite and mullite. It appears that the mullite at lower temperatures may contain excess silicon distributed among

[83] Wahl, F. M., Effect of Impurities on Kaolinite Transformations as Examined by High-temperature X-ray Diffraction, Advances in X-ray Analysis, 10th Conference, pp. 264–275, Plenum Press, New York (1962).

interstitial sites of the mullite lattice which is also discarded as the temperature is increased.

Comer,[84] using electron diffraction, found the spinel phase developing with a definite structural relationship to that of metakaolin and the mullite with a definite relation to the spinal, both in agreement with the structural transformation suggested by Brindley and Nakahira.[70] Later Von Gehlen[85] presented similar data for a metakaolin-spinel-mullite transformation.

Vaughan[86] found that the evolution of heat at the abrupt exothermic reaction is from 8 to 40 cal/g for various kaolinites, with usual values in the range of 16 to 25 cal/g.

Fig. 9-20. Thermal expansion of the clay minerals. One vertical division is 0.80 percent expansion or contraction. (A) Kaolinite. china clay, Cornwall, England; (B) halloysite, Missouri; (C) beidellite (Putnam soil), Missouri; (D) illite, Fithian, Illinois; (E) illite, Gilead, Illinois. (After Hyslop and McMurdo.[87])

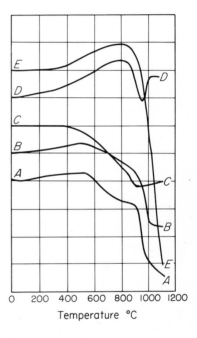

Temperature °C

Hyslop and McMurdo[87] presented expansion-contraction curves for the various clay minerals (Fig. 9-20). Kaolinite shows an initial slight expansion up to about 500°C, followed by contraction to the limits of the experiments

[84] Comer, J. J., New Electron-Optical Data on the Kaolin-Mullite Transformation, *J. Am. Ceram. Soc.*, **44**: 561–563 (1961).

[85] Von Gehlen, K., Oriented Formation of Mullite from Al-Si Spinel in the Transformation Series Kaolinite-Mullite, *Ber. Deut. Keram. Ges.*, **39**(6): 315–320 (1962).

[86] Vaughan, F., Energy Changes of Heated Kaolins, *Clay Minerals Bull.*, **2**(13): 265–274 (1955).

[87] Hyslop, J. F., and A. McMurdo, The Thermal Expansion of Some Clay Minerals, *Trans. Ceram. Soc. (Engl.)*, **37**: 180–182 (1938).

at 1100°C. The beginning of the contraction accompanies the loss of OH water. There is a flexure at about 900°C, corresponding to an interval of relatively reduced contraction; this is the temperature of the exothermic reaction that is attributed to the formation of γ-Al_2O_3 or mullite.

McKinstry[88] recently determined the thermal expansion of kaolinite and other clay minerals by high-temperature X-ray diffractometry.

Harmon and Parmelee[89] stated that

> . . . only minor changes in true specific gravity take place up to the dehydration of the kaolinite. At this point the specific gravity drops suddenly as would be expected from the theory that dehydrated kaolinite is a mixture of amorphous alumina and silica. The specific gravity then slowly increases apparently indicating that a small amount of amorphous oxides are recombining to form a compound. The increase is suddenly increased at about 900°C.

Harmon and Parmelee did their work with ball clays in which the kaolinite crystallinity was probably low. Rieke and Mauve[90] presented similar data (Table 9-1) for a kaolinite from Zettlitz in Czechoslovakia, which is well

Table 9-1   *Specific gravity of clay minerals after heating to various temperatures for 12 hr*

(After Rieke and Mauve[90])

Mineral	Raw clay	400°C	450°C	500°C	600°C	700°C	800°C	900°C	950°C	1000°C
Kaolinite, Zettlitz	2.64	2.64	2.49	2.47	2.50	2.53	2.60	2.62	2.69	2.74
Mica, Sárospatak	2.81	2.80	2.71	2.63	2.58	2.56	2.51	2.52	2.53	2.55
Halloysite	2.62	2.57	2.46	2.45	2.48	2.52	2.58	2.65	2.70	2.58
Muscovite	2.86	2.89	2.91	2.92	2.84	2.81	2.71	2.64	2.59	2.58
Montmorillonite, Wyoming	Could not be determined because of swelling				2.59	2.58	2.49	2.52	2.55	2.52

crystallized; their data are in essential agreement with those of Harmon and Parmelee.

[88] McKinstry, H. A., Thermal Expansion of Clay Minerals, *Am. Mineralogist,* **50:** 212–222 (1964).

[89] Harmon, C. G., and C. W. Parmelee, Testing and Classification of Ball Clays—Thermal History, *Bull. Am. Ceram. Soc.,* **21:** 280–286 (1942).

[90] Rieke, R., and L. Mauve, Zur Frage des Nachweises des mineralischen Bestandteile der Kaoline, *Ber. Deut. Keram. Ges.,* **23:** 119–150 (1942).

According to Norton,[91] kaolinites fuse from about 1650 to 1775°C. The fusion point is preceded by a vitrification interval, which may extend over several hundred degrees, during which a glassy component is an increasingly abundant phase. Small amounts of associated impurities may cause large changes in the determined fusion point and vitrification range.

Rehydration. Grim and Bradley[65] showed that well-crystallized kaolinite, after heating to 600°C slowly (70 hr), picks up a small but appreciable amount of OH water at room temperature, and that poorly crystallized kaolinite similarly heated apparently does not regain any OH water under the same conditions. They interpreted this as evidence that, on dehydration of the well-crystallized mineral at 600°C, some structure remains which has sufficient regularity to take up at least minor amounts of hydroxyl water.

Van Nieuwenberg and Pieters[92] showed that, after heating to 850°C, kaolinite can be completely rehydrated under steam pressure of 100 atm. Schachtschabel[93] stated that after heating to 800°C kaolinite rehydrates slowly at 110°C but that at 175 to 205°C under pressure it completely rehydrates in about 100 hr.

Roy and Brindley[94] also showed the conversion of metakaolin to kaolinite by hydrothermal treatment. The rate and completeness varied with temperature, pressure, and time, with pressure being less important than temperature. Metadickite showed a slight conversion to kaolinite.

Saalfeld[95] and Dietzel and Dhekne[96] rehydrated metakaolin hydrothermally, but these investigators reported the reconversion to poorly crystallized kaolinite, with the first author also reporting that the product had relatively higher plasticity than the original kaolinite. Hill[97] also rehydrated metakaolin and reported that fired test pieces had greatly increased strength as compared with natural kaolinite. Further, the endothermic dehydroxylation peak was about 70°C lower and the activation energy was reduced from

[91] Norton, F. H., "Refractories," McGraw-Hill, New York (1949).

[92] Van Nieuwenberg, C. J., and H. A. J. Pieters, Rehydration of Metakaolin and the Synthesis of Kaolin, *Ber. Deut. Keram. Ges.,* **10:** 260–263 (1929).

[93] Schachtschabel, P., Dehydration and Rehydration of Kaolin, *Chem. Erde,* **4:** 395–419 (1930).

[94] Roy, R., and G. W. Brindley, A Study of the Hydrothermal Reconstruction of the Kaolin Minerals, *Natl. Acad. Sci. Publ.* 456, pp. 125–132 (1956).

[95] Saalfeld, H., The Hydrothermal Formation of Clay Minerals from Metakaolin, *Ber. Deut. Keram. Ges.,* **32:** 150 (1955).

[96] Dietzel, A., and B. Dhekne, Rehydration of Metakaolin, *Ber. Deut. Keram. Ges.,* **34(11):** 366–377 (1957).

[97] Hill, R. D., The Rehydration of Fired Clay and Associated Minerals, *Trans. Brit. Ceram. Soc.,* **52:** 589–613 (1953).

55 to 40 kcal/mole in the rehydrated material in comparison with natural kaolinite. Hill[98] suggested that these results indicate a splitting up of the original crystals into smaller domains on rehydration and later firing.

Halloysite

Dehydration and phase changes on heating. Dehydration curves published by Nutting[7] and Ross and Kerr[9] (Figs. 9-1 and 9-3) show significant loss of water below 100°C for some halloysites and not for others, depending, as is now known, on whether or not the halloysite is the hydrated form ($4H_2O$). The curves show that at temperatures between 100 and 400°C there is a very slight gradual loss of water, with the amount increasing somewhat between about 300 and 400°. From 400–430°C to 500°C the hydroxyl water is lost rapidly and abruptly. Above about 500°C there is again a gradual loss of water up to about 800°C, where dehydration is essentially complete. Ross and Kerr pointed out that halloysite loses its OH lattice water at a temperature 60 to 80°C lower than that for kaolinite.

The water lost below 100°C is pore water and that occurring between basal plane surfaces of adjacent unit layers. Most of this interlayer water can be lost at room temperature under conditions of low humidity and over moderate periods of time. Brindley and Goodyear[99] showed that drying at room temperature does not completely remove all interlayer water, and that the formula for air-dried halloysite is $Al_2O_3.2SiO_2.nH_2O$ with $n = 2\frac{1}{4}$ to $2\frac{3}{4}$, which is equivalent to about one layer of water to every four silicate layers. Drying to about 400°C is required to remove the interlayer water completely. The moisture gradually lost between 100 and 400°C undoubtedly corresponds to the residue of interlayer water, which is driven off with difficulty. Brindley and Goodyear also showed that the complete loss of the interlayer water is not accompanied by a reorganization of the randomly displaced silicate layers.

Harrison and Greenberg[100] reported that the loss of interlayer water takes place randomly, but instantly, and that a partially dehydrated particle is a mixed assemblage of hydrated and dehydrated layers.

[98] Hill, R. D., Studies on Rehydrated and Refired Kaolinite Minerals, *Trans. Brit. Ceram. Soc.,* **55**: 441–456 (1956).

[99] Brindley, G. W., and J. Goodyear, X-ray Studies of Halloysite and Metahalloysite, II, The Transition of Halloysite to Metahalloysite in Relation to Relative Humidity, *Mineral. Mag.,* **28**: 203–215 (1948).

[100] Harrison, J. L., and S. S. Greenberg, Dehydration of Fully Hydrated Halloysite from Lawrence County, Indiana, Proceedings of the 9th National Clay Conference, pp. 374–377, Pergamon Press, New York (1962).

Brindley and Nakahira[59] studied the kinetics of the dehydroxylation of halloysite and concluded that it is a first-order reaction with an activation energy of 55 kcal/mole. As in the case of kaolinite, the rate process is limited by the outward diffusion of water molecules.

Differential thermal curves for halloysites are given in Fig. 9-8. The hydrated form shows an initial V-shaped endothermic peak resulting from the loss of the interlayer water. This peak is, of course, small or completely absent in the lower-hydration forms of the mineral. The shape of the initial endothermic peak is unlike the one sometimes yielded by poorly crystalline kaolinite at about the same temperature. The peak given by kaolinite is less intense and less abrupt than that of halloysite.

The differential thermal curves above about 200°C are essentially like those for kaolinite. The sharp endothermic reaction due to loss of OH water tends to develop a peak at a slightly lower temperature than that in the case of kaolinite. Grimshaw, Heaton, and Roberts[52] found in their differential thermal analyses that the endothermic peak for halloysite was at 500°C, for poorly crystalline kaolinite at 550 to 562°C, and for well-crystallized kaolinite at 583°C. Kerr, Kulp, and Hamilton[101] stated that this peak for halloysite is asymmetrical, being more abrupt on the high-temperature side than the similar peak for kaolinite.

For the temperature interval above the dehydration reaction for OH lattice water, the differential thermal curves of some halloysites are like that of well-crystallized kaolinite, suggesting that some structure is retained after dehydration and up to the temperature of the slight endothermic reaction at about 900°C. It is not known whether or not this is a characteristic of all halloysites.

Following the loss of hydroxyl water, a poorly ordered phase develops which probably is similar in some ways to that of metakaolin. According to Roy et al.,[102] this phase persists only to temperatures of the order of 675 to 700°C where it is lost and γ-alumina is simultaneously developed. According to these authors, the γ-alumina phase persists until about 950°C where it is lost with the nucleation of mullite. Johns[78] findings support the conclusion that the nucleation of mullite is the cause of the exothermic reaction at about 950°C; however, Richardson[76] and Wahl and Grim[103] showed that mullite cannot be detected by X-ray diffraction below 1100°C.

[101] Kerr, P. F., J. L. Kulp, and P. K. Hamilton, "Differential Thermal Analyses of Reference Clay Mineral Specimens," Rept. 3, American Petroleum Institute Project 49, Columbia University, New York (1949).

[102] Roy, R., D. M. Roy, and E. E. Francis, New Data on Thermal Decomposition of Kaolinite and Halloysite, *J. Am. Ceram. Soc.,* **38:** 198–205 (1955).

[103] Wahl, F. M., and R. E. Grim, High Temperature DTA and X-ray Diffraction Studies of Reactions, Proceedings of the 12th National Clay Conference, pp. 69–81, Pergamon Press, New York (1964).

Richardson[76] and Glass[77] arrived at somewhat different conclusions. According to these authors, γ-alumina does not develop until about 950°C, and its development is responsible for the sharp exothermic reaction at this temperature. Visconti[104] pointed out that the tubular form of halloysite persists to 970°C, which is at least into the range of the high-temperature exothermic reaction. Stubican and Gunthard[105] indicated, on the basis of infrared analyses, that adsorbed water is present in halloysite following its heating to 850°C, whereas in well-ordered kaolinite no such water is present in material heated above 650°C. It can only be concluded that further research is necessary to elucidate the nature of halloysite after it is dehydrated and heated into the range of the intense exothermic reaction at about 950°C.

Richardson[76] reported that cristobalite developed at about 1100 to 1300°C. According to Wahl and Grim,[103] cristobalite does not form from pure halloysite even at temperatures of about 1400°C. However, with traces of certain impurities (e.g., phosphates) cristobalite may form at about 1200°C.

The presence of traces of elements appears to have about the same effect on high-temperature-phase development in halloysite as it does in kaolinite except that, as shown by Wahl,[83] the additives are relatively more effective in halloysite.

Hyslop and McMurdo[87] presented an expansion-contraction curve for halloysite (Fig. 9-20). The curve shows the halloysite to expand slightly up to the dehydration of the mineral at about 500°C. Above this temperature there is a gradual contraction up to about 900°C, followed by a steep contraction up to about 1000°C. From 1000°C to the end of the experiments at 1100°C there is a further small amount of contraction. The difference in the character of the curves for halloysite and kaolinite may reflect differences in the high-temperature phases noted above. When halloysite is fired to temperatures just short of complete vitrification, masses of the mineral crack into small pieces, whereas masses of kaolinite remain intact. This difference in physical character must also reflect a difference in high-temperature-phase reactions. Although halloysite is very refractory, it is difficult to use in ceramic products because of this high-temperature characteristic. The tendency of halloysite to crack into pieces at high temperatures makes it difficult to determine its vitrification range and fusion point. It seems that the mineral fuses at a slightly higher temperature than kaolinite and that the fusion point is preceded by a very short vitrification interval.

[104] Visconti, Y. S., New Observations Relative to Tubular Kaolin by Means of Chemical Dispersion and Electron Microscopy, *2nd Conf. Ceram., Brazil* (1956).

[105] Stubican, V., and H. H. Gunthard, Infrared Spectra of High-temperature Phases of Kaolinite and Halloysite, *Nature,* **179:** 542 (1957).

Rehydration. The interlayer water lost at room temperature is not ordinarily regained; i.e., the reaction is not reversible. Bradley[106] showed that certain organic molecules, e.g., glycol, will penetrate between the layers of the dehydrated mineral and that, when the glycol is removed by a solvent such as alcohol, mixed with water, water remains between the layers. According to MacEwan[107] and Brindley,[108] this reintroduction of interlayer water is effective only if the original interlayer water has not been completely removed. Brindley suggested that the glycol molecules cannot enter between the layers unless some water is present to keep them partly open.

Grim and Bradley[65] showed that halloysite, after being heated to 600°C for 1 hr, regains a very small amount of OH lattice water in the course of standing at room temperature for 70 days. A similar result was found for well-crystallized kaolinite, but not for poorly crystalline kaolinite. They interpreted this to mean that some structural organization is left in halloysite after the loss of OH lattice water. The similarity in this attribute between halloysite and well-crystallized kaolinite rather than poorly crystalline kaolinite suggests that the difference between the kaolinite and halloysite structures is based on more than the variation in the stacking of adjacent silicate sheets. Bates et al.[109] emphasized this point in considering the rolled tube-like shape of the halloysite minerals revealed in electron micrographs.

Roy and Brindley[94] attempted to reconstitute by hydrothermal treatment halloysite that had been dehydrated at 610°C. At their highest temperatures and pressures (375°C and 20,000 psi) after several weeks there was much less change than for kaolinite.

Smectite

Dehydration and phase changes on heating. Dehydration curves, after Ross and Hendricks,[11] are given in Fig. 9-4. The curves show considerable water

[106] Bradley, W. F., Diagnostic Criteria for Clay Minerals, *Am. Mineralogist,* **30**: 704–713 (1945).

[107] MacEwan, D. M. C., Complexes of Clays with Organic Compounds, I, Complex Formation between Montmorillonite and Halloysite and Certain Organic Liquids, *Trans. Faraday Soc.,* **44**: 349–367 (1948).

[108] Brindley, G. W., The Kaolin Minerals, "X-ray Identification and Structures of Clay Minerals," chap. II, pp. 32–75, Mineralogical Society of Great Britain Monograph (1951).

[109] Bates, T. F., F. A. Hildebrand, and A. Swineford, Morphology and Structure of Endellite and Halloysite, *Am. Mineralogist,* **35**: 463–484 (1950).

loss at low temperatures (100 to 200°C), and, as has been shown (pages 283–288), the amount of this water, which occurs mostly as interlayer water between the silicate sheets, is contingent upon the nature of the adsorbed ions and the pretreatment of the sample, e.g., amount of drying, relative humidity, etc. Mackenzie[110] suggested that the amount of interlayer water depends on the hydration energy of the adsorbed cations and on the hydration of the surface, and that thermogravimetric curves can be interpreted to give relative values for each of these amounts of water. For most divalent cations (ranging in size from Cu^{++} to Ca^{++}) the ion is more important than the layer surface, but for larger divalent cations as well as for the monovalent cations the influence of the surface is dominant.

Rowland, Weiss, and Bradley[111] concluded, on the basis of X-ray-diffraction oscillating-heating data, that calcium, magnesium, manganese, lithium, and hydrogen montmorillonites have octahedral coordination of their exchange cations and upon heating they pass through stable hydrates with c-axis spacings of about 14.5 and 11.5Å. Montmorillonite with sodium or potassium has a regular one-layer configuration at 12.4 Å. They further concluded that the foreshortened oxygen-oxygen approaches of the lower hydrate (11.5 Å) may be accounted for by assuming that the exchange ions take up coordination positions between silica-oxygen water levels or, alternatively, that some of the silica tetrahedrons are inverted. Glaeser, Mantin, and Mering[112] showed that polyvalent ions tend to detach themselves from the silicate surface and incorporate themselves in the water layer. The monovalent ions, on the contrary, remain on the silicate surface. Van Olphen and Deeds[113] have shown that a sequence of discrete hydration states develop when certain montmorillonite-organic complexes are allowed to hydrate.

The dehydration curves are S-shaped, usually showing no distinct break between the temperature of loss of the last interlayer water and the beginning of loss of (OH) lattice water. Ross and Hendricks[11] arbitrarily took the temperature of 300°C as the break point, and this is probably the best single value that can be named. The complete loss of the interlayer water is accompanied by a reduction in the c dimension to 9.4 to 10 Å, with the exact value depending on the size of the interlamellar ions.

On further heating up to the dehydroxylation temperature, the diffraction

[110] Mackenzie, R. C., Hydration Characteristics of Montmorillonite, *Ber. Deut. Keram. Ges.*, **41**(12): 696–707 (1964).

[111] Rowland, R. A., E. J. Weiss, and W. F. Bradley, Dehydration of Monoionic Montmorillonites, *Natl. Acad. Sci, Publ.* 456, pp. 85–95 (1956).

[112] Glaeser, R., I. Mantin, and J. Mering, Étude sur l'acidité de la montmorillonite, *Intern. Geol. Congr. 21st Session, Norden, France*, pp. 28–34 (1960).

[113] Van Olphen, H., and C. T. Deeds, Stepwise Hydration of Clay Organic Complexes, *Nature*, **194**: 176 (1962).

data show little change, indicating that the silicate structure is about unchanged (Warshaw et al.[114]; Rowland et al.[111]).

The curves of various smectites show wide variation of temperature for the loss of OH lattice water. For normal montmorillonites, i.e., those with moderate to low substitutions of iron and magnesium for aluminum, rapid loss of OH water begins at about 500°C, and the dehydration is practically complete at 800°C. For nontronites the rapid loss of hydroxyl water begins at about 400°C, and the dehydration is substantially complete at 800°C. Hectorite is unusual in that the rapid loss does not begin until about 700°C and the loss is still not complete at 930°C. In hectorite the hydroxyls are partially replaced by fluorines, which are possibly more tightly held within the lattice structure than the hydroxyls.

The loss of (OH) water in smectites, unlike that in kaolinite and halloysite, does not begin or end abruptly, except in the case of hectorite. Another point of difference from kaolinite and halloysite, as Caldwell and Marshall[115] and Kelley et al.[10] showed, is the small variation in the dehydration characteristics of smectites with particle size. This is to be expected, since a reduction in particle size of the smectites is essentially a reduction of aggregates and not of primary components; i.e., it is a separation of flakes along basal planes of easy separation rather than a cross breaking of flakes.

The dehydration curves of some of the smectites show flexures in the temperature interval of the loss of OH water; the significance of these will be discussed when the differential curves are considered.

Differential thermal curves for a series of smectites are given in Fig. 9-9. At low temperatures they show variable endothermic peaks due primarily to the loss of interlayer water. The size and character of these peaks are contingent upon the nature of the adsorbed cation and on the pretreatment of the sample (see pages 254–261).

Most dioctahedral smectites with relatively small amounts of iron and magnesium replacing aluminum show an endothermic reaction due to loss of OH lattice water, beginning rather gradually at about 450 to 500°C, ending at about 750°C, and with a peak temperature of about 700°C. Large replacement of aluminum by iron causes a reduction in the temperature of the reaction, and for nontronite the peak temperature is 500 to 600°C. Some smectites which have a moderately high magnesium content, e.g., the sample from Otay, California, also show a lowering of the dehydration temperature. On the other hand, the magnesium-rich hectorite dehydrates at a higher tempera-

[114] Warshaw, C. M., P. E. Rosenberg, and R. Roy, Changes Effected in Layer Silicates by Heating Below 550°C, *Clay Minerals Bull.*, 4(23): 113 (1960).

[115] Caldwell, O. G., and C. E. Marshall, A Study of Some Chemical and Physical Properties of the Clay Minerals Nontronite, Attapulgite and Saponite, *Missouri Univ., Coll. Agr., Res. Bull.* 354 (1942).

ture than the aluminous smectites. In general, trioctahedral smectites show a gradual loss of hydroxyls and a broad slight endothermic peak.

Some smectites of relatively low iron and magnesium contents show unusually low temperatures for the hydroxyl-loss endothermic reaction, and some samples (e.g., smectite from Palmer, Arkansas) show a dual endothermic peak for hydroxyl water loss. In some cases the dual peak can be explained tentatively by a mixing of smectites, e.g., high-iron and aluminous varieties. In some other examples the possibility of small amounts of kaolinite or halloysite as impurities cannot be eliminated. Kerr et al.[101] reasoned that the explanation may reside in the geometry of the replacement and population of the octahedral positions in the structure. According to these authors, variation in the position of the cations in the octahedral packing and hence variation in the nearness to hydroxyl ions should affect the strength of the bonding of the OH and hence the energy necessary for its release. Grim and Kulbicki[116] showed that many montmorillonite clays are very intimate mixtures of two forms of montmorillonite varying in composition and perhaps also in structure, especially in the octahedral layer. Greene-Kelly[117] pointed out that defect structures might explain variations in dehydroxylation. Earley et al.[118] stated that particle size might be important in the temperature of the loss of hydroxyls, although it is difficult to see how this could be a significant factor in explaining dual dehydroxylation peaks.

Numerous investigators have shown that the dehydroxylation of smectite and the accompanying endothermic reaction may vary substantially, depending on the nature of the exchangeable cation. Ferrandis et al.[119] gave the following sequence for the rate of hydroxyl loss for a given montmorillonite, $Li > K > Na > Mg > Ba > Ca > Sr$. Vivaldi et al.[120] reported that hydroxyl loss for an ammonium montmorillonite may begin as low as 300°C and be completed at 400°C.

Jacobs and De Fre[121] showed that the dehydroxylation of montmorillonite

[116] Grim, R. E., and G. Kulbicki, Montmorillonite: High Temperature Reaction and Classification, *Am. Mineralogist*, **46**: 1329–1369 (1961).

[117] Greene-Kelly, R., The Montmorillonite Minerals (Smectite), "The Differential Thermal Investigation of Clays," chap. V, pp. 140–164, Mineralogical Society, London (1957).

[118] Earley, J. W., I. H. Milne, and W. J. McVeagh, Thermal, Dehydration, and X-ray Studies on Montmorillonite, *Am. Mineralogist*, **38**: 770–783 (1953).

[119] Ferrandis, V. A., and M. C. Rodriguez-Pascual, Influence of the Exchange Cations on the Rate of Dehydration of Montmorillonite, *Anales Edafol. Fisiol. Vegetal (Madrid)*, **17**: 257–288 (1958).

[120] Vivaldi, J. L. M., F. G. Vilchez, M. H. B. de Castro, and M. R. Gallego, The Thermal Decomposition of Ammonium Montmorillonites, *Clay Mineral Bull.*, **4**(22): 81–87 (1959).

[121] Jacobs, T., and M. de Fre; Kinetics of the Dehydroxylation of Montmorillonite by Thermogravimetric Methods, *Silicates Ind.*, **26**: 363 (1961).

is a first-order reaction with activation energies of 54.2, 50.3, and 40.7 kcal/mole for Na^+, Ca^{++}, and H^+ montmorillonites, respectively. Earlier Ferrandis et al.[119] gave the following sequence of activation energies: $Li > K > Na > Ca > Ba > Mg > Sr$.

Thilo and Schunemann[122] and Grim and Bradley[123] showed that the general layer type of structure illustrated by smectite and the micas is able to survive the elimination of OH water with only moderate readjustments. According to Bradley and Grim,[124] the removal of hydroxyl water is correlative with increases of 0.1 to 0.3 Å in c-axis periodicity and involves the explusion of about one-sixth of the oxygens of the octahedrally coordinated portion of the structure. This correlation is apparently confined to the dioctahedral types of smectites. An idealized rearrangement of the octahedral layer which adequately accounts for the meager data observable is illustrated in Fig. 9-21. It represents merely the lifting of the adjacent oxygen layer out of the packed position, with the immediate consequences that the vertical height is increased and the intensity of the 4.5-Å diffraction line is augmented. A diagram of a typical dehydration specimen is compared with its normal dry state in Fig. 9-21. Diffraction data for the rearranged material, after Bradley and Grim, are given Table 9-2. The idealized scheme shown in Fig. 9-21 involves shared pairs of octahedral coordination faces and is probably unstable. Applications of the same scheme to a trioctahedral type would demand sharing of three such pairs of faces, which is presumably impossible. No rearranged anhydrite was observed for hectorite and talc, which were the only two certain trioctahedral structures studied by Bradley and Grim.[124]

Jonas[125] suggested a somewhat different structure for the anhydrous modification in which the remaining oxygens of the O-OH sheet lie in the same planes as the octahedral cations, causing a rearrangement of the octahedral cations. This rearrangement is presumably impossible for trioctahedral forms. Heller et al.[126] raised objections to both suggested structures of Bradley and Grim[124] and Jonas.[125]

[122] Thilo, E., and H. Schunemann, Chemical Studies of Silicates, IV, Behavior of Pyrophyllite on Heating and the Existence of a "Water-free" Pyrophyllite, *Z. Anorg. Allgem. Chem.*, **230**: 321–325 (1937).

[123] Grim, R. E., and W. F. Bradley, Investigation of the Effect of Heat on the Clay Minerals Illite and Montmorillonite, *J. Am. Ceram. Soc.*, **23**: 242–248 (1940).

[124] Bradley, W. F., and R. E. Grim, High Temperature Thermal Effects of Clay and Related Materials, *Am. Mineralogist*, **36**: 182–201 (1951).

[125] Jonas, E. C., The Reversible Dehydroxylization of Clay Minerals, *Natl. Acad. Sci., Publ.* 395, pp. 66–73 (1955).

[123] Heller, L., V. C. Farmer, R. C. Mackenzie, B. D. Mitchell, and H. F. W. Taylor, The Dehydroxylation and Rehydroxylation of Trimorphic Dioctahedral Clay Minerals, *Clay Minerals Bull.*, **5**(28): 56–72 (1962).

Heller and Kalman[127] concluded that magnesium and calcium as well as lithium enter the vacant octahedral positions of dioctahedral forms on dehydroxylation (see also pages 77–88). Lindquist[128] believes that dioctahedral layer silicates, when heated, are covered to trioctahedral modifications.

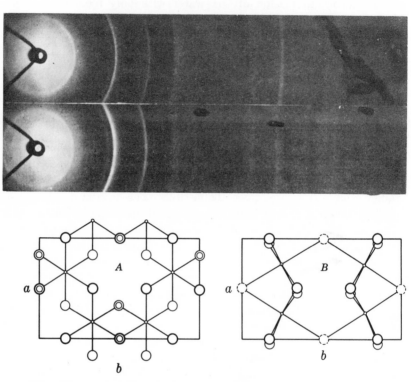

Fig. 9-21. Probable rearrangement in heptaphyllite-type clay minerals upon loss of hydroxyl water. The diffraction diagrams illustrate the Tatatilla montmorillonite after heating to 550° (upper) and 725°C (lower). The schematic sketches represent the arrangement within an octahedral layer before and after dehydroxylation. (After Bradley and Grim.[124])

The structure of many montmorillonites persists to temperatures of the order of 800 to 900°C, and Grim and Bradley[123] correlated the destruction

[127] Heller, L., and Z. H. Kalman, An Approximate Determination of the Position of Some Cations in Dehydroxylated Montmorillonite, *Clay Minerals Bull.*, **4**(25): 213–220 (1961).

[128] Lindquist, B., Polymorphic Phase Changes During Heating of Dioctahedral Layer Silicates, *Geol. Foren. Forh.*, **84**(2): 224–229 (1962).

of the lattice with the endothermic reaction frequently shown at about this temperature interval. This correlation probably applies only to dioctahedral smectites, for, as has been shown, an anhydride structure should not persist in the trioctahedral members. Some smectites which have been considered to be dioctahedral, e.g., some nontronites, do not produce this high-temperature endothermic peak, and therefore it appears that the anhydrite structure may not persist in all dioctahedral types.

Table 9-2   *Diffraction data for montmorillonite*
(Tatatilla, Mexico) ignited to 725°C
(From Bradley and Grim[124]*)*

Indices as orthohexagonal	d, Å	Observed relative intensities	Calculated relative intensities for idealized scheme of Fig. 9-21
001	9.7	m	2
002	4.85	m	2
110, 020	4.48	ss	18
003	3.22	s	10
200, 130	2.59	m	4
	2.5	Diffuse m	
220, 040	2.23	m	2
	2.1	Diffuse w	
005	1.93	w	1
240, 150	1.72	w	2
	1.67	Diffuse m	
330, 060	1.51	m	2
	1.49	Diffuse w	
400, 260	1.31	w	3
420, 350	1.25	w	1

ss = very strong; s = strong; m = moderate, w = weak.

Page[129] suggested another explanation for the third endothermic peak of the montmorillonites. According to him, it is due to the loss of OH water which is bonded with magnesiums in octahedral coordination, rather than to lattice destruction. It has not been established that there is a correlation between the magnesium content and the presence or size of this peak. McConnell[130] attributed this high-temperature endothermic reaction to the loss

[129] Page, J. B., Differential Thermal Analyses of Montmorillonite, *Soil Sci.,* **56:** 273–283 (1943).

[130] McConnell, D., The Crystal Chemistry of Montmorillonite, *Am. Mineralogist,* **35:** 166–172 (1950).

of hydroxyls which are in the silica layer in tetrahedral configuration. He attempted to correlate the peak with the amount of water lost at high temperatures, but the data are not completely convincing. An amount of water of the order of 0.5 to 1 percent persists beyond the temperature of the major loss of hydroxyl water at 500 to 700°C. The amount of water to be lost at high temperatures is small in comparison with the size of the endothermic peak, and furthermore the loss of these hydroxyls does not seem to take place abruptly, as would be required for the development of a peak of the character frequently shown on the thermal curves. Earley et al.[118] showed that there is no relationship between the amount of water loss and the third endothermic peak.

Following the third endothermic reaction, the differential thermal curves show a wide variety of thermal reactions, indicating that there is considerable variation in the phases formed when various montmorillonites are heated to elevated temperatures (Fig. 9-12). Data given in Table 9-3 from Bradley and Grim[124] show that this is the case.

Bradley and Grim[124] arrived at some general conclusions, based on a careful study of many samples, and the following discussion is taken from their work. The high-temperature portions of the DTA curves of montmorillonites tend to differentiate into two separate types: those which present an S-shaped feature and those which exhibit a distinct shoulder directly above the third endothermic reaction. Clays showing the S-shaped curves yield spinel following the first exothermic reaction; it is proportional in amount to the magnitude of the thermal effect. In the second group the first prominent new phase appears to be quartz. The new phase is thus seen to develop in the one case from the octahedral region in the clay and in the other case from the tetrahedral region.

From the point of view of the composition of the mineral, montmorillonite seems to develop quartz if there is substantially no substitution in the tetrahedral sheet, and spinel if there is substantial substitution of Al^{3+} in tetrahedral positions. The regularity in the composition of the tetrahedral scheme appears to *endow* it with ability to maintain its utility under conditions that disorganize substituted layers. Furthermore the montmorillonites developing the second type of curve, with the delayed exothermic reaction, are relatively low in iron. In these low-iron montmorillonites, enstatite, mullite, or anorthite develops at about 1000°C, in addition to the β-quartz, probably depending on the amount of magnesium present and possibly also on the amount of exchangeable Ca^{++}.

Clays of either type, which have undergone either of the foregoing syntheses, may show an additional, fairly prominent reaction near 1200°C which seems to be due to either mullite or cordierite with about equal frequency. Bradley and Grim[124] pointed out that the first high-temperature phases appear to be largely determined by the structural attributes of the montmorillonite itself.   The initial phases are not the ones which necessarily would be expected from phase-diagram considerations and bulk chemical composition but are due primarily to "inheritance" from the original structure.   The later phases developed at higher temperatures are dependent primarily on bulk composi-

Table 9-3   High-temperature phases developed on firing montmorillonites
(From Bradley and Grim[124])

Sample location	900°C	1000°C	1100°C	1200°C	1300°C
Otay, California	. . . .	β-Quartz (1) Enstatite (3)	Cristobalite (1) β-Quartz (1) Enstatite (2)	Cristobalite (1) Cordierite (1)	Cristobalite (1) Cordierite (1) Periclase (3)
Tatatilla, Vera Cruz, Mexico	. . . .	β-Quartz (2)	. . . .	. . . .	Cristobalite (1) Mullite (2) Cordierite (2)
Upton, Wyoming	Spinel (1)	Spinel (1) α-Quartz (2)	Spinel (2) Cristobalite (1)	. . . .	Mullite (2)
Cheto, Arizona	. . . .	β-Quartz (1) Anorthite (?)(3)	β-Quartz (1) Cristobalite (3) Anorthite (?)(3)	. . . .	Cristobalite (1) Cordierite (1)
Palmer, Arkansas	. . . .	Spinel (1)	Spinel (1) Quartz (3)	Cristobalite (1) Spinel (1) Mullite (1)	
Sierra de Guadalupe, Mexico	. . . .	Spinel (1)	Cristobalite (1) Spinel (1)	Cristobalite (1) Spinel (1) Cordierite (2)	Cristobalite (1) Cordierite (1)
Harris County, Texas	. . . .	Spinel (2) Cristobalite (3)	Cristobalite (1) Spinel (1)	Cristobalite (1) Spinel (1) Mullite (2)	Cristobalite (1) Mullite (2)
Pontotoc, Mississippi	. . . .	Spinel (1) α-Quartz (2)	Cristobalite (1) Spinel (1)	. . . .	Cristobalite (2) Cordierite (2)
Fairview, Utah	. . . .	. . . .	. . . .	Mullite (1) Cristobalite (1)	Mullite (1)
Wagon Wheel Gap, Colorado	Spinel (2)	Cristobalite (1) Mullite (3) Spinel (2)	. . . .	. . . .	Cristobalite
Nashville, Arkansas (woody nontronite)	. . . .	. . . .	. . . .	. . . .	Mullite Cristobalite Spinel

Parenthetic numbers signify (1) important, (2) moderate, (3) minor.

tion, and here "inheritance is of little importance.   Thus, in the case of the development of mullite or cordierite, the controlling factor is probably the amount of magnesium present.

The spinel-forming specimens all develop more or less cristobalite around

1000 or 1100°C without going through the quartz intermediate stage. No separate features appear in the differential thermal curves to indicate any energy effect associated with the growth of the cristobalite. Since this is true also of opal fired alone, it is assumed that any change involved is not picked up by the thermal method. Cristobalite is actually effectively somewhat less dense than the condensed-layer configuration derived from the articulation of the silica layers (see page 323) and might even require an energy input under the conditions of the analyses.

The spinel tends to disappear at temperatures above about 1200°C in all samples studied except the nontronite sample.

Bradley and Grim[124] considered the structual changes involved in the development of high-temperature phases when montmorillonites are heated, and the following statements are also from their work.

As a spinel phases first appear, X-ray-diffraction lines are notably diffuse, and probably the phase does not have any particular composition. Rather, it is merely an irregularly constituted assemblage of small cations, some octahedrally coordinated and others tetrahedrally coordinated, with only the oxygen packing actually approaching crystalline regularity. The spinels grow rapidly to a clearly crystalline powder affording a clean diffraction diagram. No noticeable difference in relative intensities is observed between the diffuse and the sharp patterns, although cube-edge parameters may change somewhat (with changing composition), and no erratic appearance of new lines has been observed.

The cryptocrystalline quartz produced on firing many montmorillonites at about 1000°C exhibits many vagaries with respect to its apparent relation to high-low inversion. Synthesized quartz, when observed by Bradley and Grim at room temperature, was found in some cases as the α, in others as the β modification, and in one case apparently intermediate. Both the α and β forms make the normal slow transition to cristobalite at more elevated temperatures. Bradley and Grim[124] suggested that the failure to invert is caused by the trapping of occasional extraneous ions in the channels of the structure.

The synthesized quartz develops suddenly with a correlative sharp exothermic effect, and the crystallites are subject to moderate gain growth in soaking periods of a few hours before eventually transforming into cristobalite. The original concept of the layer silicates was drawn from observations of the dimensional compatibility of brucite and the cristobalite-tridymite type of tetrahedral layer. One might postulate that a quick crystallization of silica would lead to one of these forms but that does not seem to be the case. Bradley and Grim[124] pictured the synthesis of the quartz as illustrated in Fig. 9-22, which shows two sets of linked tetrahedrons, representing adjacent montmorillonite surfaces, in which oxygens are partitioned into two kinds. Suitable articulation of these pairs actually simulates the quartz arrangement of height equivalent to one complete unit cell, and the new grouping effects a localized gain in density of the order of about 10 percent. This postulation of the articulation of montmorillonite surfaces into the quartz arrangement is a consequence of extended clean clay-mineral surfaces. To articulate two such surfaces into a segment of the cristobalite structure would require the introduction of additional oxygen between surfaces to complete the coordination of any silicons which might provide valence bonds between layers.

The fine-grained cristobalite resulting from syntheses in heterogeneous matrices is noticeably variable and of uncertain inversion character. Such material gradually becomes typical cristobalite upon ignition to sufficiently high temperatures, apparently without any sudden energy effects.

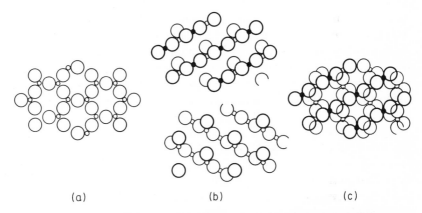

(a)   (b)   (c)

Fig. 9-22. The articulation of tetrahedrally arranged montmorillonite surfaces to form quartz. (a) A montmorillonite surface; (b) two distorted arrangements of (a); (c) a superposition of the two levels b. (After Bradley and Grim.[124])

The $MgSiO_3$ formed in the firing of some montmorillonites is comparable with the "mesoenstatite" of Thilo and Rogge.[131] This material is probably to be looked upon as a rudimentary nucleation rather than as a normal cry tallization. The structural relations of the layer silicates and of the pyroxenes are such that projections onto the respective orthohexagonal axes are approximate equivalents. In the imperfectly developed diffraction diagrams of the fired montmorillonites, the lines which appear are only those which relate to the pseudohexagonal nature. In fact they constitute an identification of $MgSiO_3$ only in the sense that they correspond to equivalent lines which appear in better and better crystallized materials of higher mag-

---

[131] Thilo, E., and G. Rogge, Chemische Untersuchungen von Silikaten, VIII, *Ber. Deut. Chem. Ges. B.,* **72:** 341–362 (1939).

nesium content for gradations up to and including the typical enstatite crystallization obtainable on firing tremolite asbestos.

Bradley and Grim[124] also discussed the development of forsterite from chrysotile and clinochlore. They concluded that the transition from the old to the new phase requires no chaotic disintegration followed by the reconstitution of a new assemblage, but simply the freedom of a few Mg ions to migrate to the nearest equivalent interstices. In other words, the oxygen packing remains substantially the same, and the transformation is a matter of the shift of a few magnesium ions. They concluded further that the development of mullite from pyrophyllite, which is probably similar to its development from montmorillonite, is carried forward to a considerable extent along the same lines as the development of the forsterite.

Grim and Kulbicki[116,132] studied the high-temperature-phase reactions of a large number of montmorillonites by continuous X-rays. They confirmed the finding of Bradley and Grim[124] that the high-temperature reactions of various smectites were somewhat different. According to them, iron-rich forms of the mineral and those carrying potassium as exchangeable cations showed very little, if any, development of new crystalline phases at elevated temperatures. These investigators reported the same high-temperature phases identified by Bradley and Grim[124] except that in many instances they did not record the initial development of spinel, perhaps because a longer time of firing at about 1000°C was required to develop this species than that used by Grim and Kulbicki in their continuous-heating experiments. Grim and Kulbicki[132] divided the aluminous smectites into two types on the basis of their high-temperature reactions (Fig. 9-23). One type, represented by a specimen of bentonite from Cheto, Arizona, loses all diffraction characteristics between about 800 and 850°C. At about 900°C β-quartz appears; it inverts to β-cristobalite beginning at about 1000°C. The cristobalite continues to develop until about 1200°C and then slowly disappears. A small amount of anorthite is present from about 1000 to 1150°C. Cordierite appears at about 1260°C. Both cordierite and cristobalite disappear at about 1450°C, above which temperature no crystalline phases are present.

The other type of montmorillonite, represented by samples of bentonite from Mendoza, Argentina, and from Wyoming, loses all diffraction characteristics at about 900°C. No diffraction effects are shown until about 1150°C, when mullite appears. Cristobalite develops above 1200°C, and the mullite and cristobalite persist to about 1500°C, above which temperature there are no crystalline phases (Fig. 9-23).

Grim and Kulbicki[116] showed that the aluminous smectites of many bentonites can be characterized as Wyoming type or Cheto type or as a mixture of these types. The differential thermal characteristics of these two types,

---

[132] Grim, R. E., and G. Kulbicki, Étude des réactions de hautes températures dans les minéraux argileux au moyen des rayons, I, *Bull. Soc. Franc. Ceram.*, **36**: 21–28 (1957).

in general, conform to the two groups distinguished by Bradley and Grim.[124] However, detailed chemical analyses did not show any significant variation in the population of the tetrahedral layers of these two types; hence the suggestion that the amount of replacement of aluminum for silicon was a major distinction between the forms of aluminous montmorillonites could not

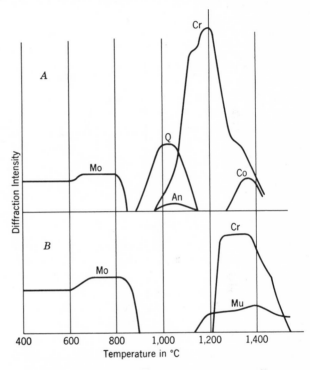

Fig. 9-23. Phase changes on heating montmorillonite to elevated temperatures as shown by continuous X-ray diffraction. (A) Cheto type; (B) Wyoming type; Mo, montmorillonite anhydrite; Q, β-quartz; Cr, β-cristobalite; An, anorthite; Co, cordierite; Mu, mullite. (After Grim and Kulbicki.[132])

be confirmed. Grim and Kulbicki[116] presented data to indicate that a major difference resides in the population of the octahedral positions, with the amount of magnesium being larger in the Cheto than in the Wyoming type. These authors further suggested that the inversion of a small number of the silica tetrahedrons, as suggested by Edelman and Favejee,[133] might be a

[133] Edelman, C. H., and J. C. L. Favejee, On the Crystal Structure of Montmorillonite and Halloysite, *Z. Krist.*, **102**: 417–431 (1940).

distinguishing characteristic of the two forms. They presented evidence suggesting that there may be an appreciable amount of such inversion in the Cheto form and not in the Wyoming form.

Grim and Kulbicki[132] showed the great influence of the exchangeable-cation composition of montmorillonite on the development of high-temperature phases (Fig. 9-24). An investigation of a sample of Cheto-type montmorillonite showed an enhanced development of the high-temperature phases when

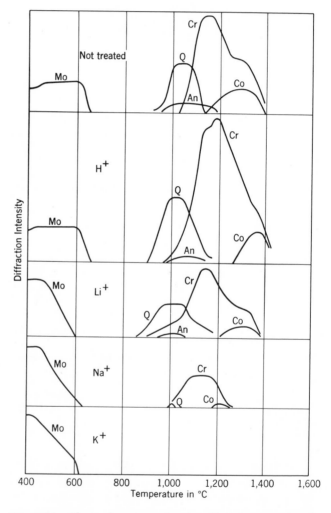

Fig. 9-24. Phase changes on heating Cheto-type montmorillonite with various exchangeable cations to elevated temperatures as shown by continuous X-ray diffraction. See Fig. 9-23. (After Grim and Kulbicki.[132])

the exchangeable cations were replaced by hydrogen (and possibly also by some aluminum). The lithium form of the mineral showed a slight reduction in the intensity of the diffraction characteristics for the high-temperature phases. The sodium form showed a very great reduction in the development of high-temperature phases and also reduced the maximum temperature to which they persisted from about 1400°C to slightly above 1250°C. In the potassium form of the mineral, no high-temperature crystalline phases developed. These data show that additives or the presence of nonmontmorillonite components in a natural montmorillonite clay would be likely to have a very great influence on the nature of the high-temperature phases formed.

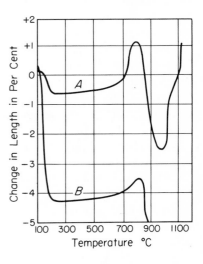

Fig. 9-25. Expansion and contraction of montmorillonite (Wyoming bentonite) on firing. (A) Preheated to 110°C; (B) without preheating. (After Steger.[135])

Byrne[134] showed that adsorbed interlayer organic molecules may influence the formation of high-temperature crystalline phases even though the organic molecules are completely eliminated at far lower temperatures.

Steger[135] investigated the expansion-contraction of montmorillonite. The specimen examined by him showed a very slight gradual expansion from 200 to about 700°C. Between 700 and 800°C there was a rapid expansion, possibly to be correlated with the development of the anhydride form. This expansion was immediately followed by a large contraction up to about 950°C, when the sample again expanded (Fig. 9-25).

Rieke and Mauve[90] showed (Table 9-1) that the specific gravity of a sample of montmorillonite decreased on heating from 600 to 800°C, followed

[134] Byrne, P. J. S., Some Observations on Montmorillonite, Organic Complexes, *Natl. Acad. Sci., Publ.* 327, pp. 241–253 (1954).

[135] Steger, W., Die Längenänderungen von Kaolinen und einigen anderen Tonmineralien beim Brennen bis 1100°C, *Ber. Deut. Keram. Ges.,* **23**: 46–92, 157–174 (1942).

by a slight increase when it was heated to 950°C. There was a further slight decrease on firing to 1000°C, the maximum temperature of the experiments.

As would be expected from the large variations in the chemical composition of the smectites, the fusion points of smectite minerals exhibit wide variations. Iron-rich members of the group may fuse below 1000°C, whereas iron-poor members may not fuse until about 1200 to 1300°C. The presence of alkalies and alkaline earths renders smectites nonrefractory. The vitrification range of smectite minerals is frequently not large.

Rehydration. Hofmann and Endell[136] showed that the property of montmorillonite to rapidly regain interlayer water and expand is lost for Li montmorillonite after heating to 105 to 125°C, for H or Ca montmorillonite after heating to 300 to 390°C, and for Na montmorillonite after heating to 390 to 490°C. Hofmann and Klemen[137] stated that the pronounced effect of lithium is due to its migration into octahedral sites, neutralizing the charge on the lattice. Greene-Kelly[138] agreed that such migration takes place but found that it is possible only if the preponderance of charge on the silicate layer is due to octahedral substitutions. Greene-Kelly[139] reported the temperature for the loss of rapid regaining of interlayer water for montmorillonites with a variety of exchangeable cations.

Grim and Bradley[65] showed that Na montmorillonite, after being heated for 1 hr at 600°C, very slowly picked up interlayer water, so that a small amount was shown at the end of 268 days. Samples heated to 800°C showed no pickup of interlayer water after the same length of time. The same authors presented data for a Ca montmorillonite showing that it also regained interlayer water after heating to 600°C and that the rehydration was slightly slower than for the Na montmorillonite.

According to Grim and Bradley, samples heated to 600°C for 1 hr regained about one-fourth of the original OH lattice water in 268 days. As shown in Fig. 9-26, the endothermic reaction corresponding to the removal of this rehydration water yields a double peak, suggesting that the regained water is of two kinds or forms in the lattice. Some of the rehydration water is removed at about the same temperature as the original hydroxyl water,

[136] Hofmann, U., and J. Endell, Die Abhängigkeit des Kationaustausches und der Quellung bei Montmorillonit von der Vorerhitzung, *Ver. Deut. Chem. Beih.*, **35**: 10 (1939).

[137] Hofmann, U., and R. Klemen, Effect of Heating on Li-Bentonite, *Z. Anorg. Chem.*, **262**: 95–99 (1950).

[138] Greene-Kelly, R., Dehydration of the Montmorillonite Minerals, *Mineral. Mag.*, **30**: 604–615 (1955).

[139] Greene-Kelly, R., Irreversible Dehydration in Montmorillonite, *Clay Minerals Bull*, **2**(9): 52–56 (1953).

and the remainder at a temperature about 150°C lower. Both kinds or forms of water are regained at about the same rate. The rehydration of the OH lattice water does not afford any clear-cut X-ray-diffraction interference, but in the light of the anhydrous configuration illustrated (Fig. 9-21), it seems reasonable to assume that only a small part is recombined into the original configuration, with an additional significant amount attaining some other bonding of lower energy.

Jonas[125] studied rehydroxylation of montmorillonite on cooling from 700°C in an atmosphere of water vapor. He found almost complete regaining

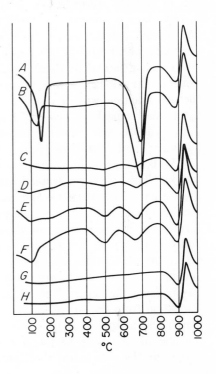

Fig. 9-26. Differential thermal curves of sodium montmorillonite (bentonite), Clay spur, Wyoming. (*A*) Not heated; (*B*) heated, 500°C for 1 hr, curve after standing 13 days; (*C*) heated, 600°C for 1 hr, curve after standing 11 days; (*D*) heated, 600°C for 1 hr, curve after standing 68 days; (*E*) heated, 600°C for 1 hr, curve after standing 146 days; (*F*) heated, 600°C for 1 hr, curve after standing 268 days; (*G*) heated, 800°C for 1 hr, curve after standing 76 days; (*H*) heated, 800°C for 1 hr, curve after standing 268 days. (After Grim and Bradley.[65])

of hydroxyls with an accompanying exothermic reaction. In general, the loss of hydroxyls from the rehydrate took place at a lower temperature than that of the original mineral. Jonas suggested that the rehydrate formed with a structure similar to that for the anhydrite (see page 317) but that it quickly changed to that of the original montmorillonite.

Experience in drying bentonite commercially has shown that, as a practical matter, regaining of interlayer water is difficult if the last trace of interlayer water is removed. So long as some water remains between the layers, swelling is generally relatively easy.

Vermiculite

Dehydration and phase changes on heating. A dehydration curve published by Nutting[7] (Fig. 9-2) shows a large water loss below about 100°C and a gradual loss from this temperature to about 850°C, where dehydration is substantially complete. The curve shows flexures indicating relatively more rapid dehydration in the vicinity of 400°C and above 700°C. Walker[140] also published dehydration curves for vermiculite. His data show the water to be lost mostly in three steps: below 100°C, from 250 to 400°C, and from 600 to 850°C. According to Walker, the water lost in each of these steps expressed as percentage of total water content of the mineral is about 48, 25, and 27 percent, respectively. Keay and Wild[141] published dehydration curves for vermiculite with various interlayer cations.

Barshad[142] and Walker[140] published differential thermal curves for the mineral (Fig. 9-12). Natural vermiculites show a large, dual, initial endothermic reaction with a peak at about 150 to 200°C, immediately followed by a smaller peak at about 250 to 275°C. Walker[143] later showed that the initial endothermic reaction varies, depending on the equilibrated relative humidity, and that three peaks are shown frequently on DTA curves.

Walker[143] explained the character of the initial endothermic peaks as follows: The interlayer water in natural vermiculites consists of a hydration shell around the Mg ions and also "unbound" water in the space between the hydrated Mg ions (see pages 106–112). The initial endothermic peak corresponds to the removal of about four of the water molecules not in immediate contact with the cation in the normally hydrated structure; the end of this stage coincides with the displacement of the 14.4-Å spacing to 13.8 Å and almost immediately afterward to 11.6 Å. The second endothermic peak results from the removal of about five water molecules not in contact with the cation from the completely filled single-sheet structure; there is no accompanying structural change. The third endothermic peak corresponds to the removal of about $2\frac{1}{4}$ of the 3 remaining water molecules which constitute a part of the octahedral shell around the cation; the basal reflections shift

[140] Walker, G. F., Vermiculite and Some Related Mixed-layer Minerals, "X-ray Identification and Structures of Clay Minerals," chap. VII, pp. 199–223, Mineralogical Society of Great Britain Monograph (1951).

[141] Keay, J., and A. Wild, Hydration Properties of Vermiculite, *Clay Minerals Bull.,* 4(25): 221–228 (1961).

[142] Barshad, I., The Effect of Interlayer Cations on the Expansion of the Mica Type of Crystal Lattice, *Am. Mineralogist,* 35: 225–238 (1950).

[143] Walker, G. F., The Vermiculite Minerals, "The Differential Thermal Investigation of Clays," chap. VII, pp. 191–206, Mineralogical Society, London (1957).

to about 20.6 Å indicating a regular interstratified structure of 11.6 Å and 9.6 Å (completely dehydrated) layers. The remaining interlayer water is lost gradually at higher temperatures without an accompanying thermal reaction and may not be complete until about 650°C.

As shown in Fig. 9-12, the part of the differential curves corresponding to the loss of interlayer water varies with the exchangeable cation carried by the vermiculite. This variation is to be expected, since Barshad[142] showed that, when Ca^{++} and Mg^{++} are the exchangeable ions, there is a double layer of water; when Ba^+, Li^+, and Na^+ are the exchangeable ions, there is a single layer of water; and when NH_4^+, K^+, Rb^+, and Cs^+ are the exchangeable ions, the mineral has no interlayer water. The sizes and complexities of the initial peaks for vermiculite with any of these ions vary in accordance with the hydration characteristics of the particular ion. Thus vermiculite shows only a very slight low-temperature endothermic reaction, and Na vermiculite shows only a single initial endothermic peak.

It appears from the data given by Nutting,[7] Barshad,[142] and Walker[140,143] that the OH lattice water from the silicate part of the structure is lost gradually from about 500 to 850°C and hence does not tend to cause a sharp corresponding thermal reaction. The endothermic reaction between 700 and 800°C is probably due to the destruction of the silicate structure, accompanied by the loss of the last hydroxyl water. The endothermic peak at 600°C for the sample from Pennsylvania (Fig. 9-12) is probably due to some chlorite in mixed layers with the vermiculite.

According to Barshad,[142] the initial stages of dehydration are attended by a contraction of the lattice, with the development of diffraction characteristics like those of biotite. When vermiculite is heated abruptly to about 300°C, the mineral exfoliates, with the production of a material of low bulk density of considerable economic importance. The exfoliation does not take place on heating to about 150°C, from which fact it may be concluded that all interlayer water must be driven off for it to occur.

The high-temperature phases formed on the ignition of vermiculite have not been studied in detail. Walker[140] showed the enstatite is formed, corresponding to the sharp exothermic reaction at about 800°C in samples studied by him. The high-temperature phases could be expected to show considerable variation, depending on lattice substitutions and exchangeable ions, and probably would develop the general composition and high-temperature phase relations shown by montmorillonites (see pages 317–328). Indeed, in the high-temperature parts of the thermal curves shown by Barshad for vermiculite prepared with various exchangeable cations (Fig. 9-12), the variations must mean a considerable variety in high-temperature phases.

No information is available concerning the expansion-contraction of vermiculite as it is heated to elevated temperatures. Also there are no data on specific-gravity changes with temperature. The temperature of fusion of the mineral and its vitrification range would be similar to those of montmorillonite.

Rehydration. Walker[146] also showed that the unbound water driven off at about 100°C is regained almost instantly on cooling to room temperature, with expansion of the lattice to its original dimensions. Barshad[142] showed that, after all the interlayer water is driven off at temperatures up to 550°C, the mineral will still rehydrate rapidly and regain its original lattice dimensions and dehydration characteristics.

The ability to rehydate disappears gradually above 550°C, and according to Walker it is completely lost at about 700°C. Apparently the loss of the rehydration ability accompanies the loss of hydroxyl water. Since the silicate unit is octaphyllitic, a loss of structure would probably accompany the loss of hydroxyls, and the ability to rehydrate after heating to the higher temperatures would not be expected.

Illite

Dehydration and phase changes on heating. As currently defined, the illite group includes clay minerals having either the muscovite type or the biotite type of structure, and it is necessary to consider the dehydration properties of muscovite and biotite, as well as of the clay-mineral micas. Dehydration curves for muscovite, biotite, and phlogopite are given in Fig. 9-27 after Roy.[144] The dehydration curve shows that muscovite has a slow, relatively gradual loss of water up to about 800°C without any restricted interval of large water loss. Phlogopite shows virtually the same dehydration characteristics as muscovite. Biotite loses water gradually up to about 400°C; it shows very little loss between about 400 and 850°C and considerable loss between 850 and 1000°C. Walker[145] also presented a dehydration curve for biotite, and, unlike Roy's curve (Fig. 9-27), it shows an almost uniform rate of water loss up to about 850°C, where the dehydration is essentially complete. Kelley, Jenny, and Brown[10] presented dehydration curves for muscovite which show an interval of relatively rapid dehydration between about 450 and 850°C. The variation in dehydration data obtained by various investigators for the micas undoubtedly results to a considerable extent from the factor of time. Very great differences in water loss are obtained by heating at a given temperature for varying intervals of time.

[144] Roy, R., Decomposition and Resynthesis of the Micas, *J. Am. Ceram. Soc.*, **32**: 202–210 (1949).

[145] Walker, G. F., The Decomposition of Biotite in Soils, *Mineral. Mag.*, **28**: 693–703 (1949).

Roy[144] investigated the length of time required for dehydration of muscovite to be completed at various temperatures, and his results are given in Fig. 9-28. His data show that about 10 hr is required for complete dehydration at 300°C, that is, for removal of all the water which can possibly

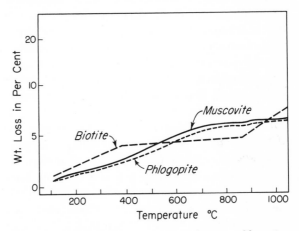

Fig. 9-27. Dehydration curves of biotite, phlogopite, and muscovite. Samples held at respective temperatures for 24 hr. (After Roy.[144])

Fig. 9-28. Dehydration curves for muscovite, showing relation of water loss to time of heating. (After Roy.[144])

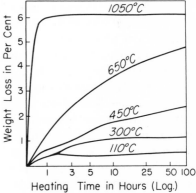

be removed at this temperature under ordinary pressure. At 450 and 650°C some water was still being lost at the end of 100 hr, and at 650°C the loss of water in 1 hr is only 20 percent of the total lost in 100 hr. Obviously, tremendously variable dehydration results can be obtained for muscovite micas unless the time factor is carefully considered. The same conclusion probably holds also for the biotite micas. There may well be some variation in dehydration properties within particular types of micas due to variations in composition and structure, and some of the differences found in the literature may be due to this cause.

Dehydration curves for illites published by Grim, Bray, and Bradley[12] (Fig. 9-5) show a considerable water loss below 100°C, a gradual loss from 100 to about 350°C, a relatively abrupt, large loss from 350 to about 600°C, and a gradual loss above 600°C. These illites have a muscovite type of crystallization. A sample of weathered biotite which probably is somewhat representative of an illite with biotite attributes is shown by Walker[145] to have a small loss of water below 100°C, a gradual, very slight loss from 100 to about 300°C, a rather sharp but gradual loss from 300 to 600°C, and a slow, gradual loss above 600°C. Thus, the dehydration properties of the illites with muscovite and biotite attributes for which data are available are about the same. The illites differ from muscovite and biotite by having a water loss below 100°C and a more abrupt loss of much of the OH lattice water between about 300 and 600°C. The low-temperature water loss is to be expected, since the clay-mineral micas differ from the well-crystallized micas in having some interlayer water, which is a consequence of fewer interlayer cations, less bond between layers, less uniform orientation of successive layers, smaller particle size, and/or a difference in composition within the silicate layer itself. The difference in loss of hydroxyl water cannot be so completely explained but is probably largely the result of the difference in particle size, as shown in the following discussion of differential thermal curves.

Differential thermal curves (Fig. 9-10), after Grim and Bradley,[146] for coarse-grained muscovite (10 to 20 μ) show only a single dehydration reaction, beginning at about 800°C with a peak at about 900°C. In the rapid heating of the differential thermal procedure there would be little loss of hydroxyls below about 800°C, and then, in accordance with Roy's[144] data, an abrupt loss of the OH water should occur. Grim and Bradley showed that a decrease in the particle size of the muscovite is accompanied by a reduction in the temperature of the beginning of the endothermic dehydration reaction and by an increase in the temperature interval during which it takes place. Differential thermal analyses of biotite yield a curve without any pronounced deflections up to 1000°C, indicating an absence of any abrupt dehydration in this temperature range under the conditions of the analyses (heating rate, 10°C/min).

Differential thermal curves for illite presented by Grim and Bradley[146] show an initial endothermic reaction corresponding to the loss of interlayer water, a second endothermic reaction beginning at about 450 to 500°C with

[146] Grim, R. E., and W. F. Bradley, The Mica Clay Minerals, "X-ray Identification and Structures of Clay Minerals," chap. V, pp. 138–172, Mineralogical Society of Great Britain Monograph (1951).

a peak at about 550 and 650°C, a third, slight endothermic reaction between about 850 and 950°C, and frequently an exothermic reaction between about 900 and 1000°C (Fig. 9-10). The size and temperature interval of the second endothermic peak, which corresponds to the loss of OH lattice water, vary in different samples, and the final part of the curve also shows considerable variation. Information is not yet available on the precise variation of thermal characteristics of illites with differences in their structural and composition attributes. Grim and Bradley (Fig. 9-10) showed one trioctahedral illite that yields a differential thermal curve like those of dioctahedral illites. On the other hand, another sample of glauconite showed a relatively smaller initial endothermic reaction and a relatively higher temperature for the reaction accompanying the loss of hydroxyl water than the dioctahedral illites. Mackenzie, Walker, and Hart[147] described an illite showing a double endothermic reaction in the hydroxyl dehydration interval, with peaks at 550 to 600°C and 713°C, and for which there is as yet no satisfactory explanation.

The initial water loss and the accompanying endothermic peak due to interlayer water shown by many samples described as illite are undoubtedly due in part to a mixed-layer expandable component. However, work[148] in the author's laboratory has shown that samples of illite giving no evidence of any mixed layering have the low-temperature dehydration and endothermic peak indicating interlayer water.

Holt[149] reported an activation energy for the dehydration of muscovite of 90.3 kcal/mole. Earlier Holt et al.[150] concluded that the dehydration of muscovite must be a first-order reaction but that such an equation is difficult to derive mathematically. They suggested a rate equation that explains the observed logarithmic time dependence based on a model involving a systematic relief of strain during dehydration. An increase of activation energy is required with time, which means that the energy barrier for hydroxyls to form water becomes higher as dehydration proceeds. They postulated that the normal muscovite lattice is strained and that the strain is relieved somewhat as the hydroxyl ions leave the lattice, thereby making the escape of the remaining hydroxyls more difficult. Holt et al.[150] provided infrared data to support their conclusion. They also proposed that the aluminum ions in the octahedral layer shift from 6- to 4-coordination positions on dehydration.

[147] Mackenzie, R. C., G. F. Walker, and R. Hart, Illite in Decomposed Granite at Ballater, Aberdeenshire, *Mineral. Mag.*, **28:** 704–713 (1949).

[148] Gaudette, H. E., J. L. Eades, and R. E. Grim, The Nature of Illite, Proceedings of the 13th National Clay Conference, pp. 33–48, Pergamon Press, New York (1966).

[149] Holt, J. B., Rate of Thermal Dehydration of Muscovite, Pyrophyllite, and Kaolinite, *Brit. Ceram. Abstr.* 60, p. 78A (1961).

[150] Holt, J. B., I. B. Cutler, and M. E. Wadsworth, Rate of Thermal Dehydration of Muscovite, *J. Am. Ceram. Soc.*, **41:** 242–246 (1958).

According to Roy,[144] muscovite shows no marked structural change on heating to 940°C, and the loss of the structure takes place between 940 and 980°C. The expulsion of at least most of the OH lattice water is accompanied by a slight expansion of the lattice in the c dimension and not by complete destruction of the lattice. On heating, phlogopite exhibits the same characteristics as muscovite, except that its structure is lost at a temperature about 50°C lower. The biotite structure, according to Roy, persists to about 1100°C, and the biotite structure, therefore, is also not lost simultaneously with the expulsion of the OH lattice water. This is surprising in view of the reasoning of Bradley and Grim[124] that trioctahedral anhydrite-mica structures are unlikely (see page 335).

Grim and Bradley[123] showed that the loss of OH water in the illite studied by them is not accompanied by a loss of structure. It is accompanied by a slight structural change of the kind described for montmorillonite (see page 318). According to these investigators, the structure of these illites is not destroyed until at least 850°C, and Grim and Bradley would correlate its destruction with the third endothermic reaction in the differential thermal curves. The illites studied by them had the muscovite type of crystallization. Maegdefrau and Hofmann[151] also reported the continuation of an illite-like structure beyond the loss of OH water, and, in at least one of their samples, it was retained to 1000°C.

Roy[144] reported that muscovite on firing at 1000°C develops γ-Al_2O_3 and/or spinel (the X-ray data were too poor for a positive determination); at 1200°C the presence of γ-Al_2O_3 is definitely established, and α-Al_2O_3 appears; and above 1400°C corundum (α-Al_2O_3) and glass are the only phases present. Zwetsch[152] reported that muscovite ignites at 1050°C to γ-Al_2O_3, α-Al_2O_3, and leucite, and finally at 1300°C the crystalline phases are leucite and α-Al_2O_3. On firing phlogopite, according to Roy,[144] spinel is formed at 1000°C, and this persists as the only crystalline phase to 1550°C. On firing biotite to about 1100°C, the phases developing are a high-iron magnetic spinel, leucite, and mullite; at about 1300°C the phases are the high-iron spinel and leucite; and at 1500°C only olivine and glass are to be found. In a biotite studied by Grim and Bradley, leucite, γ-Fe_2O_3, and a spinel were identified after firing to 1200°C.

On heating several illites, Grim and Bradley[123] found spinel appearing in all of them at about 850°C, and the spinel increased in amount and particle size up to about 1200°C. These investigators suggested that the octahedral sheet of the illite lattice, carrying the alumina, magnesia, and iron, goes into

[151] Maegdefrau, E., and U. Hofmann, Glimmerartige Mineralien als Tonsubstanzen, *Z. Krist.*, **98**: 31–59 (1937).

[152] Zwetsch, A., Röntgenuntersuchungen in der Keramik, *Ber. Deut. Keram. Ges.*, **14**: 2–14 (1934).

the formation of the spinel and that the alkalies and the silica from the tetrahedral layers yield an amorphous glass. They found that at 1300°C the spinel had dissolved in the glass and that mullite became apparent at 1100°C and persisted to at least 1400°C. The amounts of spinel and of mullite appeared to be complementary, and the specimens richer in spinel were subject to fusion at a somewhat lower temperature. These investigators found no clear evidence of a quartz phase developing from their illites at elevated temperatures.

Hill[97] reported that the structure of illite persists to about 900°C and that at 1000°C spinel and mullite are developed. Sundius and Bystrom[153] stated that the X-ray pattern of muscovite is still recognizable at 1000°C and that

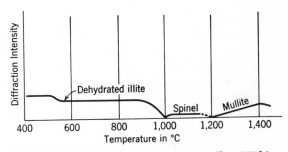

Fig. 9-29. **Phase changes on heating an illite (Fithian, Illinois) to elevated temperatures as shown by continuous X-ray diffraction.** (After Grim and Kulbicki.[132])

needles of mullite and minute crystals of corundum can be identified microscopically at this temperature. Mullite becomes abundant at 1200°C. Attempts to identify the corundum by X-ray diffraction were not very successful, and its presence appears doubtful.

Grim and Kulbicki[132] studied the high-temperature reactions of illite by continuous X-ray diffraction. In some samples which contained relatively large amounts of iron and alkalies, they were unable to detect the development of any crystalline phases following the complete loss of the dehydrated illite structure. In all cases the diffraction characteristics of the high-temperature phases were very poor, indicating only a small development of new crystalline phases, extremely small particle size, and/or a very low order of crystallinity. In a sample from Fithian, Illinois, they found (Fig. 9-29) that the dehydrated illite structure persisted to about 1000°C and that this was followed immediately by the development of a spinel-type crystallization which persisted to

[153] Sundius, N., and A. M. Bystrom, Decomposition Products of Muscovite at Temperatures between 1000° and 1260°C, *Trans. Brit. Ceram. Soc.*, **52**: 632–642 (1953).

about 1200°C, when mullite developed. Verduch and Estrada[154] reported mullite formed on heating a sericite mica to 1000°C. Eberhart[155] stated that a spinel and small amounts of mullite and leucite were formed on heating a muscovite to 1050°C and that the (111) and (110) planes of the spinel lattice were parallel to the original (001) and (110) mica planes, respectively.

It is reasonable to expect that there would be considerable variation in the high-temperature phases developed from illites of different structural and composition attributes, and indeed the variations in the high-temperature portions of the differential thermal curves show that this must be true. Considerably more study of ignited illites is necessary before general conclusions can be reached regarding high-temperature-phase transformations in these minerals.

Hyslop and McMurdo[87] published expansion-contraction curves for two illite samples (Fig. 9-20). Both samples showed considerable expansion (1.1 and 0.8 percent, respectively) up to 800°C. Above 800°C one of the samples showed an abrupt contraction, which was continuous up to the limit of the experiments at 1100°C. The other sample showed slight contraction between 800 and 950°C, followed by a secondary expansion from 950 to 1000°C. The secondary expansion could well be explained by a reaction producing gas from an impurity in the sample and probably is not a characteristic of the mineral.

The fusion point of the illites varies between wide limits. Some glauconites fuse below 1000°C, whereas iron-poor varieties relatively low in alkalies do not fuse until temperatures of the order of 1300°C. In general, illites with the higher fusion point have a moderately long vitrification range (200°C).

Rehydration. Grim and Bradley[65] studied the rehydration properties of three illites. In one of them (Fig. 9-30), after heating to 600°C there was a rapid pickup of adsorbed water and hydroxyl lattice water, so that a considerable part (25 percent) of both types of water was regained within a few hours. Additional amounts of adsorbed and hydroxyl water were regained slowly for a period of many months. The loss of rehydration hydroxyl water on heating again began at a lower temperature (400°C) than that of the original water (475°C). After heating to 800°C there was also a gradual regaining of adsorbed water and OH lattice water. The data suggest that, on heating to either 600 or 800°C, there was an initial, relatively rapid rehydration followed by a very slow and gradual rehydration. A second sample of illite

[154] Verduch, A. G., and D. A. Estrada, The Formation of Mullite from Sericite and Its Mixtures with Alumina and Kaolin, *Sci. Ceram.,* **1**: 285–294 (1962).

[155] Eberhart, J. P., Formation of the Spinel-type Phase in Muscovites Heated from 1050° to 1200°C, *Compt. Rend.,* **256**: 3860–3863 (1963).

showed nearly the same rehydration characteristics as those just described, except that the rehydration was slower and there was no lowering of the temperature required to remove the water regained on rehydration. The third sample studied by Grim and Bradley contained some chlorite mixed with the illite. After heating to 600°C, this sample showed a slow gradual pickup of adsorbed water, and after 217 days a larger amount was adsorbed than had been present in the original sample. Also in this sample, some hydroxyl lattice water was regained in material heated to 800°C. Roughly the same amount of hydroxyl water was taken up by samples heated to both 500 and

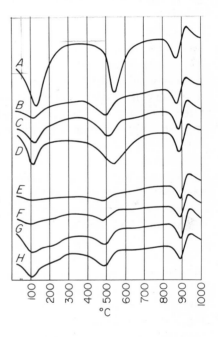

Fig. 9-30. Differential thermal curves of illite, Fithian, Illinois: (A) not heated; (B) heated, 600°C for 3 hr, curve after standing 2 hr; (C) heated 600°C for 1 hr, curve after standing 1 day; (D) heated, 600°C for 1 hr, curve after standing 70 days: (E) heated, 800°C for 1 hr, curve after standing 13 days; (F) heated, 800°C for 1 hr, curve after standing 70 days; (G) heated 800°C for 1 hr, curve after standing 147 days; (H) heated, 800°C for 1 hr, curve after standing 275 days. (From Grim and Bradley.[65])

800°C. The total amount of OH lattice water taken up by each of the illites in about 9 months was close to one-third of the original amount. The nature of the probable pairing of the layers in the illites investigated is such that data similar to that obtained for montmorillonites cannot be obtained (see page 317). However, it seems likely that the structural changes are much the same for both minerals.

Roy[144] succeeded in re-forming muscovite and phlogopite from material decomposed at 1050°C by heating the latter to 400 to 650°C for at least 12 hr under pressures varying between 1,000 and 10,000 psi. No success was attained in a similar treatment of decomposed biotite.

Dehydration and phase changes on heating. Nutting[7] presented dehydration curves for several chlorites (Fig. 9-2). All Nutting's curves for the chlorite minerals show very little water loss prior to about 500°C and a sharp loss of a good deal of water between about 500 and 550°C. The curves above about 600°C show considerable variation; some of them show a gradual continuous loss of weight from about 600 to 850°C, where dehydration is essentially complete, whereas others show a relatively slight loss between 600 and about 750 to 800°C, and then a relatively abrupt loss from 800° to essentially complete dehydration at 850 to 900°C.

Orcel,[156] Orcel and Renaud,[157] Barshad,[19] and the writer have obtained differential thermal curves for a larger number of chlorite samples (Fig. 9-12). The samples that have been studied show little or no low-temperature endothermic reaction, indicating an absence of interlayer or adsorbed water. Further work may well reveal chlorites yielding such a reaction, as it seems likely that very fine-grained chlorites with some interlayer water would occur in some clay materials. All the samples exhibited a distinct, sharp endothermic reaction between 500 and about 700°C. Most of them showed a second endothermic reaction at about 800°C, and this peak was frequently followed immediately by a sharp exothermic peak. Orcel concluded from his work that the hydroxyl water of the chlorites is driven off in two stages, and when a second endothermic peak is not shown, the two stages of dehydration occur within such a narrow temperature interval as to be unresolved by differential analyses. Unpublished work by the writer and studies by Barshad[19] show that extremely sensitive differential analyses of a series of chlorites (Fig. 9-12) tend to resolve the endothermic reaction into two parts. Nutting's[7] dehydration curves show definitely that the chlorites he investigated lost their OH water in two stages, but that the second stage may immediately follow the first stage without an interval during which no water is lost, and further that the second stage may take place gradually. Differential curves of chlorites in which the second-stage loss of water was very gradual would be expected to show an overlap of thermal reactions or even the complete absence of a second endothermic reaction.

Phillips[158] indicated that septechlorites show only one large dehydration stage which leads directly to the structural breakdown.

[156] Orcel, J., Analyse thermique des chlorites, *Bull. Soc. Franc. Mineral.,* **50:** 278–322 (1927).

[157] Orcel, J., and P. Renaud, Dehydration of Ferromagnesian Chlorites, *Compt. Rend.,* **212:** 918–920 (1941).

[158] Phillips, W. R., A Differential Thermal Analysis Study (DTA) of the Chlorites, *Mineral. Mag.,* **33:** 404–414 (1963).

All the available dehydration and differential thermal data are for well-crystallized chlorites, and the variations that would be found in clay-mineral chlorites are a matter of surmise. Reasoning from analogies with illites as compared with muscovite and biotite, it appears likely that clay-mineral chlorites would show the same thermal reaction as the well-crystallized chlorites, but perhaps at somewhat lower temperatures. The smaller particle size of the chlorite clay minerals and possibly some defects of orientation of layers would perhaps permit the loss of some of the OH water of the brucite sheet at a relatively lower temperature. Also, the clay-mineral chlorites could well have some interlayer water, providing a small initial endothermic reaction as in many illites.

In general, the lower temperature reaction can be correlated with the loss of hydroxyls from the brucite part of the structure and the second one from the mica part which varies in temperature and abruptness, depending on whether the structure is dioctahedral or trioctahedral. The detailed dehydration characteristics vary with particle size (Sabatier[159]) as well as composition. Caillere and Henin[160] gave data on dehydration temperature for chlorites with various cations in the octahedral layers of both the brucite and mica part of the structure (Table 9-4).

Table 9-4   Temperature of loss of water from chlorites

(After Caillere and Henin[160])

	Temperature, °C	
Nature of cation	*Brucite layer*	*Mica layer*
Mg	640	820
Al	500	750
Fe_2	430	530
Fe_3	250	

Brindley and Ali[161] investigated the changes taking place in some chlorites (penninite, clinochlore, and sheridanite) after heating to successively higher temperatures. According to these investigators, in the temperature range from

[159] Sabatier, G., Influence of Particle Size on DTA Curves of Chlorites, *Bull. Soc. Franc. Mineral.*, **73**: 43–48 (1950).

[160] Caillere, S., and S. Henin, Relationship between the Crystallochemical Constitution of Phyllites and Their Dehydration Temperature. Application in the Case of Chlorites, *Bull. Soc. Franc. Ceram.*, **48**: 63–67 (1960).

[161] Brindley, G. W., and S. Z. Ali, Thermal Transformations in Magnesium Chlorites, *Acta Cryst.*, **3**: 25–30 (1950).

400 to 700°C, the intensities of the (00l) reflections progressively change; in particular, the (001) reflection becomes several times more intense, and the (002) and (003) become several times weaker. Also the lattice constants change, c_0 showing a small contraction and a_0 and b_0 showing small expansions. These changes are attributed to the gradual breakdown of the brucite layer, while the mica layer remains unaffected. The dehydration of the brucite layer corresponds to the first endothermic reaction of the differential thermal curves. At 700°C about two-thirds of the hydroxyls of the brucite layer have been removed, and Fourier syntheses of (00l) intensities show that the magnesium atoms move away from their original position in a plane midway between the mica sheets to positions directly adjacent to the mica-unit surface. This change precedes the dehydration of the mica-layer portion of the chlorite structures. In some chlorites the loss of water by the mica layer is abrupt, whereas in others it is gradual. In either case it is accompanied by the total destruction of the chlorite structure, which in turn is immediately followed by the crystallization of olivine. The endothermic reaction frequently shown at 750 to 800°C is correlated with the dehydration of the mica layer and the loss of its structure. The exothermic reaction at about 800 to 900°C is correlated with the development of the olivine.

Brindley and Ali[161] showed that the newly formed olivine phase has a high degree of preferential orientation relative to the structure of the original chlorite, and they considered in detail the probable shifts in the atomic lattice accompanying the formation of the olivine. They showed that the temperature of lattice destruction and the temperature of the formation of olivine vary for different chlorites. They also showed that other phases in addition to olivine form when various chlorites are heated and that the temperature of formation of the additional phases is also variable. Thus for penninite, olivine first appears at about 800°C, for clinochlore at 850°C, and for sheridanite at 950°C. According to these workers, penninite at 950°C shows olivine and spinel, and at 1180°C olivine, spinel, and enstatite; clinochlore at 850°C shows olivine and spinel; and sheridanite at 1100°C shows olivine, spinel, and enstatite. Brindley and Ali[161] also presented differential thermal curves for these same chlorites carried to high temperatures, and they were unable to correlate the thermal reactions satisfactorily in each case with the phase developed as shown by X-ray-diffraction analysis.

The explanation for the apparent lack of correlation between the high-temperature phases revealed by X-ray analysis and the reactions shown by differential thermal curves for chlorites and for some other clay minerals is probably as follows: In some cases the new phase may develop slowly and/or involve only a very slight structural shift, so that it would not be accompanied by an intense or sudden energy change, and consequently no thermal reaction would be expected on the differential thermal curve. In other cases, the reaction may begin abruptly by a somewhat crude nucleation of the new phase, which would involve a structural change of considerable magnitude and a definite thermal reaction but would not yield discernible diffraction evidence of the new phase. Further heating or more time would be required to develop the new phase to the point where distinct diffraction effects become obvious. In the writer's experience, the thermal reaction for a new phase frequently precedes its distinct appearance in X-ray reflections, and the X-ray reflections further precede its appearance in optical measurements.

Shirozu[162] found that for iron chlorites the first stage of dehydroxylation takes place at 560 to 600°C, accompanied by a modification of the structure. At 700 to 800°C, dehydroxylation was essentially complete, with the loss of structure followed immediately by the development of a spinel-type phase. Subsequently at higher temperatures, enstatite, cristobalite, and hematite are reported. A small amount of Al_2O_3 is inferred in the hematite. Earlier Brindley and Youell[163] had arrived at similar conclusions deduced from a study of products formed on heating berthierine (kaolinite-type chamosite). They also reported variations depending on whether the mineral was heated in a vacuum or in air.

No data are available for the expansion-contraction properties of chlorite on heating or for the changes in specific gravity of the mineral on firing. Also no precise data are available for the fusion temperature and vitrification range of these minerals. It appears likely that they would be closely similar to the biotite type of mica in fusion and vitrification characteristics.

Rehydration. No experimental data could be found on the possible rehydration of the chlorite minerals. Reasoning from structural analogies with the micas, it is likely that the chlorite clay minerals would show a slow, perhaps relatively slight pickup of hydroxyl water, at least in the mica-layer part of the structure, after heating to temperatures of the order of 700°C.

Sepiolite-attapulgite-palygorskite

Dehydration and phase changes on heating. The dehydration and high-temperature characteristics of only a few samples of these minerals have been studied, and there is no assurance that the results obtained have wide application to the group as a whole.

A dehydration curve for a sepiolite given by Nutting[7] (Fig. 9-2) shows a large water loss below about 100°C and gradual continuous loss above

[162] Shirozu, H., Thermal Reaction of Iron Chlorites, *Clay Sci.,* **1**(5): 20–25 (1962).

[163] Brindley, G. W., and R. F. Youell, Chamosite, *Mineral. Mag.,* **30**: 57–70 (1953).

that temperature to complete dehydration at about 800°C. There is a relatively slight increase in water loss between about 200 and 400°C and above 675°C. Migeon[13] presented (Fig. 9-6) a continuous, i.e., not equilibrium, weight-loss curve for sepiolite which is generally similar to Nutting's curve in that it shows a large loss below 100°C and a gradual loss above that temperature, with flexures indicating relatively rapid loss at about 275 to 325°C, 550 to 600°C, and slightly above 800°C.

A differential thermal curve for a sepiolite is given in Fig. 9-13. It shows a large initial endothermic peak beginning below 100°C, followed by a series of slight endothermic peaks between about 200 and about 700°C. The exact temperature of these peaks is known to vary from sample to sample. Between 700 and 800°C there is generally a fairly sharp endothermic reaction, followed immediately by a sharp exothermic reaction. Caillere[20] published a series of differential thermal curves for many sepiolites, and they show the same general characteristics and variations noted above.

In general (Nagy and Bradley[164]), it can be concluded that adsorbed water in the channels is lost at about 100°C, water in the channels attached to the silicate units is lost between about 275 and 375°, and hydroxyl water is lost from about 550 to 800°C.

According to Migeon[13] and Longchambon,[165] there is no change in the structure of sepiolite on heating to 350°C. Above this temperature, the loss of water is accompanied by a slight change in the dimensions of the structure. The modified form of sepiolite is said to persist to 700°C, when it passes into a transient amorphous phase, according to Caillere.[20] Differential thermal curves of the mineral show an exothermic reaction at about 800°C, which is probably due to the development of a magnesium silicate phase. At 1000°C, Longchambon[165] recorded the presence of enstatite and, at 1300°C, a mixture of enstatite and cristobalite.

Preisinger[166] showed that the ribbons of sepiolite take a tilted position below 800°C, as indicated in Fig. 9-31. Bradley and Kulbicki[167] found that the tilting takes place at 350 to 450°C following the loss of water attached to the sides of the silicate units, and that the original structure is regained with difficulty after heating to 450°C. According to Kulbicki,[168] the sepiolite

[164] Nagy, B., and W. F. Bradley, Sepiolite Structure, *Am. Mineralogist,* **40:** 885–892 (1955).

[165] Longchambon, H., Sur certaines caractéristiques de la sepiolite d'Ampandandrava et la formule des sepiolites, *Bull. Soc. Franc. Mineral.,* **60:** 232–276 (1937).

[166] Preisinger, A., An X-ray Study of the Structure of Sepiolite, Proceedings of the 6th National Clay Conference, pp. 61–67, Pergamon Press, New York (1959).

[167] Bradley, W. F., and G. Kulbicki, Structural Changes on Dehydration of Sepiolite and Attapulgite, manuscript (1959).

[168] Kulbicki, G., High Temperature Phases in Sepiolite, Attapulgite, and Saponite, *Am. Mineralogist,* **44:** 752–764 (1959).

structure is lost at about 800°C, followed immediately by the nucleation of enstatite which develops slowly to a maximum at about 1350°C. β-cristobalite forms in some samples at 1075°C; in others only above 1350°C. Both phases persist to above 1500°C, and the elongate units may persist to above 1400°C.

In a series of dilation curves for sepiolite, Longchambon[165] showed that there are two intervals of contraction perpendicular to the fiber axis, one interval of about 1 percent at about 350°C, and another of about 2 percent taking place abruptly at about 800°C. Parallel to the fiber axis, there is a slow gradual contraction of about 0.5 percent up to 800°C. The contraction during this temperature interval becomes more prominent as the temperature increases. At 800°C, a further contraction, exceeding 2 percent in the direction parallel to the fiber axis, takes place suddenly.

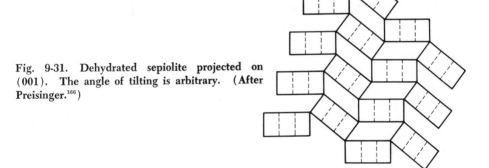

Fig. 9-31. Dehydrated sepiolite projected on (001). The angle of tilting is arbitrary. (After Preisinger.[166])

Nutting[7] presented a dehydration curve for a palygorskite (Fig. 9-2), showing a sharp water loss below 100°C, abrupt water losses at 150 to 200°C and 375 to 425°C, and a gradual loss above 425°C, with complete dehydration at about 700°C. A continuous weight-loss curve (Fig. 9-6), given by Longchambon[14] shows similar results.

Differential thermal curves for a palygorskite and an attapulgite (Fig. 9-13) show an initial endothermic reaction below about 200°C and additional endothermic reactions at about 225 to 350°C and 400 to 525°C, corresponding approximately to the intervals of abrupt water losses shown by the dehydration curves. There is no definite-higher-temperature endothermic peak, and apparently the small water loss above about 550°C causes no thermal reaction adequate for recording.

Caillere and Henin[169] presented a series of differential thermal curves

[169] Caillere, S., and S. Henin, Application de l'analyse thermique différentielle a l'étude des argiles des sols, *Ann. Argon.,* 17: 23–72 (1947).

for palygorskites, and in general they have similar characteristics (Fig. 9-13), except for larger initial endothermic reactions. The various palygorskites that have been analyzed show additional variations in detail, particularly in the high-temperature part of the curves; this is to be expected because of the probable differences in composition, with resulting variations in high-temperature phases.

Bradley[170] stated that the water molecules in the channel-like interstices of the structure are lost below 100°C, a temperature which would correspond to the initial endothermic reaction. The water coordinated about the magnesium ions at the sides of the silicate units is lost at a slightly higher temperature, and this dehydration is probably correlated with the second endothermic reaction. The loss of the hydroxyl water from the silicate unit is probably correlated with the third endothermic reaction. Apparently most of the hydroxyl water is lost at about 400°C, with the remainder being lost slowly between 450 and about 700°C.

The palygorskites-attapulgites differ in dehydration from the sepiolites in the loss of the hydroxyl water more abruptly and generally at a lower temperature. Also in at least some sepiolites, there is little or no indication of so-called zeolitic water, which in attapulgite is that coordinated with the magnesiums.

According to Longchambon,[171,172] the loss of water up to about 400°C is not accompanied by any structural changes. At about 400°C, there is a rapid contraction of the crystal lattice normal to the length of the fiber. The modified structure persists to about 775°C, when the structure is destroyed and further shrinkage takes place.

According to Caillere and Henin,[173] a comparison of the results of Longchambon and Bradley indicates that above 400°C the $(hk0)$ reflections are replaced by halos. This is explained by dislocations in the octahedral units of the amphibole chains, which lead to a closing of the cavities or channels with the subsequent disappearance of the water-adsorbing property. Bradley and Kulbicki[167] found that the ribbons of attapulgite take a tilted position at 300 to 350°C, similar to that found for sepiolite. The original structure is readily regained by attapulgite unless heating is continued considerably above about 350°C. Kulbicki[168] stated that the attapulgite structure is lost

[170] Bradley, W. F., The Structural Scheme of Attapulgite, *Am. Mineralogist,* **25:** 405–413 (1940).

[171] Longchambon, H., On the Characteristic Properties of Palygorskite, *Compt. Rend.,* **203:** 672–674 (1936).

[172] Longchambon, H., *Rec. Commun. Congr. Tech. Ind. Ceram.,* pp. 75–141 (1938).

[173] Caillere, S., and S. Henin, Palygorskite-Attapulgite, "X-ray Identification and Structures of Clay Minerals," chap. IX, pp. 234–243, Mineralogical Society of Great Britain Monograph (1951).

at about 800°C, followed immediately by the nucleation of mullite which develops slowly as the temperature is raised. β-quartz appears at about 1100°C, and both phases disappear at about 1200°C with the appearance of β-cristobalite, which persists above 1400°C. The needle-like morphology begins to disappear at about 1200°C. Exchangeable cations have about the same influence on the high-temperature phases developed from attapulgite as from aluminous montmorillonites (see pages 324–327).

According to Grim,[174] attapulgite fuses at approximately the same temperature as montmorillonite. Data are not available for the fusion point of other members of this group, but probably they would be similar to attapulgite.

Rehydration. According to Migeon,[13] the water lost below 350°C by sepiolite is easily regained, whereas water lost at higher temperatures is not taken up again.

As indicated above, attapulgite rehydrates readily, regaining its original structure following heating to about 350°C. Sepiolite rehydrates slowly and regains its original structure slowly, following heating to about 450°C. Above these temperatures, the minerals lose hydroxyl water, and there appears to be slight rehydration.

Studies of the surface properties of attapulgite and sepiolite have not always recognized the reversible structural changes that take place at relatively low temperatures, and hence the results have limited value.

Clay-mineral mixtures

Many clay materials are made up of mixtures of several different clay minerals. The mixtures may be of two types: a combination of discrete particles of the separate minerals or an interstratification of layers of the separate minerals in which the thickness of the layers is of the order of one or a few unit layers.

In mixtures of discrete particles the dehydration characteristics tend to reflect those of the individual components. Numerous investigations[21,47] have published differential thermal curves of prepared mixtures of various clay minerals. As shown in Fig. 9-32, such mixtures may show the thermal effects of the separate clay minerals with intensities proportionate to their relative abundance.

Grim[39] published a large series of differential thermal curves of clay-min-

[174] Grim, R. E., Relation of the Composition to the Properties of Clays, *J. Am. Ceram. Soc.*, **22**: 141–151 (1939).

eral mixtures prepared in different ways. He showed that the thermal characteristics depend on the intimacy of mixing and on the nature of the clay
minerals themselves. In mixtures prepared by combining dry powders without subsequent wetting, the dehydration characteristics of the individual components are maintained in proportions roughly equal to their relative abundance. If the mixing is done by stirring together wet suspensions of the

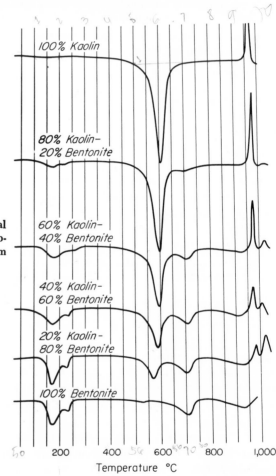

Fig.   9-32. Differential   thermal
curves of discrete mixtures of kaolinite and montmorillonite. (From
Spiel et al.[21])

separate minerals, and particularly if the mixture is allowed to dry and then
is rewetted and remixed several times, the dehydration characteristics of the
individual components may be modified significantly. As shown in Fig. 9-33,
the modifications may be such as to mask completely the presence of relatively
minor components. Amounts of the order of 25 percent could easily be hidden
by such modifications of dehydration characteristics.

In such mixtures of discrete particles the modification of dehydration properties takes place more readily and to a greater degree as the crystallinity decreases. Thus, the effect is more pronounced for poorly crystalline kaolinite than for the well-crystallized mineral.

In general, it seems that, as the intimacy of mixing ,increases, the temperature for the loss of hydroxyl water is reduced and the loss of this water

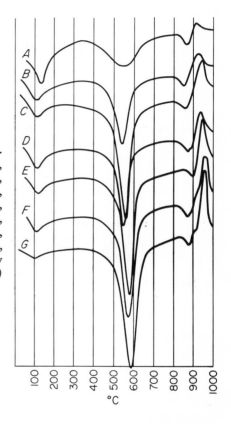

Fig. 9-33. Differential thermal curves of prepared mixtures. (A) Illite, Grundy County, Illinois; (B) illite, 0.504 g, and kaolinite K_1, 0.252 g; (C) illite, 0.371 g, and kaolinite K_1, 0.371 g; (D) illite, 0.518 g, and kaolinite K_2, 0.259 g; (E) illite, 0.371 g, and kaolinite K_3, 0.371 g; (F) illite, 0.502 g, and kaolinite K_3, 0.251 g; (G) illite, 0.363 g, and kaolinite K_3, 0.363 g. Kaolinite K_1 and K_2 relatively poorly crystalline; K_3 well crystallized. (After Grim.[39])

is more gradual, so that its thermal effect is spread out over a wider temperature interval.

With regard to interlayer mixtures, the published curves show that in some cases the individual components are registered in the differential thermal data and that in other cases they are not. Grim and Rowland[47] showed that a series of samples listed as beidellite provided differential thermal curves indicating that they are mixtures. Grim and Bradley[146] showed (Fig. 9-11) that the so-called sarospatite and bravaisite from the type localities are mixtures' of mica and montmorillonite and that both components are indicated by the thermal curves. These same investigators pointed out that other samples composed of mixtures of illite and montmorillonite provide thermal data suggestive of a single mineral (Fig. 9-11) ,and also that mixtures of illite and chlorite may give no definite indication of more than one component.

In unpublished work by the author and his colleagues, it was found that some bentonites from Texas composed of mixtures of kaolinite and low-iron montmorillonite, in nearly equal amounts, yielded a single endothermic peak at about 600°C for the loss of hydroxyl water. In contrast to this finding, some of the kaolins from the Georgia area contain 15 to 20 percent of montmorillonite, and differential thermal curves reveal distinct thermal reactions for both the kaolinite and the montmorillonite.

The factors determining whether or not the individual components are revealed by dehydration data are not completely known. The tendency for a separate reaction decreases as the relative abundance of the component decreases. Also a further factor probably is the nature of the interlayering; e.g., a component present in layers of nearly unit-cell thickness would not be revealed, whereas if the same mineral were present in the same amount but concentrated in layers several unit cells thick, its presence might be shown. Variations in composition may be a further influencing factor. Thus, in mixtures of illites and montmorillonites, a montmorillonite of one composition may be more readily concealed than one with a different composition.

Differential thermal data are used widely for clay-mineral identification, and in some cases quantitative estimates have been made solely on such data. The foregoing discussion illustrates the fact that such identifications must be made with caution. Differential thermal data are extremely valuable for clay-mineral identifications, but they cannot be used blindly, basing identifications solely on data derived from a few pure clay minerals. Frequently, trustworthy identifications can be made only on the basis of several types of analytical data, e.g., differential thermal plus X-ray-diffraction plus chemical data. Cole and Hosking[175] published DTA curves for various clay-mineral mixtures which substantiate the conclusions of work summarized above.

A considerable amount of work has been done in recent years on the crystalline phases that develop when clay-mineral mixtures are heated to elevated temperatures (e.g., Brindley and Maroney[176] and Cole and Segnit[177]).

[175] Cole, W. F., and J. S. Hosking, Clay Mineral Mixtures and Interstratified Minerals, "The Differential Thermal Investigation of Clays," chap. X, pp. 248–274, Mineralogical Society, London (1957).

[176] Brindley, G. W., and D. M. Maroney, High Temperature Reactions of Clay Mineral Mixtures and Their Ceramic Properties, II, Reactions of Kaolin-Mica-Quartz Mixtures Compared with the K_2O-Al_2O_3-SiO_2 Equilibrium Diagram, *J. Am. Ceram. Soc.*, 43: 511–516 (1960).

[177] Cole, W. F., and E. R. Segnit, High Temperature Phases Developed in Some Kaolinite-Mica-Quartz Clays, *Trans. Brit. Ceram. Soc.*, 62: 375–395 (1963).

Such mixtures are likely to have a complex chemical composition, and a wide variety of phases have been reported. In general, it can be concluded that bulk chemical composition is likely to play a more important part in determining the high-temperature phases that develop than is the case for pure clay minerals.

Additional references

Allaway, W. H., Differential Thermal Analyses of Clays Treated with Organic Cations as an Aid in the Study of Soil Colloids, *Soil Sci. Soc. Am. Proc.*, 13: 183–188 (1948).

Augustnik, A. I., P. Z. Tandura, and L. I. Sverchkova, Mechanism of Reactions in Talc upon Heating, *J. Appl. Chem. USSR* (English Transl.), 22: 1156–1159 (1949).

Bohor, B. F., High-temperature Phase Development in Illitic Clays, Proceedings of the 12th National Clay Conference, pp. 233–246, Pergamon Press, New York (1964).

Donnay, G., J. Wyart, and G. Sabatier, Structural Mechanism of Thermal and Compositional Transformations in Silicates, *Z. Krist.*, 112(1): 161–168 (1959).

Dubois, P., Balance-analyseur thermique à enregistrement photographique, *Bull. Soc. Chim. France,* 3: 1178–1181 (1936).

Eberhart, J. P., Étude des transformations du mica muscovite par chauffage entre 700 et 1200°C, *Bull. Soc. Franc. Mineral. Crist.*, 86: 213–251 (1963).

Eitel, W., H. O. Muller, and O. E. Rodzewski, Übermikroscopische Untersuchungen an Tonmineralien, *Ber. Deut. Keram. Ges.*, 20: 165–180 (1939).

Endell, K., and M. Ardenne, Demonstration of Sintering and Fusion of Ceramic Raw Material in a Heated Electron Microscope, *Kolloid-Z.*, 104: 223–231 (1943).

Faust, G. T., Thermal Analysis of Quartz and Its Use in Calibration in Thermal Analysis Studies, *Am. Mineralogist*, 33: 337–345 (1948).

Forkel, W., Qualitative und Quantitative Untersuchungen an Dilatometerkurven von Tonen, *Heidelberger Beitr. Mineral. Petrog.*, 5(1): 1–22 (1955).

Fripiat, J. J., P. Rouxhet, and H. Jacob, Proton Delocalization in Micas, *Am. Mineralogist,* 50: 1937–1958 (1965).

Gastuche, M. C., F. Toussaint, J. J. Fripiat, R. Touilleaux, and M. Van Meersche, Study of Intermediate Stages in the Kaolin-Metakaolin Transformation, *Clay Minerals Bull.*, 5(29): 227–236 (1963).

Gilard, P., Autoclave Treatment of Kaolin and Metakaolin, *Verre Silicates Ind.*, 13: 57–59 (1948).

Hendricks, S. B., S. S. Goldich, and B. A. Nelson, A Portable Differential Thermal Analysis Unit for Bauxite Exploration, *Econ. Geol.*, 41: 64–76 (1946).

Johns, W. D., Topotactic Development of High-temperature Phases from Two-layer Silicates, *Am. Ceram. Soc. Bull.*, 44: 682–686 (1965).

Larsen, E. W., and G. Stieger, Dehydration and Optical Studies of Alunogen, Nontronite, and Griffithite, *Am. J. Sci., ser.* 15, 5: 1–19 (1928).

McCarter, W. S. W., K. A. Krieger, and H. Heinemann, Thermal Activation of Attapulgus Clay, *Ind. Eng. Chem.*, 42: 529–533 (1950).

Menshutkin, B. M., History of Thermal Analysis, *Ann. Secteur Anal. Phys.-Chim. Inst. Chim. Gen. (USSR)*, 8: 373–406 (1936).

Mervielli, J., and A. Boureille, Identifications des argiles céramique par la thermo-balance, *Bull. Soc. Franc. Ceram.*, 7: 18–27 (1950).

Norton, F. H., and W. G. Lawrence, Notes on Anomalous Heat Adsorption of Kaolinite, *J. Am. Ceram. Soc.,* **26**: 388–389 (1943).

Okuda, H., S. Kato, and T. Iga, High Temperature Phase Change in Kaolinitic Clays, *J. Ceram. Assoc. Japan,* **71**: 119–128 (1963).

Rios, E. G., and J. L. Vivaldi, Silicates of Laminar Structure, I, Hydration, *Anales Fis. Quim. B,* **45**: 291–342 (1949).

Rode, E. Y., and T. V. Rode, Physicochemical Nature of Nontronite, *Akad. Nauk SSSR,* **15**: 273–278 (1947).

Roy, R., Metastable and Stable Dehydration Reactions in Clays and Zeolites, *Proc. Conf. Silicates Ind.,* **7**: 141–154 (1963).

Russell, J. D., and V. C. Farmer, Infrared Spectroscopic Study of the Dehydration of Montmorillonite and Saponite, *Clay Minerals Bull.,* **5**(32): 443–464 (1964).

Stringham, B., and A. O. Taylor, Nontronite at Bingham, Utah, *Am. Mineralogist,* **35**: 1060–1066 (1950).

Tscheiswili, L., W. Bussem, and W. Weyl, Metakaolin, *Ber. Deut. Keram. Ges.,* **20**: 249–276 (1939).

Urbain, P., Introduction à l'étude pétrographique et géochemique des roches argileuses, *Actualites Sci. Ind.,* **50** (1935).

Vivaldi, J. L. M., and J. Cano-Ruiz, Contribution to the Study of Sepiolite, III, The Dehydration Process and the Types of Water Molecules, *Natl. Acad. Sci., Publ.* **456**, pp. 177–180 (1956).

Vivaldi, J. L. M., D. M. C. MacEwan, and M. R. Gallego, Effect of Thermal Treatment on the *c*-axial Dimension of Montmorillonite as a Function of the Exchange Cation, *Proc. Intern. Clay Conf., Stockholm,* **1**: 45–51 (1963).

Zvanut, F. J., and L. J. Wood, X-ray Investigation of the Pyrochemical Changes in Missouri Halloysite, *J. Am. Ceram. Soc.,* **20**: 251–255 (1937).

Ten
Clay-Mineral—
Organic Reactions

Introduction

Investigations of clay materials in many different fields for many years have produced evidence of some kind of a reaction between the fundamental components of clay and organic material. For example, early literature[1] on decolorizing oil with clays refers to the adsorption of some components of the oil during the process and to changes in the nature of the oil as a consequence of decolorization.

Students of soils in the field of agriculture have presented evidence for some sort of inorganic-organic combination in many soils. For example, Demolin and Barbier[2] showed a definite fixation of humic acid and protein by clay, and Mattson[3] demonstrated a reduction in base-exchange capacity by complexing clay with protein. Sedletsky[4] investigated the matter in con-

[1] Gilpin, J. E., and M. P. Cram, Fractionation of Crude Oil by Capillary Diffusion, *U.S. Geol. Surv., Bull.* 365 (1908).

[2] Demolin, A., and G. Barbier, Conditions de formation et constitution du complexe argilo-humique des sols, *Compt. Rend.*, **188**: 654–656 (1929).

[3] Mattson, S., The Laws of Soil Colloidal Behavior, VII, Proteins and Proteinated Complexes, *Soil Sci.*, **23**: 41–72 (1932).

[4] Sedletsky, I. D., "X-ray Study of Soils," Moscow (1939).

siderable detail and concluded that many soils contain clay-mineral–organic complexes.

Further evidence for such reactions is provided by the color reactions produced when certain clays and organic compounds are mixed (see pages 407–410). The development of color or a change in color requires some kind of reaction between the clay and the organic compound added. The fact that smectite and halloysite experience changes in their optical properties when immersed in certain oils (see pages 423–426) also indicates a reaction between these clay minerals and some organic compounds.

Prior to the determination of the exact nature of the fundamental constituents of clays and soils, there could be no precise concept of the reaction between the components of clay and organic compounds and of the nature of clay-organic complexes. The finding that clay materials are composed of small crystalline particles and the determination of the structure of the particles provided a new basis for studying clay-organic reactions. Since the early 1930s a large amount of work has been done on this problem, and the character of clay-mineral-organic reactions has now been fairly well established. Thus, in 1934, Smith[5] reacted organic bases and their salts with montmorillonites and presented evidence that the reaction was one of ion exchange. Gieseking and Jenny[6] in 1936 showed that methylene blue replaced sorbed cations in certain clays, and later Gieseking[7] and his colleagues[8,9] showed definitely that organic ions enter into cation-exchange reactions with the clay minerals, particularly with smectite.

Many common organic cations contain basic amino groups, and it is the reaction between these compounds and the clay minerals that has been studied in most detail. These cations are ammonium ions in which one or more of the hydrogens have been substituted by organic groups. Other onium ions, such as phosphonium, sulfonium, and oxonium, also replace inorganic cations.

Hofmann and his colleagues,[10] in 1934, showed that the c-axis dimension

[5] Smith, C. R., Base Exchange Reactions of Bentonites and Salts of Organic Bases, *J. Am. Chem. Soc.*, **56**: 1561–1563 (1936).

[6] Gieseking, J. E., and H. Jenny, Behavior of Polyvalent Cations in Base Exchange, *Soil Sci.*, **42**: 273–280 (1936).

[7] Gieseking, J. E., Mechanism of Cation Exchange in the Montmorillonite-Beidellite-Nontronite Type of Clay Minerals, *Soil Sci.*, **47**: 1–14 (1939).

[8] Ensminger, L. E., and J. E. Gieseking, The Adsorption of Proteins by Montmorillonite Clay, *Soil Sci.*, **48**: 467–473 (1939).

[9] Ensminger, L. E., and J. E. Gieseking, Adsorption of Proteins by Montmorillonite Clays and Its Effect on Base-Exchange Capacity, *Soil Sci.*, **51**: 125–132 (1941).

[10] Hofmann, U., K. Endell, and D. Wilm, Röntgenographische und kolloidchemische Untersuchungen über Ton, *Angew. Chem.*, **47**: 539–547 (1934).

of montmorillonite varied following treatment with alcohol, acetone, and ether. In 1939, Ensminger and Gieseking[8] showed definitely that organic ions could be sorbed on the basal plane surfaces of montmorillonite.

The early work of Hofmann et al.[10] in 1934, and of Myers[11] and Sharov[12] in 1937, suggested that organic compounds with polar active groups could be adsorbed by the clay minerals. Bradley[13,14] in 1945 and MacEwan[15,16] quite independently at about the same time showed definitely that the nonionic organic molecules of polar character could be adsorbed by the clay minerals. These investigators also considered the nature of the bond between the nonionic organic molecules and clay minerals.

The initial work on clay-mineral–organic reactions was done largely on smectite, and such investigations have continued. Recently, a considerable variety of organic derivatives have been prepared with other clay minerals. Reactions with the two-layer minerals, kaolinite and halloysite especially, have been studied in detail (Wada[17] and Weiss et al.[18]). In recent studies particular attention has been given to the orientation of the organic molecules when they are intercalated in the silicate structures and to the nature of the bond tying them to the structure.

Various clay minerals may serve to catalyze certain organic reactions (Robertson[19]), and they may also act as initiators of polymerization reactions (Hauser and Kollman[20] and Friedlander[21]). The foregoing clay-mineral

[11] Myers, H. E., Physicochemical Reactions between Organic and Inorganic Soil Colloids as Related to Aggregate Formation, *Soil Sci.*, **44**: 331–359 (1937).

[12] Sharov, V. S., Action of Substances Having a Hydrophobic and a Gel-forming Effect on Clays, *Grozhensk. Neftyanik*, **6**: 10–15 (1936); *Chem. Abstr.*, **31**: 2764 (1937).

[13] Bradley, W. F., Molecular Associations between Montmorillonite and Some Polyfunctional Organic Liquids, *J. Am. Chem. Soc.*, **67**: 975–981 (1945).

[14] Bradley, W. F., Diagnostic Criteria for Clay Minerals, *Am. Mineralogist*, **30**: 704–713 (1945).

[15] MacEwan, D. M. C., Identification of the Montmorillonite Group of Minerals by X-rays, *Nature*, **154**: 577–578 (1944).

[16] MacEwan, D. M. C., The Identification and Estimation of the Montmorillonite Group of Clay Minerals, with Special Reference to Soil Clays, *J. Soc. Chem. Ind. (London)*, **65**: 298–305 (1946).

[17] Wada, K., Lattice Expansion of Kaolin Minerals by Treatment with Potassium Acetate, *Am. Mineralogist*, **46**: 78–91 (1961).

[18] Weiss, A., W. Thielepape, W. Ritter, H. Schäfer, and G. Goring, Zur Kenntnis von Hydrazin-Kaolinit, *Z. Anorg. Allgem.*, **320**(1–4) (1963).

[19] Robertson, R. H. S., Clay Minerals as Catalysts, *Clay Minerals Bull.*, no. 2, pp. 38–42 (1948).

[20] Hauser, E. A., and R. C. Kollman, Clay Complexes with Conjugated Unsaturated Aliphatic Compounds of Four to Five Carbon Atoms, U.S. Patent 2,951,087 (1960).

properties are of considerable economic importance. Also, clay-mineral–organic reaction products have found wide application in industry. Grim[22] and Nahin[23] have reviewed briefly the very extensive literature in this area.

Ionic reactions

Smectite. Ionic reactions with smectite have been studied in greater detail than those with any other clay minerals. The relatively rapid rate of the reaction and the expandable structural characteristic of the mineral, permitting the reaction to be studied by X-ray diffraction, make it most amenable to such investigations.

The data in Fig. 10-1, after Ensminger and Gieseking,[9] show that the

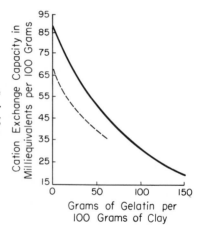

Fig. 10-1. **Effect of gelatin adsorption on the cation-exchange capacity of montmorillonite. Montmorillonite (——); colloid fraction of Hartsburg soil (- - - - - -). (After Ensminger and Gieseking.[9])**

adsorption of gelatin may have a very marked effect on the cation-exchange capacity of montmorillonite and a montmorillonitic soil. The gelatin-clay complexes were prepared by mixing alkaline suspensions of gelatin and clay together and then acidifying to pH 2.6 with acetic acid. These authors

---

[21] Friedlander, H. Z., Two Methods Give Polybutene-2. Homopolymer Forms by Contact Polymerization on Oxides or with Ziegler Catalyst on Clay, *Chem. Eng. News,* p. 42 (Feb. 10, 1964).

[22] Grim, R. E., "Applied Clay Mineralogy," McGraw-Hill, New York (1962).

[23] Nahin, P. G., Perspectives in Applied Organo-Clay Chemistry, Proceedings of the 10th National Clay Conference, pp. 257–271, Pergamon Press, New York (1963).

showed that the protein did not reduce the cation-exchange capacity in an alkaline medium. As the hydrogen-ion concentration is increased, however, the cation-exchange capacity decreases. They concluded, from the fact that an increase in hydrogen-ion concentration increases the basic properties of the protein, that it is adsorbed as cations.

The data in Table 10-1, after Hendricks,[24] show that the cation-exchange

Table 10-1 Cation-exchange capacity of a montmorillonite determined with certain organic cations
(*After Hendricks*[24])

Base	*Cation-exchange capacity, meq/100 g*
Benzidine	91
p-Aminodimethylaniline	90
p-Phenylenediamine	86
α-Napthylamine	85
2,7-Diaminofluorene	95
Piperidine	90
Barium	90–94

capacities determined with relatively small organic molecules are nearly the same as that found by displacement of the hydrogen ion with the barium ion. Hendricks showed also that extremely weak bases like the *o*- and *m*-nitroanilines do not form salts with montmorillonite, and that large molecules like the alkaloids brucine (*strychnos* group) and codeine (*morphine* group) neutralize less of the hydrogen on the clay, even though they are stronger bases than aniline and benzidine. Thus codeine, in concentrations of 4 to 0.5 meq of acid clay in 25 cc of water, neutralized 0.63 meq of hydrogen per gram of clay, and brucine similarly neutralized 0.65 meq of hydrogen per gram of montmorillonite. The difference, namely, about 0.30 meq/g, between these amounts and the exchange capacity of the montmorillonite (see Table 10-1) represents the amount of hydrogen so covered by the large alkaloid molecules as to be unavailable for neutralizing other molecules (cover-up effect). Organic ions larger than the area per exchange position, i.e., about 80 A², would be required to produce the cover-up effect. Hendricks showed that both groups of an organic diamine are effective in neutralizing an acid clay.

Hendricks[24] indicated that the organic ions are held by van der Waals forces in addition to the coulombic force. In general, the larger ions are more strongly adsorbed because of the greater van der Waals forces, and

[24] Hendricks, S. B., Base-Exchange of the Clay Mineral Montmorillonite for Organic Cations and Its Dependence upon Adsorption Due to van der Waals Forces, *J. Phys. Chem.*, **45**: 65–81 (1941).

larger organic ions are difficult or impossible, to replace with smaller ions. Grim, Allaway, and Cuthbert[25] showed that small ions are adsorbed only up to the cation-exchange capacity, whereas larger ions may be adsorbed in excess, and that these excess molecules are not dissociated and probably are adsorbed by van der Waals forces. Bradley and Grim[26] believed that the coulombic forces are supplemented by C—H· · ·O bonds (see page 373) between the organic molecule and the clay-mineral surface.

The amount of organic ion adsorbed varies with its orientation on the clay-mineral surface as well as with its size. For example, Greenland and Quirk[27] investigated the adsorption of 1-(*n*-alkyl) pyridinium bromides by montmorillonite and found that, in compounds with up to eight carbon atoms in the alkyl chain, adsorption was approximately limited to the exchange capacity. With larger ions, adsorption occurred beyond the exchange capacity and extended to double with the cetyl derivative on calcium montmorillonite and triple with the same derivative on sodium montmorillonite. For the short chains and for fractional saturations with longer chains, the organic molecules lay parallel to the clay surface, but for excessive saturation, especially with the cetyl compounds, the chains stood erect to the surface.

McAtee[28] showed that the primary amine, octadecylamine, has enough basic character so that it will replace some of the sodium ions from an exchange position in sodium montmorillonite. The replacement is not as rapid nor as complete as with the corresponding amine hydrochloride but takes place to an extent corresponding to its relative basic character. McAtee also showed that a large quaternary amine salt replaces sodium much more easily than does octadecylamine hydrochloride (a simple primary amine ion). The sodium of montmorillonite is replaced stoichiometrically while calcium and magnesium require greater than the stoichiometric amounts of organic for replacement. Because of the larger groups of flocculated clay in calcium and magnesium systems the quaternary cation probably cannot effectively exchange onto the clay or even have room to become adsorbed after approximately 75 meq have been added. The author's data indicate that the cations on hectorite are somewhat easier to replace than those on a Wyoming montmorillonite. The author's explanation is that in hectorite the charge on the

[25] Grim, R. E., W. H. Allaway, and F. L. Cuthbert, Reaction of Different Clay Minerals with Organic Cations, *J. Am. Ceram. Soc.*, **30**: 137–142 (1947).

[26] Bradley, W. F., and R. E. Grim, Colloid Properties of Layer Lattices, *J. Phys. Chem.*, **52**: 1404–1413 (1947).

[27] Greenland, D. J., and J. P. Quirk, Adsorption of 1-(*n*-alkyl) Pyridinium Bromides by Montmorillonite, Proceedings of the 9th National Clay Conference, pp. 484–499, Pergamon Press, New York (1962).

[28] McAtee, J. L., Jr., Inorganic-Organic Cation Exchange on Montmorillonite, *Am. Mineralogist*, **44**: 1230–1236 (1959).

lattice is a consequence of replacements within the octahedral part of the structure, whereas in the Wyoming montmorillonite there are replacements in both the octahedral and tetrahedral parts of the structure.

Weiss and Michel[29] and Weiss[30] investigated the reaction of alkylammonium ions and mica-type layer silicates in great detail. They showed that the longer the *n*-alkyl chain, the steeper the exchange isotherm corresponding to bond strength. Equilibrium is achieved after only a few hours for low-charged smectites but requires up to 14 months for highly charged materials such as the micas. Surprisingly, with the latter, attainment of equilibrium takes less time with long-chain, that is, larger, ions than with short-chain ones; this behavior is consistent with the mechanism for change of lattice sites within the interior of the crystal.

According to Weiss,[30] the bonding strengths of the alkyl compounds decrease sharply in the series RNH_3^+, $R_2NH_2^+$, and R_3NH^+. Quaternary alkyl-ammonium ions R_4N^+ behave quite differently; asymmetrical ions such as trimethylcetyl and dimethyl-di-*n*-octadecylammonium ions must be classified among the primary ammonium ions, whereas symmetrical ions, on the other hand, rank behind the secondary ammonium ions. Because of the bulkiness of the groups, the bonding strengths of $R_2NH_2^+$, R_3NH^+, and symmetrical R_4N^+ also decrease rapidly with increasing charge on the montmorillonite layers. The behavior of sulfonium and oxonium ions is similar to that of R_3NH^+.

Doehler and Young[31] investigated the adsorption of quinoline by sodium and calcium montmorillonite, illite, and kaolinite in water suspension under varying conditions of quinoline concentration, pH, salinity, time, and temperature. In the experiments, quinoline existed both in ionic and nonionic forms, depending upon the pH. It was found that quinoline was adsorbed by ion exchange and molecular adsorption. Of the variables investigated, pH was found to be critical.

Bodenheimer and others[32] investigated the sorption of various diamines from aqueous solutions by montmorillonite in the presence of copper. They showed that the clays sorbed copper diamine complexes from solutions and that the sorbed complex can be returned into solution by ion exchange. Peter-

[29] Weiss, A., and E. Michel, Uber Kationenaustausch und innerkrystallines Quellungs-vermögen bei kettenformigen Polyphosphaten, *Z. Anorg. Allgem. Chem.*, **296:** 313 (1958).

[30] Weiss, A., Organic Derivatives of Mica-type Layer-Silicates, *Angew. Chem.*, **2:** 134–143 (1963).

[31] Doehler, R. W., and W. A. Young, Conditions Affecting the Adsorption of Quinoline by Clay Minerals in Aqueous Suspensions, Proceedings of the 9th National Clay Conference, pp. 468–483, Pergamon Press, New York (1962).

[32] Bodenheimer, W., and L. Heller, Organo-Metallic Clay Complexes, Part 2, *Clay Minerals Bull.*, **5**(28): 145–154 (1962).

son and Freeman[33] also reported the preparation of organometallic derivatives of clays.

Palmer and Bauer[34] reported the sorption of gaseous amines. They found the order of adsorption was $MeNH_2 > ETNH_2 > Me_2NH$ and concluded that this might be the result of stearic hindrance.

The adsorption of the organic ions on the basal plane surface of montmorillonites is shown conclusively by the shift in the *c*-axis spacing of the montmorillonite units accompanying the adsorption (Fig. 10-2). Hendricks[24] found that the organic ions he studied were oriented between the montmorillonite layers so as to cause the minimum expansion of the layers. Thus, flat-shaped ions were found to lie with their flat surfaces parallel to the basal surfaces of the montmorillonite flake (Fig. 10-2).

Hendricks,[24] Bradley and Grim,[26] and Jordan[35] showed that for some organics more than one molecular layer may be adsorbed on the basal surfaces of the montmorillonite planes. Jordan determined the *c*-axis spacing of montmorillonite from Wyoming bentonite treated with aliphatic amines of increasing chain lengths. With the total internal surface area per cation-exchange position calculated as 165 A^2, the *c*-axis spacings are plotted in Fig. 10-3 as a function of amine chain length. The stepwise separation of the flakes in increments of 4 Å, which is about the van der Waals thickness of a methyl group, indicates that the chains lie flat along the clay flake surfaces, with the planes of the zigzag carbon chains parallel to the plane of the mineral. On this basis the areas covered by the amine chains have been computed from atomic dimensions and included in Fig. 10-2. When the organic chains occupy no more than half of the available area per exchange position, the organic molecules on the top surface of one layer fit into the gaps between those on the bottom surface of the layer directly above it, so that the resulting separation of the two layers is 4 Å, or the thickness of one hydrocarbon chain. When the chains occupy more than 50 percent of the surface area per exchange position, adjacent laminae are unable to approach more closely than 8 Å, which is the thickness of two hydrocarbon chains. With long-chain quaternary amines McAtee[36] reported the formation of up to six-layer complexes.

Numerous investigators, e.g., Brindley and Hoffmann,[37] have studied the

[33] Peterson, W. H., and P. K. Freeman, Organometallic Derivatives of Acid Clays, U.S. Patent 3,067,223 (Dec. 4, 1962).

[34] Palmer, J., and N. Bauer, Sorption of Amines by Montmorillonite, *J. Phys. Chem.,* **65:** 894–895 (1961).

[35] Jordan, J. W., Organophilic Bentonites, I, Swelling in Organic Liquids, *J. Phys. Colloid Chem.,* **53:** 294–306 (1949).

[36] McAtee, J. L., Jr., Random Interstratification in Organophilic Bentonite, *Natl. Acad. Sci., Publ.* 566, pp. 308–317 (1958).

[37] Brindley, G. W., and R. W. Hoffmann, Orientation and Packing of Aliphatic Chain Molecules on Montmorillonite, Proceedings of the 9th National Clay Conference, pp. 546–556, Pergamon Press, New York (1962).

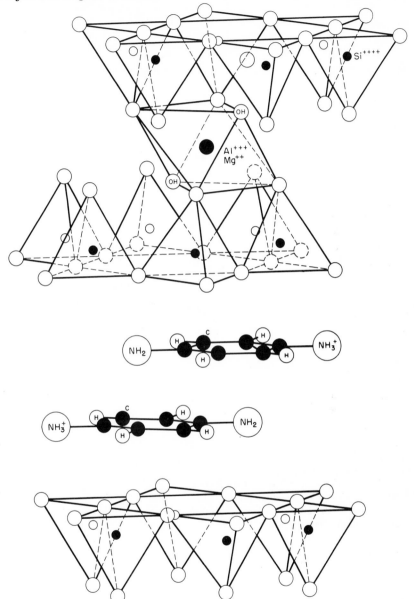

Fig. 10-2.   Schematic drawing showing position of organic cations between montmorillonite layers.   (After Hendricks.[24])

orientation and packing of organic molecules on various clay minerals. Weiss[30] recently summarized data on the intercalation of inorganic ions in mica-type layer silicates, and the following is taken largely from his work.

On intercalation of *n*-alkylammonium ions the resulting distance between

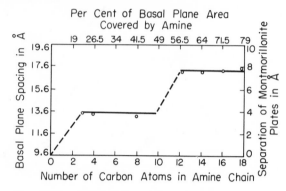

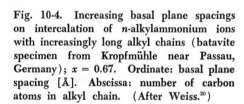

Fig. 10-3. Effect of amine chain length on montmorillonite basal spacing. (After Jordan.[35])

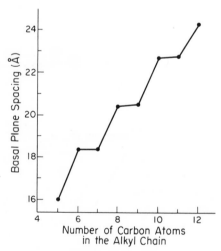

Fig. 10-4. Increasing basal plane spacings on intercalation of *n*-alkylammonium ions with increasingly long alkyl chains (batavite specimen from Kropfmühle near Passau, Germany); $x = 0.67$. Ordinate: basal plane spacing [Å]. Abscissa: number of carbon atoms in alkyl chain. (After Weiss.[30])

the silicate layers depends upon the length of the alkyl chain and the layer charge. With increasing length of the alkyl chain, this distance increases alternately by a greater or lesser amount, depending on whether the resulting number of carbon atoms in the intercalated chain is even or odd (Fig. 10-4).

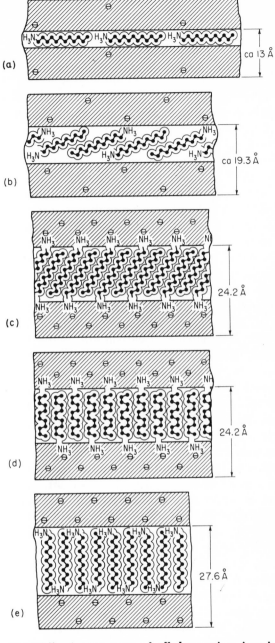

Fig. 10-5. Arrangements of alkylammonium ions in mica-type layer silicates with different layer charges. (a) Very low charge; (b) medium charge; (c) preferred arrangement with the formation of three hydrogen bonds per RNH_3 + ion and with trans, trans-configuration; (d) as (c) but with cis, trans-configuration (for reason of clarity the energetically unfavorable, planar cis, trans-configuration is shown here); (e) high charge. Hatched areas: silicate layers. (After Weiss.[30])

Thus, the alkyl chains can be present neither in the form of the statistical coil nor with their longitudinal axes parallel or perpendicular to the silicate layers. The favored arrangement of the chains is illustrated in Fig. 10-5c and d which shows the pronounced stepwise changes in the interplanar distance encountered when the alkyl chain is lengthened. According to Fig. 10-4, on passing from an odd to the next higher even number of carbon atoms, the interplanar distance increases by 2.0 to 2.1 Å, but on passing from an even to the next higher odd number the increase is only 0.0 to 0.1 Å. The value of 2.0 to 2.1 Å per carbon atom exceeds the length of the single C—C bond (= to 1.54 Å). The additional distance originates from the coordination of the terminal CH_3 group with a six-membered SiO ring. According to Fig. 10-5c, one would expect the greatest variation in the interlayer distance to occur on transition from an even to an odd number of carbon atoms. The opposite is, in fact, observed because the terminal CH_3 group may dip into the six-membered SiO ring only when there is an even number of carbon atoms. Brindley and Hoffmann[37] and Rowland and Weiss[38] have shown that CH_3 groups can penetrate far into the six-membered SiO ring. On transition to an odd number of carbon atoms, the distance between the silicate layers must increase by the entire van der Waals radius of a CH_3 group.

On replacing the methylammonium with ethylammonium cations, the basal plane spacing becomes about 2 Å larger. If the alkyl chain is made even longer, the increase in the interplanar distance begins to depend also on the charge on the layer (Weiss[30]). In highly charged silicate layers ($x = 1\pm$, with x = charge per half unit cell), the basal plane spacing increases, starting with the ethyl compound; if a somewhat lower charge density ($x = 0.66$) is present, it increases beginning with the n-butyl compound; with a still lower charge density ($x = 0.33$), it remains practically unchanged at 13 to 14 Å from the ethyl to the n-decylammonium derivative. With an interlayer distance of 13 to 14 Å, there is available a free space with a height of 4 to 5 Å which is equal to the van der Waals diameter of the chain. Evidently when the layer charge is low, the alkyl chains remain flat until the layers are densely covered (Fig. 10-5a). As the charge is increased, the chains begin increasingly to slip over one another (Fig. 10-5b) and stand more erect, until finally the arrangement in Fig. 10-5e is achieved. These arrangements are especially favored because the hydrogen atoms of the RNH_3 group are able to form hydrogen bonds with the oxygen atoms of the six-membered SiO rings without distortion of their valence angle. Weiss and Kantner[39] pointed out that the relation between the length of the alkyl chain, basal

[38] Rowland, R. A., and E. J. Weiss, Bentonite-Methylamine Complexes, Proceedings of the 10th National Clay Conference, pp. 460–468, Pergamon Press, New York (1963).

[39] Weiss, A., and I. Kantner, Uber eine einfache Möglichkeit zur Abschätzung der Schichtladung glimmerartiger Schichtsilikate, Z. *Naturforsch.* **15b:** 804–807 (1960).

spacing, and charge density may be used to determine the charge on the layers of smectite minerals.

Talibudeen[40] studied the intercalation of proteins and found fairly low spacings, concluding that the organics lay flat and were "stretched." On the other hand, Ensminger and Gieseking[8] found for some proteins very large spacings which were not related to molecular size and configuration. Weiss[30] showed that alkaloid cations are very strongly bound by mica-type layer silicates, and thus they are adsorbed from even very dilute solutions. Intracrystalline swelling capacity is largely blocked by alkaloid cations. However, zeolitic uptake of liquid is always possible whenever closest packing of the cations does not occur.

At low pH values, proteins are exchanged almost quantitatively. With increasing pH values, the numbers of cationic positions in the protein drops, and the exchange becomes incomplete (Weiss and Koch[41]). The reactions with protamines proceed especially rapidly. Albumins, globulins, and prolamines are also intercalated. With serum globulins, the protein seems to penetrate the lattice to only about 20 to 20 Å from the outer edge of the silicate crystal.

The water-adsorbing properties of montmorillonite are gradually reduced as the basal surfaces of the mineral are coated with the organic ions. In general, the larger the organic ion, the greater is the reduction in the water-adsorbing capacity. Figure 10-6 shows differential curves for untreated montmorillonite from Wyoming bentonite and for butyldodecyl- and octadecylamine complexes of the same clay after exposure to an atmosphere of 75 percent relative humidity for 4 days. These curves show a progressive decrease in the size of the low-temperature endothermic peak; this decrease results from the loss of adsorbed water with an increase in the size of the aliphatic chain. Gieseking,[7] Hendricks,[24] and Grim, Allaway, and Cuthbert[25] presented data on the relation of water adsorption to amine adsorption which indicate the same conclusion.

With the exclusion of a few still unexplained exceptions (Walker[43]), n-alkylammonium ions and smectite do not build diffuse ionic double layers with water since the alkyl groups are hydrophobic and the N—H· · ·O hydrogen bridges with the silicate sheets are too strong. There is, however, some

[40] Talibudeen, O., Complex Formation between Montmorillonoid Clays and Amino Acids and Proteins, *Trans. Faraday Soc.,* **51**: 582–590 (1955).

[41] Weiss, A., and G. Koch, Ph.D. thesis, Technische Hochschule, Darmstadt, Germany (1960).

[42] Jordan, J. W., Alteration of the Properties of Bentonite by Reaction with Amines, *Mineral. Mag.,* **28**: 598–605 (1949).

[43] Walker, G. F., Macroscopic Swelling of Vermiculite Crystals in Water, *Nature,* **187**: 312–313 (1960).

swelling of n-alkylammonium montmorillonites in water, and Fig. 10-7 (after Weiss[30]) shows the relationship that exists between the charge density, length of the alkyl chains, and the swelling in water. A sample with x (layer charge) = 0.42 swells in water only when the n-alkylammoniun ion contains

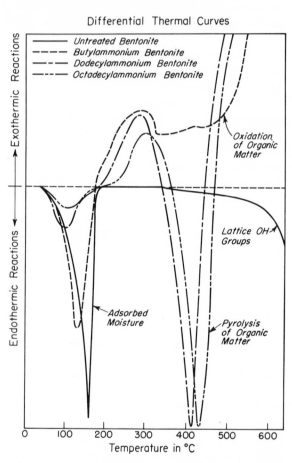

Fig. 10-6. Differential thermal curves of amine-treated and untreated montmorillonite. (After Jordan.[42])

more than eight carbon atoms. With a higher charge even the n-butyl and ethylammonium derivatives, respectively, swell in water. In contrast, the lower-charged montmorillonites swell only with n-decylammonium ions or still longer chains.

Cowan and White[44] investigated the organophilic properties of a series of montmorillonite complexes with *n*-aliphatic amines of varying chain length. They found that, in a series from ethylammonium to octadecylammonium montmorillonite, the dodecylammonium compound is most active in the adsorption of phenols from dilute aqueous solution. They concluded that several factors affect the extent of adsorption, but the main criteria appear to be (1) the clay possesses both hydrophilic and organophilic properties, with a balance between the two types of surface being critical, and (2) the adsorbate molecule possesses a localized center of high electron density while the remainder of the molecule is essentially nonpolar. It is supposed that

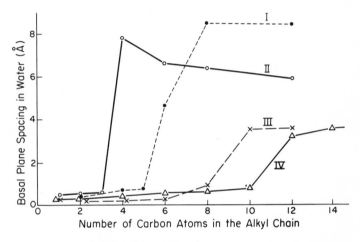

Fig. 10-7. **Relationship between the length of the alkyl chain, the layer charge** *x*, **and the swelling power of** *n*-alkylammonium **montmorillonites in water: curve I:** $x = 0.55$; **curve II:** $x = 0.66$; **curve III:** $x = 0.42$; **curve IV:** $x = 0.33$. **Ordinate: basal plane spacing in water (Å). Abscissa: number of carbon atoms in the alkyl chain. (After Weiss.[30])**

adsorption occurs simply by two complementary mechanisms: first, by electrostatic hydrogen bonding between polar groups and, second, by van der Waals bonding between organophilic centers.

White and Cowan[45] found that there was no fundamental difference between the cation-exchange reaction of aromatic and aliphatic amines with montmorillonite. They found that in certain cases chemical changes probably

[44] Cowan, C. T., and D. White, Adsorption by Organo-Clay Complexes, I, Proceedings of the 9th National Clay Conference, pp. 459–467, Pergamon Press, New York (1962).

[45] White, D., and C. T. Cowan, Aromatic Amine Derivatives of Montmorillonite, *Trans. Brit. Ceram. Soc.*, **59:** 16–21 (1960).

initiated by the clay surface result in the formation of color compounds. They did not investigate the possible cause of this color formation.

The swelling behavior in electrolyte solutions is also unexpected. If organic cations are intercalated between the silicate layers, the basal plane spacings are decreased when an electrolyte is added to the aqueous swelling agent (Weiss[30]). However, in the case of montmorillonite derivatives with long-chain n-alkylammonium ions, the interplanar distances increase when an electrolyte is dissolved in the swelling agent.

Jordan[42] showed that the viscosity of aqueous bentonite suspensions varies with the amount of treating amine, and that the maximum effect is obtained in the region of complete cation exchange. As shown in Fig. 10-8, maximum

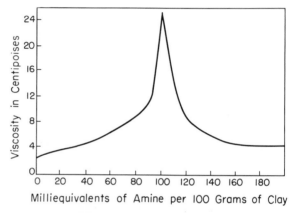

Fig. 10-8. Effect of octadecylamine acetate on the viscosity of montmorillonite (Wyoming bentonite). Concentration of suspension, 1.9 percent. (After Jordan.[42])

flocculation with maximum viscosity is reached in an octadecylamine-montmorillonite when the amount of the added amine is nearly equal to the cation-exchange capacity.

The thermodynamic aspects of the reactions of organic cations and montmorillonite have been studied by various investigators, and this matter may be pursued by consulting the work of Barrer and Reay,[46] Slabaugh and Kupka,[47] and Cowan and White.[44,48]

[46] Barrer, R. M., and J. S. S. Reay, Sorption of NH_4^+ and Cs^+ Montmorillonite and Ion Fixation, *J. Chem. Soc.*, pp. 3824–3830 (1959).

[47] Slabaugh, W. H., and F. Kupka, Organic Cation Exchange Properties of Calcium, *J. Phys. Chem.*, **62**(5): 599 (1958).

[48] Cowan, D. T., and D. White, The Mechanism of Exchange Reactions Occurring Between Sodium Montmorillonite and Various n-Primary Aliphatic Amine Salts, *Trans. Faraday Soc.*, **54**: 691–697 (1958).

Other Clay Minerals. *Vermiculite.* Among the clay minerals other than montmorillonite, vermiculite is most amenable to study because of its high cation-exchange capacity and structural expansion properties. Walker[49] first reported complexes with organic ions and showed that the amount of displacement of vermiculite basal reflections depended on the concentration and nature of the organic ion. Garrett and Walker[50] studied the reaction of vemiculite and a series of *n*-alkylammonium compounds ranging from one to eighteen carbon atoms. They observed that the three- and four-carbon members exhibit distinctive unidimensional swelling in water. Shorter molecules apparently lie flat on the silicate surfaces, whereas chains having three or four carbon atoms and longer stand at high angles of inclination. They found that intercalation is facilitated if the interlayer magnesium ions are first replaced by sodium and that complete or almost complete exchange can be obtained in that way. They found also that the swelling of butylammonium-vermiculite crystals decreases as the particle size of the vermiculite decreases and further that similar compounds with montmorillonite do not show swelling water.

Haase, Weiss, and Steinfink[51] determined, on the basis of single-crystal analysis of a hexamethylene-diamine-vermiculite complex, that the organic molecule was inclined to the basal plane of the silicate layer at an angle of 32°, and the projection of the organic molecule onto the *ab* plane was at an angle of 56° to the *a* axis. The atomic distances suggested that, at one end of the hexamethylene-diamine chain, the nitrogen ion forms an N· · ·H· · ·O bond, whereas the other end the N· · ·O distance indicates a van der Waals contact only. In the space group *C*2 determined for the complex, there should be four hexamethylene-diamine molecules, whereas nitrogen analysis indicates only one organic molecule per unit cell. The molecules, therefore, must be statistically distributed over the available crystallographic sites.

Illite. Grim, Allaway, and Cuthbert[25] showed that illite reacted with organic ions up to its exchange capacity. The ions were thought to be limited to the exterior surface of the illite particles; i.e., they did not replace the potassium ions between the silicate layers. Weiss and Michel[29] showed that *n*-alkylammonium ions may replace K⁺ in micas, producing intercalation complexes. The rate of reaction decreases as the charge on the lattice increases and may require up to 14 months before equilibrium is attained. Surprisingly, larger long-chain ions require less time than smaller short-chain ions to attain equilibrium.

[49] Walker, G. F., Vermiculite-Organic Complexes, *Nature,* **166:** 695 (1950).

[50] Garrett, W. G., and G. F. Walker, Swelling of Some Vermiculite Organic Complexes in Water, Proceedings of the 9th National Clay Conference, pp. 557–567, Pergamon Press, New York (1962).

[51] Haase, D. J., E. J. Weiss, and H. Steinfink, The Crystal Structure of a Hexamethylene-diamine-vermiculite Complex, *Am. Mineralogist,* **48:** 261–270 (1963).

Chlorite.   No data are available on the reaction of organic ions and chlorite minerals, but it seems certain that the reaction would be similar to illite.

Kaolinite and Halloysite.   In general, kaolinites enter into exchange reactions with organic ions up to their exchange capacity (Grim, Allaway, and Cuthbert[25]).   These authors showed that large organic ions added in excess of the exchange capacity tend to be adsorbed by van der Waals forces.   Halloysite minerals would be expected to exhibit similar reactions.   It appears that the organic ions do not penetrate between the silicate layers unless the layers have been pried apart by some other molecule (see pages 380–383).

Mortensen[52] studied in some detail the adsorption of hydrolized polyacrylonitrile on kaolinite and found that this molecule apparently is adsorbed on kaolinite, with the formation of ionic bonds between ionized carboxyl groups of the organic and unsatisfied valence bonds of exposed lattice aluminum.   The adsorption is increased by the presence of polyvalent exchange cations, probably by a bridging mechanism.   Blocking of silanol groups did not affect adsorption, but blocking of exposed lattice aluminum completely inhibited adsorption.   Infrared-absorption spectra also gave evidence that lattice aluminum was involved in the adsorption.

In recent years organic-clad kaolinite clays have found substantial use in industry.   For information concerning them, the patent literature (Albert and Wilcox;[53] Wilcox[54]) and trade brochures of the producing companies should be consulted.

Attapulgite and Sepiolite.   The fibrous clay minerals should react with organic ions around their edges like kaolinite, but there are substantially no reported researches on this subject.   The patent literature, e.g., Sawyer,[55] indicates that such organic-clay attapulgite clays are prepared for commercial use.

Adsorption of neutral molecules

Smectite and halloysite.   Debye[56] showed that many nonionic and organic molecules are dipoles as a result of the lack of symmetry of electron distribution within individual molecules.   Such molecules act as if they carried both

---

[52] Mortensen, J. L., Adsorption of Hydrolyzed Polyacrylonitrile on Kaolinite, Proceedings of the 9th National Clay Conference, pp. 530–545, Pergamon Press, New York (1962).

[53] Albert, C. G., and J. R. Wilcox, Surface Modified Kaolin Clay, U.S. Patent 2,948,632 (Aug. 9, 1960).

[54] Wilcox, J. R., Modified Kaolin Clay, U.S. Patent 2,982,665 (May 2, 1961).

[55] Sawyer, E. W., Jr., Agent for Gelling Organic Liquids and Organic Liquids Gelled Therewith, U.S. Patent 3,049,498 (Aug. 14, 1962).

[56] Debye, P., "Polar Molecules," Reinhold, New York (1929).

centers of positive charges and centers of negative charges. Since the clay-mineral structures are also polar, when they are in contact with liquids, solutions, or suspensions containing other polar substances, the negative centers of the clay minerals attract the positive centers on polar substances in surrounding liquids, and positive centers on clay minerals attract negative centers on surrounding polar particles. The foregoing notion of dipole character is valuable as a general concept. However, it should be borne in mind that more than one polar group can enter many molecules, and some quite stable associations occur with molecules which have no dipoles.

The water molecule is polar, and it is by far the most important polar compound sorbed by the clay minerals. Bradley[13] in 1945 showed that glycols, polyglycols, and polyglycol ethers were adsorbed on the basal plane surfaces of montmorillonite, displacing water from these surfaces. About contemporaneously and quite independently, MacEwan[15,16] showed that a large variety of polar organic molecules can be adsorbed on the basal surfaces of both montmorillonite and halloysite (Table 10-2). The c-axis spacing of both minerals varies with the organic molecule adsorbed. The inorganic cations present on the surfaces of the clay minerals are not necessarily displaced by the adsorption of the organic molecules.

In recent years reactions with a considerable number of organic compounds have been studied. Thus, Greenland[57] showed that a large variety of sugars reacted with montmorillonite, forming complexes with one or two molecular layers. He pointed out that sodium montmorillonite yielded two-layer complexes more readily than did the clay with other exchangeable cations. Emerson[58] also found that sugars were adsorbed, but according to him the reaction was favored if Ca++ and Sr++ were the exchangeable cations. Lynch, Wright, and Cotnoir[59] found that montmorillonite adsorbed various carbohydrates. Of those studied, the greatest adsorption was with cornstarch (22.8 percent), and little difference was noted between calcium and hydrogen saturation of the mineral. X-ray-diffraction analyses demonstrated that the carbohydrate materials were adsorbed between the basal planes of the montmorillonite clays, and the results of infrared analysis suggested that hydrogen bonding rather than cation or anion exchange played a role in the adsorptive mechanism. Barrer and McLeod[60] showed that montmorillonites selectively

[57] Greenland, D. J., Adsorption of Sugars by Montmorillonite, *J. Soil Sci.,* **7**: 319–328 (1956).

[58] Emerson, W. W., Complexes of Calcium-Montmorillonite with Polymers, *Nature,* **186**: 573–574 (1960).

[59] Lynch, D. L., L. M. Wright, and L. J. Cotnoir, The Adsorption of Carbohydrates and Related Compounds on Clay Minerals, *Soil Sci. Soc. Am. Proc.,* **20**: 6–9 (1956).

[60] Barrer, R. M., and D. M. McLeod, Intercalation and Sorption by Montmorillonite, *Trans. Faraday Soc.,* **50**: 980–989 (1954).

Table 10-2   Data for montmorillonite- and halloysite-organic complexes
(*After MacEwan*[62])

Complex-forming substances	Montmorillonite					Halloysite			
	Method of formation	Δ obs.	Δ calc.	No. of layers	Base saturation of clay	Method of formation	Δ obs.	Δ calc.	No. of layers
Normal monohydric alcohols									
Methanol	a	7.4	6.9	2	NH₄⁺	a	3.4	...	1
	b	7.2	....	...	Ca⁺⁺				
Ethanol	a	7.9	8.1	2	NH₄⁺	a	2.8d		
1-Propanol	b	4.5	4.5	1	NH₄⁺	a	0.3	...	0
1-Butanol	b	4.6	4.5	1	NH₄⁺	a	0.3	...	0
1-Pentanol	b	4.6	4.5	1	NH₄⁺				
1-Hexanol	b	4.1	4.5	1	NH₄⁺				
1-Heptanol	b	4.1	4.5	1	NH₄⁺				
1-Hexadecanol	K	4.1d	4.5	1	NH₄⁺				
Other monohydric alcohols									
4-Heptanol	b	3.7	4.5	1	NH₄⁺				
2-Octanol	b	4.1d	4.5	1	Ca⁺⁺				
2-Ethyl-1-butanol	b	4.2	4.5	1	NH₄⁺				
2-Methyl-2-butanol	b	5.4	5.4	1	NH₄⁺				
Cyclohexanol	b	5.6	5.5	1	NH₄⁺				
Polyhydric alcohols									
Ethylene glycol	a	7.6	7.3	2	NH₄⁺, Ca⁺⁺	a, e	3.7	...	1
1,3-Propanediol	a	8.6	8.6	2	NH₄⁺, Ca⁺⁺	a, e	4.4	4.5	1
1,3-Butanediol	...	.....	....	...	....	a	4.0d	4.5	1
Glycerol	a	8.3	8.3	2	NH₄⁺,	a, e	3.8	3.9	1
Alcohol ethers and ethers									
1,4-Dioxane	a, b	5.6	5.9	1	Ca⁺⁺	a	d		
Methyl cellosolve	...	.....	....	...	....	a	3.5d	...	1
Ethyl cellosolve	...	.....	....	...	....	a	3.6d	...	1
Butyl cellosolve	...	.....	....	...	....	a	3.1d	...	1
Diethylene glycol	...	.....	....	...	....	a	3.6	...	1
Triethylene glycol	...	.....	....	...	....	a	3.5	...	1
Hydrocarbons									
n-Hexane	b	1–2d	....	0	Ca⁺⁺				
n-Heptane	b	0.6d	....	0	NH₄⁺				
Benzene	b, c	4.6d	....	2	NH₄⁺				
Naphthalene	b	3.5d	....	1	NH₄⁺				
Tetrahydronaphthalene	b	4.6d	....	2	NH₄⁺				
Decahydronaphthalene	b	0.6	....	0	NH₄⁺				
Other compounds									
Chloroethanol	b	8.1	7.8	2	NH₄⁺	a, e	3.6	...	1
Ethanediamine	b	4.3	4.4	1	Ca⁺⁺	a	4.5	4.4	1
1,2-Propanediamine	...	.....	....	...	....	a	d		
Diethylamine	...	.....	....	...	....	a	0.3	...	0
Acetone	a, b	8.2	8.1	2	Ca⁺⁺	a, f	3.9	...	1
Acetaldehyde	...	.....	....	...	....	a	3.6	...	1
Acetonitrile	a, b	10.2	10.6	3	Ca⁺⁺	a	3.4	...	1
Nitromethane	b	10.4	10.5	3	Ca⁺⁺	a, g, h	3.3	...	1
Nitrobenzene	b	5.7	....	2	NH₄⁺	i	d		

Δ = "clear space" between mineral sheets measured from the surface of the O or OH layers, in k*X*.

a = treated with large excess of cold liquid.

b = boiled down to half volume with excess of liquid.

c = dehydrated at 100°C over P₂O₅ in vacuum, then covered with liquid.

e = covered with excess of liquid, and heated to 100°C for 1 to 2 hr.

f, g, h, i = like a, starting from complex with ethanediamine (f), methyl cellosolve (g), methanol (h), or chloroethanol (i).

K = treated with the molten alcohol at about 200°C.

d = diffuse or double line.

adsorb various *n*-paraffins. Weiss and Hofmann[61] showed that unexpected derivatives with neutral, colored chelate complexes are formed if the exchangeable cations between the silicate sheets react with organic components capable of forming complexes, for example, Mg ions with quinalizarine. In this case, the hydronium ions balance the negative charges of the silicate layers.

On the basis of one-dimensional Fourier syntheses of diffraction data, Bradley[13] concluded that the polar molecules are held to the clay surface through a C—H· · ·O (clay-mineral surface) bond. McEwan[62] came to the same conclusion regarding the nature of the bond. In general, the sorption energy for the organic dipolar molecules is greater than for water molecules, for montmorillonite dehydrated at 400 to 500°C will not rehydrate in water but will take up glycol and expand. They concluded that the organic molecules they studied were oriented between the basal surfaces of the clay minerals in positions as flat as possible.

Bradley showed that some organic dipoles were adsorbed in multiple molecular layers by montmorillonite. MacEwan confirmed this finding for montmorillonite and indicated that a single molecular layer is the general result for all organic polar molecules with halloysite. The number of layers intercalated when in contact with an excess of liquid appears to increase with the molecular dipole moment and molecular size of the organic.

MacEwan[62] explained the multiple-layer adsorption by smectite as compared with the single-layer adsorption by halloysite as follows: Each of the surfaces of the smectite sheets is formed by a layer of oxygen ions, and thus they behave rather like sheets with a uniform diffuse negative charge. Each sheet of smectite will tend to collect a layer of dipoles on both its surfaces, so that between such sheets there will be two layers. In the case of nonpolar molecules or groups, only the van der Waals force of attraction between them and the montmorillonite sheets need be considered. Since this force is nondirectional, the question of orientation does not arise and only a single molecular layer could be held in common by two neighboring montmorillonite sheets. Thus markedly polar molecules would form a two-layer structure, and less-polar molecules a one-layer structure. Very strongly polar liquids may form even more than two layers (e.g., acetonitrile and nitromethane). According to MacEwan,[63] the number of layers in montmorillonites in contact with excess of the liquid appears to increase with the function

[61] Weiss, A., and U. Hofmann, Batavite, *Z. Naturforsch.*, **6b:** 405–409 (1951).

[62] MacEwan, D. M. C., Complexes of Clays with Organic Compounds, I, *Trans. Faraday Soc.*, **44:** 349–367 (1948).

[63] MacEwan, D. M. C., The Montmorillonite Minerals, "X-ray Identification and Structures of Clay Minerals," chap. IV, pp. 86–137, Mineralogical Society of Great Britain Monograph (1951).

$\mu/[P]$, where μ is the molecular dipole moment and $[P]$ is molecular size (parachor). *Some* polar groups appear to be necessary for intercalation because saturated hydrocarbons do not form such complexes (they may, however, be adsorbed on the surface; Barrer and McLeod[60]). Presumably, therefore, as the hydrocarbon chain of a monohydric alcohol is made longer and longer, the adsorption would gradually diminish. The necessity of polar groups for adsorption is no doubt bound up with the fact that the montmorillonite sheets are charged and are held together by positive ions between the sheets. The energy of formation of a purely van der Waals adsorption complex is presumably insufficient to cause the breakup of the charged-sheet–ion–charged-sheet complex. But if a few polar groups are present, these will tend to congregate about the charged spots on the sheets and about the positive ions. The result will be an effective increase of the radius of the ions and a consequent diminution in electrical stress. The resulting structure may be more stable than the dehydrated montmorillonite structure, in which the base cations occur between the structural sheets.

With halloysite, only highly polar molecules are adsorbed between the unit layers. In this mineral the adjacent sheets of successive layers carry positive and negative charges, respectively, and there is no tendency for two layers of polar molecules to form. Figure 10-9, after MacEwan,[62] shows schematically the process of complex formation.

On the basis of observed distances between successive clay-mineral sheets in the complexes from c-axis measurements and computed values based on van der Waals dimensions (Table 10-2), MacEwan concluded that each layer of dipoles occupies its full "clearance space" in the c-axis direction, i.e., the space necessary to allow its molecules to move freely over a plane parallel to the surface of the structural sheets. This suggests that the adsorbed layers are in the state of a two-dimensional liquid, firmly adhering to the clay surface but able to move over it in a random manner. This conclusion is supported by the extremely labile nature of the complexes, which is such that one complex is changed to another by washing with excess of another miscible liquid, and by the absence of a simple integral number of molecules per unit cell, which would be expected for a crystalline arrangement of the adsorbed molecules on the clay surface.

Hoffmann and Brindley[64] pointed out that, if the concept of a C—H· · ·O—Si bond is true, differences in CH activity of molecules should lead to different adsorption patterns. They found that increasing CH activity leads to higher molar adsorption. They found the two main factors that influence adsorption, namely, chain length and CH activity, appeared to oper-

---

[64] Hoffmann, R. W., and G. W. Brindley, Adsorption of Non-ionic Aliphatic Molecules from Aqueous Solutions on Montmorillonite, Clay-Organic Studies, II, *Geochim. Cosmochim. Acta,* **20:** 15–29 (1960).

ate independently; for example, a lack of CH activity can be overcome by higher chain length.

Hoffman and Brindley[64] studied the *c*-axis spacing of their montmorillonite-organic complexes and found that precise and significant values were difficult to obtain because the hydroscopic character of the montmorillonite was not diminished by adsorption of the organic substances, and because the organic material tended to evaporate from the complexes on drying. They recorded one-, two-, and three-layer complexes and considered that aliphatic

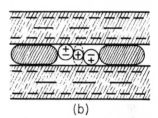

(a)

Fig. 10-9. The process of complex formation in montmorillonite and halloysite. (*a*) Polar molecules with montmorillonite; (*b*) mainly nonpolar groups with some polar groups with montmorillonite; (*c*) polar molecules with halloysite. The specific role of hydrogen-bond formation, which is important, has been neglected. (After MacEwan.[62])

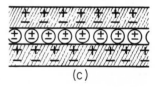

(b)

(c)

molecules orient with the plane of their chain both parallel and perpendicular to the clay surface. Their data agreed with that presented by Greene-Kelly[65] suggesting that straight-chain aliphatic compounds, in general, are oriented with the plane of the carbon chain perpendicular to the clay surface. In the case of ketones and acetoaceticethylester, Hoffmann and Brindley[64] suggested that an orientation coplanar to the montmorillonite surface was probable in order to place the negative carbonyl oxygen at an equal distance from both negative silicate surfaces. In the case of two other groups investi-

[35] Greene-Kelly, R., Sorption of Saturated Organic Compounds by Montmorillonite, *Trans. Faraday Soc.,* **52:** 1281–1286 (1956).

gated (ethernitriles and polyethylene glycoldiacetates) the spacings were borderline between what would be expected for perpendicular and from coplanar orientation. They stated that the orientation of two-layer complexes was too uncertain to be discussed.

Tensmeyer, Hoffmann, and Brindley[66] obtained infrared-absorption spectra of one- and two-layer complexes of several ketones with calcium montmorillonite. They found that the infrared spectrum of the calcium montmorillonite itself had not changed discernibly upon adsorption of the ketone; however, the spectrum of the adsorbed water was altered appreciably, for example, the band at 1,632 cm^{-1} was shifted to 1,653 cm^{-1}. Upon absorption they found significant changes in the carbonyl stretching frequency, and the methyl and methylene deformation frequencies of the ketone. Their concept is that in the one-layer complex the ketone molecules are lying flat between the clay plates with the carbonyl groups providing bonding to adjacent molecules. They concluded that for the one-layer complexes a type of molecular packing with a high degree of order, possibly even higher than that of the solid state of these organic compounds, must be postulated. In a sense the clay particles have induced the ketone to "solidify" at a temperature above its normal melting point. They further concluded that their data suggest that organic-organic interaction predominates and that the adsorption is of a physical rather than chemical nature. In their opinion, the formation of the clay-organic complex does not proceed through random orientation but, once a few molecules are adsorbed as nuclei, succeeding molecules are adsorbed in a manner similar to a crystallization process. The two-layer complexes offer the adsorbed molecules more possibilities of interaction and arrangement. The spectra show that the kind of arrangement is similar to that of the one-layer complex, but somewhat less ordered.

Brindley and Hoffmann[67] investigated the orientation and packing of aliphatic chain molecules on montmorillonite. They concluded that aliphatic molecules with strongly polar groups prefer an orientation in which the zigzag plane is parallel to the silicate sheet and those molecules without such groups prefer an arrangement in which the zigzag plane lies perpendicular to the silicate sheet. These authors found that the c-axis dimension of complexes was appreciably less than the sum of the van der Waals cross section of the organic molecules plus the thickness of the montmorillonite layer. There is a contraction of about 0.6 to 0.8 Å for complexes with the organic molecules

[66] Tensmeyer, L. F., R. W. Hoffmann, and G. W. Brindley, Infrared Studies of Some Complexes Between Ketones and Calcium Montmorillonite, Clay-Organic Studies, III, *J. Phys. Chem.*, **64:** 1656–1662 (1960).

[67] Brindley, G. W., and R. W. Hoffman, Orientation and Packing of Aliphatic Chain Molecules in Montmorillonite, Proceedings of the 9th National Clay Conference, pp. 546–556, Pergamon Press, New York (1962).

in a perpendicular position and about 0.4 to 0.8 Å for complexes with the organic in a parallel position. They concluded that chemical bonding seems unlikely as the cause of this shortening, but they showed that an appreciable contraction can occur at the clay-organic interfaces, owing to a keying of the molecules into the surface structure, and that this is equally possible for parallel and perpendicular orientation of the organic molecule.

Greene-Kelly[68] studied the sorption of aromatic organic compounds by montmorillonite and found that at low surface concentrations the plane of the ring of the organic was parallel to that of the silicate sheets, but at higher surface concentrations the organics reoriented so that their planes were perpendicular to that of the silicate sheet. He found that the C—H· · ·O distances in the organic complexes were substantially shorter than expected if normal van der Waals binding was operative.

Emerson[69] reported that polyvinyl alcohol and simple alcohols form complexes with montmorillonite in which the carbon chains are perpendicular to the 001 plane. Hydrogen bonding of two alcoholic OH groups to a clay oxygen atom to satisfy tetrahedral coordination, with van der Waals contact at the other clay oxygen surface, determines the orientation of the interlayer molecules.

In agreement with observations by Grim and coworkers[25] and Cowan and White,[48] Hoffmann and Brindley[64] found that the molar adsorption increased with increasing chain length. They found that a chain length of from five to six units was necessary to start adsorption from aqueous solution. The chain length had a strong influence up to about 10 units, but beyond this point differences in chain length did not seem to affect the adsorption very appreciably.

Brunton, Tettenhorst, and Beck[70] investigated the intercalation of a series of polyalcohols with montmorillonite. Complexes with polyalcohols having OH groups at the end of the chain had mean d spacings of about 13.7 Å, whereas with polyalcohols with interior OH groups the mean d spacing was about 14 Å. They concluded that the difference of 0.3 Å is probably due to a steric effect provided by OH groups, and the difference is reasonable if the plane of the zigzag carbon chains is parallel to the basal surfaces of the montmorillonite.

MacEwan[62] indicated that, when glycol complexes are made in the presence of water, less glycol is needed to form the complex. He stated that

[68] Greene-Kelly, R., Sorption of Aromatic Organic Compounds by Montmorillonite, I, Orientation Studies, *Trans. Faraday Soc.*, **51**: 412–424 (1955).

[69] Emerson, W. W., Organo-Clay Complexes, *Nature*, **180**: 48–49 (1957).

[70] Brunton, G., R. Tettenhorst, and C. W. Beck, Montmorillonite-Polyalcohol Complexes, Proceedings of the 9th National Clay Conference, pp. 500–519, Pergamon Press, New York (1962).

regardless of the water present the c-axis spacing is the same, and he concluded that the glycol molecules must set the pattern into which the water molecules fit. Mackenzie[71] found that the c-axis spacing of a montmorillonite–ethylene glycol complex can be maintained by adding water if the glycol content is low. The additional water is proportional to the glycol deficiency in the molecular ratio of about 1:6. Glaeser[72] showed that this conclusion does not hold for all complexes. For example, in the case of acetone-montmorillonite complexes, when the complex is formed in the absence of water, the distance between the montmorillonite layers is 3.7 to 3.9 Å; at 5 percent relative humidity, 8.0 to 8.2 Å; and at 100 percent relative humidity, 12.5 to 15.1 Å. Apparently a water-acetone complex develops between the montmorillonite layers. The data given in Table 10-2 therefore may be subject to variations in some cases, depending on the water present during the preparation of the complexes. Glaeser[73] showed also that the character of the adsorbed cation is only of slight importance in the formation of dipole complexes if they are formed in the absence of water or at low relative humidities. However, if they are formed at high relative humidities, the thickness of the adsorbed layer may vary greatly according to the cation present. Thus for acetone-montmorillonite complexes formed at 100 percent relative humidity, the distance between the units is about 12.5 Å, if Na^+ is the adsorbed ion, and 15.1 Å, if it is Ca^{++}. The explanation probably resides in the hydration tendencies of the cation. Ruiz-Amil and MacEwan[74] showed that sodium montmorillonite may swell more readily in mixtures of water and acetone than in pure water for a given electrolyte concentration. Norrish[75] also showed that the swelling of montmorillonite in sodium chloride–acetone–water systems is related to the concentration of the sodium chloride.

The effect of exchange cations on the sorption of organic molecules tends to be similar to the effect found with water though the organic complexes are often less sensitive to variation in cation composition. Ethylene glycol and glycerol, for example, give two-layer complexes with a wide range of cations of all valences including sodium and calcium, but large monovalent cations such as potassium, cesium, and rubidium may give one-layer complexes

[71] Mackenzie, R. C., Complexes of Clays with Organic Compounds, II, *Trans. Faraday Soc.,* **47**: 368–375 (1948).

[72] Glaeser, R., Sur le mécanisme de formation des complexes montmorillonite-acetone, *Compt. Rend.,* **226**: 935–937 (1948).

[73] Glaeser, R., On the Mechanism of Formation of Montmorillonite-Acetone Complexes, *Clay Minerals Bull.,* no. 3, pp. 88–90 (1949).

[74] Ruiz-Amil, A., and D. M. C. MacEwan, Interlamellar Sorption of Mixed Liquids by Montmorillonite, *Kolloid-Z.,* **55**: 134–135 (1957).

[75] Norrish, K., Swelling of Montmorillonite, *Discussions Faraday Soc.,* **18**: 120–134 (1954).

(Greene-Kelly[76]). Acetone commonly gives a two-layer complex with calcium and may give a one-layer complex with sodium. According to Greene-Kelly, it appeared, in general, that ions with a low electric field (ve/r^2, where v = valency and r = radius) tend to give "lower" complexes though no rigid rule can be stated.

Brindley and Ruston[77] investigated the adsorption of a polyethylene glycol ester of oleic acid on montmorillonite in the presence of water. They found that the basal spacing of wet clay with the adsorbed organic prior to any drying was 18 to 22 Å, largely irrespective of the exchangeable cations and the amount of adsorbed organic material, suggesting that prior to drying the lattice spacing is determined principally by water remaining between the silicate sheets. On drying the clay-organic complexes at 110°C, lattice spacings are found with stationary values of about 14 and 17.5 Å, with the value increasing as the total organic content is increased. There is a marked tendency toward the formation of ordered complexes containing either one layer or two layers of organic material between successive silicate sheets. The one-layer phase with a spacing of 13.9 Å develops when the adsorbed organic per gram of clay is in the range of 12 to 30 percent of the saturated value; when 30 percent saturation is reached, some two-layer complex makes its appearance; and the two phases coexist until the adsorbed organic exceeds 42 percent of saturation when the two-layer phase alone is formed. This means that, even though the organic material is not quite sufficient to fill one layer to maximum capacity, a regular two-layer structure is formed. Repeated washings with water reduce the two-layer to the one-layer type of sequence, but a single organic layer between silicate sheets is firmly held.

Other clay minerals. Kaolinite. Bradley[14] pointed out that adsorption of polar molecules would take place on the surfaces of kaolinite. Lynch, Wright, and Cotnoir[59] found that calcium-saturated kaolinite adsorbed various carbohydrates in amounts up to 3.7 percent. Mortensen[78] showed that the nature of the anion and the cation on kaolinite influenced the amount of adsorption of hydrolized polyacrylonitrile. He concluded that the organic was apparently adsorbed as a monolayer on a single set of sites, which are most likely positive spots on the edge of the kaolinite crystals.

Weiss[79] pointed out that it is known from protein chemistry that hydrogen

[76] Greene-Kelly, R., Identification of Montmorillonoids, *J. Soil Sci.*, **4**: 233–237 (1953).

[77] Brindley, G. W., and M. Ruston, Adsorption and Retention of an Organic Material by Montmorillonite in the Presence of Water, *Am. Mineralogist*, **43**: 627–640 (1958).

[78] Mortensen, J. L., Adsorption of Hydrolized Polyacrylonitrile (HPAN) on Kaolinite, I, Effect of Exchange Cation and Anion, *Soil Sci. Soc. Am. Proc.*, **21**: 385–388 (1957).

[79] Weiss, A., Eine Schichteinschlussverbindung von Kaolinit mit Harnstoff, *Angew. Chem.*, **73**: 736 (1961).

bonds between macromolecules can be broken by certain low-molecular-weight substances, such as urea or formamid, which have outstanding tendencies to form hydrogen bonds. He showed[18,80] that formamid, acetamid, hydrazine, and urea react with kaolinite to form intercalation compounds in which the c-axis dimension ranges between 9.5 and 11 Å. The intercalation compounds are formed by reacting concentrated solutions in water or alcohol or, in the case of acetamid, by a molten compound. Weiss and his colleagues studied

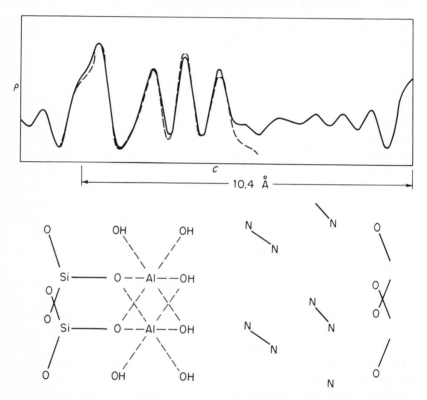

Fig. 10-10. Fourier projection and crystal structure of hydrazin-kaolinite. (After Weiss.[18])

these reactions in detail and presented data regarding the rate of reaction, influence of temperature, etc. Also, by means of one-dimensional electron-density projection, they indicated the arrangement of the organic molecules relative to the silicate layers (Fig. 10-10).

Similar experiments using poorly crystallized kaolinite rather than the well-crystallized mineral showed little or no intercalation. No completely

[80] Weiss, A., W. Thielepape, W. Ritter, H. Shafer, and G. Goring, Kaolinit-Einlaherungs-Verbindungen, *Proc. Intern. Clay Conf., Stockholm.,* **1**: 287–304 (1963).

satisfactory explanation for the failure of such kaolinite to react presents itself, and it is not established that this is generally true.

Weiss and his colleagues[80] pointed out that, by using hydrazine, urea, etc., as wedges to pry apart the kaolinite-silicate layer, intercalation compounds can be formed with various materials which do not react directly with kaolinite. Hydrazine was shown to be a particularly good wedge, and intercalation compounds were formed, for example, with lithium and sodium acetate, alkali salts of amino acids, benzidine, octylamine, etc., starting with solutions of these compounds in hydrazine and water or hydrazine and alcohol.

Kukharskaya et al.[81] suggested that mixing kaolinite and certain organics in the presence of ultrasonic vibrations causes the formation of kaolinite-organic derivatives. With the organics used there were no measurable changes in interplanar spacing, and it was concluded that only surface layers were affected.

Wada[17] showed that washing complexes of potassium acetate and nacrite and dickite with ethylene glycol forms well-defined complexes with a 10.8-Å basal spacing. Fifty percent conversion occurred within 3 days, following which the reaction slowed down and was not complete in 3 months. Earlier Miller and Keller[82] had produced ethylene glycol complexes with various halloysite minerals, using a similar technique, but had failed in the case of kaolinite.

Illite, Chlorite. Substantially no investigation has been made of the possible intercalation of neutral molecules with illites and chlorites. It has been considered that such molecules would be adsorbed only on the exterior surfaces of particles of the mineral. Recent work with kaolinite (see above) makes this conclusion less certain, particularly if the reactions are studied over long periods of time.

Vermiculite. Bradley[14] pointed out that the ideal structure for vermiculite suggests that the interlayer water could be displaced by organic molecules. Barshad[83] showed that Mg, Ca, Ba, H, Li, and Na vermiculites take up a single layer of glycol molecules, with a c-axis spacing of 14.3 Å, when immersed in hot anhydrous glycol. Under similar conditions smectites take up two molecular layers of glycol, with a c-axis spacing of 17.7 Å. Also under similar conditions NH_4, K, Rb, and Cs vermiculites remain contracted, taking up no glycol molecules between unit layers. Barshad[84] related the

[81] Kukharskaya, E. V., Yu. I. Skornik, and N. G. Boiko, Organic Derivatives of Kaolin, *Dokl. Akad. Nauk SSSR,* **148:** 350–352 (1963).

[82] Miller, W. D., and W. D. Keller, Differentiation Between Endellite-Halloysite and Kaolinite by Treatment with Potassium Acetate and Ethylene Glycol, Proceedings of the 10th National Clay Conference, pp. 244–253, Pergamon Press, New York (1963).

[83] Barshad, I., The Effect of the Interlayer Cations on the Expansion of the Mica Type of Crystal Lattice, *Am. Mineralogist,* **35:** 225–238 (1950).

[84] Barshad, I., Factors Affecting the Interlayer Expansion of Vermiculite and Montmorillonite with Organic Substances, *Soil Sci. Soc. Am. Proc.,* **16:** 176–182 (1952).

extent of interlayer expansion to the magnitude of the dipole moment and the dielectric constant of the liquid.

Walker[85] showed that ethylene, glycol, methane, cyclohexanol, acetone, and acetonitrile show limited penetration of Mg vermiculite, or none at all, but that penetration is easy if Li vermiculites are used. Probably because of the generally larger size of the vermiculite flakes and the higher perfection of stacking orientation, penetration of organic molecules is more difficult in vermiculites than in smectites.

Walker[85] emphasized that the *c*-axis spacing of glycol-vermiculite (14.28 Å) is very close to that of natural magnesium-saturated vermiculite (14.36 Å), and therefore the penetration of the glycol molecule is difficult to observe. Walker also pointed out that in the smectite complex the glycol molecule orients itself with a height of 3.9 Å, whereas in vermiculite the organic molecules either (1) attach themselves to the clay-mineral surfaces by a weaker bonding force than that in a smectite complex or (2) orient themselves differently with respect to the adjacent silicate surfaces. Walker stated that vermiculite exerts a catalytic effect on the decomposition of glycerol. Vermiculite boiled a few minutes in glycerol acquires a black metallic coating on interlamellar as well as boundary surfaces. The basal spacing of such a vermiculite complex is 14.06 Å.

Walker[86,87] later showed that vermiculite may form stable single- or double-layer complexes with glycerol and glycol. For natural vermiculites with magnesium as the exchangeable cation, single-layer complexes are formed; with other exchangeable cations, double-layer complexes may be formed. The ease of formation of the complexes varies with the particle size of the vermiculite and with the charge on the lattice, but in general complexes are formed less readily than with smectites.

Walker[87] concluded that the rate and degree of complex formation of vermiculite with glycerol and ethylene glycol are related to the surface charge of the silicate layers, to the type of interlayer cation present, and to crystal size. With glycerol, but not with ethylene glycol, the hydration state of the mineral immediately prior to complex formation is of some importance.

Bradley et al.[88] determined, on the basis of one-dimension Fourier synthesis, that the glycol molecules in a glycol–sodium-vermiculite complex were oriented with the plane of the zigzag normal to the silicate face.

[85] Walker, G. F., Vermiculite-Organic Complexes, *Nature,* **166**: 695–697 (1950).

[86] Walker, G. F., On the Differentiation of Vermiculites and Smectites in Clays, *Clay Minerals Bull.,* 3(17): 154–163 (1957).

[87] Walker, G. F., Reactions of Expanding Lattice Clay Minerals with Glycerol and Ethylene Glycol, *Clay Minerals Bull.,* 3(20): 302–313 (1958).

[88] Bradley, W. F., E. J. Weiss, and R. A. Rowland, A Glycol–Sodium Vermiculite Complex, Proceedings of the 10th National Clay Conference, pp. 117–122, Pergamon Press, New York (1963).

Attapulgite. The fibrous clay minerals belonging to the sepiolite-attapulgite-palygorskite group should adsorb polar molecules around their edges. Bradley[14] pointed out that the dimensions of the cross section of the channels in attapulgite are about 3.7 to 6.0 Å which is the approximate cross-sectional dimension for single strings of ethylene glycol molecules. The presence of organic molecules in the channels would be difficult to detect by X-ray diffraction, because only minor intensity variations in the reflections would be expected. However, careful determinations of the change in the optical properties of attapulgite on treatment with glycol suggest that some organic molecules penetrate the channels. Nederbragt and deJong[89] presented evidence, based on the preferential adsorption of some paraffins, indicating that some aliphatic chains are adsorbed in the open channels or in the gutters around the edges of particles and perhaps also in the interior channels.

Barrer and Mackenzie[90] showed that, in addition to substantial adsorption upon the external surfaces, H_2O and NH_3, and possibly (although to a lesser extent) CH_3OH and C_2H_5OH, may enter the intercrystalline channels of attapulgite. The channels are not, however, equally available to N_2, O_2, or CO_2; thus, polarity rather than molecular dimension governs the results. These authors pointed out that the channels in attapulgite are of considerable size and perfect crystals should be able to accommodate small nonpolar species. They proposed that cations sparsely distributed along the channels, together with water molecules, could provide a high energy barrier opposing diffusion into the channels.

Barrer, Mackenzie, and McLeod[91] showed that attapulgite and sepiolite selectively adsorb various *n*-paraffins. In some instances there is a loss of selectivity when the attapulgite is heated above 88°C, and they pointed out that this change might be due to structural changes influencing the surface following heating to this temperature.

Thermodynamic aspects, selective adsorption, etc. The literature on thermodynamic and selective-adsorption aspects of clay-mineral–organic reactions is too abundant to be reviewed in detail. The following remarks may illustrate the kind of data and conclusions that are available.

Granquist, Mitch, and Edwards[92] presented adsorption isotherms for nor-

[89] Nederbragt, G. W., and J. J. deJong, The Separation of Long-chain and Compact Molecules by Adsorption, *Rec. Trav. Chim.*, **65:** 831–834 (1946).

[90] Barrer, R. M., and N. Mackenzie, Sorption by Attapulgite, I, Availability of Intracrystalline Channels, *J. Phys. Chem.*, **58:** 560–568 (1954).

[91] Barrer, R. M., N. Mackenzie, and D. M. McLeod, Sorption by Attapulgite, II, Selectivity Shown by Attapulgite, Sepiolite, and Montmorillonite, *J. Phys. Chem.*, **58:** 568 (1954).

[92] Granquist, W. T., F. A. Mitch, and C. H. Edwards, Adsorption of Normal Saturated Hydrocarbons in Fuller's Earth, *Ind. Eng. Chem.*, **46:** 358–362 (1954).

mal saturated hydrocarbons and attapulgite. Jurinak and colleagues,[93-97] in a series of papers, presented adsorption isotherms for kaolinite and montmorillonite carrying various exchangeable cations and degassed at various temperatures with various organics. Their data permit certain conclusions regarding the character of the clay-mineral surfaces. For example, they concluded that montmorillonite systems present a porous aggregate whose capillary dimensions restrict adsorption while kaolinite systems present a relatively unrestrictive planar surface for vapor adsorption. Varying the degassing temperature of aluminum kaolinite from 25 to 315°C indicated an intimate relation between surface acidity and the moisture associated with the clay surface.

It is known that various organic derivatives have selective absorbent properties for other organic compounds. For example, White[98] showed that dimethyldioctadecylammonium montmorillonite is a very effective absorbent in the gas phase for the separation of the aromatic compounds from the paraffins and for the separation of the disubstituted isomers of benzene. He showed that derivatives identical except for the method of preparation can have widely different characteristics. White disagreed with the conclusion of Barrer and Kelsey,[99] who suggested that the cohesive energy density of the adsorbent is responsible for this phenomenon. White concluded that the aromatic and aliphatic compounds are being adsorbed on a similar type of surface but that the interlamellar space is not available to the aliphatic compound. He indicated as an alternative explanation that the free organic base in the clay is oriented with the hydroxyl groups at the edges of the platelets, and this is a barrier which prevents the aliphatic compounds from penetrating the lattice.

Aleixandre and Rodrigues[100] studied in detail the modification of the

---

[93] Jurinak, J. J., Effect of Clay Minerals and Exchangeable Cations on the Adsorption of Ethylene Dibromide Vapor, *Soil Sci. Soc. Am. Proc.,* **21:** 599–602 (1957).

[94] Jurinak, J. J., and D. H. Volman, Thermodynamics of Ethylene Dibromide Vapor Adsorption by Ca-Montmorillonite and Ca-Kaolinite, *Soil Sci.,* **86(1):** 6 (1958).

[95] Jurinak, J. J., and D. H. Volman, Acid-Base Interaction in the Adsorption of Olefins on Aluminum Kaolinite, *J. Phys. Chem.,* **63(9):** 1373 (1959).

[96] Jurinak, J. J., and D. H. Volman, Adsorption of Butane by Kaolinite and Montmorillonite. Phase Changes in the Monolayer Region, *Soil Sci. Soc. Am. Proc.,* **23:** 25–28 (1959).

[97] Jurinak, J. J., The Effect of Pretreatment on the Adsorption and Desorption of Water Vapor by Lithium and Calcium Kaolinite, *J. Phys. Chem.,* **65(1):** 62 (1961).

[98] White, D., The Structure of Organic Montmorillonites and Their Adsorptive Properties in the Gas Phase, Proceedings of the 12th National Clay Conference, pp. 257–267, Pergamon Press, New York (1964).

[99] Barrer, R. M., and K. E. Kelsey, Thermodynamics of Interlamellar Complexes, *Trans. Faraday Soc.,* **57:** 452–462 (1961).

adsorption properties of various clay minerals carrying various cations toward alcohols that takes place as a consequence of heating to temperatures up to 700°C.

Thermal stability of organic complexes

Several investigators have studied the thermal stability of intercalated organic molecules. Carthew[101] found that the intensity and temperature of the DTA exothermic combustion reaction for piperidine-saturated montmorillonite, nontronite, and illite vary for each mineral. Montmorillonite and nontronite give peaks of relatively high intensity, whereas illite gives peaks of low intensity. The exothermic peaks are complex, with a component at 260 to 300°C representing the combustion of hydrogen released by the cracking of the piperidine, and additional peaks representing the combustion of the carbon in interlayer surfaces.

McAtee and Concilio[102] found that dimethyldioctadecylammonium chloride reacted with sodium montmorillonite to form a four-layer complex with the c-axis spacing of about 28 Å. On heating to temperatures up to 170°C an expansion to 32 Å took place, which is interpreted as indicating an untangling of the hydrocarbon chains in the middle two layers with the consequent development of a higher order. With further heating, there is a very rapid collapse to a single organic layer between the silicate sheets. Heating in an oxygen atmosphere does not cause the change to take place at a lower temperature, nor does a nitrogen atmosphere allow the sample to be heated to a higher temperature before a collapse takes place. The authors concluded that the interaction between the CH_2 groups and the oxygen on the clay surface must be greater than simple van der Waals forces and that probably there is an augmented attraction of the nature of a C—H· · ·O—Si bond as suggested by Bradley.[14]

McAtee and Hawthorne[103] showed, on the basis of heating-oscillating

---

[100] Aleixandre, F. V., and M. C. Rodrigues, Modification of the Adsorptive Properties of Clay Minerals by Thermal Treatment and Change of Cation, *Anales Edafol Fisiol. Vegetal (Madrid)*, 16: 799–856 (1957).

[101] Carthew, A. A., Use of Piperidine Saturation in the Identification of Clay Minerals by Differential Thermal Analysis, *Soil Sci.*, 80: 337–347 (1955).

[102] McAtee, J. L., Jr., and C. B. Concilio, Effect of Heat on an Organo-Montmorillonite Complex, *Am. Mineralogist*, 44: 1219–1229 (1959).

[103] McAtee, J. L., and J. M. Hawthorne, Heating-Oscillating X-ray Diffraction Studies of Some Organo-Montmorillonites, *Am. Mineralogist*, 49: 247–257 (1964).

X-ray studies, that in some ionic organic-montmorillonite complexes there is a rotation of the aromatic portion of the organic ion from a vertical to a parallel position on the clay surface on heating to temperatures of the order of 100 to 175°C. As the temperature is increased still further, the organic slowly decomposes, with the organic ions undergoing a catalytic decomposition followed by a thermodesorption from the clay surface.

Beck and Brunton[104] investigated the thermal stability of a series of hectorite- and montmorillonite-guanidine complexes. They found that temperatures as high as 600°C were necessary for the complete decomposition of the intercalated organic material, and further that the hectorite complexes were slightly more stable thermally than the analogous montmorillonite complexes.

Byrne[105] showed that the adsorption of organics may influence the formation of high-temperature phases, from which he concluded that the adsorbed organic influences the oxygen network surface of the silicate units which in turn is related to the high-temperature phase formation.

The enhanced stability of some intercalated organics is shown by the increased radiation stability when clay-mineral fillers are incorporated in plastics (Metz[106] and Caffrey and Allen[107]). Allen and Caffrey[108] found that the radiolysis of normal pentane was profoundly altered in the presence of montmorillonite.

Kinter and Diamond[109] showed that glycerol-treated montmorillonite gradually loses one layer of the normal two-layer complex on moderate heating at 110°C, forming a monolayer complex with a sharp X-ray-diffraction peak at about 14 Å, and in the presence of free glycerol vapor this complex is quantitatively stable on further heating. With this as a basis, these authors presented a method for determining the internal and external surface area of the montmorillonites. They concluded that cation saturation affects the values of both internal and external glycerol retention for montmorillonite.

[104] Beck, C. W., and G. Brunton, X-ray and Infrared Data on Hectorite-Guanidines and Montmorillonite-Guanidines, Proceedings of the 8th National Clay Conference, pp. 22–39, Pergamon Press, New York (1960).

[105] Byrne, P. J. S., Montmorillonite Organic Complexes, *Natl. Acad. Sci., Publ.* 327, pp. 241–253 (1954).

[106] Metz, D. J., Graft Polymers—A New Technology? *Nucleonics,* 16(4): 73–77 (1958).

[107] Caffrey, J. M., and A. O. Allen, Radiolysis of Pentane Adsorbed on Mineral Solids, *J. Phys. Chem.,* 62: 33–37 (1958).

[108] Allen, A. O., and J. M. Caffrey, Irradiation Method of Converting Organic Compounds, U.S. Patent 2,955,997 (1960).

[109] Kinter, E. B., and S. Diamond, Gravimetric Determination of Monolayer Glycerol Complexes of Clay Minerals, *Natl. Acad. Sci., Publ.* 566, pp. 318–333 (1958).

Bradley et al.[110] arrived at the same conclusion, using the oscillating-heating X-ray-diffraction technique.

The collapse of montmorillonite-organic complexes to a *c*-axis spacing of 10 Å requires temperatures as much as 275°C higher than those necessary when water alone is present, thereby attesting to the stability of the inter-calated organic molecules (Beck and Brunton[104]).

Brunton, Tettenhorst, and Beck[70] determined the temperature of collapse of one-layer polyalcohol-montmorillonite complexes by the oscillating-heating X-ray-diffraction technique. They found that the mean *d* spacing of the one-layer complexes was not significantly different with respect to the inter-layer cation composition, with the possible exception of the potassium complex, for which the *d* spacing was 0.1 to 0.2 Å below that of the other com-plexes. The mean temperature of final collapse for all polyalcohols is related to the valence of the interlayer cations and inversely related to their ionic radii. Investigation using a single montmorillonite showed that the difference in temperature between the collapse of the clay-organic complex and the boiling point of the polyalcohol was proportional to the length of the carbon chain and was also a function of the number and position of the OH groups.

Resistance of adsorbed organic molecules to biological decomposition

The work of several investigators suggests that the resistance of some organic materials to biological decomposition may be increased when they are ad-sorbed by the clay minerals. Thus, Waksman[111] reached this conclusion be-cause it is easier to build up the humus content in a clay-rich soil than in a clay-poor soil. Mattson[3] believed that adsorbed proteins might be more resistant to microbial decomposition than free proteins.

Ensminger and Gieseking[112] and Erickson[113] investigated the matter of the decay of adsorbed organics and showed that the decay characteristics are often quite different in free as compared with clay-mineral-adsorbed or-

[110] Bradley, W. F., R. A. Rowland, E. J Weiss, and C. E. Weaver, Temperature Stabilities of Montmorillonite- and Vermiculite-Glycol Complexes, *Natl. Acad. Sci., Publ.* 566, pp. 348–355 (1958).

[111] Waksman, S. A., "Humus; Origin, Chemical Composition and Importance in Nature," Williams & Wilkins, Baltimore (1936).

[112] Ensminger, L. E., and J. E. Gieseking, Resistance of Clay-adsorbed Proteins to Proto-lytic Hydrolysis, *Soil Sci.,* **53**: 205–209 (1942).

[113] Erickson, A. E., Ph.D. thesis, University of Illinois (1948). Reference from J. E. Gieseking, The Clay Minerals of Soils, *Advanc. Agron.,* **1**: 598–605 (1949).

ganic material. Ensminger and Gieseking observed that protein-montmoril-lonite complexes prepared by ion exchange, in which relatively large amounts of protein were adsorbed as cations, showed only a slight tendency to putrefy. These same authors showed that the adsorption of albumin and hemoglobin by ion exchange in montmorillonite interfered with the enzymatic hydrolysis of these proteins in both acid pepsin suspensions and alkaline pancreatin suspensions. In kaolinite complexes there was no significant effect on the hydrolysis of albumin and hemoglobin. Apparently sorption between basal units, as in the case of montmorillonite, rather than around particle edges, as in kaolinite, is necessary to retard chemical change. Ensminger and Gieseking[112] concluded that the hydrolysis is reduced because the protein mole-cules, when adsorbed by the clay, are oriented in such a way that the active groups are inaccessible to the enzyme. Another possible explanation is that the enzyme is partly adsorbed and thereby rendered partly inactive.

Erickson[113] showed that the stability of amine-montmorillonite complexes, or their tendency to decomposition, depends on the enzyme and on the amine used in the complex. Actually in some cases, e.g., in tyrosine-montmorillonite complexes, after an initial lag in oxidation, the clay seems to catalyze the enzymatic reaction.

Lynch and his colleagues[114] investigated the relation of the breakdown by the action of soil of certain carbohydrate compounds in the soil to the presence of various clay minerals. They found that the breakdown by soil microorganisms is retarded in some cases by the presence of montmorillonite and attapulgite whereas illitic and kaolinitic clay minerals have very little protective influence.

Pinck[115] studied the adsorption of proteins, enzymes, and antibiotics by montmorillonite. He found that proteins react with montmorillonite forming mono- and polylayer complexes and that there is a striking difference between the rates of decomposition of protein in clay complexes and in protein alone. In monolayered complexes, there is relatively little microbial decomposition, whereas in polylayer complexes the protein undergoes extensive decomposition, with a corresponding shrinkage of the *c*-axis spacing. The enzyme urease was found to be adsorbed completely by hydrogen montmorillonite and only partially by basic montmorillonite and hydrogen kaolinite. It was shown that hydrogen montmorillonite inactivated about three-fourths of the urease, and untreated montmorillonite about one-third. The activity of the urease adsorbed on hydrogen montmorillonite was about half as great as that ad-

[114] Lynch, D. L., and L. J. Cotnoir, Jr., The Influence of Clay Minerals on the Breakdown of Certain Organic Substrates, *Soil Sci. Soc. Am. Proc.,* **20:** 367–370 (1956).

[115] Pinck, L. A., Adsorption of Proteins, Enzymes, and Antibiotics by Montmorillonites, Proceedings of the 9th National Clay Conference, pp. 520–529, Pergamon Press, New York (1962).

sorbed on kaolinite. Pinck found that basic and amphoteric antibiotics formed interlayer complexes with montmorillonite, whereas there was little such formation with acid antibiotics. Biological assays showed that in the interlayer complexes the antibiotics were relatively inactive.

Catalysis and polymerization

It has been known for a long time that the clay minerals possess in varying degrees catalytic properties toward some organic liquids (Grim[22]). Robertson[19] listed a large number of reactions catalyzed by various clay minerals. Kayser and Bloch[116] showed that montmorillonite catalyzes the inversion of sucrose and the formation of sulfanilic acid, and Alberada et al.[117] pointed out that kaolinite and montmorillonite clays catalyze the oxidation of ethyl alcohol.

Weiss[30] stated that chemical reactions occur between the silicate layers of the clay minerals that do not take place under similar conditions in free solutions. For example, if the aromatic rings lie parallel to the silicate sheets, intercalated acrylammonium ions are oxidized by atmospheric oxygen. The type of product formed depends on the charge density of the silicate layer. Thus, with low charge ($0.33\pm$) aniline is oxidized to a red ion; with a somewhat higher charge ($0.55\pm$), to a bluish-purple ion; and with a still higher charge, finally to aniline black. In highly charged micas with H_3O^+ ions, proteins are cleaved into peptides and amino acids. The small fragments are not bound as strongly as the intact protein ions; therefore, they can be exchanged again by intact protein molecules. Thus, the reaction is a catalytic one.

The variability of catalytic properties of the clay minerals causes difficulties in the use of color reaction for the identification of clay minerals (see pages 407–410). Thus, Dodd and Ray[118] pointed out that semiquinone cations derived from aromatic diamines by one-electron oxidation in aqueous solutions possess characteristic colors but depend on hydrogen-ion concentration. The colored free radicals are formed by oxidation on the surface of certain montmorillonite clays where they are stabilized by adsorption appar-

[116] Kayser, F., and J. M. Bloch, Some Catalytic Properties of Montmorillonite, *Compt. Rend.*, **234**: 1885–1887 (1952).

[117] Alberada, J., M. Alexandre, and T. Fernandez, Influencia de la composicion mineralogica de las arcillas de los cationes de cambio en la oxidacion catalitica del alcohol etilico en fase vapor, *Anales edafol. Fisiol. Vegetal (Madrid)*, **12**: 89–140, 281–308 (1953).

[118] Dodd, C. G., and Satyabrata Ray, Semiquinone Cation Adsorption on Montmorillonite as a Function of Surface Acidity, Proceedings of the 8th National Clay Conference, pp. 237–252, Pergamon Press, New York (1960).

ently in cation-exchange positions, but many clays react anomalously when the procedure is employed as a diagnostic test for montmorillonite, probably because of variations in the chemical nature of various semiquinones and variations in the clay-mineral crystal structure. These authors assumed that some property related to the clay-mineral surface was responsible for its power to oxidize the organic compounds to their semiquinone form but were unable to determine precisely the nature of this property.

The general idea of putting polymerizable organic groups on clays and then polymerizing the resultant derivatives with themselves or with other organics is discussed in the patent literature by Hauser and Dannenberg,[119] Te Grotenhuis,[120] and others. Organoclay may be considered as "fillers" or complex formers in polymeric systems. If the organoclay forms a chemically cross-linked complex with the polymer matrix, it will insolubilize or gel the system in a polymer solvent; if it behaves simply as extender of the polymer, extraction in a solvent for the polymer will result in a phase separation.

Van der Watt and Bodman[121] discussed three types of clay to polymer linkages for the case of a vinyl acetate maleic anhydride–aluminum-montmorillonite complex.

Emerson[58] investigated the clay polymer bonds formed by the interaction of calcium montmorillonite with a wide range of polymers. He found three types of associations: one in which the polymer is readily removable by 1 N sodium chloride, which is interpreted to mean that the bond is formed by a precipitation complex with the exchangeable calcium ions (sodium alginate is such a polymer); another in which the polymer is removable only by a strong sequestering agent such as sodium pyrophosphate in which the bond is made by forming a coordination compound with the less readily exchangeable calcium ion (polyacrylamide is such a polymer); and a third type of association in which the polymer is not removable by either of the above reagents in which the bond is the hydrogen from polymer OH to plate SiO and is independent of the valence of the exchange ion (polyvinyl alcohol is such a polymer).

Dekking[122] prepared derivatives with hydrogen kaolinite and hydrogen

[119] Hauser, E. A., and E. M. Dannenberg, Molding Composition, Molded Products, and Method of Making, U.S. Patent 2,401,348 (1946).

[120] Te Grotenhuis, T. A., High Polymers with Chemically Bonded Reinforcing and Method of Making Same, U.S. Patent 2,841,566 (1958).

[121] Van der Watt, H. v. H., and G. B. Bodman, Viscosimetric Constants of Suspensions of Clay-Organic Complexes, Proceedings of the 9th National Clay Conference, pp. 568–584, Pergamon Press, New York (1962).

[122] Dekking, G. G., Preparation and Properties of Some Polymer-Clay Compounds, Proceedings of the 12th National Clay Conference, pp. 603–616, Pergamon Press, New York (1964).

montmorillonite and amine-terminated polystyrene and polymethancryloni-trile. He presented evidence to indicate there is a definite reaction between the clay and the polymer, which he considered to be a neutralization of the acid clays by the basic polymers. In the case of montmorillonite, the organic molecule forms a single molecular layer between the silicate sheets.

Hauser and Kollman[123] showed that certain conjugated dienes are slowly converted to a monolayer of polymer on some clay minerals, while Friedlander[21] noted a similar conversion of butadiene and *cis*-butene-2 on sodium montmorillonite. This same author also briefly noted the vigorous exothermic polymerization of styrene on sodium montmorillonite; this is in contrast with the findings of Blumstein,[124] who failed to polymerize styrene with this mineral.

Solomon and Rosser[125] investigated in detail the polymerization of styrene by various minerals. They concluded that the polymerization of styrene by sodium montmorillonite and attapulgite shows the characteristics of both a radical and an ionic mechanism. They proposed the formation of a radical—carbonium ion as the initiating step, followed by a rapid dimerization and then cationic propagation. Termination could then proceed by proton transfer, if necessary to monomer, and this could result in the initation of some polymerization by a conventional cationic mechanism. Evidence to support termination by proton transfer comes from the presence of approximately two double bonds for every three styrene units in the methanol-soluble product. No evidence for stereospecificity was found with NMR or X-ray-diffraction techniques. Their results indicate that the activity can be related to the presence of aluminum in octahedral coordination; for example, pyrophyllite was active whereas talc was not, while montmorillonite was more active than hectorite. However, the presence of aluminum in octahedral coordination does not necessarily result in catalytic activity, as shown by the failure of gibbsite to yield polystyrene. The strong activity shown by kaolinite suggests that the reactivity of the alumina sheet is influenced by the presence of the silicate sheet. Attempts to form complexes with the styrene between the silicate layers were not successful, and the catalytic sites are not, therefore, between the silicate sheets. The greater activity shown by attapulgite as compared with sodium montmorillonite would suggest that crystal edges are involved in the catalysis since attapulgite has a much greater edge-to-surface-area ratio because of its structure. It is suggested, therefore, that the cata-

[123] Hauser, E. A., and R. C. Kollman, Clay Complexes with Conjugated Unsaturated Aliphatic Compounds of Four to Five Carbon Atoms, U.S. Patent 2,951,087 (1960).

[124] Blumstein, A., Etude des polymerisations en couche adsorbée, I, *Bull. Soc. Chem. France,* pp. 899–905 (1961).

[125] Solomon, D. H., and M. J. Rosser, manuscript (1965).

lytic activity is related to the aluminum in octahedral coordination that is situated at crystal edges.

Instances of interlayer polymerization are known, and they would appear to be either free-radical or anionic although the exact mechanism by which the clay initiates these reactions is still obscure (Friedlander and Fink[126]). In other examples, clay minerals inhibit known free-radical-initiated polymerization (Solomon and Swift, personal communication), and the mineral edge appears to be the site responsible for this inhibiting effect. Clay minerals also appear to act as stabilizers for some degradation reactions (Blumstein[127]).

In the majority of cases of polymerization studied, it has been necessary to activate the clay minerals by acid treatment, which in effect converts the clay to a Bronsted acid; by heat treatment, which removes the water, particularly water adsorbed at aluminum edges, and creates Lewis acid sites; and/or by mechanical grinding, resulting in bond fracture and the formation of either free ions or radicals.

Organophilic clay-mineral complexes

Hauser[128] appears to have discovered that certain organic compounds of montmorillonite have the property of swelling and dispersing in organic liquids. Jordan[35,42] first investigated this matter in great detail and the following statements are based on his work.

The gel volumes of a series of normal primary aliphatic amine-montmorillonite complexes plotted as a function of the number of carbon atoms are given in Fig. 10-11. The gel volumes were determined in nitrobenzene, benzene, and isoamyl alcohol. The data show that the organophilic properties are negligible until an amine with a chain of 10 carbon atoms is reached and that 12 carbons are required for maximum swelling. An amine with a 12-carbon chain covers slightly more than half of the available basal surface of the montmorillonite units per exchange position and initiates the interlayer distance of about 8 Å, which is the thickness of two molecular layers (Fig. 10-3). Apparently this condition is necessary for maximum swelling.

The swelling behavior of dodecylamine complexes in a wide variety of liquids is given in Table 10-3. These data suggest that swelling is low in liquids of a nonpolar nature, such as the aliphatic and aromatic hydrocarbons,

[126] Friedlander, H. Z., and C. R. Fink, Organized Polymerization, III, Monomers Intercalated in Montmorillonite, *J. Polymer Sci. Letters,* **2:** 475–479 (1964).

[127] Blumstein, A., Photoelastic Properties of Tightly Crosslinked Networks, *J. Polymer Sci.,* **A2:** 769–784 (1964).

[128] Hauser, E. A., Reference in J. W. Jordan, *Mineral. Mag.,* **28:** 598–605 (1949).

and that generally the gel volume increases with the dielectric constant of the liquid, although the correlation is not perfect. It appears that the most effective liquids are those which combine highly polar with highly organophilic characteristics, good examples being nitrobenzene and benzonitrile. It is thought that, in the case of an incompletely clad amine-montmorillonite complex, adsorption of a highly polar organic liquid on the uncoated portion of the clay surface would greatly enhance solvation of the micelle in the remainder of the liquid.

Jordan found that, for complexes prepared with single-chain primary amines of less than sufficient size to coat the montmorillonite particles completely with a layer of hydrocarbon chains, swelling can be developed in nonpolar hydrocarbons if a second organic liquid that is highly polar, e.g.,

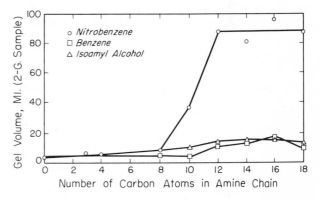

Fig. 10-11. **Effect of amine chain length on swelling of organic-ammonium–montmorillonite in organic liquids. (After Jordan.[35])**

an alcohol, ester, ketone, or aldehyde, is added to the system. Figure 10-12 shows the great influence of small additions of alcohols on the swelling characteristics of octadecylammonium-montmorillonite in toluene.

It is postulated that the highly polar additive is adsorbed on the uncoated surface of the montmorillonite, thereby rendering the individual flakes entirely organophilic and compatible with the hydrocarbon portion of the solvating liquid. In Fig. 10-13 data are presented for toluene-ester systems with octadecylammonium-montmorillonite; these data indicate that the acid portion of the ester is more important than the alcohol portion. For example, ethyl formate is more effective than methyl acetate; methyl, ethyl, and butyl acetate are approximately equal in effect; and butyl stearate is practically without influence.

Jordan showed that montmorillonite complexes with quarternary ammonium salts having two long aliphatic chains are organophilic in single-hydro-

Table 10-3   Gel volumes of 2-g samples of dodecylammonium-bentonite in various liquids
(After Jordan[35]*)*

Liquid	Gel volume, ml	Liquid	Gel volume, ml
Water (untreated bentonite)	31	Butyl carbitol	12.5
Water (dodecylammonium-bentonite)	2.0	n-Butyl phthalate	12.5
Petroleum oil, Gulfpride SAE 10	2.5	Isophorone	12.5
Petroleum oil, Gulfpride SAE 40	2.5	Benzyl alcohol	13.0
Dow Corning Fluid 200	2.5	Bromoform	13.0
Petroleum ether	3.0	Ethyl acetate	13.0
Piperidine	3.0	Tricresyl phosphate	13.0
Naphtha	3.5	Acetone	13.5
Carbon disulfide	4.0	Ethanol (95%)	13.5
Carbon tetrachloride	4.0	Nitroethane	13.5
Dibutylamine	4.0	Acetonitrile	14.0
Glycerol	4.5	Isoamyl acetate	14.0
Tributylamine	4.5	Castor oil	14.0
Amyl nitrate	6.0	Linseed oil	14.0
α-Butylene bromide	6.0	Oleic acid	14.5
Eucalyptol	6.5	n-Butylaldehyde	15.0
Styrene	6.5	Cyclohexanone	15.0
Toluene	6.5	Dodecylamine	15.5
Bromobenzene	7.0	Ethyl bromide	15.5
Linoleic acid	7.0	n-Butyl tartrate	16.5
Cymene	7.5	n-Heptaldehyde	18.0
Aniline	8.0	Methyl iodide	18.0
Cyclohexanol	8.0	γ-Picoline	18.0
Ethylene dichloride	8.0	Acetophenone	19.0
Benzene	9.0	Tetraethyl orthosilicate	19.0
Paraldehyde	9.0	Coconut oil	20.0
Acetic anhydride	10.0	Dodecyl alcohol	20.0
Chloroform	10.0	Methyl ethyl ketone	20.0
o-Cresol	10.0	Diethyl ketone	21.0
Ethyl malonate	10.0	Hexadienal	21.0
Formamide	10.0	Pyridine	28.0
Furfuryl alcohol	10.0	Benzaldehyde	31.0
Toluidine	10.0	Benzoyl chloride	33.0
Phenol	10.5	Crotonaldehyde	34.0
Butyl stearate	11.0	Ethyl ether	35.0
2-Nitropropane	11.0	Furfural	35.0
Acetic acid, glacial	12.0	Benzonitrile	50.0
Isoamyl alcohol	12.0	Nitrobenzene	88.0
1-Nitropropane	12.0		

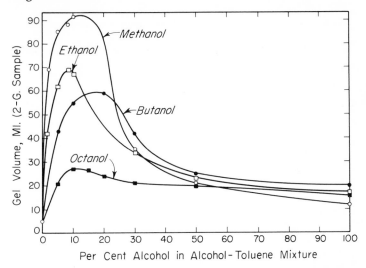

Fig. 10-12. Swelling of octadecylammonium-montmorillonite in binary mixtures. (After Jordan.[35])

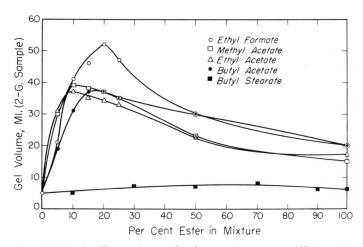

Fig. 10-13. Swelling of octadecylammonium-montmorillonite in toluene-ester mixtures. (After Jordan.[35])

carbon liquid systems. Dispersion in toluene takes place much more readily and completely with the double-chain complexes than with the single-chain complexes, probably because the completely coated particles are more compatible with the hydrocarbon. In the case of double-chain complexes, there is a tendency for an initially greater separation of the montmorillonite units than in single-chain complexes.

Jordan concluded that the swelling process involves two factors for an incompletely clad clay complex: (1) high adsorption energy of liquid for uncoated clay and (2) high solvation energy of liquid for the organic coating. Both conditions are fulfilled by a single liquid like nitrobenzene or by a binary liquid system of the toluene-alcohol type. Adsorption of polar molecules by the clay must tend to separate the plates, thereby rendering the

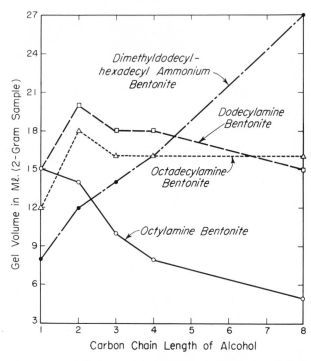

Fig. 10-14. **Effect of chain length on swelling of organic-ammonium–montmorillonite in alcohol. (After Jordan.[35])**

micelle more organophilic and the organic coating more accessible. The dialkyl quarternary complexes are solvated by toluene alone because of the complete coating of organic matter and the initially higher separation of the plates. Data for this conclusion are given in Fig. 10-14, which shows gel volumes for several organic-ammonium-montmorillonite complexes in several primary aliphatic alcohols. In agreement with this conclusion, octylammonium-montmorillonite solvates most in the most polar alcohol, while dimethyllaurylcetylammonium-montmorillonite solvates least in methanol and to a progressively greater extent in the less-polar members of the series.

Weiss[30] reviewed the complex-forming characteristics of the layer silicates with various onium ions, and the following is largely from his work. Swelling adopts a much simpler pattern in nonaqueous liquids than in water or aqueous electrolytes. In *n*-alkyl alchols or *n*-alkyl amines of the same chain length as the intercalated *n*-alkyammonium ion, the charge density no longer has any effect on the interplanar distance; that is, they are the same for montmorillonites, vermiculites, and micas. The sum of the number of *n*-alkylammonium ions and the molecules of swelling liquid is nearly the same in all mica-type silicates if compounds with unbranched alkyl chains are used as swelling agents. This number is largely independent of the nature of the functional groups, as the results of experiments with alcohols, amines, aldehydes, nitriles, carboxamides, and carboxylic acids have shown. Additional neutral molecules evidently fill the vacant cation positions in low-charge montmorillonite. In all these compounds, the alkylammonium ions and the molecules of swelling fluid are arranged in a bimolecular layer between the silicate anions, as shown in Fig. 10-15. Determinations of basal plane spacings confirm the concept that the alkyl chains must be densely packed, with their longitudinal axes perpendicular to the silicate layer.

Saturated hydrocarbons cannot be intercalated to *n*-alkylammonium montmorillonites. On the other hand, alkenes and aromatic hydrocarbons are absorbed vigorously, even when present at very high dilution of saturated hydrocarbon. The aromatic rings arrange themselves parallel to the silicate layers, insofar as the steric conditions permit.

The factors determining the swelling and orientation of compounds containing branched alkyl chains, such as ethers, esters, ketones, and similar compounds, are not clear. In general the size and form of the intercalated molecule and the position of the polar group play a role in their orientation. The polar group is usually in a position in direct contact with the silicate layer so that the arrangements formed have various packing densities, depending on the molecular shape. It is especially perplexing that, owing to their effect on the hydrogen bonds, molecules with strongly negative groups cause the preferably prone alkyl chains of *n*-alkylammonium cations to stand erect and thereby cause sheet separation that has no direct relation to their dimensions.

Montmorillonites with unsymmetrical cations (e.g., di- and tri-alkylammonium ions) have properties similar to those of *n*-alkylammonium–montmorillonite. With large symmetrical cations of this type, the situation is quite complicated even in unswollen samples. Here, as a result of the bulky structure of such cations, there are formed cavities open to the exterior which can be filled with gases or liquids of small molecular size with no apparent change in lattice dimensions.

No peculiarities appear on intercalation of small quarternary alkylammonium ions. Gases can be adsorbed in a zeolite-type exchange with no change in the basal spacings (Barrer and McLeod[60]). These montmorillonites form thixotropic gels with high contents of liquid with the majority of slightly polar liquids, excluding saturated hydrocarbons. Because of their colloidal

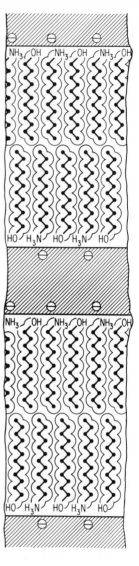

Fig. 10-15.   Arrangement of *n*-alkylammonium ions and of *n*-alkyl compounds intercalated as swelling agents between silicate layers (hatched areas).   The total (*n*-alkylammonium ions plus molecules of *n*-alkyl compound) = 2.00 to 2.07 per $(Si,Al)_4O_{10}$ unit.   Here, an *n*-alkyl alcohol is drawn as an example of a swelling agent.   (After Weiss.[30])

chemical properties, derivatives with unsymmetrical ions are of special importance and are the chief constituents of some commercial Bentone.*   Intensive gel formation of these onium montmorillonites is always associated with the presence of traces of water (Damerell and Milberger[129]).

* Trade name for montmorillonite-organic compounds produced by the National Lead Company.

[129] Damerell, V. R., and E. Milberger, Organophilic Montmorillonite Gels, *Nature,* **178:** 200 (1956).

Greene-Kelly[130] published data on the swelling characteristics and *c*-axis dimensions for a large variety of montmorillonite-organic complexes. Numerous authors have presented similar data which were assembled by MacEwan[131] and are given in Table 10-4.

Barrer and his colleagues (Barrer and Perry[132]) studied in considerable detail the selective-adsorption capacity of a variety of montmorillonite-organic complexes for a considerable series of organic compounds.

Very little information is available for organophilic complexes of the clay minerals other than montmorillonite. Vermiculite forms such complexes, and Weiss, Mehler, and Hofmann[133] showed that if alkylammonium ions are

Table 10-4   *X-ray data for a selection of organic complexes of montmorillonite*

(From MacEwan[131])

1*	2†	3	4	5‡	6§	7
Cation (and amount, meq/g if not saturated)	Sorbed molecules	Spacing (Å)	No. of molecular sheets	Orientation	Method	Reference
Na	Aniline	15.0	1	u	s	68, 137
Aniline HCl also Me-, Et-, Pr-, Bu-, diMe-	H_2O	15.0–15.1		u	dw	68, 137
Aniline, 0.62		12.8	1?	f?	dw, P_2O_5	24
Aniline, 0.91		14.2	1?	u?	dw, P_2O_5	24
Aniline HCl	H_2O	14.8–15.0	1	u	dw	7
Aniline HCl	H_2O excess	15.2	1	u	dw	7
Na	Benzidine	15.2	1	u	s, et	68, 137
Na	Nitrobenzene	15.2	1	u	s	68, 137
Na	Nitrobenzene	12.5	1	f	d	68, 137
Na	o-toluonitrile	16.2	1	u	s	68, 137
Na	m-toluonitrile	15.6	1	u	s	68, 137
Na	p-toluonitrile	15.0	1	u	s	68, 137
Pyridinium chloride	H_2O	12.5	1	f	dw	68, 137
Na	Pyridine	12.5	1	f	d	68, 137
Na	Pyridine	14.8	1	u	d	68, 137
Na	Pyridine (+H_2O tr.)	23.3	3	u, int	s	68, 137
Na	a:a′-dipyridyl	18.6	1	u	sc	68, 137

[130] Greene-Kelly, R., The Swelling of Organophilic Montmorillonites in Liquids, *J. Colloid Sci.*, **11**: 77–79 (1956).

[131] MacEwan, D. M. C., Montmorillonite Minerals, "The X-ray Identification and Crystal Structures of Clay Minerals," chap. IV, pp. 143–207, Mineralogical Society of Great Britain Monograph (1961).

[132] Barrer, R. M., and G. S. Perry, Sorption of Mixtures and Selectivity in Alkylammonium Montmorillonites, Part I, Monomethylammonium Bentonite, *J. Chem. Soc.*, pp. 842–849 (1961); Part II, Tetramethylammonium Montmorillonite, *J. Chem. Soc.*, pp. 850–858 (1961).

[133] Weiss, A., A. Mehler, and U. Hofmann, Zur Kenntnis von organophilem Vermikulit, *Z. Naturforsch.*, **11b**: 341, 435 (1956).

Table 10-4   X-ray data for a selection of organic complexes of montmorillonite (Continued)

1* Cation (and amount, meq/g if not saturated)	2† Sorbed molecules	3 Spacing (Å)	4 No. of molecular sheets	5‡ Orienta- tion	6§ Method	7 Reference
a:a'-Dipyridyl chloride	H_2O	12.6	1	f	dw	68, 137
Na	γ:γ'-dipyridyl	15.1	1	u	sc	68, 137
Na	Quinoline	16.9	1	u	s	68, 137
Na	6-Me-quinoline	18.5	1	u	s	68, 137
Na	8-Me-quinoline	17.1	1	u	s	68, 137
Na	6-nitroquinoline	12.6	1	f	s	68, 137
Na	8-nitroquinoline	17.3	1	u	s	68, 137
Brucine, 1.30	H_2O	16.9			dw	24
Brucine	H_2O	16.5			dw	7
Brucine	H_2O excess	16.7			dw	7
Magdala Red	H_2O	29.0			dw	7
$N(CH_3)_4$		13.5	1		o	134
$N(C_2H_5)_4$		13.9	1		o	134
$N(CH_3)_3H$		13.0	1		o	135
$N(CH_3)_2H_2$		12.2	1		o	135
$N(CH_3)H_3$		12.0	1		o	135
Cetyl tri-methylamine bromide, eq amount	H_2O?	20.5	2	f		136
bromide, 2.4 × eq amount	Neutral molecules?	38.6	2	i, int		136
Ca	Acetone	17.6	2	f	s	62
Na	Acetone	13.1	1	f	v (dry)	72, 138
Ca	Acetone	13.3	1	f	v (dry)	72, 138
Na, Ca	Acetone	17.4–17.6	2	f	v (50)	72, 138
Na, Ca	Acetone, H_2O	24.5	3–4?		v (100)	72, 138
Ca	Acetone	17–17.5	2	f	s	72, 138
Na	Acetone	13.3 or 17.5	1 or 2	f	s	72, 138
Ca	Acetonitrile	19.6	3	f	s	62
Ca	Nitromethane	19.8	3	f	s	62
Ca	1, 4 dioxane	15.0	1	u	s	62
H	Glucose	18.5	2	f	e (5, 2.5)	57
Na	Glucose	18.1	2	f	e (5, 2.5)	57
K	Glucose	14.3	1	f	e (5, 2.5)	57
Ca	Glucose	14.5	1	f	e (5, 2.5)	57
Ca	Glucose	18.1	2	f	e (5, 5)	57
Na	Glucose	18.5	2	f	e (2.5, 1)	57
Mg	Glucose	{ 17.9 { 14.3	2 1	f	e (2.5, 1)	57
	Ethoxy-trimethylene glycol	16.7	2	f		13
	Ethoxy-pentamethyl- ene glycol	14.3	1	f		13
	Ethylene glycol, dimethyl ether	17.0	2	f		13
	Triethylene glycol, dimethyl ether	17.5	2	f		13
Gelatin, edestin, pepsin		18.0	2		pH 2.5, conc. >1 percent	139
Gelatin, edestin, pepsin		14.2	1		pH 2.5, conc. <½ percent	139
Gelatin		26.3	4?		slow deposition	139
Glycine, arginine		12.9	1		1 percent solution	139

* eq   = equivalent quantity.

† tr   = trace.

introduced in interlayer positions a wide variety of organic molecules can then enter readily.

The patent literature indicating the use of organic-clad clay minerals in organic systems shows the organophilic nature of such materials. Thus, Sawyer[57] showed that attapulgite coated with certain fatty alkanolamides is an agent for gelling certain organic liquids.

Wilcox[140] described the coating of kaolinite particles by certain aliphatic alkylene polyamines by an exchange reaction and physical adsorption. Such coated particles may then be incorporated in a variety of organic vehicles, imparting unique flow properties. Similarly Albert[141] described kaolinite particles coated with rosin amine stearate, which could be incorporated in certain organic systems.

The attapulgite and kaolinite complexes are not intercalation compounds, but they may have somewhat similar properties in organic vehicles.

Structural implications of montmorillonite-organic complexes

Studies of the reactions between montmorillonite and organic molecules have led some investigators to certain conclusions regarding the atomic structure

[140] Wilcox, J. R., Modified Kaolin as a Polyester Resin Filler, U.S. Patent 2,999,080 (1961).

[141] Albert, C. G., Surface Modified Kaolin Clay, U.S. Patent 2,948,632 (1960).

---

‡ f   = flat, u = upright, i = inclined, int = interpenetrating.

§ d   = first saturated, then washed with solvent or evaporated.

s   = saturated in concentrated solution or liquid.

o   = outgassed.

e   = suspended in solution which is then evaporated.

c, w, et = cyclohexane, water, ether solution.

P_2O_5 = dried over P_2O_5 *in vacuo*.

v (dry)   = sorption of vapor by dry clay.

v (50)   = sorption of vapor in 50 percent relative humidity.

v (100)   = sorption of saturated vapors of organic substance and water.

(5, 2.5) = are the percent concentrations of clay and sugar, respectively, in aqueous solution (or suspension).

[134] Barrer, R. M., and D. M. McLeod, Sorption Complexes of Tetraalkyl Ammonium Montmorillonites, *Trans. Faraday Soc.*, **51**: 1290–1300 (1955).

[135] Barrer, R. M., and J. S. S. Reay, Methylammonium Montmorillonites, *Trans. Faraday Soc.*, **53**: 1253–1261 (1957).

[136] Franzen, P., Montmorillonite-Lissolamine Complexes, *Clay Minerals Bull.*, **2**(13): 223–225 (1955).

[137] Greene-Kelly, R., Unusual Montmorillonite Complex, *Clay Minerals Bull.*, **2**(13): 225–232 (1955).

[138] Glaeser, R., Organic-Montmorillonite Complexes, *Mem. Serv. Chim. Etat.* **36**(1): 69–76 (1951).

[139] Talibudeen, O., Interlamellar Sorption of Proteins, *Nature*, **166**: 236–237 (1950).

of the mineral. Thus, Berger[142] concluded that methylation takes place when dry hydrogen montmorillonite is treated with diazomethane, and that this requires the presence of acidic hydroxyls on the exterior of the montmorillonite structure. Berger further concluded from quantitative determinations of the methylation that the Hofmann structure would not provide an adequate number of such hydroxyls. Since the Edelman-Hofmann structure (see page 88) postulates an abundance of such hydroxyls, the methylation was believed to be evidence for this structure. Edelman[143] pointed out that the capacity for methylation of montmorillonite determined by Berger requires the inversion of only about one-third of the number of silica tetrahedrons (see page 89) originally postulated as inverted by Edelman and Favejee.[144]

Gieseking[145] expressed the opinion that the reaction of acetyl chloride with air-dried montmorillonite, producing a lyophobic material which no longer swells in water, affords evidence for the occurrence of OH groups on the basal surfaces of montmorillonite.

Deuel et al.[146] studied the esterification of hydrogen montmorillonite by 1,2-epoxides and the reaction of sodium montmorillonite and mustard gas, with the production of hydrophobic clay esters of largely reduced cation-exchange capacities. The clay esters produced in the reactions can be saponified. These investigators also found that hydrogen clays can be transformed by reaction with $SOCl_2$ into clay chlorides, which in turn will react with alcohols to form esters. Deuel et al. concluded that these reactions require the presence of a sufficiently large number of surface hydroxyls to provide evidence for the Edelman-Favejee structure.

Numerous investigators have pointed out that the Edelman-Favejee structure is not supported by diffraction data for montmorillonite. For example, the value of $d(001)$ in completely dried alkali and alkaline-earth montmorillonites is less than 10 Å, which is too low for the inverted silica tetrahedron structure of Edelman and Favejee. Projecting silica tetrahedrons from the basal planes of the mineral would require a $d(001)$ value of at least 11 Å in the dried condition.

Deuel[147] published additional experimental data which led him to the

[142] Berger, G., The Structure of Montmorillonite, *Chem. Weekblad,* **38:** 42–43 (1941).

[143] Edelman, C. H., Relation entre les propriétés et la structure de quelques minéraux argileux, *Verre Silicates Ind.,* **21:** 3–6 (1947).

[144] Edelman, C. H., and J. L. Favejee, On the Crystal Structure of Montmorillonite and Halloysite, *Z. Krist.,* **102:** 417–431 (1940).

[145] Gieseking, J. E., The Clay Minerals in Soils, *Advan. Agron.,* **1:** 59–204 (1949).

[146] Deuel, H., G. Huber, and R. Iberg, Organische Derivate von Tonmineralien, *Helv. Chim. Acta,* **33:** 1229–1232 (1950).

[147] Deuel, H., Organische Derivate von Tonmineralien, *Ber. Deut. Keram. Ges. B.,* **31**(1): 1–7 (1954).

conclusion that there were active hydroxyls on the basal surfaces of the clay minerals, and Gentilli and Deuel[148] claimed to have shown the presence of free Si-OH groups in investigations of the preparation and degradation of a phenol-montmorillonite derivative.

The existence of complexes of the type described by Deuel and collaborators are far from being accepted. Brown, Greene-Kelly, and Norrish[149] examined certain of these complexes by X-ray and chemical analysis and considered them to be merely sorption complexes of hydrogen montmorillonite with impurities such as thiopene, or with alteration products of the reaction. Greenland and Russell[150] arrived at essentially the same conclusion. Slabaugh[151] studied reactions similar to those of Deuel, and he did not feel that Deuel's structural interpretation is essential.

Analytical techniques based on clay-mineral–organic reactions

The reactions of clay minerals and organic molecules have been used as the basis for analytical techniques for the identification of the clay minerals, for the determination of certain properties of the clay minerals themselves, and for the determination of the geometry and properties of the organic molecules.

X-ray techniques. The smectite minerals commonly give diffuse reflections and a nonintegral series of basal reflections. Bradley[14] and Mac-Ewan[15] pointed out that organic-smectite complexes tend to have a high degree of regularity in the *c* spacing, giving an integral series of very sharp reflections. Thus, the identification of smectite by X-ray diffraction is greatly simplified by preliminary treatment with certain organic reagents. It is desirable to use an organic liquid of low volatility which

[148] Gentilli, R., and H. Deuel, Organische Derivate von Tonmineralien, 5, Mitteilung, Abbau von Phenylmontmorillonit, *Helv. Chim. Acta,* **40:** Fasciculus primus 12, 106–113 (1957).

[149] Brown, G., R. Greene-Kelly, and K. Norrish, Organic Derivates of Montmorillonite, *Clay Minerals Bull.,* **1**(7): 214–220 (1952).

[150] Greenland, D. J., and E. W. Russell, Organoclay Derivatives and the Origin of the Negative Charge on Clay Particles, *Trans. Faraday Soc.,* **51:** 1300–1307 (1955).

[151] Slabaugh, W. H., The Synthesis of Organo-Bentonite Anhydrides, *J. Phys. Chem.,* **56:** 748–751 (1952).

is readily miscible with water, so that the natural hydrated smectite may readily be changed into an organic complex. Glycerol and ethylene glycol are used extensively. The hydration state of the natural mineral does not alter appreciably the position of intensity of the reflections given by the organic-smectite complex. It is common practice to add enough glycerol to air-dried material to make it visibly moist. Any smectite present will give rise to a periodicity along *c* of about 17.7 Å, with a series of higher orders, most of which will be adequately separated from reflections given by the other clay minerals. By the use of the organic treatment technique, relatively small amounts (5 percent) of smectite can be identified. Prior to the development of this technique, such small amounts of smectite would not have been detected.

Glycerol treatment of halloysite ($4H_2O$) causes a shift in the basal reflections to 11 Å, and first, third, and fifth orders from this spacing are apparent. A similar shift takes place for the dehydrated form of this mineral if the dehydration is not essentially complete. Thus, air-dried samples will show the shift, whereas samples dried at 75°C for about 12 hr or at higher temperatures for shorter intervals do not form the complex and exhibit no shift.

Glycerol forms complexes with vermiculite with a spacing of 14 Å, which is so close to the spacing of the fully hydrated natural mineral that it has little diagnostic value.

The glycol- or glycerol-treatment technique is of tremendous value in resolving very intimate interlayer mixtures of some of the clay minerals. For example, the reflections from a mixture of smectite and illite are likely to be diffused and complex because of the variable *c* spacing of the former mineral. Treatment with glycol or glycerol develops a uniform spacing of the smectite and thereby sharpens the reflections so that they can be interpreted more readily. Use of the organic treatment technique has shown that many materials described as single species are actually interlayer mixtures. Thus, the type bravaisite and many materials described as beidellite were found to be mixtures of illite and smectite (see pages 43 and 47). For a detailed discussion of the glycol-glycerol diffraction technique the works of MacEwan[15,63] and Bradley[14] should be consulted.

Differential thermal techniques. When clay-mineral–organic complexes are heated, the organic material is oxidized, and an exothermic reaction takes place. Allaway,[152] working with piperidine-montmorillonite complexes, showed that the detailed features of the exothermic reactions varied with

[152] Allaway, W. H., Differential Thermal Analyses of Clays Treated with Organic Cations as an Aid in the Study of Soil Colloids, *Soil Sci. Soc. Am. Proc.,* **13:** 183–188 (1948).

the composition of the montmorillonite. Triethanolamine, monoethanolamine, and *n*-butylamine complexes gave results similar to those for piperidine complexes. Allaway found that the exothermic reaction was always multiple and that a peak at 700°C suggested a high-magnesium montmorillonite, a peak between 450 and 500°C suggested an iron-rich montmorillonite, and a peak at about 600°C suggested a member of the montmorillonite group containing some aluminum in tetrahedral coordination. Similar variations in the exothermic reactions were found for several illites, but they were of relatively lower intensity than for the montmorillonites. Piperidine treatment had little effect on the differential thermal curves of kaolinites.

Allaway concluded that the piperidine held by the montmorillonite breaks down on firing, with a loss of hydrogen, leaving a coating of carbon on the external surfaces and between the unit layers of the clay mineral. The carbon is burned off rather slowly as long as the clay-mineral structure remains intact, but when the structure breaks down, the carbon is quite rapidly oxidized. The use of organic cations is essentially a technique for magnifying the thermal effects of the breakdown of such structures as can be coated with carbon, with strong exothermic reactions replacing weaker endothermic reactions. Certain lattice changes which appear in the ordinary differential thermal procedure to be slow gradual breakdown of the clay are resolved into two separate peaks of thermal activity by the use of the piperidine treatment. Thus, the treatment procedure may resolve mixtures or permit the identification of certain mineral components which would not be revealed by the usual method of thermal study.

Optical methods. The adsorption of organic molecules between the basal plane surfaces of smectite and to a lesser extent of halloysite causes a slight but definitely determinable change in their indices of refraction (see pages 423–426). Therefore, when these clay minerals are mounted in certain liquids for optical determinations, the indices of refraction can be seen to change as the liquid is adsorbed by the clay mineral. This characteristic can be used as a diagnostic criterion for the identification of smectite and halloysite. No specific data are available for vermiculite, but its indices of refraction should also change, since it possesses the property of interlamellar adsorption of organic molecules.

Cation-exchange capacity. Robertson and Ward[153] described a rapid method for estimating the cation-exchange capacity of clay materials based on the

[153] Robertson, R. H. S., and R. M. Ward, The Assay of Pharmaceutical Clays, *J. Pharm. Pharmacol.*, **3:** 27–35 (1951).

adsorption of methylene blue. The change in color of a standard methylene blue solution produced by the adsorption of the clay material permits a determination of the amount of the organic molecule adsorbed, from which the cation-exchange capacity can be calculated. For some clays the results are in good agreement with those obtained by other procedures. Thus, Ramachandran et al.,[154] using malachite green and methyl violet as well as methylene blue, obtained good agreement for kaolinite and illite but not for several smectites.

Surface-area determination. Dyal and Hendricks[155] pointed out that the total surface, the external surface, and, by difference, the internal surface of clays can be determined from adsorption reactions with polar molecules, e.g., ethylene glycol. Total surface can be determined by the retention of ethylene glycol in an evacuated system. After the clay has been heated to about 100°C for halloysite and 600°C for smectite, the organic molecule does not penetrate between the unit layers, and the values for retention on external surfaces alone are obtained. The difference between total retention and external retention expresses the amount of interlayer surface, and from this value the amount of halloysite or smectite can be estimated. Morin and Jacobs[156] recently presented a method for measuring surface area based on the adsorption of ethylene glycol vapor.

White and Cowan[45] used the methylene blue adsorption for the determination of surface area and obtained satisfactory results for montmorillonite but overestimated the area of kaolinite clays by about 50 percent.

Geometry and properties of organic molecules. The study of clay-mineral–organic reactions and the resulting complexes provides an approach to the investigation of the shape and certain bonding characteristics of the adsorbed organic molecules themselves. Thus, the space required for the organic molecules between the montmorillonite layers, as revealed by determination of $d(001)$ spacings, provides information on the thickness, areal dimensions, and in some cases the bond distribution in the organic molecules. This subject is aside from the field of this volume, and for information concerning it, the works of Hendricks,[24] MacEwan,[62] Bradley,[13] Brindley and Hoffmann,[67] and Weiss et al.[18] should be consulted.

[154] Ramachandran, V. S., K. P. Kacker, and N. K. Patwardhon, Adsorption of Dyes by Clay Minerals, *Am. Mineralogist*, **47**: 165–169 (1962).

[155] Dyal, R. S., and S. B. Hendricks, The Surface of Clays in Polar Liquids as a Characteristic Index, *Soil Sci.*, **69**: 421–432 (1950).

[156] Morin, R. E., and H. S. Jacobs, Surface Area Determination of Soils by Adsorption of Ethylene Glycol Vapor, *Soil Sci. Soc. Am. Proc.*, **28**(2): 190–194 (1964).

Staining tests for clay minerals

According to Faust,[157] who investigated the history of staining techniques in general and summarized much of the early work, Behrens[158] in 1871 appears to have first applied staining methods to mineralogy. Staining techniques were used for biological material beginning about 1838. An extensive literature has grown up regarding the use of staining methods, and many persons have attempted to apply them to clays.

The adsorption of various organic substances by natural and chemically altered or heat-treated clays can produce color changes in the clay. Frequently such changes vary, depending on the identity of the clay mineral and its composition. The color changes, therefore, provide a possible basis for identifying the clay-mineral components of clay materials. A staining test has the advantage of being rapid and simple to perform, even in the field.

In general, the reaction and resulting color change are slight or absent for clay minerals of low adsorptive capacity, such as kaolinite, but typically are pronounced for highly adsorptive clay minerals, such as smectites. Several mechanisms have been proposed to explain the color reactions obtained with clay minerals and various reagents: an acid-base reaction in which the natural or acid-treated clay reacts as an acid; an oxidation-reduction phenomenon in which certain ions, mainly ferric iron contained in the clay-mineral lattice, cause an oxidation of the reagent; and by electron transfer from the reagent to the clay mineral (Hasegawa[159]). Substances probably causing color changes by the acid-base reaction include triphenylmethane, azine, and azo dyes. Substances probably changing the color of clays by the oxidation-reduction reaction include benzidine and other aromatic amines (Dodd[160]).

The general applicability of staining tests for the identification of the clay minerals is restricted, because the development of characteristic color reactions can be inhibited, augmented, or masked by several common ingredients of clay materials other than the clay minerals. Manganese dioxide may cause the oxidation-reduction reaction with benzidine, even in the absence

[157] Faust, G. F., Staining of Clay Minerals as a Rapid Means of Identification in Natural and Beneficiation Products, *U.S. Bur. Mines, Rept. Invest.* 3522 (1940).

[158] Behrens, H., Mikrochemische Methods fur Mineralanalyse, *Verslag. Mededeel. Koninkl. Akad. Wetenschap. Amsterdam,* **1:** 17 (1881).

[159] Hasegawa, H., Spectroscopic Studies on the Color Reaction of Acid Clay with Amines, *J. Phys. Chem.* **65:** 292–296 (1961); see also **66:** 834–836 (1962).

[160] Dodd, C. C., Dye Adsorption as a Method of Identifying Clays, *Dept. Nat. Resources, Calif., Div. Mines, Bull.* 169, pp. 105–111 (1955).

of adsorptive clays, and ferrous iron and other reducing agents may prevent the color-forming reaction from taking place. In techniques that require the preparation of hydrogen clays for the staining reaction, the acid treatment may destroy the structure of the clay minerals. Thus iron- and magnesium-rich smectites, and consequently some nontronites and hectorite, might not appear to react with the staining solutions like other members of the smectite group. Also, the strong original color of a sample particularly rich in iron might tend to mask the color change that would result from the adsorption of the staining reagent.

Because of the above factors, caution must be used in applying any staining test indiscriminately to clay materials. Under restricted conditions, when the samples to be studied are relatively pure clay-mineral materials, or where the range of composition is restricted and known in general terms, staining techniques may be entirely satisfactory. For example, some of the sedimentary kaolins in the Georgia area contain smectite in addition to kaolinite. There are no other constituents of these kaolins that might cause a color reaction, and a staining test quickly shows the presence of smectite.

A very large number of reagents have been suggested and tried for clay-mineral staining tests. In recent years the tendency has been to use several reagents separately in a test, so that the deficiencies of one can be compensated by the value of another for a given material. Such a test was described by Mielenz, King, and Schieltz[161] and appears to give quite satisfactory results. In this procedure a pulverized portion of the natural clay is treated with benzidine, and portions of the clay, after treatment with strong hydrochloric acid and washing to remove the excess chloride, are treated separately with a saturated solution of safranine Y in nitrobenzene and with a saturated solution of malachite green in nitrobenzene. In addition to the visual determination of the color developed, the samples are examined with a petrographic microscope to determine the development of any pleochroic colors and other changes in optical characteristics. The results of the use of these reagents are summarized in Table 10-5 taken from the work of Mielenz, King, and Schieltz.[161] Even with such an elaborate staining technique, caution must be used; anomalous results are sometimes obtained because various nonclay substances may cause or inhibit staining reactions; and some clays fail to react in a characteristic manner for as yet inexplicable reasons.

Fahn and Gennrich[162] proposed a method using dyestuff with strong specific fluorescent colors. Only dyestuffs containing nitrogen were found very useful, with kaolinite having the lowest adsorption and illite, calcium

[161] Mielenz, R. C., M. E. King, and M. C. Schieltz, "Staining Tests," Rept. 7, American Petroleum Institute Project 49, Columbia University, New York (1950).

[162] Fahn, R., and M. Gennrich, Staining of Clay Minerals with Fluorescent Dyes, *Tonind.-Ztg. Keram. Rundschau,* **79**(15/16): 233–236 (1955).

Table 10-5   Characteristic staining of clay minerals
(After Mielenz et al. [161])

Clay mineral	Untreated clay	Acid-treated clay	
	Benzidine	Safranine Y	Malachite green
Kaolinite	No reaction	Red–purple red.* Strong to weak pleochroism from reddish purple parallel to cleavage to yellowish red perpendicular to cleavage	Blue–green blue and blue–green.* Strong to weak pleochroism from yellowish green parallel to cleavage to blue perpendicular to cleavage
Halloysite	No reaction	Blotchy stain: purple, purple–blue purple, and red–purple red. Not pleochroic	Blotchy stain: yellow green–yellow, blue–green, and green–yellow. Not pleochroic
Dickite	No reaction	Crystals not stained. Very weak pleochroism from reddish purple or purple parallel to cleavage to reddish yellow perpendicular to cleavage	Crystals not stained. Very weak pleochroism from yellowish green or colorless parallel to cleavage to light blue perpendicular to cleavage
Nacrite	No reaction	Crystals not stained. Weak pleochroism from reddish purple parallel to cleavage to yellowish red perpendicular to cleavage	Crystals not stained. Weak pleochroism from yellowish green parallel to cleavage to blue perpendicular to cleavage
Montmorillonite	Purple–blue	Purple–blue	Yellow–red yellow
Nontronite	Blue–green	Red–purple red*	Green blue–green and blue–green blue*
Hectorite	Purple–blue	Red–purple red*	Blue–green blue*
Illite	No reaction	Red–purple red*	Green blue–green*
Attapulgite	No reaction	Red–purple red*	Blue–green and blue–green blue*
Pyrophyllite	No reaction	Not stained	Not stained

* Dye absorbed without change in color. Samples of nontronite included in these tests did not change the color of the dyes in these tests, but specimens of nontronite reacting in a manner similar to montmorillonite have been examined.

montmorillonite, and sodium montmorillonite having increasingly better responses.

Russian investigators have given much attention to staining techniques in recent years (Vedeneeva,[163] Vedeneeva and Rateev,[164] Vikulova[165]).

[163] Vedeneeva, N. E., *Tr. Sovesch. Vses. Khromatografi, Akad. Nauk SSSR*, pp. 144–146 (1952). (See Eitel[166].)

Eitel[166] presented a very valuable summary of this literature from the U.S.S.R.

A large amount of work on color reactions with clay minerals and organics, largely unpublished except in the patent literature (Green[167] and Cormack and Thacker[168]) has resulted from the use of clays in duplicating papers, e.g., NCR paper (no carbon required).

Additional references

Armstrong, D. E., and G. Chesters, Properties of Protein-Bentonite Complexes as Influenced by Equilibration Conditions, *Soil Sci.,* 98(1): 39–52 (1964).

Bangham, D. H., and S. Mosallam, The Adsorption of Vapours at Plane Surfaces of Mica, *Proc. Roy. Soc. (London) A,* pp. 558–571 (1938).

Caillere, S., R. Glaeser, and J. Esquerin, Preparation of Halloysite with Interplanar Distances of 14 Å and 17 Å, *Compt. Rend.,* 230: 308–310 (1950).

Cowan, C. T., Adsorption by Organo-Clay Complexes, II, Proceedings of the 10th National Clay Conference, pp. 226–234, Pergamon Press, New York (1963).

Diamond, S., and E. B. Kinter, Characterization of Montmorillonite Saturated with Short-chain Amine Cations, 1, Interpretations of Basal Spacing Measurements, Proceedings of the 10th National Clay Conference, pp. 163–173, Pergamon Press, New York (1963).

Fripiat, J. J., A. Servais, and A. Leonard, Etude de l'adsorption des amines par les montmorillonites, III, la nature de la laison amine-montmorillonite, Extrait, *Bull. Soc. Chim. France,* pp. 635–644 (1962).

Grandjean, M. F., Coloration des argiles par les couleurs d'analine, *Bull. Soc. Franc. Mineral.,* 32: 408–419 (1909).

Greenland, D. J., R. H. Laby, and J. P. Quirk, Adsorption of Glycine and Its Di-, Tri-, and Tetra-Peptides by Montmorillonite, *Trans. Faraday Soc.,* 58(472): 829–841 (1962).

Heller, L., Sorption of Glycol and Glycerol by Preheated Monoionic Montmorillonite, *Proc. Intern. Clay Conf., Stockholm,* 2: 105–113 (1965).

Jordan, J. W., B. J. Hook, and C. M. Finlayson, Organophilic Bentonites, II, *J. Phys. Colloid Chem.,* 54: 1196–1208 (1950).

Leonard, A., A. Servais, and J. J. Fripiat, Etude de l'adsorption des amines par les montmorillonites, II, la structure des complexes, Extrait, *Bull. Soc. Chim. France,* pp. 625–635 (1962).

[164] Vedeneeva, N. E., and M. A. Rateev, *Dokl. Acad. Nauk SSSR,* 100: 559–562 (1955). (See Eitel[166].)

[165] Vikulova, M. F., General Characteristics of Clays, *Tr. Vses. Nauchn.-Issled. Geol. Inst. Acad. Sci. USSR,* pp. 7–90, 229–269 (1957).

[166] Eitel, W., "Silicate Science," vol. I, Academic, New York (1964).

[167] Green, B. K., Manifold Record Material and Process for Making It, U.S. Patent 2,548,366 (1951).

[168] Cormack, M. J., and N. A. Thacker, Desensitized Clay-coated Record Sheets, U.S. Patent 2,777,780 (1957).

McAtee, J. L., Random Interstratification in Organophilic Bentonites, *Natl. Acad. Sci., Publ.* 566, pp. 308–317 (1958).

McAtee, J. L., Organic Cation Exchange on Montmorillonite as Observed by Ultraviolet Analysis, Proceedings of the 10th National Clay Conference, pp. 153–162, Pergamon Press, New York (1963).

McAtee, J. L., and J. R. Hackman, Exchange on Montmorillonite Involving Organic Cations, *Am. Mineralogist,* **49**: 1569–1577 (1964).

Moll, W. F., The Orientation of Some Cyclic Amine Cations on Vermiculite, Univ. Microfilms (Ann Arbor, Mich.), order no. 64-2444 (1964).

Rateev, M. A., Investigation of Clays by Spectrophotometry with the Use of Organic Dyes, *Materialy po Geol., Mineralog. i Ispol'z. Glin Sb.,* pp. 98–104 (1958).

Servais, A., J. J. Fripiat, and A. Leonard, Etude de l'adsorption des amines par les montmorillonites, I, Les processus chimiques, Extrait, *Bull. Soc. Chim. France,* pp. 617–625 (1962).

Sieskind, O., The Interaction of Clay and Organic Matter: Adsorption of Amino Acids by Montmorillonite, *Mem. Serv. Carte Geol. Alsace Lorraine,* **22**: 1–90 (1962).

Skawinski, R., Method of Gradual Staining of Clay Minerals, *Bull. Acad. Polon. Sci., Ser. Sci. Geol. Geograph.,* **13**(1): 39–42, 43–48 (1965).

Van Olphen, H., and C. T. Deeds, The Stepwise Hydration of Clay-Organic Complexes, *Nature,* **194**: 176–177 (1962).

Eleven
Optical Properties

The optical properties of the clay minerals as reported in the literature are summarized in Table 11-1. In general it is necessary to give a range of optical values instead of a single value, because of inherent variations in the clay minerals themselves and because of difficulties in measuring precisely the optical properties of the extremely small clay-mineral particles. Also, it is usually possible to determine only some of the optical characteristics.

Inherent variations in the clay minerals which cause changes in optical properties are replacements within the lattice. Also, variations in the amount of any interlayer water in the expanding-lattice minerals are reflected by changes in optical properties. In addition, the clay minerals from time to time may exhibit strain or warping within the lattice or deviations from perfection of crystallinity, which would cause some alteration in optical characteristics. The variation due to strain or imperfections in crystallinity is likely to be small. Further, the optical values of some clay minerals are different in different liquids which may be used as index liquids to measure them (see pages 423–426).

Correns and Piller[1] described in detail all aspects of the various micro-

---

[1] Correns, C. W., and H. Piller, Mikroskopie der feinkörnigen Silikatminerale, "Mikroskopie

scopic procedures that are applicable to fine-grained silicate minerals. Particular attention was given to the clay minerals and the possible use of a phase-contrast microscope.

Allophane

The values given for this mineral in Table 11-1 are for air-dried samples. White[2] reported an index of 1.398 to 1.423 for a sample with a moisture content of 34.69 percent from Lawrence County, Indiana, and that the index increased to 1.473 to 1.480 when the sample was dried at 100°C, reducing the moisture content to 10.58 percent. On heating to 500°C for 1 hr there was substantially no further change in the index.

Very few determinations of optical properties for allophane are reported in the literature. Because of the rather wide range in chemical composition of allophane, a wide variation in indices would be expected.

Kaolinite

As shown in Table 11-1, there is a measurable and diagnostic difference in the optical properties of kaolinite, dickite, and nacrite. The optical characteristics reported for anauxite are substantially the same as those of kaolinite.

Because of the absence of replacements within the kaolinite lattice and the low adsorptive properties of the mineral, kaolinite exhibits narrowly restricted optical values. Kaolinite frequently occurs in particles of sufficient size to permit fairly complete and precise optical measurements. As the degree of crystallinity decreases, the poorly crystallized variety has slightly decreasing indices of refraction and birefringence. Frequently only a mean index of refraction can be determined. For the "fireclay" mineral, values between 1.552 and 1.563 are common.

The data in Fig. 11-2, after Mehmel,[3] show that the mean index of refraction of kaolinite increases slightly on heating up to the point of loss of OH lattice water slightly above 400°C. Accompanying this dehydration reaction,

der Silikate," pt. 1, pp. 699–780, "Handbuch der Mikroskopie in der Technik," vol. IV, Mikroskopie der Gesteine, Frankfurt (1955).

[2] White, W. A., Allophanes from Lawrence County, Indiana, *Am. Mineralogist,* **38**: 634–642 (1953).

[3] Mehmel, M., Beiträge zur Frage des Wasserhaltes der Minerale Kaolinit, Halloysit und Montmorillonit, *Chem. Erde,* **11**: 1–16 (1937).

Table 11-1 *Optical properties of the clay minerals*

Mineral	α	γ	$\gamma - \alpha$	Sign	2V	Dispersion	Remarks
Kaolinite	1.553–1.563	1.560–1.570	0.006–0.007	−	24°–50° 42° mean value	$p > v$ weak	Optic plane and $Z \perp$ to (010); $X \wedge \perp$ (001) = 3° ±
Dickite	1.560–1.562	1.566–1.571	0.006–0.009	+	52°–80°	$p < v$ weak	$Z = b$; $X \wedge C = +15$–20°
Nacrite	1.557–1.560	1.563–1.566	0.006	−	40°	$p > v$	$Z = b$; $X \wedge \perp$ (001) 10–12°
				+	90°	$p < v$	
Allophane	$n = 1.468$–1.512						
Halloysite 4H$_2$O	Mean value: 1.526–1.532						
Halloysite 2H$_2$O	Mean value: 1.548–1.556		0.002–0.001				
Smectite	1.480–1.590	1.515–1.630	0.025–0.040	−	0–30° ±	……………	X about ⊥ (001)
Hectorite	1.485	1.516	0.031	−	Small		
Saponite	1.480–1.490	1.510–1.525	0.030–0.035	−	Moderate		
Nontronite	1.565–1.60	1.600–1.640	0.035–0.040	−	Moderate	……………	Pleochroic; yellow–brown–green
Volkhonskoite	1.551	1.585	0.034	−	Small		
Sauconite	1.550–1.575	1.592–1.615	0.035–0.042	−	Small		
Illite	1.545–1.63	1.57 –1.67	0.022–0.055	−	Small	……………	X about ⊥ (001); some pleochroic
Glauconite	1.545–1.63	1.57 –1.66	0.022–0.030	−	0–20° ±	$p > v$	Pleochroic; yellow green
Hydrobiotite	1.59 –1.62	1.64 –1.67	0.045–0.055	−	Small	……………	Pleochroic; green–brown–yellow
Chlorite	1.57 –1.64	1.575–1.645	0.003–0.007	±	Small	……………	Pleochroic; green brown
Vermiculite	1.525–1.56	1.545–1.585	0.020–0.030	−	Small	……………	X about ⊥ (001); pleochroic; green brown
Sepiolite	1.483–1.520	1.505–1.530	0.009–0.027	−	0–60°	……………	$Z \approx c$ (elongation)
Attapulgite	1.510–1.520	1.540–1.555	0.025–0.032	−	Small		
Palygorskite	1.524–1.502	1.533–1.512	0.011–0.020	−	51–61°		$Z \parallel$ elongation

there is a sharp reduction in the mean index from about 1.575 to about 1.525 and probably also a complete loss of birefringence.

Halloysite

Usually only the mean indices of refraction can be determined for the halloysite minerals because of their low degree of crystallinity. As shown in

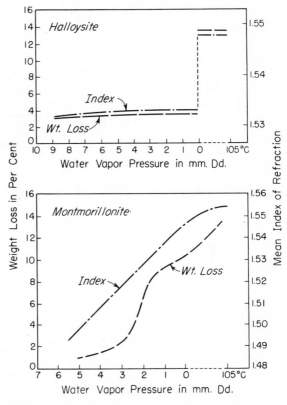

Fig. 11-1. **Relation of the mean index of refraction, weight loss, water-vapor pressure, and drying at 105°C for halloysite and montmorillonite. (After Mehmel.[3])**

Fig. 11-1, the transition from the $4H_2O$ to $2H_2O$ form is accompanied, according to Mehmel[3] and Correns and Mehmel,[4] by an increase in index from

[4] Correns, C. W., and M. Mehmel, Über den optischen und röntgenographischen Nachweis von Kaolinit, Halloysit, und Montmorillonit, *Z. Krist.*, **94**: 337–348 (1936).

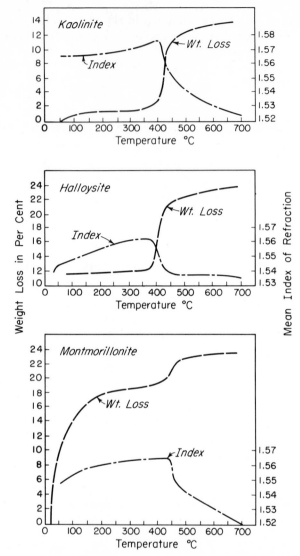

Fig. 11-2. Relation of the mean index of refraction to weight loss and firing temperature for kaolinite, halloysite, and montmorillonite. (After Mehmel.[3])

about 1.532 to almost 1.550. Alexander and colleagues[5] computed the theoretical value of the $4H_2O$ form at 1.490, assuming the structure for the mineral suggested by Hendricks[6] and using the molecular refractivity of kaolinite and water. No indices nearly as low as 1.490 appear to have been reported for the $4H_2O$ form, which may be taken as lack of support for this structural concept of the mineral.

On heating to temperatures above 100°C, halloysite shows a slight gradual increase in index of refraction, until the OH water is driven off at about 400°C. The dehydration of the mineral is accompanied by a sharp reduction in the index of refraction from about 1.560 to 1.535, as shown in Fig. 11-2.

Smectite

The smectite clay minerals show large variations in optical properties as a consequence of substitutions within the lattice. The data in Fig. 11-3,

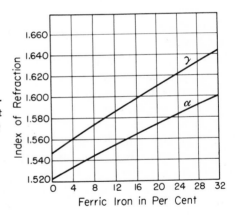

Fig. 11-3. Indices of refraction and birefringence of smectite in relation to content of ferric iron. (After Ross and Hendricks.[7])

after Ross and Hendricks,[7] show that an increase in the Fe_2O_3 content from 0 to 32 percent is accompanied by an increase in the γ index from about

[5] Alexander, L. T., G. T. Faust, S. B. Hendricks, H. Insley, and H. F. McMurdie, Relationship of the Clay Minerals Halloysite and Endellite, *Am. Mineralogist*, **25**: 1–18 (1943).

[6] Hendricks, S. B., On the Structure of the Clay Minerals: Dickite, Halloysite, and Hydrated Halloysite, *Am. Mineralogist*, **23**: 295–301 (1938).

[7] Ross, C. S., and S. B. Hendricks, Minerals of the Montmorillonite Group, *U.S. Geol. Surv., Profess. Paper* 205-B, pp. 23–79 (1945).

1.545 to 1.645 and by an increase in the birefringence from about 0.025 to 0.045. Martinet and Martinet[8] discussed theoretically the reason for the influence of ferrous and ferric iron on the optical properties of natural silicates. Mehmel[3] showed that the indices increase as the content of magnesium increases. In the samples of smectite studied by him, the mean index increased from about 1.530 to 1.548 as the MgO content increased from about 2 to 6 percent.

As shown in Fig. 11-1, the mean index of refraction increases in the range from about 1.49 to 1.55 as the thickness of the interlayer water decreases. The identities of the exchangeable cations affect the indices indirectly under certain conditions of humidity because of their control of the thickness of the interlayer water. Thus air-dried Na montmorillonite tends to have interlayer water one molecular layer thick, whereas Ca montmorillonite under the same conditions tends to have interlayer water two molecular layers thick. Under such conditions the Ca montmorillonite will have the lower indices. There is no positive evidence that the exchangeable cations themselves directly influence the optical properties, but the data by Marshall[9] on birefringence in an electric field (see page 428) suggest that such an influence exists.

Mehmel (Fig. 11-3) and Grim and Bradley[10] independently showed that the mean index of refraction of a smectite relatively low in iron increases from about 1.54 to about 1.570 when all the interlayer water is eliminated on heating to 250 to 300°C. At higher temperatures there is no appreciable change in indices until about 450 to 500°C, when the OH lattice water is lost, accompanied by a sharp drop in mean index to about 1.52. This drop in index is accompanied by a reduction in birefringence. Caldwell and Marshall[11] presented data for saponite (Fig. 11-4) showing somewhat the same effect of heating, except that the decrease in indices accompanying the loss of OH water is more gradual. These investigators showed also that at higher temperatures, above that at which the hydroxyl water is lost, there is an increase in indices and birefringence. A sample of nontronite investigated by Caldwell and Marshall[11] showed (Fig. 11-4) a gradual increase in indices on heating to about 850°C, with only a slight flattening of the curve accompanying loss of OH water. These authors also showed that the birefringence

[8] Martinet, J., and A. Martinet, Pleochroism and Structure of Natural Silicates, *Bull. Soc. Chim. France,* 46(21): 563–565 (1952).

[9] Marshall, C. E., The Orientation of Anisotropic Particles in an Electric Field, *Trans. Faraday Soc.,* 26: 173–189 (1930).

[10] Grim, R. E., and W. F. Bradley, The Effect of Heat on the Clay Minerals Illite and Montmorillonite, *J. Am. Ceram. Soc.,* 23: 242–248 (1940).

[11] Caldwell, O. G, and C. E. Marshall, A Study of Some Physical and Chemical Properties of the Clay Minerals Nontronite, Attapulgite, and Saponite, *Missouri Univ., Coll. Agr. Res. Bull.* 354 (1942).

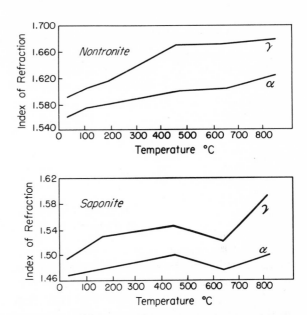

Fig. 11-4. Effect of heating on the refractive indices and birefringence of nontronite and saponite. (After Caldwell and Marshall.[11])

of nontronite increases as the mineral is dehydrated, up to about 700°C. At temperatures above 700°C, the birefringence decreases moderately.

Illite

The exact variations of the optical properties of the illites accompanying substitutions within the lattice are not known, but they are probably of the same order of magnitude as those for the smectite minerals. The illites with the muscovite type of structure have somewhat lower indices of refraction and lower birefringence than muscovite itself, and the same relation apparently holds for the illites with the biotite type of structure. Thus, in the case of illite with the muscovite structure, the γ value is frequently of the order of 1.570 to 1.580, as compared with 1.595 for muscovite itself. For biotite-type illite the γ value is frequently 1.62 to 1.63, as compared with 1.64 to 1.66 for biotite. This difference must result, at least in part, from the lesser population of interlayer cations and from the presence of some water molecules between the silicate layers of the illites. It probably also is somewhat dependent on the crystallinity of the clay minerals, i.e., on defects

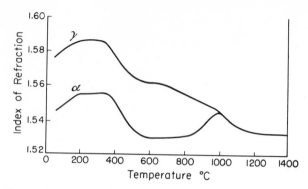

Fig. 11-5. Effect of heating on the refractive indices and birefringence of illite. (After Grim and Bradley.[10])

in the stacking of the silicate layers and on some variations in the population of the silicate units.

Grim and Bradley[10] showed (Fig. 11-5) that on heating there is frequently a slight increase in indices at very low temperatures (200°C); e.g., the γ value increases from about 1.575 to 1.585. Above about 200°C there is little change until about 400°C, when there is a drop in indices accompanying the loss of OH water. At this temperature the γ index decreases from about

Table 11-2 Index, percent expandable, and total iron, calculated as Fe_2O_3 for seventeen glauconites

(*After Toler and Hower*[12])

	Formation	Index	Percent expandable	Percent Fe_2O_3
1	Sundance	1.584	29	8.9
2	Colorado Shale	1.584	28	9.4
3	Byram	1.590	36	
4	Moody's Branch	1.596	30	14.2
5	Carrizo	1.606	25	
6	Kinkaid	1.612	14	
7	Folkestone	1.614	13	14.8
8	"B" New Jersey	1.622	10	
9	Park Shale	1.622	8	
10	Gros Ventre (1)	1.626	13	
11	Gros Ventre (2)	1.626	10	
12	Bashi	1.626	10	22.7
13	Franconia	1.628	13	21.5
14	Boone Terre	1.630	6	
15	Birkmose	1.630	8	20.8
16	Reno	1.636	7	
17	Tonto	1.638	0	19.4

1.585 to 1.560. At still higher temperatures, there is a further decrease in the γ index, and the birefringence also decreases.

It has been shown (see Chap. 4) that glauconite frequently contains expandable layers in addition to illite in mixed-layer assemblages. Toler and Hower[12] showed that there is a general decrease in the index of refraction (n_z) of glauconite with an increase in the percent of expandable layers, as shown in Table 11-2 and Fig. 11-6. Further, as the content of iron increases, the index increases and the amount of expandable layers decreases. These

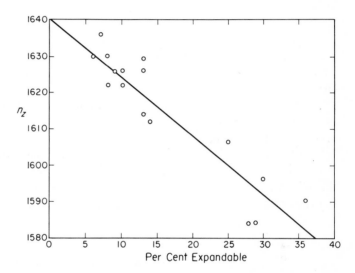

Fig. 11-6. Relation between index of refraction and percent expandable layers in glauconites. (After Toler and Hower.[12])

authors concluded that the decrease in index with increasing percentages of expandable layers is caused in part by the presence of water molecules between the layers and in part because the iron content decreases with increasing amounts of expandable layers.

Chlorite

There is substantially no information available in the literature on the optical properties of the chlorite clay minerals. Unpublished data obtained by the author indicate that the chlorite clay minerals have somewhat lower indices

[12] Toler, L. G., and J. Hower, Determination of Mixed Layering in Glauconites by Index of Refraction, *Am. Mineralogist,* **44:** 1314–1320 (1959).

of refraction than the well-crystallized chlorite minerals and that the difference is of about the same order of magnitude as that for the illites as compared with muscovite and biotite. There are no experimental data on the changes in optical properties on heating the chlorite clay minerals, but it is probable, by analogy with the illite minerals, that they would show an increase in indices accompanying any loss of adsorbed water and a decrease in indices at higher temperatures, accompanying the loss of hydroxyl water.

Vermiculite

There are no precise data available on the optical properties of the vermiculite clay minerals. However, they probably have substantially the same optical characteristics as those of coarse-grain vermiculite. Because of the structural similarity of vermiculite and smectite, it would be expected that the optical properties of vermiculite would vary with the degree of hydration and with substitutions within the lattice in a similar manner to that discussed for the smectite clay minerals. Also, it would be expected that heating vermiculite would cause changes in optical characteristics similar to those found for the smectites.

Sepiolite-attapulgite-palygorskite

The optical properties of only a few specimens of these clay minerals have been measured, and the data reported in the literature show wide variations. This is particularly true for sepiolite for which reported values for birefringence vary from 0.009 to 0.027, and $2V$ from 0 to 60°. Preisinger[13] described a sepiolite that is pleochroic, the mineral being dark yellow in the direction of the fiber and lemon yellow normal to it, with the γ index in the direction of the fiber. Caillere[14] described a nickeliferous sepiolite with γ of 1.57 and α of 1.55. According to Von Engelhardt,[15] the indices increase in the palygorskites as the magnesium content increases. The similarity between the indices and birefringence of some of these clay minerals and those

[13] Preisinger, A., X-ray Study of the Structure of Saponite, Proceedings of the 6th National Clay Conference, pp. 61–67, Pergamon Press, New York (1959).

[14] Caillere, S., Nickeliferous Sepiolite, *Bull. Soc. Franc. Mineral.,* **59:** 163–166 (1936).

[15] Von Engelhardt, W., Ueber silikatische Tonminerale, *Fortschr. Mineral. Krist. Petrog.,* **21:** 276–337 (1937).

of the smectites is noteworthy and has led sometimes to their erroneous identification as smectites.

Caillere[16] stated that the indices of sepiolite increase from a mean value of 1.49 to 1.54 when the mineral is dehydrated above about 465°C. At still higher temperatures the mean index increases further to about 1.61 on calcination.

Caldwell and Marshall[11] showed that attapulgite behaves differently from the other clay minerals on heating (Fig. 11-7). It shows an increase in indices

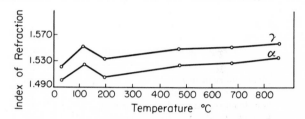

Fig. 11-7. **Effect of heating on the refractive indices and birefringence of attapulgite. (After Caldwell and Marshall.[11])**

up to about 100°C, the γ value increasing from 1.52 to 1.55; there follows a decrease to about 200°C, γ decreasing to 1.53. Above 200°C there is a slight gradual increase up to about 850°C, with γ attaining a value of about 1.56.

Influence of immersion media on optical properties

Larsen and Wherry[17] in 1917 found that the indices of refraction of a clay material from Beidell, Colorado, now known to contain a smectite mineral, increased gradually when the material was immersed in a mixture of cinnamon oil and oil of clove. This appears to be the first mention in the literature of variations in the optical properties of the clay minerals resulting from the media in which they are immersed. Correns and Mehmel[4] showed that the indices of montmorillonite and halloysite ($4H_2O$) varied in different liquids. In general, the indices of the mineral increased with time over a period

[16] Caillere, S., Etude de quelques silicates magnésiens à facies asbestiforme ou papyrace n'appartenant pas du groupe de l'antigorite, *Bull. Soc. Franc. Mineral.*, **59**: 353–386 (1936).

[17] Larsen, E. S., and E. T. Wherry, Leverrierite from Colorado, *J. Wash. Acad. Sci.*, **7**: 208–217 (1917).

of about 2 hr, and it is now known that much longer time, perhaps several days, would be necessary for equilibrium to be reached. In the case of montmorillonite, changes in optical values were noted for both air-dried samples and material dried at 105°C. Correns and Mehmel believed that the effect was due to the penetration of the organic molecules between the silicate layers

Table 11-3   Index of refraction in relation to immersion liquid
(After Van Baren[18])

Liquid and index of liquid	Kao-linite	Hal-loysite	Non-tronite	Smec-tite	Smec-lite (HCl treated)
Mean index of mineral	1.57	1.575	1.565	1.50	1.50
Chlorobenzene, 1.524	+	+	+	−	−
Ethylenebromide and chlorobenzene, 1.535	+	+	+	−	−
Benzylchloride, 15.41	+	+	+	−	−
Benzylamine, 1.546	+	+	+	+	+
o-Nitrotoluene, 1.547	+	+	+	−	−
Nitrobenzene, 1.552	+	+	+	−	−
Monobromobenzene, 1.560	+	+	+	−	−
Xylidine, 1.560	+	+	+	+	+
o-Anisol, 1.562	+	+	+	−	−
o-Toluidine, 1.572	=	+	x	+	+
Anisidine, 1.576	−	−	+	x	+
Aniline, 1.585	−	−	−	+	+
Bromoform, 1.596	−	−	−	−	−
Iodobenzene and bromobenzene, 1.605	−	−	−	−	−
Iodobenzene and bromobenzene, 1.616	−	−	−	−	−
Chinoline, 1.624	−	−	+	x	x
α-Tetrahydro-ar-napthylamine, 1.631	−	−	−	?	−

+ = index of mineral higher than liquid
− = index of mineral lower than liquid
= = index of mineral and liquid the same
x = index of mineral increased but not to a value higher than liquid

and thought that no effect would be produced by organic molecules too big for such penetration.

Van Baren[18] investigated the influence of a variety of organic liquids on the indices of refraction of kaolinite, halloysite ($2H_2O$), and montmorillonite (Table 11-3). He found no changes in the indices of kaolinite or halloysite

[18] Van Baren, F. A, Ueber den Einfluss verschiedener Flüssigkeiten auf den Brechungsindex von Tonmineralien, *Z. Krist.*, **95:** 464–499 (1936).

(2H₂O) but significant increases for montmorillonite for some immersion liquids. It is probable that under somewhat different conditions, and perhaps with different liquids, halloysite material with only slightly more than $2H_2O$ would show some variation in optical properties. Van Baren concluded that organic molecules with NH_2 radicals are most likely to cause significant variations in optical characteristics and that the time necessary for the liquid to obtain its maximum influence increased as the size of the organic molecule increased. He showed further that the indices of a smectite, when immersed in a liquid of higher index than the highest index of the mineral, sometimes increase to even higher values than that of the liquid itself. That is, the indices of the mineral not only may increase to equal that of the liquid but may become higher than that of the liquid. Thus a smectite with a mean index of 1.50 increased to a value above that of benzylamine (1.546) when immersed in this liquid.

Vendel[19] studied in detail the characteristics which are likely to be found in liquids causing a variation in indices of the clay minerals. According to him, a liquid should have the following properties to permit a determination of the true optical values of the clay minerals:

1. The liquid must not chemically attack the clay mineral.

2. It must not be soluble in water, and water must not be soluble in it; otherwise a change in hydration state might develop with an attendant change in optical values.

3. It must not enter into any ion-exchange reaction with the clay.

4. It must not permit the dissociation of any of the adsorbed ions of the clay.

5. It must be nonpolar.

6. Its index must be approximately that of the mineral.

7. If a mixture is used, the components must be completely miscible and must have nearly the same boiling point.

8. The liquids should be well-defined compounds.

Smectite, halloysite, and vermiculite to a limited extent (see Chap. 10) have the property of adsorbing certain organic molecules on their basal plane surfaces. The types of organic molecules adsorbed are also those which influence the optical properties. The variation in optical properties is, therefore, undoubtedly largely a consequence of this adsorption. This was confirmed by the work of Greene-Kelly,[20] who found that organic complexes of montmorillonite and aromatic molecules having the plane of the aromatic ring perpen-

[19] Vendel, M., Zur Bestimmung der Lichtbrechung silikatischer Tonminerale, *Chem. Erde,* **15:** 325–370 (1945).

[20] Greene-Kelly, R., Birefringence of Montmorillonite Complexes, *Nature,* **184:** 181 (1959).

dicular to the silicate sheet have a reduced negative birefringence or even a positive birefringence, whereas in cases where the plane of the aromatic ring is parallel to the silicate sheet the complexes have a strong negative birefringence.

The optical properties of smectite, halloysite, and vermiculite could also be altered by certain liquids as a consequence of an effect on the extent of interlayer hydration water. Ordinarily illite, kaolinite, and chlorite minerals would be expected to show no variation with different immersion liquids, because of their very low adsorptive power for organic molecules and because of their lack of interlayer water. As shown in Chap. 10, it is possible to intercalate some organic and inorganic molecules in these clay minerals. Certainly such intercalation would alter the optical properties, but the matter remains to be studied. The attapulgite-sepiolite minerals also should show no effect from immersion liquids except in the rare cases when an organic molecule is of the correct size and shape to enter the channels of their structure.

This attribute of the clay minerals requires that the index liquids used in their study must be carefully chosen, and, in general, when values are published, the liquids in which they were determined should be recorded. Obviously, the influence of immersion liquids on certain clay minerals and not on others can be of diagnostic value in the identification of the clay minerals.

Oriented-aggregates technique

When the flake-shaped clay-mineral particles settle from a deflocculated clay-water suspension on a horizontal surface, they tend to come to rest with one flake on top of another so that their basal plane surfaces are substantially parallel. Frequently successive flakes assume positions in which there is a considerable degree of parallelism of the a and b crystallographic axes. In other words, an aggregate orientation closely approximating crystal growth may take place in the accumulation of such material under the foregoing conditions.

Grim[21] in 1934 described a technique taking advantage of the above phenomenon for the preparation of oriented aggregates for the determination of the optical properties of the clay minerals. In this technique, a clay-water suspension is allowed to stand until some of the clay mineral has accumulated on a flat surface, either the bottom of a vessel or a glass plate hung in a horizontal position in the suspension. The flat surface with the accumulated

[21] Grim, R. E., The Petrographic Study of Clay Minerals—A Laboratory Note, *J. Sediment. Petrol.*, **4:** 45–46 (1934).

clay is then removed from the suspension and dried. Aggregate particles are sliced or cut from the dried film of clay for mounting in index-of-refraction liquids.

The α and γ indices and the birefringence of such aggregates can usually be determined with considerable accuracy. Frequently the aggregates yield reasonably good interference figures, from which the sign of the mineral and the order of magnitude of $2V$ can be determined. Measurement of the latter value implies considerable orientation parallel to the a and b crystallographic axes.

Without the use of oriented aggregates, it is often possible to determine only a mean index of refraction from the more or less randomly oriented clay-mineral particles in the crude clay. The birefringence and other optical values frequently cannot be even approximated in the crude clay. Oriented aggregates are valuable also in X-ray-diffraction techniques because data are obtainable in which characteristic basal reflections are emphasized (see Chap. 4).

Interlayer mixtures

It has been pointed out (see pages 121–122) that clay materials may be composed of several clay minerals mixed primarily by interstratification on substantially unit-cell dimensions. Such materials may be composed of aggregates made up of alternations of one or a few unit layers of the component clay minerals. Also in some clay materials the clay-mineral units are interlayered with ferric hydroxide units.

Optical data from such intimate interlayered mixtures can easily be mistaken for that of a single mineral. Thus, Grim and Rowland[22] showed that a sample from Namiquipa, Mexico, previously described as beidellite on the basis of optical values, was actually a mixture of halloysite and limonite. X-ray and differential thermal data left no doubt regarding this interpretation, whereas the optical values were suggestive of a single clay mineral of the smectite type.

An example such as that above emphasizes the fact that clay-mineral determinations based solely on optical data must be made with great caution to avoid gross errors. Frequently the character of the interference figure reveals whether a clay-mineral aggregate is made up of a single mineral or a mixture. If the interference figure is not good and the isogyres are not reasonably complete, there is probably more than one mineral present.

[22] Grim, R. E., and R. A. Rowland, Differential Thermal Analysis of Clay Minerals and Other Hydrous Material, *Am. Mineralogist,* **27**: 746–761 (1942).

Homogeneity and parallelism in the *a* and *b* directions are required for fairly complete figures of the layered clay minerals.

Orientation in an electric field

Marshall[9,23] studied in considerable detail the tendency of clay-mineral particles to orient themselves in an electric field, and the following statements are taken largely from his publications.

Freundlich, Kruyt, and their coworkers[24] showed that extremely nonequidimensional particles, such as needle-shaped particles, are oriented in a water suspension in an electric field relative to the electrical lines of force. Procopiu[25] showed that the prime cause of orientation in suspensions of various minerals was the electrical anisotropy of the particles themselves, rather than hydrodynamic effects. Thus, irrespective of the shape of the particles or the hydrodynamic conditions of flow, a turning moment operates when an electric field is applied, until the direction of greatest dielectric constant becomes parallel to the lines of force.

Table 11-4 Electrical birefringence of Putnam (beidellite) clay fractions
(After Marshall[23])

Exchange cations	Birefringence at 800 volts/cm, 60 cycles ac		
	500–200 $m\mu$	200–100 $m\mu$	100 $m\mu$
H+	0.023	0.029	0.026
Ca++	0.026	0.031	0.031
K+	0.023	0.020	0.011

Marshall[9,23] measured the birefringence developed in suspensions of various clays when subjected to an electric field, and some typical results for fractions of a soil listed as beidellite, containing a large proportion of smectite, are given in Table 11-4. It can be seen from the data in the table that the electrical birefringence is affected by the exchangeable cation. No such variation was found for kaolinite, which is evidence, as Marshall pointed

[23] Marshall, C. E., "The Colloidal Chemistry of the Silicate Minerals," Academic, New York (1949).

[24] Freundlich, H., H. R. Kruyt, et al. See Marshall.[23]

[25] Procopiu, S., Sur la biréfringence électrique et magnétique des suspensions, *Ann. Phys.*, **1:** 213–223 (1924).

out, for the exterior position of the exchange cations in kaolinite and their internal position in smectite.

Marshall[23] found that dilute suspensions of lithium and sodium smectites sometimes gave a negative instead of a positive birefringence. Muller[26,27] and Norton[28] found that, at low frequencies, negative values exist for smectite suspensions and, for high frequencies, the values are positive. The lower the concentration of the clay, the lower was the range of frequencies for which negative values were observed. These authors concluded that the negative values are due to the presence of long-range interparticle forces, the positive ones being normal for an orientation like that of the Kerr effect (see Marshall[23]).

Whiteside and Marshall[29] showed that, at 4,000 volts/cm with 60-cycle alternating current, the saturation values for Na and K clays can be obtained, and, for the Putnam material (beidellite), the Na and K clays are very significantly lower in birefringence than the hydrogen clay. Their data indicate that the optical and electrical ellipsoids are similar in character.

More recently Joly,[30] Kahn and Lewis,[31] and Shah and Hart[32] discussed the optical properties of clay suspensions, and the latter two authors especially considered the electro-optical birefringence of montmorillonite-water suspensions.

Granquist and Hollingsworth[32a] found that flow birefringence data at low shear of polydisperse systems of hectorite, montmorillonite, and attapulgite gave gradient and concentration rotational diffusion constants for such systems. They[32b] also studied the rate of relaxation of streaming birefringence after stirring for these same systems.

[26] Muller, H., Electro-Optical Effects in Colloids, *Phys., Rev.,* **55:** 508 (1939).

[27] Muller, H., and B. W. Sakmann, A New Electro-Optical Effect in Colloids, *Phys. Rev.,* **56:** 615–616 (1939).

[28] Norton, F. J., Influence of Frequency on the Electro-Optical Effects in Colloids, *Phys. Rev.,* **55:** 668–669 (1939).

[29] Whiteside, E. P., and C. E. Marshall, Mineralogical and Chemical Studies of the Putnam Silt Loam Soil, *Missouri Univ., Coll. Agr. Res. Bull.* 386 (1944).

[30] Joly, M., Theory of the Orientation of Rigid Particles Interacting in a Flowing System, *Kolloid-Z.,* **126:** 77–87 (1952).

[31] Kahn, A., and D. R. Lewis, The Size of Sodium Montmorillonite Particles in Suspension from Electro-Optical Birefringence Studies, *J. Chem. Phys.,* **58:** 801–804 (1954).

[32] Shah, M. J., and C. M. Hart, The Electro-Optical Birefringence of Polydisperse Bentonite Suspensions, *IBM J. Res. Develop.,* **7**(1): 44–57 (1963).

[32a] Granquist, W. T., and C. A. Hollingsworth, Rotational Diffusion of Some Polydisperse Clay Mineral Sols, *J. Colloid Sci.,* **18**(6): 538–554 (1963).

[32b] Granquist, W. T., and C. A. Hollingsworth, Streaming Birefringence Decay of Some Clay Mineral Sols, *Trans. Faraday Soc.,* **59**(489): 2192–2199 (1963).

Form birefringence

Wiener[33] showed that birefringence ("form birefringence") may arise in a system in which there is a parallel arrangement of plate- or rod-shaped particles when these particles are separated by a medium of different refractive index. There is no form birefringence when the solid and the liquid have the same index of refraction. The birefringence increases as the difference between the indices of the liquid and solid increases. Wiener developed his concepts for isotropic particles, but Ambronn and Frey,[34] who studied this phenomenon in great detail, showed that the effect can be superimposed on systems which contain birefringent components. Thus, the inherent birefringence of the clay-mineral particles would augment the form birefringence developed in suspensions.

In the case of the clay minerals, form birefringence would arise primarily because of the penetration of water or an organic liquid between the unit layers of the expanding clay minerals. Thus, under the usual conditions for the measurement of optical properties, it might be significant for montmorillonite, vermiculite, and halloysite but probably not for the other clay minerals. Where the ratio of volume of liquid to solid is small, the effect would be small, but when considerable fluid has been adsorbed between layers, it might be significant. Undoubtedly part of the variation of the optical characteristics of montmorillonite and halloysite minerals reported in the literature is due to form birefringence.

Obviously the phenomenon of form birefringence must be kept in mind in any optical study of fine-grained particles in a liquid medium. Unfortunately this has not always been the case.

Discussion of the application of optical methods in clay-mineral studies

The value of optical methods in studying the clay minerals has been the subject of much discussion. Some students of clays assert that optical methods are worthless for such fine-grained materials, whereas others plead for

---

[33] Wiener, O., Theory of Reaction Constants, *Abhandl. Math.-Phys. Klasse Sächs. Akad. Wiss.* (*Leipzig*), **32**: 256–276 (1912).

[34] Ambronn, H., and A. Frey, "Das Polarizationsnikroskop," Akademische Verlagsgesellschaft mbH, Leipzig (1926).

their universal use and base detailed clay-mineral determinations solely on optical studies. In the writer's opinion, optical methods are very frequently of value, but they must be used with caution, and it must be realized that they are worthless for some types of materials. An optical examination can be made in a very short time, and the chance of obtaining valuable data usually justifies the expenditure of the required time. Also, after some experience has been gained in the study of clay minerals under the microscope, more can be done with them than would be supposed at first. It is usually possible to determine with fair accuracy at least the indices of refraction and the birefringence of clay-mineral samples. The optical properties of the various clay minerals are sufficiently diverse so that an idea of the identity of the clay minerals usually can be gained from such data. As noted previously (page 427), caution must be used because of anomalous results possible from interlayer mixtures.

Under the microscope, certain of the clay minerals frequently have distinctive appearances which aid in their identification. Thus, aggregates of the smectite clay minerals often appear as a large single crystal that has been strained. The individual particles of smectite either cannot be seen or else are indistinct and tend to merge into each other. The attapulgite-sepiolite minerals sometimes have a somewhat similar appearance. Aggregates of the other clay minerals have the appearance of being made up of discrete particles, even though the individual particles are less than $1\,\mu$ in diameter, and the particles have a high degree of uniform orientation. With a little experience one can usually recognize relatively pure smectite quite easily under the microscope by its appearance without reference to optical values.

The presence of nonclay minerals, such as quartz, calcite, and feldspar, may cause considerable difficulty if these minerals are present in particles of about the same size as the clay minerals. An abundance of extremely fine quartz or calcite makes optical methods valueless in many cases. Sometimes the presence of such nonclay minerals will produce an unusual appearance or strange optical characteristics which suggest that some such material might be present. However, in many cases, the nonclay material would go undetected by optical methods.

The presence of even relatively small amounts of extremely fine-grained carbonates makes optical study of the clay minerals impossible, because of the high birefringence of these nonclay minerals. Further, it is not possible to remove the carbonates in all cases by dissolving them, because of the susceptibility of some of the clay minerals to acid attack. Certain of the clay minerals (see pages 435–439) are relatively insoluble in acid, but others are quite soluble. Unfortunately many calcareous materials also contain the more soluble clay minerals (see page 438), so that in a study of such material it is necessary to determine the clay-mineral composition by nonoptical methods.

Organic material and ferric iron compounds may mask the optical characteristics of the clay minerals. The organic material can frequently be removed by solvents or by mild oxidants, such as hydrogen peroxide, without altering the clay minerals. Some organic material, particularly that in the older sediments, is not affected by such reagents and can be removed only by vigorous oxidizing agents or by heating to elevated temperatures. Unfortunately, if there is iron present, it may be oxidized in the process, so that it may then conceal the characteristics of the clay minerals.

Sometimes the free iron compounds can be removed by reducing, e.g., by hydrogen sulfide, and then dissolving the ferrous iron. Again caution must be used because of the danger of attack on the clay minerals, particularly those of the smectite and attapulgite-palygorskite groups, by the reducing agents.

Optical methods, of course, are of great value in determining the nonclay minerals present in a clay material. In general, minerals found in particles as small as $5\,\mu$ in diameter can be identified. If the mineral has particularly distinctive optical properties, considerably finer particles can be identified. Thus with a little experience carbonates can be spotted in particles at least as small as $1\,\mu$. Quantitative determinations of such nonclay minerals are, however, very difficult for particles smaller than about $5\,\mu$.

The petrographic microscope is, of course, useful in studying the texture of clays as revealed by thin sections. Examples are the shards which are characteristic of many bentonites and the parallel orientation of the clay minerals in many shales. The textures of clay materials are not included in the objective of this text and will not be considered. The identification of the clay minerals in thin sections is usually difficult and often impossible. Optical methods of identification can be applied best by the use of immersion techniques and prepared, oriented aggregates.

Additional references

Edsall, J. T., Streaming Birefringence and Its Relation to Particle Size and Shape, *Advan. Colloid Sci.*, **1**: 269–316 (1941).

Emiliani, F., Relations Between the Chemical and Optical Properties of Muscovite, *Rend. Soc. Mineralog. Ital.*, **12**: 118–127 (1956).

Greene-Kelly, R., and B. V. Deryagin, Birefringence of Thin Liquid Films in Swollen Montmorillonite, *Issled. Obl. Poverkhn. Sil, Akad. Nauk SSSR, Inst. Fiz. Khim., Sb. Dokl. Vtoroi Konf., Moscow*, pp. 127–135 (1962). (Publ. 1964.)

Grim, R. E., Properties of Clays, "Recent Marine Sediments," pp. 466–496, *Amer. Assoc. Petrol. Geol.* (1939).

Huggins, W. H., M. V. Denny, and H. R. Shell, Properties of Palygorskite, an Asbestiform Mineral, *U.S. Bur. Mines, Rept. Invest.* 6071 (1962).

Jasmund, K., Die Silikatischen Tonminerale, *Monograph Z. Angew. Chem.* and *Chem.-Ing.-Tech.*, no. 60.2 (1955).

Marshall, C. E., Mineralogical Methods for the Study of Silts and Clays, *Z. Krist.*, **90:** 8–34 (1935).

Ross, C. S., and P. F. Kerr, The Kaolin Minerals, *U.S. Geol. Surv., Profess. Paper* 165E, pp. 151–180 (1931).

Ross, C. S., and P. F. Kerr, Halloysite and Allophane, *U.S. Geol. Surv., Profess. Paper* 185G, pp. 135–148 (1934).

Twelve
Miscellaneous Properties

Solubility of the clay minerals

General statement. The solubility of the clay minerals in acids and alkalies is important fundamentally because it reveals certain attributes of the clay minerals and certain differences between them which are not obvious from other methods of study. From a practical standpoint, solubility characteristics are important in determining the utility of various clay materials as sources of metallic aluminum and for the manufacture of alum. Such data are also of practical value in numerous other ways, e.g., in the bonding of refractories by acid media, in the fixation of certain materials such as phosphates in soils, in the preparation of clay for decolorizing oils, and in the manufacture of catalysts. The solubility characteristics of the clay minerals are important in determining the limitations of acid treatment and electrodialysis as auxiliary tools in the analyses of complex clay materials containing carbonate, and in determining their ion-exchange capacity.

Nutting,[1] who studied the solubility of the clay minerals in great detail, arrived at the following general conclusions:

---

[1] Nutting, P. G., The Action of Some Aqueous Solutions on Clays of the Montmorillonite Group, *U.S. Geol. Surv., Profess. Paper* 197F, pp. 219–235 (1943).

The solubility of a number of clays in acids and alkalies of a wide range of concentrations indicates that the solution process is of at least three kinds for every clay, dependent on the concentration of solvent.

At equilibrium, a fraction of the acid or alkali remains free, and a fraction of each clay remains undissolved regardless of the proportions present. Hence the reactions appear to be reversible as indicated by varying equilibrium conditions.

Over a range of low acid concentrations, 0.05 to 0.2 normal, the "solution" is essentially a silicate hydrosol similar in composition to the clay dissolved. At higher concentrations it contains also salts in solution; at lower concentrations excess silica. At acid concentrations of 20 per cent and over, bases but no silica go into solution and no hydrosols are formed.

Over a range of low alkali concentrations, 0.002 to 0.005 normal (0.02 to 0.05 per cent) sodium carbonate, the solution essentially is an alkali silicate hydrosol, only silica being removed from the clay. At concentrations of alkali below the critical range, no carbonate is left in solution.

With montmorillonite clays the critical anion concentration is around 0.1 mole per liter while the critical cation (Na^+) concentration is around 0.001 mole per liter. Within the critical acid range, the anions may be considered as competing on substantially equal terms with silicic acid for the bases of the clay. Within the critical alkali range, the alkali in solution is competing on equal terms with the bases of the clay for the silica.

In general, above certain minimum concentrations, acids remove alkali metals, alkaline earths, iron, and aluminum from the clay minerals, and alkalies dissolve the silica. The literature contains a large amount of information on acid solubility, because of the general economic importance of the acid-clay reaction, as noted previously, but relatively little data on solubility in alkalies.

Solubility of clay minerals in acid. The solubility of the clay minerals in acids varies with the nature of the acid, the acid concentration, the acid-to-clay ratio, the temperature, and the duration of treatment. Also the solubility of the various clay-mineral groups is quite different, and there is great variation in solubility characteristics of members of some individual groups. Thus, in general, a magnesium-rich smectite is much more soluble than one that is rich in aluminum, with an iron-rich member somewhere in between. In the case of the clay minerals showing variations in the degree of crystallinity, such as kaolinite, the solubility increases as the degree of crystallinity decreases. The solubility would, of course, also increase as the particle size decreases. Calcining the clay minerals changes their solubility characteristics and their relative solubility with respect to each other.

Pask and Davies[2] (Table 12-1), using 0.5 g of clay dried at 130°C in 30 cc of 20 percent solution of H_2SO_4 and boiling for $\frac{1}{2}$ hr, showed that only 3 percent of the total alumina is dissolved from kaolinite and only 9 percent from anauxite. Halloysite is moderately soluble, showing solution of from 50 to 90 percent of the total alumina under similar conditions. Illite is slightly soluble, with 11 percent of total alumina going into solution in one sample. Smectite ranges from low to high solubility, showing 33 to 87 percent of total alumina dissolved. The attapulgite-sepiolite clay minerals would probably have characteristics similar to those of the smectite minerals.

With acid of the same concentration and a similar acid-to-clay ratio, but with digestion under pressure at 155°C, the solubility of all the clay minerals increased (Table 12-1). All the alumina of halloysite became soluble, and almost all of it was dissolved from the smectites (85 to 93 percent). Even the ordinarily relatively insoluble kaolinite and illite lost the major part of their alumina, 70 and 87 percent, respectively, going into solution.

Under similar conditions of extraction, but after calcining the clay minerals to 800°C, all or substantially all the alumina of the kaolinite, anauxite, and halloysite is soluble, whereas the solubility of the alumina in the illite is increased only moderately, and for the smectites it actually decreases (Table 12-1). The explanation resides in the fact that the kaolinite and halloysite lattices are probably destroyed or at least considerably disrupted at this temperature, and no new high-temperature phases have yet formed. In the case of illites and smectites the lattice has been changed slightly but not destroyed; hence the solubility of the alumina has been changed, but not so that it can be completely removed. When kaolinite is heated to still higher temperatures (975°C), new crystalline phases develop, and the solubility of the alumina decreases. Similarly, when the other clay minerals are heated to near this temperature (975°C), new high-temperature phases form, which would affect the solubility in acid. In the case of some clay minerals, such as the smectites, the high-temperature phase forming (see page 321) varies considerably with substitutions in the original smectite lattice, and correlative variations in solubility in acid would be expected.

[2] Pask, J. A., and B. Davies, Thermal Analysis of Clays and Acid Extraction of Alumina from Clays, *U.S. Bur. Mines, Tech. Paper* **664**, 56–78 (1945).

A comparison of the data obtained by Thiebaut[3] with those of Pask and Davies[2] shows the great variation in results obtained under different treatment conditions. Using 50 percent hydrochloric acid, a treating temperature of 80 to 85°C, a treating time of 2 hr, and clay dried at 105 to 108°C, Thiebaut found that the solubility of biotite was 100 percent, muscovite 5 to 32 percent, kaolinite 10 percent, halloysite 6 to 15 percent, and smectite 62 percent, based on cation solubility. Using sulfuric acid of similar concentration and the same minerals but evaporating to dryness, Thiebaut found

Table 12-1   *Extraction of alumina from various clay minerals with*
sulfuric acid

(From Pask and Davies[2])

Clay mineral	Aluminum extracted, percent total aluminum present		
	Clay dried at 130°C*	*Clay calcined at 800°C**	*Pressure digestion at 155°C†*
Kaolinite	3	100	70
Anauxite	9	95	
Halloysite	63	100	98
	50	96	
	90		
	78	100	100
Illite	11	52	87
Smectite	87	28	85
	33	19	93
Muscovite	17	29	

* Determinations made after boiling 0.5 g of sample in 35 cc of 20 percent solution of sulfuric acid for 1 hr.

† Clay dried at 130°C, then digestion of 0.5-g sample in 35 cc of 20 percent solution of sulfuric acid for 1 hr.

that the solubility of all the above minerals was complete, again based on their cation solubility. The work of Thiebaut[3] indicates the generality that the clay minerals are more soluble in sulfuric than in hydrochloric acid and also that, for hydrochloric acid, both kaolinite and halloysite are less soluble than the three-layer clay minerals. Biotite occupies an anomalous position in being about equally soluble in both acids.

The solubility of the clay minerals in other acids is not well known. However, investigations of anion exchange indicate that the solubility is likely to be considerable, especially for the acids with an anion having a size and geometry approximating that of the component parts of the clay-mineral

[3] Thiebaut, J. L., "Sédiments argilo-calcaires du bassin de Paris," Nancy, France (1925).

lattice. As a consequence, some relatively weak acids may strongly attack certain clay minerals. Murray[4] showed, for example, that phosphoric acid attacks the kaolinite lattice under some conditions with greater vigor than sulfuric acid.

Wolf[5] presented interesting data showing the effect of concentration of acid on solubility. Using 10 g of Zettlitz kaolinite and cooking for 2 hr, he found that 0.02 N HCl dissolved 3.1 mg of Al_2O_3 in 100 cc of acid; that in similar amounts of acid 0.5 N HCl dissolved 53.5 mg of Al_2O_3; and that 5 N HCl dissolved 124.4 mg of Al_2O_3.

Carroll and Starkey[6] percolated water saturated with carbon dioxide at 25°C through layers of limestone alternating with montmorillonite, illite, and kaolinite for long periods of time. They found only silica in the leachate, and the minerals lost SiO_2 in the order montmorillonite > kaolinite > illite.

Ostrom[7] studied the solubility of clay minerals in various acids in order to develop a method for freeing clay minerals in carbonate sediments without seriously altering the character of the clay minerals. He found that, for hectorite, an acetic acid solution of 0.33 M concentration or hydrochloric acid solution of 0.19 M concentration was sufficient to produce a slight change in the character of the X-ray-diffraction data. Well-crystallized chlorite was unaffected by 16.6 M acetic or 10 M hydrochloric acid solutions over periods of 72 hr. For poorly crystallized chlorite, 0.8 M acetic acid or 1.3 M hydrochloric acid solutions slightly altered the structure whereas 0.1 M acetic or 0.3 M hydrochloric acid concentrations produced no detectable effects. Well-crystallized illite was not affected by acetic or hydrochloric acid solutions of 16.6 M or 10 M concentrations. In mixed-layer-illite-montmorillonite assemblages, 0.8 M acetic and 1.9 M hydrochloric acid concentrations were sufficient to alter the diffraction data whereas no detectable effect was produced by 0.17 M acetic and 0.36 M hydrochloric acid concentrations. Well-crystallized kaolinite was unaffected after treatment with 16.6 M acetic and 10 M hydrochloric acid concentrations for 70 hr.

Brindley[8] pointed out that fine-grained chlorites decompose in warm dilute hydrochloric acid, whereas kaolinite is not ordinarily decomposed by this treatment, and that this might be a method for differentiating between

[4] Murray, H., The Structure of Kaolinite and Its Relation to Acid Treatment, Ph.D. thesis, University of Illinois (1951).

[5] Wolf, L., Zur Chemie des Kaolins, *Ber. Deut. Keram. Ges.,* **14**: 393–403 (1933).

[6] Carroll, D., and J. C. Starkey, Leaching of Clay Minerals in a Limestone Environment, *Geochim. Cosmochim. Acta,* pp. 83–87 (1959).

[7] Ostrom, M. E., Separation of Clay Minerals from Carbonate Rocks by Using Acid, *J. Sediment. Petrol.,* **31**: 123–129 (1961).

[8] Brindley, G. W., Chlorite Minerals, "The X-ray Identification and Crystal Structures of Clay Minerals," chap VI, pp. 242–296, Mineralogical Society of Great Britain Monograph (1961).

these clay minerals. He pointed out further that chemical composition of the chlorite and particle size, as well as acid concentration, time, and temperature, are important in acid solubility so that such a procedure would have to be used with caution.

Oberlin et al.[9,10] found that kaolinite subjected to prolonged treatment with sulfuric acid solutions at pH 2 resulted in the appearance of a poorly crystallized kaolinite ("fireclay mineral"). Their early study by microdiffraction of kaolinite in pure water showed a tendency for the bonds between the silicate layers to be ruptured, followed by gliding along both a and b axes, forming a halloysite-type structure. In the reaction with the acid, the gliding is only along the b axis. The authors attributed the difference to the presence of a considerable number of protons between silicate sheets which are responsible for a large amount of the bonding.

Escard et al.[11] discussed the X-ray-diffraction characteristics of montmorillonite that develop with increasing attack by acid. Figure 12-1 shows the relation of the surface area developed in the acid-treated product with increasing amounts of free silica, i.e., the silica readily soluble in dilute caustic solution. For a calcined sodium montmorillonite and air-dried sodium montmorillonite, the surface area increases up to the development of about 50 percent free silica and then gradually declines. The amount of free silica developed is dependent on the acid, its concentration, and the temperature and time of treatment. According to these authors, the increased surface area can be attributed both to the formation of the amorphous silica and to the subdivision of the remaining particles of montmorillonite into smaller units.

Haden[12] showed that pugging kaolinite with concentrated sulfuric acid and then roasting at about 300°C result in the formation of an aluminum sulfate compound from which the aluminum may be readily leached with water.

Nature of the acid reaction. In the attack of acid on the smectite clay minerals and probably also on the illite and sepiolite-attapulgite clay minerals, it appears that the alkalies and alkaline earths are removed relatively more

[9] Oberlin, A., and M. C. Tchoubar, Mineralogie—étude en microscopie et microdiffraction électroniques de l'altération des cristaux de la kaolinite par une solution acide, *Compt. Rend.*, **250**: 728–729 (1960).

[10] Oberlin, A., M. C. Tchoubar, C. Schiller, H. Pezerat, and S. Kovacevic, Fireclay Produced by Alteration of Kaolinite and Some Natural Fireclays, *Colloq. Intern. Centre Natl. Rech. Sci. (Paris)*, **105**: 45–55 (1962).

[11] Escard, J., J. Mering, and I. Perrin-Bonnet, Activation de la montmorillonite, *J. Chim. Phys.*, **47**(3–4): 234–237 (1950).

[12] Haden, W. L., Jr., Method for Producing High Purity Silica from Kaolin Clay, U.S. Patent 3,116,973 (1964).

rapidly than the aluminum or iron, and that the iron is removed more rapidly than the aluminum. Mathers et al.,[13] however, did not check this early conclusion and reported that aluminum, iron, and magnesium were removed from the smectites they studied in the same proportions as they occurred in the mineral. On the other hand, Nikolaev et al.[14] found that repeated treatments of vermiculite with normal nitric acid reduced the content of magnesium and iron at a much higher rate than that of aluminum. Further, they found

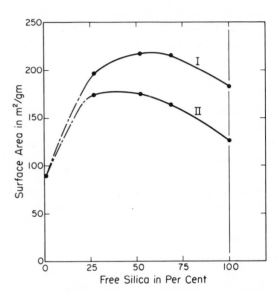

Fig. 12-1. Relation of free silica to surface area developed in montmorillonite by reaction with mineral acid. I, calcined; II, air-dried. (After Escard et al.[11])

that the removal of magnesium and iron had only a small effect on the cation-exchange capacity whereas in the case of aluminum the cation-exchange capacity was reduced in proportion to the reduction of the aluminum content. Hofmann and Endell,[15] Glaeser,[16] and Mering[17] showed that 75 to 85 percent

[13] Mathers, A. C., S. B. Weed, and N. T. Coleman, The Effect of Acid and Heat Treatment on Montmorillonoids, *Natl. Acad. Sci., Publ.* 395, pp. 403–413 (1956).

[14] Nikolaev, V. M., V. F. Bagretsov, and Yu. A. Kalmykov, The Effect of Repeated Treatments with Acid on the Cation-exchange Capacity of Vermiculite, *Zh. Prikl. Khim.,* 36: 692–693 (1963).

[15] Hofmann, U., and K. Endell, Mitteilung über die Aktivierung der rohen Bleicherde, *Z. Angew. Chem.,* 48: 187–192 (1935).

[16] Glaeser, R., Effect of Acid Treatment on the Base-Exchange Capacity of Montmorillonite, *Compt. Rend.,* 222: 1241–1242 (1946).

[17] Mering, J., Reactions of Montmorillonite, *Bull. Soc. Chim. France,* pp. 218–223 (1949).

of the total alumina must be removed from the montmorillonite lattice before it is completely destroyed. Acid attack begins around the edges of the particles and works inward. In relatively dilute alkaline solutions, the solubility of the silica residue of the structure left around the edges of the particles is increased as a consequence of the attack by acids. If much aluminum replaces silicon, the tetrahedral part of the structure as well as the octahedral part tends to disintegrate on acid attack. It is clearly the tetrahedral sheet that tends to preserve the morphology on solution of the metallic cations. Gastuche and Fripiat[18] observed that, after the octahedral layer of the clay minerals has been removed, the (001) reflections persist, indicating retention of order in the tetrahedral layers.

It appears that the removal of aluminum is stepwise; i.e., it moves first from octahedral positions to exchange positions and then to complete solubility. If the sample is dried before all the aluminum is removed, at least some of it appears to move back from exchange positions to octahedral positions. Mathers et al.[13] found that hydrogen montmorillonites changed spontaneously into aluminum clays. The rate of conversion was slow at 0°C but at temperatures at about 100°C, moist hydrogen montmorillonite changed to aluminum-saturated montmorillonite within 24 hr. It appeared that octahedral aluminum moved more rapidly from the lattice to exchange positions than tetrahedral aluminum.

Brindley and Youell[19] presented data for a chlorite (penninite) indicating that the aluminum in octahedral coordination is more soluble in HCl than the aluminum in tetrahedral coordination.

Osthaus[20] investigated the digestion of several montmorillonites and nontronite in hydrochloric acid at constant temperature for various periods of time and concluded from the interpretation of the dissolution curves that it was possible to determine the components in octahedral versus tetrahedral coordination. Rate-of-solution curves were plotted as the logarithm of the percent of residual components against time. In general, the data resulted in straight lines or curves that could be resolved into two straight lines (Fig. 12-2). Cloos et al.[21] were also able to differentiate between octahedral and

[18] Gastuche, M. C., and J. J. Fripiat, Acid Solution Techniques Applied to the Determination of the Structure of Clay and Controlled by Physical Methods, *Sci. Ceram.*, **1**: 121–138 (1962).

[19] Brindley, G., and R. F. Youell, A Chemical Determination of Tetrahedral and Octahedral Aluminum Ions in a Silicate, *Acta Cryst.*, **4**: 495–496 (1951).

[20] Osthaus, B., Kinetic Studies on Montmorillonites and Nontronite by the Acid-Dissolution Technique, *Natl. Acad. Sci., Publ.* **456**, pp. 301–321 (1956).

[21] Cloos, P., M. C. Gastuche, and M. Croegaert, Cinetique de la destruction de la glauconite par l'acide chlorhydrique étude préliminaire, *Intern. Geol. Congr.*, 21st, *Rept. Session, Norden*, pt. 24, pp. 35–50 (1960).

tetrahedral components on the basis of acid solubility. Gastuche and Fripiat[18] found that the differential attack on octahedral layers as compared with tetrahedral layers was improved by using solutions of acid saturated with silica.

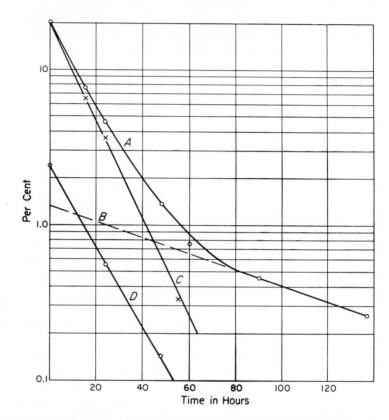

Fig. 12-2.   Dissolution curves of aluminum and magnesium in montmorillonite, Clay Spur, Wyoming. (*A*) Total aluminum, (*B*) tetrahedral aluminum, (*C*) octahedral aluminum, (*D*) octahedral magnesium. (After Osthaus.[20])

Osthaus[20] found that the solution of ions from the lattice of montmorillonites, as determined at several different temperatures in one sample and at two acid concentrations in another sample, was in agreement with chemical kinetics and was a first-order reaction. The rate constant for the acid-solution reaction increased proportionately with the acid concentration. Activation energies for single samples of montmorillonite and nontronite were 17 to 18 kcal/mole in both cases.

Recent studies by Langston and Jenne[22] and Miller[23] indicated that the order rate law is not necessarily the same for all clay minerals or even for members of the same species. Wide variations in rate constants were found for specimens of the same clay mineral, depending on particle size, crystallinity, etc.

Kerr et al.[24] studied the degradation of hectorite by hydrogen ions and concluded that two consecutive first-order reactions occurred: (1) Strong acid underwent a rapid spontaneous reaction to yield weak acid, and (2) the resulting weak acid underwent a slow spontaneous reaction to yield a neutral clay. The following tentative mechanism was suggested:

The magnesium and silicon atoms at the crystal edge are each monohydroxylated. The rate determining step in the first reaction series consists of a proton attacking the magnesium hydroxyl resulting in the formation of water. The magnesium ion is instantly released in the lattice and a second proton becomes quickly attached to the highly nucleophyllic silicon-oxygen system. The resulting di-hydroxylated silica is assumed to be a mono-basic weak acid. This acid then undergoes a relatively slow first-order hydrolysis or depolymerization to be released from the edge of the lattice as the low molecular weight silicate or silicic acid. Thus, the exchangeable cation is ultimately the magnesium ion. Upon release of magnesium and silicon, the freshly exposed edge is identical with the crystal edge before attack; i.e., the new terminal magnesiums and silicon atoms are each monohydroxylated.

Cloos et al.[21] found that the heat of solution of a glauconite in $2\ N$ hydrochloric acid saturated with respect to silica was 20 kcal/mole and that the reaction was first-order. They reported similar values for kaolinites and montmorillonites. Neumann[25] studied the degradation of montmorillonite

[22] Langston, R. B., and E. A. Jenne, NaOH Dissolution of Some Oxide Impurities from Kaolins, Proceedings of the 12th National Clay Conference, pp. 633–647, Pergamon Press, New York (1964).

[23] Miller, R. J., Mechanisms for Hydrogen to Aluminum Transformations in Clays, *Soil Sci. Soc. Am. Proc.,* **29**: 36–39 (1965).

[24] Kerr, G. T., R. H. Zimmerman, H. A. Fox, Jr., and F. H. Wells, Degradation of Hectorite by Hydrogen Ion, *Natl. Acad. Sci., Publ.* 456, pp. 322–329 (1956).

[25] Neumann, B. S., The Thermal Stability of Acid-extracted Montmorillonites, *Proc. Intern. Clay Conf.,* Stockholm, **1**: 85–96 (1963).

caused by the extraction of structural cations by acids, using thermogravimetric analysis. She considered two hypotheses: one based on the coexistence of residual montmorillonite and silica gel as separate entities and another on the assumption that the acid attack takes place uniformly throughout the structure, leading to montmorillonite with various concentrations of vacant octahedral positions. She concluded that the evidence strongly favored the second hypothesis.

Murray[4] showed that under similar conditions of treatment the actual sulfate or phosphate compound resulting from treatment with sulfuric and phosphoric acid, respectively, is *almost* the same for all kaolinites and halloysite, regardless of variations in degree of crystallinity. There were, however, some differences which suggested to him that the structural attributes of the parent clay mineral exerts some influence on the nature of the resulting reaction product.

Decomposition by electrodialysis. Numerous investigators have indicated that electrodialysis of certain of the clay minerals may cause their decomposition. Thus Kelley[26] showed that, as the cations are replaced by H^+, aluminum moves from octahedral positions to exchange positions, and Hofmann and Giese[27] showed that, in general, unexchangeable cations are lost from within the lattice *before* all the exchangeable cations are replaced by H^+.

The amount of loss of unexchangeable cations during electrodialysis varies with the clay mineral. Magnesium-rich smectite minerals seem to be most susceptible, and Nutting[28] showed that all the magnesium can be removed from hectorite by electrodialysis with complete breakdown of the structure. Caldwell and Marshall[29] showed that the same results are obtained with saponite. Other smectites are quite susceptible to such alteration, but less so than saponite and hectorite. The iron-rich smectites are more susceptible than those high in aluminum.

Attapulgite-sepiolite, vermiculite, chlorite, and some of the biotite micas are quite susceptible to disintegration by electrodialysis. Roy[30] showed that biotite suffers appreciable loss of cations after only a few hours of electrodialy-

[26] Kelley, W. P., The Agronomic Importance of Calcium, *Soil Sci.*, **40:** 103–109 (1935).

[27] Hofmann, U., and K. Giese, Ueber den Kationenaustausch an Tonmineralien, *Kolloid-Z.*, **87:** 21–36 (1939).

[28] Nutting, P. G., A Study of the Association of Magnesia with Silica in a Pure Magnesium Clay, *J. Wash. Acad. Sci.*, **30:** 233–237 (1940).

[29] Caldwell, O. G., and C. E. Marshall, A Study of Some Chemical and Physical Properties of the Clay Minerals Nontronite, Attapulgite, and Saponite, *Missouri Univ., Coll. Agr. Res. Bull.* 354 (1942).

[30] Roy, R., Decomposition and Resynthesis of the Micas, *J. Am. Ceram. Soc.*, **32:** 202–210 (1949).

sis and that 80 to 90 percent are lost at the end of 28 days without, however, complete destruction of the biotite structure. After a similar length of time, muscovite and phlogopite micas lost only a small amount of their iron and alkalies and substantially no aluminum. It appears that kaolinite is relatively little affected by electrodialysis, and data are not at hand for halloysite.

Electrodialysis has frequently been used to prepare H clays for cation-exchange studies and for use as the starting point for investigations of the physical properties of clays carrying specific exchangeable cations. Because of the likelihood of significant amounts of disintegration of the clay mineral, the procedure must be used with caution. Many of the investigations which have used this procedure are of little value, because the clay minerals were altered, and the cation composition was not what it was thought to be, as a result of very difficultly replaceable aluminum moving from the lattice to exchange positions.

Cation liberation by neutral salts. Mukherjee and his colleagues[31,32] showed that Al^{3+} and Fe^{3+} may be removed from certain hydrogen clays by repeated treatment with $BaCl_2$. They determined the amount of H^+, Al^{3+}, and Fe^{3+} in the leachate after leaching the clay minerals with $BaCl_2$. The clay minerals had been rendered acid by prior treatment with $0.02\ N$ HCl. After each leaching with $BaCl_2$, the clay minerals were rendered acid again before repeated leaching with $BaCl_2$. These investigators found that some Al^{3+} and Fe^{3+} continued to come out of the lattice of montmorillonite through at least eight cycles of acid treatment and leaching, whereas for kaolinite there was little effect after the first leaching. The removal of the Al^{3+} and Fe^{3+} from the montmorillonite was accompanied by a reduction in cation-exchange capacity, whereas the capacity of kaolinite was substantially unaffected. The work of Mukherjee shows the extreme sensitivity of some of the clay minerals to chemical treatment.

Rational analysis. It is frequently desirable, particularly in the industrial use of clays, as in the ceramic industry, to have some rapid, simple method of determining variations in the composition and properties of clays. Before the development of X-ray-diffraction, differential thermal, and improved optical methods about 1930, the only methods available were chemical procedures based on the difference in the resistance of various clay substances to chemical attack. There developed, therefore, so-called "rational methods of analysis"

[31] Mukherjee, J. N., B. Chatterjee, and B. N. Baverjee, Liberation of H^+, Al^{+++} and Fe^{+++} from Hydrogen Clay by Neutral Salts, *J. Colloid Sci.*, **2**: 247–256 (1947).

[32] Mukherjee, J. N., B. Chatterjee, and A. Roy, Liberation of H^+, Al^{+++} and Fe^{+++} from Pure Clay Minerals on Repeated Soil Treatment and Desaturation, *J. Colloid Sci.*, **3**: 437–446 (1948).

based primarily on the difference in solubility of various clay substances in acids. Such methods were widely used, e.g., in the ceramic industry, for raw-material control.

Most methods of rational analysis were based on the difference in solubility of clay components in hot concentrated hydrochloric and sulfuric acids. The part of a clay soluble in hydrochloric acid was frequently called allophaneton, or allophane clay, and the part insoluble in hydrochloric but soluble in sulfuric acid was called kaolinton, or kaolin clay. The allophaneton often was largely the smectite portion of a clay, together with other components soluble because of extremely fine grain size. The kaolinton was likely to include primarily the kaolinite and illite components. Frequently the method was elaborated to obtain some determination or estimation of the potash content as a basis for an evaluation of a possible mica component. Also in some instances a value for silica was obtained to indicate the quartz content. The wide variations in the methods of rational analysis that have been used were summarized by Harkort and Harkort.[33]

Washington[34] and later Correns[35] pointed out the great fallibility of rational analytical procedures because of the tremendous range in solubility of the components of clays as a consequence of particle-size variations. Since about 1930, with the development of other analytical methods, the use of rational analysis has gradually decreased. It should be pointed out, however, that such methods may be quite satisfactory for a specific limited purpose if used with caution and only for a limited range of clays. Thus an industry using only a particular kind of clay can set up an arbitrary set of chemical tests, which would be rapid and simple, to check certain variations in the material supplied to them.

Solubility of clay minerals in alkalies. Except for the investigations of Nutting,[1] there has been very little study of the solubility of the clay minerals in alkalies. Nutting studied particularly the solubility of montmorillonite and attapulgite in dilute solutions of sodium carbonate. His results for montmorillonite, given in Table 12-2 and Fig. 12-3 show that for this clay mineral the amount of silica removed in solution increases as the concentration of sodium carbonate increases to a maximum at about 0.025 percent sodium carbonate concentration. At higher concentrations, the solubility decreases to a minimum at about 0.05 percent sodium carbonate and then rises steadily

[33] Harkort, H., and H. J. Harkort, A Rapid Rational Analysis, *Sprechsaal,* **65**: 705–707, 723–726, 739–741 (1942).

[34] Washington, H. S., The Calculation of the "Rational Analysis" of Clays, *J. Am. Ceram. Soc.,* **I**: 405–421 (1918).

[35] Correns, C. W., Ueber die Bestandteile der Tone, *Z. Deut. Geol. Ges.,* **85**: 706–711 (1933).

Table 12-2   Silica dissolved from montmorillonite (Wyoming bentonite) by
solutions containing sodium carbonate
(after Nutting[1])

Percent Na_2CO_3	0.0	0.005	0.01	0.02	0.05	0.1	0.2	0.5	1.0
SiO_2 in solution, g/liter*	0.064	0.067	0.132	0.178	0.133	0.260	0.397	0.598	0.700

* After digestion of 48 g of clay in 2-liter solutions for 48 hr at 90°C.

to 0.70 g/liter at 1 percent sodium carbonate concentration. According to Nutting, only silica is dissolved by the alkali up to a concentration of 0.05 percent sodium carbonate. He pointed out that an extremely dilute solution of alkali is very effective in removing silica from montmorillonites and, given enough time, probably could remove all of it.

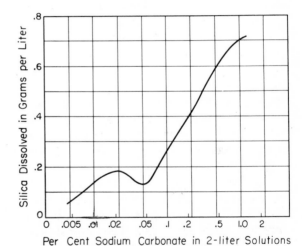

Fig. 12-3. Solubility of silica from montmorillonite (Wyoming bentonite) in sodium carbonate solutions. (After Nutting.[1])

Additional data for the removal of silica from montmorillonite by sodium carbonate are given in Table 12-3 together with comparable data for attapulgite. The results for the two clay minerals are much the same, except that the relative amount of silica dissolved is lower for the attapulgite.

Foster[36] studied the solubility of montmorillonite in sodium hydroxide

[36] Foster, M. D., Geochemical Studies of Clay Minerals, III, The Determination of Free Silica and Free Alumina in Montmorillonites, *Geochim. Cosmochim. Acta*, **3**: 143 (1953).

*Table 12-3   Milligrams of silica dissolved from 1
g of clay in various amounts of water containing
1 g of sodium carbonate*

(after Nutting[1])

Water, liters	Montmorillonite (Wyoming bentonite)	Attapulgite (Quincy, Florida)
10	169	32
5	481	253
2	215	192
1	104	79
0.5	70	87
0.2	86	50
0.1	100	59

of various concentrations and presented a method for determining free silica and alumina, based on solubility in 0.5 N solutions.

Infrared spectra of the clay minerals

General statement. Atoms grouped together in molecules do not remain at rest but are continuously in vibration. Such vibrations produce periodic displacement of atoms with respect to one another, causing a simultaneous change in interatomic distances. The frequencies of the vibrations fall within the range of 10^{13} to 10^{14} cycles per second, which is of the same order of magnitude as the frequencies of infrared radiations. Vibrations that are accompanied by a change in dipole moment give rise to the absorption of radiations in the infrared region of the electromagnetic spectrum. Several modes of vibration may occur for a particular atomic group, each at a characteristic frequency and normally independent of other modes.

If molecules of a substance whose vibrations are accompanied by a change of dipole moment are irradiated by a succession of monochromatic bands of infrared, those radiated frequencies that correspond to the intramolecular vibrational frequencies may be absorbed wholly or in part. If the percent of radiation absorbed by a substance is plotted against the incident wavelength (or frequency), the ensuing graph may be interpreted in terms of the intramolecular vibration. The graph, therefore, will be characteristic of the material and can be used in its identification. In addition, it should also provide data on the structure and bonding characteristics within the molecule.

Infrared absorption has been used extensively and successfully in the study of organic compounds. Its utility in mineralogy and particularly in clay mineralogy has been investigated extensively in recent years. For details regarding infrared absorption in general, the texts of Barnes et al.[37] and Herzberg[38] should be consulted. For the application to clay mineralogy, readers should refer to the pioneer work of Keller and Pickett[39,40] and Kerr and collaborators (see Adler et al.[41]) and the more recent work of Beutelspacher,[42] Serratosa and Bradley,[43] Tuddenham and Lyon,[44] Fripiat,[45] Stubican and Roy,[46] Miller,[47] Lyon,[48] Stubican,[49] Wolff,[50] and Farmer and Russell.[51] The latter publications have extensive bibliographies to the abundant recent work on this subject.

The range of infrared used in absorption studies in general is 4,000 to 400 cm^{-1} in frequency and 2.5 to 25 μ in wavelength. For clay-mineral investigations, wavelengths of 2 to 25 μ are used, since the characteristic spectra

[37] Barnes, R. B., R. C. Gore, U. Liddell, and V. Z. Williams, "Infrared Spectroscopy," Reinhold, New York (1944).

[38] Herzberg, G., "Molecular Spectra and Molecular Structures," Van Nostrand, Princeton, N. J. (1950).

[39] Keller, W. D., and E. E. Pickett, Absorption of Infrared Radiation by Powdered Silica Minerals, *Am. Mineralogist,* **34:** 855–868 (1949).

[40] Keller, W. D., and E. E. Pickett, Absorption of Infrared Radiations by Clay Minerals, *Am. J. Sci.,* **248:** 264–273 (1950).

[41] Adler, H., E. E. Bray, N. P. Stevens, J. M. Hunt, W. D. Keller, E. E. Pickett, and P. F. Kerr, "Infrared Spectra of Reference Clay Minerals," Rept. 8, American Petroleum Institute Project 49, Columbia University, New York (1950).

[42] Beutelspacher, H., Infrarot Untersuchungen an Bodenkolloiden, *Trans. 6th Intern. Congr. Soil Sci., Rept. B,* pp. 329–335 (1956).

[43] Serratosa, J. M., and W. F. Bradley, Determination of the Orientation of OH Bond Axes in Layer Silicates by Infrared Absorption, *J. Phys. Chem.,* **62:** 1164–1167 (1958).

[44] Tuddenham, W. M., and R. F. P. Lyon, Relation of Infrared-red Spectra and Chemical Analysis for Some Chlorites and Related Minerals, *Anal. Chem.,* **31:** 377–380 (1959).

[45] Fripiat, J. J., Application of Infrared Spectroscopy to the Study of Clay Minerals, *Bull. Groupe Franc. Argiles,* **12:** 25–41 (1960).

[46] Stubican, V., and R. Roy, Isomorphous Substitution and Infrared Spectra of the Layer Lattice Silicates, *Am. Mineralogist,* **46:** 32–51 (1961).

[47] Miller, J. G., An Infrared Spectroscopic Study of the Isothermal Dehydroxylation of Kaolinite at 470°, *J. Phys. Chem.,* **65**(5): 800–804 (1961).

[48] Lyon, R. J. P., Minerals in the Infra-Red, Stanford Research Institute, Menlo Park, Calif. (1962).

[49] Stubican, V. S., The Interaction of Ions in Complex Crystals as Shown by Infrared Spectroscopy, *Penn. State Univ., Mineral Ind.,* **33**(2): 1–7 (1963).

[50] Wolff, R. G., Structural Aspects of Kaolinite Using Infrared Absorption, *Am. Mineralogist,* **48:** 390–399 (1963).

[51] Farmer, V. C., and J. D. Russell, The Infra-red Spectra of Layer Silicates, *Spectrochim. Acta,* **20:** 1149–1173 (1964).

occur in this range. Wavelength is conventionally expressed in terms of microns and frequency in terms of wave numbers or reciprocal centimeters. The relationship is shown by

$$1\mu = 10^{-4} \text{ cm or } \frac{10^{-4}}{1 \text{ cm}^{-1}}$$

Infrared absorption by powdered material cannot be measured satisfactorily unless scattering and reflection of the radiation from the tiny particles are reduced to a low value. It has been shown that the use of particles smaller than the minimum wavelength of radiation used, about 2μ, reduces these factors to a very low level and generally improves results. Early work was done by mounting the clay in an oil medium (Nujol), which caused difficulty by adding its own spectrum to that of the clay. Currently techniques are available for mounting the clay without any additive; for example, the specimen is dispersed in KBr and pressed into disks. Care must be taken to prevent distortion of the spectra in the preparation technique (Ruthberg et al.[52]).

Oriented aggregates of clays mounted on a substrate transparent to infrared are used to study bond direction.

Data for clay minerals. Infrared-absorption curves for a variety of clay minerals are given in Figs. 12-4 to 12-10. It has been established that the spectra are sensitive both to structural and compositional variations in the minerals, and numerous attempts (see especially Stubican and Roy[46] and Farmer and Russell[51]) have been made to correlate particular bonds and compositional variations with specific absorption bands. The following discussion, taken largely from these reports, is confined to broad generalities rather than details.

Frequencies between 3,500 and 3,750 cm^{-1}. Buswell and Dudenbostel[53] first showed that absorption in this range is due to hydroxyl groups.

In dioctahedral minerals, pyrophyllite can be considered as the parent mineral and the other members as derived from it by ionic substitution. In this structure each pair of aluminum ions shares two hydroxyl groups which are related by a center of symmetry between the aluminum ions. According to Farmer and Russell,[51] coupling between the vibrations of the two hydroxyl groups can give rise to two frequencies of vibration: one in which the two hydroxyl groups are $180°$ out of phase and the other an in-phase vibration. The antisymmetric (out of phase) is infrared-active and is the origin of the

[52] Ruthberg, S., M. W. Barnes, and R. H. Noyce, Alteration of the Muscovite Spectrum by KBr Pellet Preparation, *Appl. Opt.*, **2**(2): 177 (1963).

[53] Buswell, A. M., and B. F. Dudenbostel, Spectroscopic Studies of Base Exchange Materials, *J. Am. Chem. Soc.*, **63**: 2554–2558 (1941).

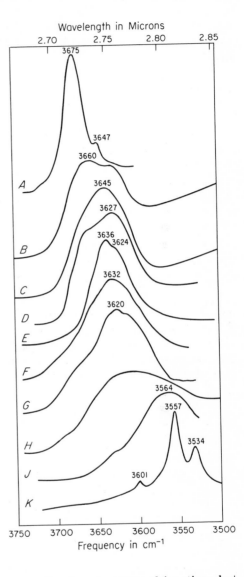

Fig 12-4. Hydroxyl absorption of randomly oriented samples of (A) pyrophyllite, (B) beidellite, (C) rectorite, (D) muscovite or paragonite, (E) margarite, (F) Wyoming montmorillonite, (G) Skrevedalen montmorillonite, (H) Woburn montmorillonite, (J) nontronite, and (K) ferric celadonite. (After Farmer and Russell.[51])

strong band at 3,675 cm^{-1}. Symmetric vibration is infrared-inactive, but the weak band at 3,647 cm^{-1} could arise from this vibration if the symmetry of the structure is not ideal. An alternative assignment is that it arises from hydroxyl groups shared between aluminum and iron as the octahedral layer may contain a small proportion of iron. Progressively increasing substitutions of aluminum for silicon in the tetrahedral layer of the pyrophyllite structure with the introduction of balancing interlayer cations gives rise to two absorption bands at 3,660 and 3,627 cm^{-1} (see Fig. 12-4B and D for beidellite and muscovite) with the lower-frequency band increasing in intensity with increasing aluminum substitution and the higher-frequency component decreasing.

The displacement of the OH stretching vibration to lower frequencies

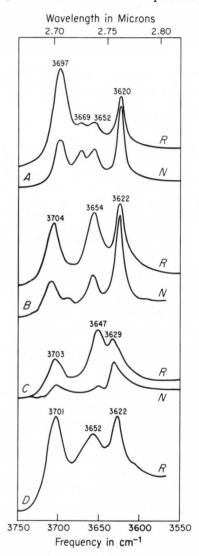

Fig. 12-5. Hydroxyl absorption of (*A*) kaolinite, (*B*)dickite, (*C*) nacrite, (*D*) Pugu kaolin. R indicates randomly oriented specimens, and N, films at normal incidence. (After Farmer and Russell.[51])

associated with band broadening in the smectites (see Fig. 12-4*F,G,H*) is commonly ascribed to hydrogen bonding, but this explanation is unlikely. The breadth of both components in these minerals compared with the narrowness of the hydroxyl absorption in pyrophyllite can reasonably be ascribed to structural irregularities arising from the random nature of the aluminum for silicon substitution, leading to a range of hydroxyl-oxygen distances. In general, substitution of magnesium and/or iron for aluminum in the octahedral position tends to cause a broad absorption band in this frequency range,

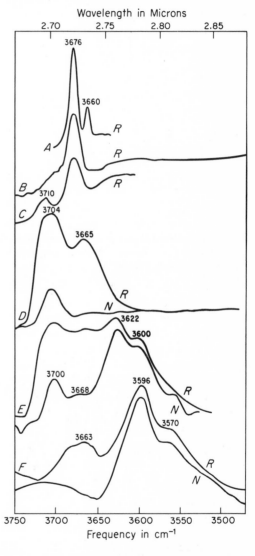

Fig. 12-6. Hydroxyl absorption of (A) talc, (B) hectorite, (C) saponite, (D) phlogopite (126 μ), (E) green biotite (115 μ), and (F) dark-brown biotite (25 μ). The mica spectra were obtained from single flakes of the thickness indicated at normal (N) and 45° incidence (R). (After Farmer and Russell.[51])

with a maximum at a lower frequency. For example, in nontronite, the displacement is about 90 cm^{-1}. According to Farmer and Russell,[51] the pattern of absorption is not affected by changing the interlayer cation and is only slightly changed by removing interlayer water.

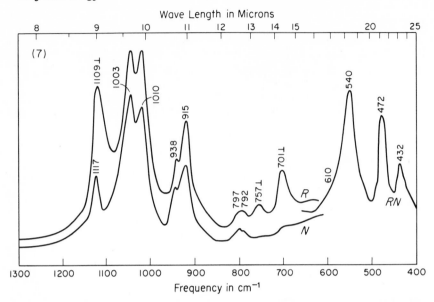

Fig. 12-7. Infrared spectrum of kaolinite, using randomly oriented samples (R) and oriented layers at normal incidence (N). RN indicates spectra independent of orientation. Bands showing perpendicular polarization are indicated. (After Farmer and Russell.[51])

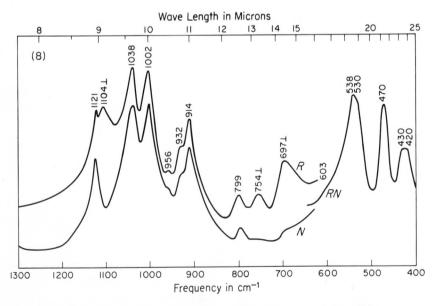

Fig. 12-8. Infrared spectrum of nacrite. (After Farmer and Russell.[51])

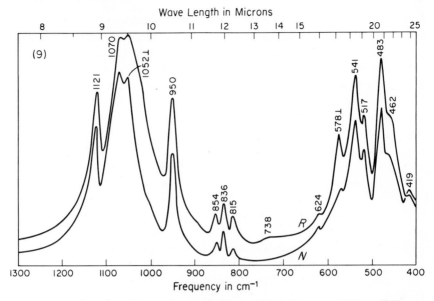

Fig. 12-9. Infrared spectrum of pyrophyllite. (After Farmer and Russell.[51])

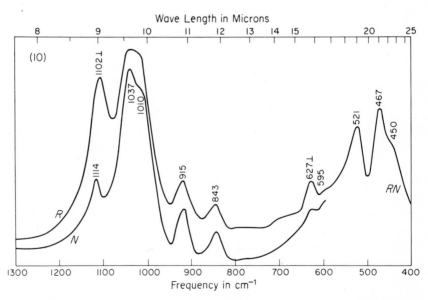

Fig. 12-10. Infrared spectrum of Skrevedalen montmorillonite. (After Farmer and Russell.[51])

Serratosa[54] thought that the absorption frequency of O—H bonds depends on the degree of association of these groups. For free groups the frequency is around 3,700 cm^{-1}, with the frequency decreasing under the influence of associations, the decrease being a measure of the strength of the hydrogen bond. As a general rule, the layer-silicate structural OH groups that are comparatively slightly associated show absorption of high frequencies of 3,600 to 3,700 cm^{-1}, whereas the adsorbed water shows adsorption at lower frequencies, 3,400 cm^{-1}, and another band around 1,640 corresponding to the deformation vibration of water. A sample of montmorillonite and nontronite showed a broad band around 3,400 cm^{-1} modified by a weak shoulder at 3,240 cm^{-1} and a simpler band at 1,640 cm^{-1}, all of which disappeared on heating to 200°C and are correlated with the adsorbed water. Both minerals showed a strong absorption band between 3,600 and 3,700 cm^{-1} which was not lost on heating to 200°C and is correlated with structural OH groups. Both minerals show several absorption bands between 1,900 and 650 cm^{-1} which are substantially the same for both minerals except with the most significant difference in the band at 915 cm^{-1} for montmorillonite and 820 cm^{-1} for nontronite, thereby indicating the influence of iron in the structure. On heating to 700°C there is a progressive decrease in the 3,600- to 3,700-cm^{-1} band and the 915- and 820-cm^{-1} bands. These two absorption bands, therefore, can be assigned the same origin, that is, a vibration of protons associated with the octahedral layer. Each octahedral hydroxyl has two cation neighbors, and the frequency difference relates to the nature of the cations occupying the octahedrons, that is, whether it is aluminum or iron.

Roy and Roy[55] came to the conclusion that there was no simple correlation of OH absorption frequencies in the 3,000- to 5,000-cm^{-1} range with types of OH bonding or the extent of hydrogen bonding in the kaolin minerals. However, Newnham[56] assigned the absorption band at 3,700 cm^{-1} in dickite to a "free" hydroxyl bond, that is, the hydroxyl groups in which the oxygen alone affects the hydrogen. The band at 3,644 cm^{-1} was assigned to a bond between basal hydroxyls of one sheet and the puckered oxygens of the next sheet, and the band at 3,617 cm^{-1} was attributed to interlayer hydrogen bonding.

Miller[47] concluded that the bands at 3,718 and 3,677 cm^{-1} for kaolinite

[54] Serratosa, J. M., Dehydration and Rehydration Studies of Clay Minerals by Infrared Absorption Spectra, Proceedings of the 9th National Clay Conference, pp. 412–418, Pergamon Press, New York (1962).

[55] Roy, D. M., and R. Roy, Hydrogen-Deuterium Exchange in Clays and Problems in the Assignment of Infra-red Frequencies in the Hydroxyl Region, *Geochim. Cosmochim. Acta*, 11: 72–85 (1957).

[56] Newnham, R. E., A Refinement of the Dickite Structure and Some Remarks on Polymorphism in Kaolin Minerals, *Mineral. Mag.*, 32: 683–704 (1961).

are due to stretching vibrations of relatively unassociated O—H groups, i.e., due to hydroxyl groups nearly free of hydrogen bonding to other atoms. The band at 3,677 cm^{-1} is attributed to intralattice OH groups. The band at 3,718 cm^{-1} is attributed to basal plane hydroxyl groups which are probably involved in at least two different kinds of associations: (1) bonds between groups within the hydroxyl plane, which should be of a strength equal to those of the intralattice hydroxyls, and (2) bonds involved in cementing the unit layers together, which would be considerably weaker because of their large O-O distance. Therefore, some of the basal plane hydroxyls would contribute to one band and others contribute to the second band, the relative contribution depending upon the regularity of the stacking of the unit layer, i.e., upon the degree of order or crystallinity of the kaolinite.

Wolff[50] studied the OH stretching vibrations for kaolinite and found that there was no difference in the intensity of the 3,704-, 3,635-, and 3,663-cm^{-1} absorption bands when samples were heated to 200°C. The 3,704-cm^{-1} band showed considerable orientation parallel to the *c* axis. Variations in the angle of incidence of the infrared beam caused substantial variation in the intensity of the 3,704-cm^{-1} band, only slight variation for the 3,663-cm^{-1} band, and substantially none for the 3,635-cm^{-1} band. Wolff interpreted his data somewhat differently from Newnham. He considered that the 3,704-cm^{-1} frequency is due to OH groups in the layer common to the dioctahedral and tetrahedral layers, with the proton end pointing toward the hole formed by the oxygens of the basal tetrahedral sheet, and also by hydroxyl groups located in the basal octahedral layer of the opposing sheet which are pointing away from the octahedral layer toward the same hole, as these hydroxyls would meet the requirements of "free" hydroxyl. The 3,633-cm^{-1} band is indicative of interlayer hydrogen bonding in which a major component of the bonding is perpendicular to the basal planes. The 3,665-cm^{-1} band is indicative of the hydroxyl bonds in which the hydrogen end is pointing into vacant octahedral sites, resulting in a bond orientation for which the major component is parallel to the basal plane.

According to Farmer and Russell[51] (Fig. 12-5), the pattern of absorption in kaolinite can be explained by ascribing the 3,697-cm^{-1} band to the symmetrical vibration of the gibbsite-like sheet, the bands at 3,669 and 3,652 cm^{-1} to the in-plane vibrations, with their degeneracy lifted, while the 3,620-cm^{-1} absorption arises principally from the hydroxyl group on the side of the tetrahedral sheet.

The application of infrared spectrometry to determine bond direction has been very useful in the case of the clay minerals. When the dipole is oriented parallel to the direction of the propagation of the infrared beam no absorption takes place. If the hydroxyl band is oriented so that the dipole has a component lying across the direction of propagation or parallel to the direction of vibration of the infrared, the absorption takes place when the frequency of vibration corresponds to the resonance frequency of the O—H bonds. Serratosa and Bradley[57] showed that in the trioctahedral mica, phlogopite, the OH stretching band at 3,700 cm^{-1} is highly sensitive to the orientation of a single crystal, indicating that the hydroxyl ions are perpendicular to the OH plane. On the other hand, the data for muscovite, which has a single absorption band at 3,600 cm^{-1}, indicated that the hydroxyl ions are inclined to the OH plane.

In the trioctahedral mineral talc, as in pyrophyllite, the main absorption band at 3,676 cm^{-1} has a weak satellite of uncertain origin at 3,660 cm^{-1}. According to Farmer and Russell,[51] partial substitution of aluminum for silicon, as in saponite, or lithium for magnesium, as in hectorite (Fig. 12-6), leaves the main absorption unchanged in frequency although it is broadened and the lower-frequency satellite is lost. With considerable substitution of aluminum for silicon, as in saponite, a higher-frequency component develops at 3,710 cm^{-1} which increases in strength as the substitution increases. With substantial substitutions of iron for magnesium, as in the case of some biotites, relatively weak absorption bands, insensitive to orientation of the flake, develop at 3,622 and 3,600 cm^{-1}. With increasing amounts of iron it appears that these bands shift to lower frequencies and are more intense. Bassett[58] also noted the tendency of hydroxyl absorption in biotites to shift to lower frequencies with increasing iron content.

Frequencies between 1,150 and 400 cm^{-1}. In this region Farmer and Russell[51] described what they called lattice vibrations. In this range the layer silicates in which aluminum-for-silicon substitution is absent or low give the sharpest spectra. Figure 12-11, after Stubican,[49] shows this investigator's assignment of bands in this region. Table 12-4, after Farmer and Russell,[51] shows their assignment of bands for two montmorillonites and a nontronite.

Minerals with low aluminum-for-silicon substitution show a broadly similar pattern of absorption in the 1,150- to 960-cm^{-1} region, arising from Si-O stretching vibrations. Comparison of the spectra of oriented and random samples show three well-defined medium or strong in-plane vibrations and one perpendicular vibration for each of the samples.

The position and sharpness of the perpendicular vibration vary with physical state; thus in the spectra of kaolinite of large crystal size, this node appears as a broad shoulder near 1,080 cm^{-1}. The band intensifies and shifts to higher frequencies in the spectra of smaller crystals until in very finely ground material the band is at 1,109 cm^{-1}. Smaller but distinct shifts in

[57] Serratosa, J. M., and W. F. Bradley, Determination of the Orientation of O—H Bond Axes in Layer Silicates by Infrared Absorption, *J. Phys. Chem.,* **62:** 1164–1167 (1958).

[58] Bassett, W. A., Role of Hydroxyl Orientation in Mica Alteration, *Bull. Geol. Soc. Am.,* **71:** 449–456 (1960).

frequency, together with sharpening and intensification of the bands as particle size decreases, also occur with the other less intense perpendicular nodes of the kaolinite minerals, but the absorption of the in-plane nodes is not affected. There is evidence too that the position and appearance of the 1,107-cm⁻¹ band are very sensitive to the regularity of the structure. Thus, it is broad

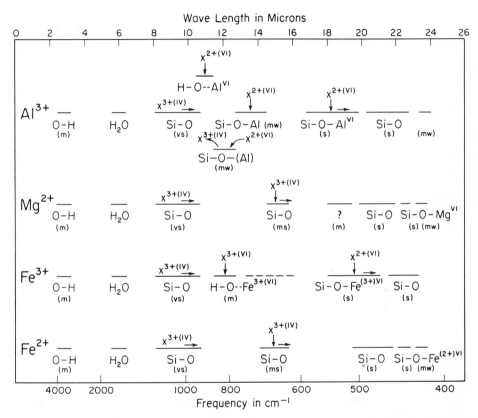

Fig. 12-11. Schematic summary of the assignments of the infrared-absorption bands in layer-lattice silicates containing basically Al^{3+}, Mg^{++}, Fe^{3+}, or Fe^{++} in the octahedral sites. Relative intensities of the bands are described as vs = very strong, s = strong, m = medium, and w = weak. (After Stubican.[49])

and diffuse in halloysite. In one of the montmorillonites examined, this same band is well defined, but this is rather unusual as the band is not so well developed in published spectra of other montmorillonites.

In pyrophyllite the frequency of the perpendicular node is at 1,052 cm⁻¹, which is considerably lower than that found in the kaolinite minerals (about 1,109 cm⁻¹). This difference cannot be related to the fact that the pyrophyllite is triformic whereas the kaolin minerals are diformic, since montmorillonite, which is triformic, has an absorption pattern in this region close to that of kaolinite. It must be concluded, therefore, that the silicate layers in pyrophyllite have a significantly different geometry from those in montmorillonite and kaolinite minerals. As the vibration giving rise to this absorption involves the oxygen shared between tetrahedral and octahedral layers, its frequency might reasonably be expected to be influenced by the nature of the cations in the octahedral layer; however, it is found that partial substitution

Table 12-4 *Frequency* (*cm⁻¹*) *of infrared-absorption bands*
(After Farmer and Russell)[51]

	Dickite	Wyoming montmorillonite	Woburn montmorillonite	Nontronite
Si—O stretching	1,120	1,120	1,117	1,091
	1,107 (p)	1,080 (sh p)	1,074 (p)	1,034 (p)
	1,034	1,048	1,038	1,017
	1,003	1,025	1,010	
	968			
R—O—H bending	936	920	919	848
	913	890	877	818
	795	849	839	786
	755 (p)	800	790 (p)	753
	699 (p)	622 (p)		680 (p)
	605 (sh)			
Si—O—R³⁺ and R³⁺—OH	540	523	568	587 (p)
	530 (sh)		521	493
	470	468	473	450
	431		430	430
	422 (sh)			

sh—shoulder; p—perpendicular polarization.

of magnesium for aluminum, as in montmorillonite, and complete replacement of aluminum by ferric, ferrous, and magnesium ions cause little change in this frequency from that found in the kaolinite minerals. Nontronite appears to be anomalous since it has a very-low-frequency perpendicular node lying at 1,034 cm⁻¹. In general, the effect of iron substitution seems to contribute to a greater diffuseness and to slightly lower frequency of the perpendicular node in some montmorillonites. In the trioctahedral minerals, talc, saponite, and hectorite, the frequency is in the range of 1,073 to 1,045 cm⁻¹.

The two strongest in-plane vibrations occur in the frequency range 1,037 to 1,018 cm^{-1} and are about the same in the kaolinite and montmorillonite minerals, but for pyrophyllite they lie at a much higher frequency (mean 1,061 cm^{-1}), again suggesting a significantly different structure.

The third high-frequency in-plane vibration in the 1,150- to 960-cm^{-1} region occurs within a relatively narrow range from 1,121 to 1,100 cm^{-1} in most of the minerals examined. Like the perpendicular node, the sharpness and definition of this absorption band are very dependent on the degree of order of the structure. For example, in halloysite as compared with kaolinite, it is less well defined.

Vibrations in the 960- to 550-cm^{-1} region are attributed to R-O-H bending vibrations. Comparison of the spectra of normal kaolinite and nontronite with synthetic minerals containing deuterium in place of hydrogen has established that the bands near 935 and 914 cm^{-1} in the kaolinite minerals and those at 848 and 819 cm^{-1} in nontronite arise from OH bending vibrations (Stubican and Roy[46]). By analogy the same assignment can be given to the absorption bands at 950 cm^{-1} in pyrophyllite, 942 cm^{-1} in beidellite, and 915 to 920 cm^{-1} in montmorillonite. Smectites with a high iron content show an absorption band at 870 cm^{-1}, and there appears to be a correlation between the intensity of this band and iron content. Also this band appears to be associated with ferric iron, as it has been found that it is reduced in intensity or eliminated in synthetic samples prepared under reducing conditions.

Possibly the 848-cm^{-1} band in nontronites, like the 870-cm^{-1} band in montmorillonites, arises from hydroxyl groups shared between ferric iron and aluminum in the octahedral layer. A bending frequency of hydroxyl groups shared between aluminum and magnesium might be expected in the spectra of montmorillonites and the band at 850 cm^{-1} may have this assignment although this is by no means certain.

Stubican[49] suggested assignments for vibrations in the region from 800 to 600 cm^{-1}. Farmer and Russell[51] also discussed the assignment of bands in this region; they concluded that further work is required before assignments can be made with confidence.

The strong absorption in the region below 550 cm^{-1} must arise principally from in-plane vibrations of the octahedral ions and their adjacent oxygen layers. These vibration couples give nodes which can equally well be described as metal-oxygen stretching or silicon-oxygen bending. Farmer and Russell[51] considered in detail the correlation of these bands with structural attributes. They stated that substitution of aluminum for silicon in the tetrahedral layers does not greatly shift the main absorption bands in this region, and it must be concluded that the pattern of absorption is largely determined by the nature and distribution of the octahedral cation. The presence of iron and magnesium replacing aluminum causes a shift of these absorption bands to lower frequency and/or the development of a new absorption band, as in the case of nontronite at 430 cm^{-1}.

Farmer and Russell[51] stated that the effect of substitutions of aluminum for silicon is particularly pronounced on the perpendicular and in-plane Si-O frequencies in the 1,150- to 950-cm^{-1} range. In nontronite and beidellite, as compared with montmorillonite, the bands are much more poorly defined. Also, these bands are very broad in muscovite and phlogopite. Farmer and Russell suggested an explanation for the variation in this region, with aluminum replacing silicon, based on the concept of the rotation of the silica tetrahedrons in the tetrahedral layer suggested by Radoslovich and Norrish.[59]

Stubican and Roy[46] noted that increasing aluminum-for-silicon substitution gives rise to increased strength of absorption in the 600- to 900-cm^{-1} region. They discussed the assignment of bands in this region, based on their study of synthetic layer silicates. The following statements are from their publications.

The spectra obtained for saponites are comparable to those for talc, but with increasing aluminum content there is an increasing intensity of the weak band at 877 cm^{-1} as well as the gradual appearance of the weak bands between 800 and 850 cm^{-1}. At the same time, the absorption bands at 527, 467, and 427 cm^{-1} become more diffuse with slight changes in their position. The most radical change can be observed with the Si-O band of talc and the frequency center at 668 cm^{-1} which decreases in intensity and moves toward lower frequencies as the aluminum content increases.

In beidellites, with increasing substitution of aluminum for silicon in tetrahedral sites, the difference in the intensities of the bands at 477 and 535 cm^{-1} decreases. These changes in intensity can be correlated with the amount of aluminum ions in tetrahedral sites, but the precision is low. In montmorillonites, with increasing amounts of magnesium, the absorption band assigned to H-O-Al at 935 cm^{-1} becomes more poorly resolved from the main Si-O band, and the absorption band at 533 cm^{-1} shows a small shift toward higher frequencies. The latter band decreases in intensity as the amount of aluminum ions in octahedral sites decreases and presumably could be generated by Si-O-Al^{VI} vibrations.

With the entry of magnesium into the octahedral sites of dioctahedral muscovite, the absorption bands with the frequency 935 cm^{-1} become more poorly defined. Also the presence of magnesium ions causes the deactivation of vibrations which give rise to absorption bands in the region 600 to 850 cm^{-1}. The displacement of the absorption band of muscovite at 543 cm^{-1} toward lower frequencies is very marked and can be correlated with the presence of magnesium ions in octahedral sites.

The substitution of the trivalent ions in the tetrahedral sites results in the displacement of the main Si-O stretching bands (900 to 1,100 cm^{-1}) toward

[59] Radoslovich, E. W., and K. Norrish, The Cell Dimensions and Symmetry of Layer-Lattice Silicates, *Am. Mineralogist*, **47**: 599–616 (1962).

lower frequencies; this is independent of the ions present in the octahedral site. The decrease is linear. Similar results with natural micas have been reported by Tuddenham and Lyon.[44] With the trioctahedral triformic family the most radical change caused by increasing amounts of aluminum ions in the tetrahedral sites is a gradual decrease in frequency and intensity in the Si-O band with the frequency center at 668 cm⁻¹. The substitution in octahedral sites of divalent ions for aluminum markedly influences the absorption bands of the type Si-O-Al in the region of 500 to 800 cm⁻¹. The strong absorption band Si-O-Al^{VI} at 538 cm⁻¹ strongly decreases in intensity and moves toward lower frequencies.

In the octahedral sites of dioctahedral minerals with increasing amounts of magnesium, the H-O-Al absorption band at 935 cm⁻¹ decreases in intensity and is more poorly resolved, and the frequency and intensity of the Si-O-Al^{VI} band at 535 cm⁻¹ decrease. With trioctahedral clay minerals the above mentioned bands are absent.

Tuddenham and Lyon[44] gave a table of the positions of infrared-absorption bands for a considerable series of chlorite samples. In a sample with low aluminum values there are two major absorption bands in the 1,100- to 900-cm⁻¹ region. As the aluminum content increases, three absorption bands appear. With further increases in aluminum there is a single absorption band, and finally at high aluminum contents two absorption bands. This relationship seems to hold for chlorites with high and low iron content. Also the absorption band nearest 1,000 cm⁻¹ increases as the total iron content increases. Spectra for septachlorite are similar to the 14-Å mineral but sufficiently different in detail to permit the differentiation of these two forms.

Surface area

Many methods have been tried to determine the surface area of clay materials. The most trustworthy results have been obtained with the method developed by Brunauer, Emmett, and Teller,[60,61] which depends on the adsorption of simple molecules, such as nitrogen, at temperatures in the neighborhood of their boiling point. A comprehensive discussion of the basic theory of physical adsorption and surface-area measurement was presented by

[60] Brunauer, S., and P. H. Emmett, The Use of Low Temperature van der Waals Adsorption Isotherms in Determining the Surface Area of Various Adsorbents, *J. Am. Chem. Soc.,* **59:** 2682–2686 (1937).

[61] Brunauer, S., P. H. Emmett, and E. Teller, Adsorption of Gases in Multimolecular Layers, *J. Am. Chem. Soc.,* **60:** 309–319 (1938).

Brunauer.[62] Emmett, Brunauer, and Love[63] and Makower, Shaw, and Alexander[64] first applied the method to soils and clay materials.

Obviously, specific surface is not a characteristic of any particular clay mineral, since it varies with particle-size distribution, particle shape, and the presence of cracks and pores in the sample. The data given in Table 12-5, taken from Nelson and Hendricks,[65] are intended to give comparative values of the order of magnitude of the surface area of some clay minerals. The low surface-area value obtained for montmorillonite is interesting in view of the fine-grain and the expanding-lattice characteristics of the mineral.

Table 12-5   Specific-surface area of some clay minerals
(From Nelson and Hendricks[65])

Mineral	Heat* treatment, °C	Loss in weight, %	Specific surface, m²/g
Kaolinite, −0.3μ	30	....	15.5
	200	0.2	15.3
	500	12.7	18.1
	700	14.3	16.5
	900	14.6	1.5
Illite, −0.3μ	30	....	97.1
	200	0.4	92.2
	500	5.8	91.6
	700	7.7	80.4
Montmorillonite	30	....	15.5
Halloysite	30	....	43.2

* Heated at temperature given until no further weight loss.

The low value for montmorillonite means that the major part of the potential surface of the mineral could not be reached by the nitrogen or other gases used in the determinations. The theoretical surface of montmorillonite on dispersion to nearly unit-cell dimensions is 8×10^6 cm²/g.

Escard,[66] using the BET method, reported the following values in square

[62] Brunauer, S., "The Adsorption of Gases and Vapours," Princeton, Princeton, N.J. (1943).

[63] Emmett, P. H., S. Brunauer, and K. S. Love, The Measurement of Surface Area of Soils and Soil Colloids by the Use of Low Temperature van der Waals Adsorption, *Soil Sci.*, **45**: 57–65 (1938).

[64] Makower, B., T. M. Shaw, and L. T. Alexander, The Specific Surface and Density of Some Soils and Their Colloids, *Soil Sci. Soc. Am. Proc.*, **2**: 101–109 (1937).

[65] Nelson, R. A., and S. B. Hendricks, Specific Surface of Some Clay Minerals, Soils, and Soil Colloids, *Soil Sci.*, **56**: 285–296 (1944).

[66] Escard, J., Adsorption des gaz, *Bull. Groupe Franc. Argiles,* **3**: 79–84 (1952).

meters per gram: attapulgite, 140; illite, 113; kaolinite, 22; montmorillonite, 82; nontronite, 72; and sepiolite, 392. The values for montmorillonite and nontronite are for external surfaces of dehydrated samples.

The values in Table 12-5 show that only a small amount of surface is lost when kaolinite and illite are dehydrated.

Hendricks and Dyal[67] used the adsorption of ethylene glycol for surface-area determinations. The organic is adsorbed on both external surfaces and inner surfaces accessible only by swelling. A combination of the Brunauer, Emmett, and Teller method with that of Hendricks and Dyal provides values for both external and swelling surfaces.

Haynes[68] determined the surface area of four samples containing kaolinite by various methods, including adsorption with various gases, air permeability, light extinction, etc., and found quite different results with different methods. The author quite accurately pointed out that it is incorrect to speak of disagreement between the various results since the different methods do not measure the same quantity. He pointed out further that this fact can be put to good use in evaluating the fundamental factors responsible for variation in the physical properties of clays.

Density

Determinations of the density of the clay minerals must be made on aggregates of fine particles. The measurement is inherently difficult, because of the likelihood of incomplete penetration of the liquid used into the interstices of the aggregate mass and because of the possibility of a chemical or physical change during the measurements as a consequence of adsorption of the liquid or an exchange reaction between the liquid and the clay mineral. It is particularly difficult to measure the density of the clay minerals in their natural state with their adsorbed water, which, in some of the clay minerals, such as smectites and halloysite ($4H_2O$), is an intregral part of their structure. For these clay minerals the density of the dehydrated material can be determined, but it does not represent the mineral as it occurs naturally.

In the case of the clay minerals which are subject to isomorphous substitution within the lattice, e.g., smectite, illite, etc., the density varies from

[67] Hendricks, S. B., and R. S. Dyal, Surface Measurement by Ethylene Glycol Retention of Clays and Its Application to Potassium Fixation, *Trans. 4th Intern. Congr. Soil Sci.*, **2:** 71–72 (1950).

[68] Haynes, J. M., The Specific Surface of Clays, *Trans. Brit. Ceram. Soc.*, **60:** 691–707 (1961).

one specimen to another. For such clay minerals, there is a range of characteristic density values rather than a single value.

Caillere and Henin[69] described a simple procedure for computing density, based on chemical composition and the dimensions of the unit cell.

Kaolinite. According to Gruner,[70] the theoretical density of kaolinite computed from the structure is 2.609. Values given in the literature vary from 2.60 to 2.68, with 2.63 frequently quoted in mineralogical textbooks. Because of the low adsorptive capacities of kaolinite and the general absence of isomorphous substitutions, the value for this mineral is a fairly definite and easily determined quantity.

Halloysite. Values of 2.55 and 2.56 are given by Makower and colleagues[64] for completely evacuated material, which was probably close to the $2H_2O$ form. The dehydrated form of the mineral ($4H_2O$) would have a lower density, and values given in the literature[71] for halloysites in general are as low as 2.0 to 2.2.

Illite. Values given by Dana[72] are 2.76 to 3.0 for muscovite and 2.7 to 3.1 for biotite. Illites would be expected to be in this range, but perhaps on the low side because of deficiencies of interlayer ions and slight replacement of such ions by hydration. DeWit and Arens[73] gave values of 2.642 to 2.688 for materials with 0.0 percent adsorbed water. These authors determined values for the decrease in density of two illites as they pick up adsorbed water in increasing relative humidities (Table 12-6). A certain amount of such change is to be expected, but a large amount of it would probably indicate the presence of interlayer montmorillonite or vermiculite.

Smectite. Makower and coworkers[64] gave a value of 2.53 for a low-iron smectite and 2.74 for one with a somewhat higher iron content (3.6 percent). DeWit and Arens[73] gave a value of 2.348, and Caldwell and Marshall[29] gave values of 2.2 to 2.7 for nontronite and of 2.24 to 2.30 for saponite. These values are for essentially dehydrated material, and it may be concluded that

[69] Caillere, S., and S. Henin, "Mineralogie des argiles," Masson et Cie, Paris (1963).

[70] Gruner, J. W., Densities and Structural Relationships of Kaolinites and Anauxites, *Am. Mineralogist,* **22:** 855–860 (1937).

[71] Bosazza, V. L., The Specific Gravity of Clays, *S. African J. Sci.,* **36:** 155–157 (1939).

[72] Dana, E. S., "Textbook of Mineralogy," W. E. Ford, ed., Wiley, New York (1921).

[73] DeWit, C. P., and P. L. Arens, Moisture Content and Density of Some Clay Minerals and Some Remarks on the Hydration Pattern of Clay, *Trans. 4th Intern. Congr. Soil Sci.,* **2:** 59–62 (1950).

the density of such montmorillonite ranges from about 2.2 to at least 2.7. Probably 2.7 is too low for the substantially pure, iron-end member of the series.

The data in Table 12-6, after DeWit and Arens, indicate the large change

Table 12-6   Density determinations of clay minerals after water adsorption at various relative humidities (After DeWit and Arens[73])

Mineral	Relative humidity	Moisture content, oven-dry basis, %	Density
Montmorillonite (thixotron)	. . . .	0.0	2.348
	0.00	. . . . .	2.608
	0.25	11.6	2.199
	0.50	16.6	2.134
	0.75	28.4	2.001
	1.00	46.0	1.772
Illite (Holland)	. . . .	0.0	2.649
	0.0	. . . . .	2.660
	0.25	3.0	2.579
	0.50	4.85	2.504
	0.75	6.9	2.422
	1.00	18.8	2.128
Illite (Maquoketa shale, Illinois)	. . . .	0.0	2.642
	0.0	. . . . .	2.688
	0.25	8.3	2.409
	0.50	12.0	2.310
	0.75	24.8	2.02
	1.00	76.0	1.48
Kaolinite "Brocades"	. . . .	0.0	2.667
	0.0	. . . . .	2.682
	0.75	0.4	2.663
	0.50	0.8	2.651
	0.75	0.9	2.642
	1.06	6.5	2.427

in density of the mineral with the increase in adsorbed water, as the measurements were made following treatment at increasing relative humidities.

Vermiculite. Precise density data are not available for clay-mineral vermiculite, but the oven-dried material would be expected to have about the density of biotite mica. Because of its hydration properties, natural material would give a lower value than biotite mica, and the density would decrease with increasing adsorbed water, as does that of montmorillonite. The minimum value would be higher for vermiculite than for montmorillonite, since the amount of adsorbed water is restricted in vermiculite.

Chlorite. Dana[72] gave values of 2.6 to 2.96 for the chlorite minerals. The clay-mineral chlorites would be expected to give values in this range. Like the illites as compared with muscovite and biotite, they would probably be in the low part of this range. The very low water-adsorptive properties of the mineral would mean that the density of air-dried material would differ little from that of natural material, and there would be little change as determinations were made following treatment at varying relative humidities.

Attapulgite-sepiolite-palygorskite. Caillere[74] gave a value of 2.08 for sepiolite and 2.29 and 2.36 for two samples of palygorskite. These values indicate the probable range for these minerals, but more data are necessary before their density characteristics are known in any detail.

Dielectric constants

Ficai[75] measured the dielectric constants (ϵ) of samples of kaolinite, montmorillonite, halloysite, and illite (containing some montmorillonite) at frequencies within the range of 30 kHz to about 3,000 kHz following heating to 75, 140, and 220°C. The results (Table 12-7) show that in this frequency range the values are characteristic of the clay minerals, with the sharpest differences between kaolinite and the minerals containing interlayer water. The dielectric constants are particularly dependent on the interlayer water, and as the amount of this water varies on drying, characteristic dielectric constants develop. The author believed that such dielectric values could be used in the identification of the various clay minerals.

Spauszus[76] measured the change in the dielectric constants of ethylene chloride, benzene, and chlorobenzene upon addition of solids containing montmorillonite and kaolinite. The two clay minerals give characteristic values, with the change for air-dried montmorillonite being larger than that for kaolinite, because of the inner crystalline swelling of the former. The results

[74] Caillere, S., S. Henin, and S. Meriaux, Xylotile, a Fibrous Variety of Sepiolite, *Compt. Rend.,* **227**: 855–856 (1948).

[75] Ficai, C., Caractéristiques des matières argileuses établies en mesurant les constantes et les pertes diélectriques dans le domaine de fréquence compris entre 30 kHz et 3 MHz, *Bull. Soc. Franc. Ceram.,* **42**: 7–16 (1959).

[76] Spauszus, S., Dielectric Measurements on Bentonites and Kaolins, *Kolloid-Z.,* **152**: 121–126 (1957).

Table 12-7   Dielectric constants of clay minerals
(From Ficai[75])

Kaolinite						Halloysite					
75°C		140°C		220°C		75°C		140°C		220°C	
kHz	ε	kHz	ε	kHz	ε	kHz	ε	kHz	ε	kHz	ε
45.58	6.45	42.22	5.30	50.17	4.11	35.01	13.91	38.94	9.91	43.04	6.63
84.17	6.15	86.61	5.18	88.24	4.10	64.68	13.51	71.88	9.51	79.40	6.65
165.80	5.88	171.21	5.14	173.10	4.06	128.66	12.64	141.96	9.25	156.35	6.21
324.50	5.64	331.09	5.02	335.40	4.06	255.00	11.73	278.50	8.62	305.80	5.99
653.0	5.41	663.90	4.90	671.80	4.04	519.20	10.88	566.40	8.02	617.40	5.71
1,387.5	5.30	1,394.7	4.79	1,408.2	4.04	1,110.9	10.09	1,204.3	7.36	1,301.4	5.52
2,879.0	5.24					2,361.5	9.42	2,569.4	6.91	3,040.3	5.40

Montmorillonite						Illite (containing some montmorillonite)					
75°C		140°C		220°C		75°C		140°C		220°C	
kHz	ε	kHz	ε	kHz	ε	kHz	ε	kHz	ε	kHz	ε
—	—	45.53	5.23	46.86	4.58			41.82	7.60	43.84	6.18
—	—	83.51	5.17	80.95	4.58	63.22	13.91	77.22	7.21	80.16	6.14
—	—	164.69	5.04	168.40	4.47	127.11	12.44	151.95	6.88	157.51	5.94
—	—	274.68	4.91	327.80	4.41	256.40	10.77	298.80	6.78	306.20	5.87
469.8	15.30	643.90	4.81	658.10	4.27	528.00	9.74	599.40	6.51	615.00	5.80
1,037.3	11.96	1,351.9	4.72	1,384.2	4.27	1,126.9	9.27	1,265.2	6.29	1,240.7	5.73
2,246.3	10.41	3,110.1	4.65	2,879.0	4.27	2,413.5	8.44	2,648.3	6.22	2,700.5	5.68

for a series of clays containing different cations are of the order of the hydration numbers of the cations, again indicating the effect of degree of swelling.

Effect of grinding

The effect of prolonged grinding on the structure and properties of the clay minerals has been studied by many persons. Murphy[77] showed that grinding kaolinite increased its capacity to fix phosphate, and Jackson and Truog[78] found that relatively short periods of grinding several minerals increased

[77] Murphy, H. F., The Role of Kaolinite in Phosphate Fixation, *Hilgardia*, **12**: 341–382 (1939).

[78] Jackson, M. L., and E. Truog, Influence of Grinding Soil Minerals to Near Molecular Size on Their Solubility and Base Exchange Properties, *Soil Sci. Soc. Am. Proc.*, **4**: 136–143 (1939).

greatly their cation-exchange capacity and the solubility of the alumina and silica in neutral salt solutions and dilute acids and bases.

Laws and Page[79] found that dry-grinding kaolinite at first increased its exchange capacity and then decreased it as grinding continued. They also reported that continued grinding produced a new mineral species since prolonged grinding induced the fixation of calcium, altered the thermal curves, and increased the solubility of alumina in solutions of bases and acids. Perkins et al.[80] ground samples of kaolinite and montmorillonite for periods up to 21 weeks and concluded on the basis of chemical characteristics, infrared spectra, X-ray diffraction, and electron micrographs that the results for both minerals were substantially the same, with the product resembling an alumina-silica gel. Bloch[81] earlier had found that prolonged grinding of montmorillonite destroyed the crystal structure of the mineral and liberated part of the alumina and magnesia as well as the crystalline (OH) groups, and he concluded that new minerals were formed.

Perkins and Dragsdorf[82] showed that intimate contact with magnesium, as compared with calcium or hydrogen, during grinding of montmorillonite hastened the decomposition of the mineral. Perkins et al.[83] reported that the rate of decomposition on dry-grinding kaolinite depended on the nature of any extraneous material that might be present. For example, decomposition was more rapid in the presence of a tribasic phosphate than calcium carbonate at equal pH values.

Takahashi[84] subjected kaolinite and halloysite to dry grinding in a mechanical mortar for several hundred hours. For both minerals he found that relatively short periods of grinding first distorted and then destroyed the crystalline structure, with a reduction in particle size. With very long periods of grinding there was a reaggregation of the amorphous material into spherical particles which had a zeolitic structure. Accompanying the development of the zeolitic structure there was an increase in cation-exchange capacity.

[79] Laws, W. D., and J. B. Page, Changes Produced in Kaolinite by Dry Grinding, *Soil Sci.*, **62**: 319–336 (1946).

[80] Perkins, A. T., R. D. Dragsdorf, E. R. Lippincott, J. Selby, and W. G. Fateley, Products of Clay Mineral Decomposition as Related to Phosphate Fixation, *Soil Sci.*, **80**: 109–120 (1955).

[81] Bloch, J. M., Effect of Grinding on the Crystal Structures and Properties of Montmorillonite, *Bull. Soc. Chim. Franc.*, pp. 774–781 (1950).

[82] Perkins, A. T., and R. D. Dragsdorf, Decomposition of Bentonite as Affected by H, Ca, Mg, and Dry Grinding, *Soil Sci. Soc. Am. Proc.*, **16**: 312 (1952).

[83] Perkins, A. T., R. D. Dragsdorf, and M. S. Bhangoo, Reactions Between Phosphates and Kaolinite Decomposition Products, *Soil Sci. Soc. Am. Proc.*, **21**: 154–157 (1957).

[84] Takahashi, H., Effects of Dry Grinding on Kaolin Minerals, I., Kaolinite, *Bull. Chem. Soc. Japan*, **32**: 235–245 (1959).

Legrand and Nicolas[85] investigated the effect of various grinding techniques on the crystal structure of kaolinite. They found that grinding had no appreciable effect in the presence of water or a nonpolar liquid such as toluene. With dry grinding, the effect varied with the type of grinding mill. Thus, with a hammer mill or an oscillating-disk vibrocrusher, there was substantially more lattice disruption than with a ball mill. Kohler et al.[86] also investigated the effect of various grinding techniques. They found that long periods of wet grinding of a coarse and fine kaolinite and a calcium montmorillonite in a ball mill caused no appreciable disturbance of the structure. The main effect was a splitting of the kaolinite particles into thinner platelets, thus causing an increase in the surface area, cation-exchange capacity, and the amount of the fraction finer than $1\,\mu$. Dry grinding, however, caused appreciable disturbance of the layer order of the kaolinite, and after long grinding the platelets of both kaolinite and montmorillonite were broken up. These investigators reported that plasticity and dry strength of the coarse kaolinite were improved by wet grinding. A short period of dry grinding improved these properties for both kaolinite and montmorillonite, but longer grinding caused their deterioration.

Gunn and Morris[87] and Duke[88] have shown that wet-grinding kaolinite in the presence of very small plastic balls or sand tends to reduce the viscosity of the clay in water suspensions.

The necessity of considering that the structure and the properties of the clay minerals are very susceptible to alteration by grinding in preparing samples for study is obvious from the work reported above. For example, it is possible that some of the divergent results reported in the literature for the infrared characteristics of kaolinite are due to variations in grinding the samples for analysis.

Intercalation complexes

In 1959, Wada[89] observed that certain inorganic compounds which do not form hydrogen bonds (see Chap. 10) could be intercalated between the silicate

[85] Legrand, C., and J. Nicolas, Contribution of X-ray Diffraction and the Electron Microscope to the Study of Ground Kaolins, *Bull. Soc. Franc. Ceram.,* **44:** 61–69 (1959).

[86] Kohler, E., U. Hofmann, E. Scharrer, and K. Fruhauf, Über den Einflus der Mahlung auf Kaolin und Bentonit, *Ber. Deut. Keram. Ges.,* **37:** 493–503 (1960).

[87] Gunn, F. A., and H. H. Morris, Delamination Domestic Sedimentary Clay Products and Method of Preparation Thereof, U.S. Patent 3,171,718 (1965).

[88] Duke, J. B., Method for Comminuting Kaolin Clay, U.S. Patent 3,097,801 (1963).

[89] Wada, K., Oriented Penetration of Ionic Compounds between the Silicate Layers of Halloysite, *Am. Mineralogist,* **44:** 153–165 (1959).

layers of halloysite and that the reactions were not ion-exchange reactions. Wada soaked hydrated halloysite with various salt solutions, dried the material at 30°C, and then finely ground the product in an agate mortar. X-ray-diffraction analyses showed variations in the basal spacing for some potassium, ammonium, rubidium, and cesium salts, and no variation for sodium, lithium, magnesium, calcium, and barium salts. In experiments with ammonium salts, Wada[90] found considerable variation in the basal spacing, depending on the accompanying anion. In such experiments he concluded that the cation and anion of the salts penetrated between the silicate layers of halloysite, replacing the interlayer water. It was of interest that no reaction occurred with lithium and sodium salts whose ionic diameters are less than about 2.7 Å, which is the approximate diameter of the hole in the surface oxygen net of the halloysite. Similar experiments with montmorillonite showed no reaction. Washing the reacted halloysites with water caused a return of the c-axis dimension to the original 10.1 Å of the hydrated halloysite.

Wada[91] also investigated the change in the basal spacing of halloysite that had been dehydrated to various degrees, followed by dry grinding with potassium acetate, ammonium chloride, and ammonium nitrate. With partially dehydrated halloysite and ammonium acetate, there was complete intercalation with the c-axis dimension at 14.3 Å. With ammonium chloride and ammonium nitrate there was partial intercalation with c-axis dimensions of 9.9 and 7.55 Å, respectively. Following heating to 300°C, producing substantially complete dehydration, only potassium acetate caused a change in the basal spacing with dry grinding. In every case, water washing following the intercalation returned the basal dimension to 10.1 Å, that is, the value for the hydrated mineral.

Garrett and Walker[92] also treated samples of hydrated halloysite with a variety of potassium salts, evaporated the solutions to dryness, and found evidence of intercalation of the salts. They found that the nature of the anion influenced the basal spacing and also that there was a relation between solubility values of the salts and intercalation. With less soluble salts, little penetration was indicated, and it may, therefore, be inferred that a high salt concentration is necessary before significant intercalation occurs. Examination of the effect of basal spacing of the chlorides of metals other than potassium failed to disclose any simple relation with salt concentration. However, they did find a correlation with hydration energies of the cation,

[90] Wada, K., An Interlayer Complex of Halloysite with Ammonium Chloride, *Am. Mineralogist,* **44**: 1237–1247 (1959).

[91] Wada, K., Lattice Expansion of Kaolin Minerals by Treatment with Potassium Acetate, *Am. Mineralogist,* **46**: 78–91 (1961).

[92] Garrett, W. G., and G. F. Walker, The Cation Exchange Capacity of Hydrated Halloysite and the Formation of Halloysite Salt Complexes, *Clay Minerals Bull.,* **4**(22): 75–80 (1959).

suggesting that, when intercalation occurs, the cation accompanies the anion into the crystal lattice and the ability of the cation to enter is influenced by its hydration shell. Very low hydration of the cation favors the formation of the complexes. These authors concluded that the ability of certain salts to form intercalation complexes with hydrated halloysite is related to the total salt concentration of the solution and to the hydration energy of the cation. Penetration of the salt into the interlayer region of the mineral evidently required the development of a concentration gradient between the external solution phase and the intracrystalline water.

Andrew, Jackson, and Wada[93] and Wada[94] showed that dickite is also expandable upon grinding with potassium acetate. Wada[91] found that mixing well-crystallized kaolinite with potassium acetate, drying, and then grinding produced no change in the basal spacing of the kaolinite. However, on grinding a mixture of well-crystallized kaolinite and potassium acetate without preliminary drying, the basal spacing increased to 14.2 Å and the diffraction pattern showed comparable higher orders, suggesting intercalation. The pattern also suggested that the intercalation was accompanied by modification in the stacking arrangements resulting from *b*-axis displacement. Washing with water removed the potassium acetate, and the structure collapsed back to about 7 Å. Heating to 100°C collapsed the spacing to 11.4 Å, but the 14.2-Å spacing was quickly regained if the material was again allowed to stand in the air, suggesting that the intercalation contains a monomolecular layer of water along with the acetate. Wada's concept of the arrangement of the intercalated acetate is shown in Fig. 12-12.

Washing the kaolinite-acetate reaction product with ammonium chloride reduced the basal dimension to about 11 Å, indicating a replacement of the potassium acetate by ammonium chloride. Grinding well-crystallized kaolinite with ammonium chloride produced a diffuse band at about 10 Å, suggesting partial random intercalation. Grinding with other potassium and ammonium salts caused no change in the basal dimension, indicating no intercalation. Wet grinding also was tried, but dry mechanical grinding was reported to be three to five times as effective. Experiments in dry-grinding montmorillonite, pyrophyllite, and talc with ammonium chloride and potassium acetate gave no indication of lattice expansion and, consequently, no indication of complex formation.

Wada[91] discussed the possible causes of the variation in the intercalation of halloysite and kaolinite with different salts, from which it can be concluded that much remains to be learned before this type of reaction is well under-

[93] Andrew, R. W., M. L. Jackson, and K. Wada, Intercalation as a Technique for Differentiation of Kaolinite from Chloritic Minerals by X-ray Diffraction, *Soil Sci. Soc. Am. Proc.*, **24**: 422–424 (1960).

[94] Wada, K., Ammonium Chloride-Kaolin Complexes, *Clay Sci. (Tokyo)*, **2**: 43–56 (1964).

stood. In addition to the influence of any interlayer water, the cation size in relation to the size of the cavity in the Si-O sheet was found to be of primary importance in the reaction.

Weiss[95] also pointed out that kaolinite can react with compounds incapable of forming hydrogen bonds. Thus, potassium acetate, ammonium acetate, and some other acetates form intercalation compounds. He indicated that the reaction is strongly dependent upon pH and occurs only in the alkaline region, which would suggest that the formation of a basic aluminum acetate plays a part in the reaction. The size of the cation is perhaps also decisive since the reaction proceeds completely only with potassium, rubidium, and

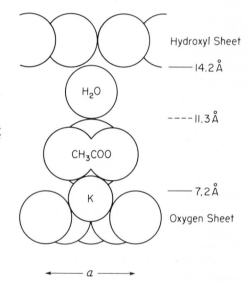

Fig. 12-12. A cross section of the 14-Å complex, showing a probable position of K[+], C₂H₃O₂[-], and H₂O. (After Wada.[91])

cesium acetates, but not with sodium, magnesium, calcium, or lithium acetates. With the propionates, however, the sodium salt reacts completely.

Wada[96] was able to produce intercalation complexes with nacrite. He showed that, when complexes of halloysite, kaolinite, dickite, nacrite, and potassium acetate and ammonium chloride are washed with water, intercalated water layers remain in the structures. As indicated by the dehydration data and *c*-axis determinations, the amount of such water is not the same for each of these minerals, with kaolinite having the smallest amount.

Numerous authors have shown how intercalation can be used to determine

[95] Weiss, A., A Secret of Chinese Porcelain Manufacture, *Angew. Chem.*, **2**: 697–703 (1963).

[96] Wada, K., Intercalation of Water in Kaolin Minerals, *Am. Mineralogist*, **50**: 924–941 (1965).

the presence and relative abundance of halloysite and kaolinite minerals in clay-minerals mixtures. Thus Wada[97] described a method based on the specific retention of NH_4Cl, and Jackson[98] and Miller and Keller[99] described techniques based on differences of the c-axis dimension of these minerals following intercalation with various reagents.

Miscellany

Corbett and Burson[100] found significant changes in the degree of crystallinity, specific-surface area, cation-exchange capacity, and slurry viscosity of two kaolinite samples after exposure to doses of 0.667-Mev gamma radiation rang-· ing from 10^{18} to 10^{22} ev per gram. Further, infrared absorption indicated a decrease in the ratio of H-bonded to unbonded OH groups, and the irradiated samples appeared to grind relatively more easily. Head et al.[101] found that the preferred orientation relationship between the structure of kaolinite and mullite formed on heating is disrupted by heavy irradiation. These authors concluded that the irradiation disrupts the ordering of the position of the aluminum atoms.

Several investigators (Ducros and Dupont,[102] Friedlander et al.,[103] and Boss and Stejskal[104]) have applied nuclear magnetic resonance (NMR) techniques to the study of the clay minerals. Results to date are concerned mainly with the mobility of the protons and the nature of the water compo-

[97] Wada, K., Quantitative Determination of Kaolinite and Halloysite by NH_4Cl Retention Measurement, *Am. Mineralogist,* **48:** 1286–1299 (1963).

[98] Jackson, M. L., Significance of Kaolinite Intercalation in Clay Mineral Analysis, Proceedings of the 9th National Clay Conference, pp. 424–430, Pergamon Press, New York (1962).

[99] Miller, W. D., and W. D. Keller, Differentiation between Endellite-Halloysite and Kaolinite by Treatment with Potassium Acetate and Ethylene Glycol, Proceedings of the 10th National Clay Conference, pp. 244–253, Pergamon Press, New York (1963).

[100] Corbett, W. J., and J. H. Burson, Delamination of Kaolin by High-energy Ionizing Radiation, *U.S. At. Energy Comm.* SRO-57 (1961).

[101] Head, C. M., W. E. Moody, L. Mitchell, and R. A. Young, Effect of Radiation Damage on Mullite Formation in Kaolinite, Proceedings of the 10th National Clay Conference, pp. 356–364, Pergamon Press, New York (1963).

[102] Ducros, P., and M. Dupont, Etude par résonance magnétique nucleaire des protons dans les argiles, *Bull. Groupe Franc. Argiles,* **8:** 59–63 (1962).

[103] Friedlander, H. Z., J. Saldick, and C. R. Frink, Electron Spin Resonance Spectra in Various Clay Minerals, *Nature,* **199**(4888)**:** 61 (1963).

[104] Boss, B. D., and E. O. Stejskal, Anisotropic Diffusion in Hydrated Vermiculite, *J. Chem. Phys.,* **43**(3)**:** 1068–1069 (1965).

nent. It is too early to know if variations in different clay-mineral groups may be determined.

It is well known that the application of ultrasonic vibrations to clay suspensions may cause large variations in their dispersion-coagulation properties. Mathieu-Sicaud and Levavasseur[105] concluded that there was particular frequency for the optimum dispersion of the particular clay minerals. Recently, Kruglitskii and his colleagues[106,107] extensively studied this matter and arrived at about the same conclusion.

Pellizzer[108] investigated the orientation of various clay minerals in a magnetic field and found a significant difference between various clay minerals; for example, in a magnetic field of 1,400 gauss kaolinite became well oriented whereas illite and smectites were not appreciably oriented. In a field of 2,800 gauss, the illite was appreciably oriented and the smectites were not appreciably oriented. At higher gauss values, the differences tended to disappear. The author attributed this phenomenon to some characteristic of the hydroxyls in the structure, as the property is lost when the clay minerals are heated above the dehydroxylation temperature.

Additional references

Solubility of the clay minerals

Gastuche, M. C., Investigation of the Alteration of Kaolinite by Different Chemical Agents, *Silicates Ind.*, **24**: 237–244, 313–320 (1959).

Ginzburg, I. I., R. S. Yashina, L. A. Mateeva, V. V Belyatskii, and T. S Nuzhdelovskaya, Decomposition of Some Minerals With Organic Acids, *Khim Zemnoi Kory, Akad. Nauk SSSR, Tr. Geochim. Konf.*, **1**: 290–305 (1963).

Hayashi, H., and K. Oinuma, Behaviors of Clay Minerals in Treatment with Hydrochloric Acid Formamide, and Hydrogen Peroxide, *Clay Sci. (Tokyo)*, **2**: 75–91 (1964).

Kallauner, O., and J. Matejka, Rational Analysis of Kaolinitic Clays, *Czech. Ceram. Soc.*, **69**: 68–70 (1928).

[105] Mathieu-Sicaud, A., and G. Levavasseur, Ultrasonic Dispersion of Clay Suspensions—Interpretation of the Results With an Electron Microscope, *Compt. Rend.*, **228**: 393–395 (1949).

[106] Kruglitskii, N. N., V. V. Simurov, F. D. Ovcharenko, and S. P. Nichiporenko, Mechanism by Which Coagulation Processes in Aqueous Clay Dispersions Are Influenced by Ultrasonic Vibrations, *Dokl. Akad. Nauk SSSR*, **159**(6): 1367–1370 (1964).

[107] Kruglitskii, N. N., and V. V. Simurov, Regulating the Processes of Coagulation Structure Formations in Aqueous Clay Dispersions by Ultrasonic Vibrations, I, Effect of Ultrasonic Vibrations on the Stability of Aqueous Clay Suspensions, *Ukr. Khim. Zh.*, **30**(8): 823–830 (1964).

[108] Pellizzer, Renato, Orientamento di minerali argillosi in campo magnetico, *Rendi. Soc. Mineral. Ital.*, **16**: 347–358 (1960).

Mering, J., Reactions of Montmorillonite, *Mem. Chim. État (Paris)*, **36**(2): 165–177 (1951).

Nutting, P. G., Solution and Dispersion of Minerals in Water, *J. Wash. Acad. Sci.*, **22**: 261–267 (1932).

Petzold, A., Determination of Clay Substance by Rational Analysis According to Kallauner-Matefka, *Silikat Tech.*, **8**: 511–513 (1957).

Polzer, W. L., and J. D. Hem, The Dissolution of Kaolinite, *J. Geophys. Res.*, **70**(24): 6233–6240 (1965).

Wilson, H., "Ceramics—Clay Technology," McGraw-Hill, New York (1927).

Infrared-absorption spectra

Buswell, A. M., K. Krebs, and W. H. Rodebush, Infrared Studies, III, Absorption Band of Hydrogels between 2.5 and 3.5 Microns, *J. Am. Chem. Soc.*, **59**: 2603–2605 (1937).

Drummond, D. G., Infrared Absorption of Quartz, *Nature*, **130**: 928–929 (1932).

Kodama, H., and K. Oinuma, Identification of Kaolin Minerals in the Presence of Chlorite by X-ray Diffraction and Infrared Absorption Spectra, Proceedings of the 11th National Clay Conference, pp. 236–249, Pergamon Press, New York (1963).

Serratosa, J. M., A. Hidalgo, and J. M. Vinas, Infrared Study of the OH Groups in Kaolin Minerals, *Proc. Intern. Clay Conf., Stockholm*, **1**: 17–26 (1963).

Vedder, W., Correlations Between Infrared Spectrum and Chemical Composition of Mica, *Am. Mineralogist*, **49**: 736–768 (1964).

Williams, V. Z., Infrared Instrumentation and Techniques, *Rev. Sci. Instr.*, **19**: 135–178 (1948).

Surface area

Emmett, P. H., The Measurement of the Surface Areas of Finely Divided or Porous Solids by Low Temperature Adsorption Isotherms, *Advan. Colloid Sci.*, **1**: 1–36 (1941).

Greene-Kelly, R., Specific Surface Areas of Montmorillonites, *Clay Minerals Bull.*, **5**(31): 392–400 (1964).

Longuet-Escard, J., The Effect of Progressive Dehydration on the Area of the Surface of Montmorillonite, *J. Chim. Phys.*, **47**: 113–117 (1950).

McAuliffe, C. D., N. S. Hall, L. A. Dean, and S. B. Hendricks, Exchange Reactions between Phosphates and Soils: Hydroxylic Surfaces of Soil Minerals, *Soil Sci. Soc. Am. Proc.*, **12**: 119–123 (1948).

Ritter, H. L., and L. C. Erich, Pore Size Distribution in Porous Materials, *Anal. Chem.* **20**: 665–670 (1948).

Schofield, R. K., Calculation of Surface Area of Clays from Measurements of Negative Adsorption, *Trans. Brit. Ceram. Soc.*, **48**: 207–213 (1949).

Spiel, S., Effect of Adsorbed Electrolytes on Properties of Monodisperse Clay-Water Systems, *J. Am. Ceram. Soc.*, **23**: 33–38 (1940).

Teichner, S., Measurement of Specific Surfaces on Certain Clays, *Compt. Rend.*, **231**: 1063–1064 (1950).

Density

Baver, L. D., "Soil Physics," Wiley, New York (1940).

Ralston, O. C., and A. G. Stern, Report of Non Metallic Division, *U.S. Bur. Mines, Rept. Invest.* 3599 (1941).

Grinding

Nemetschek, T., and S. Pollman, Influence of Milling, Ion Exchange, and Removal of Quartz on Shale, *Ber. Deut. Keram. Ges.*, **39**(9): 447–453 (1962).

Shirozu, H., Structural Changes of Some Chlorites by Grinding, *Mineral. J.,* **4**(1): 1–11 (1963).

Stanczyk, M. H., and I. L. Feld, Continuous Attrition of Coarse Kaolin, II, Closed Circuit Tests, *U.S. Bur. Mines, Rept. Invest.* (Alabama) **6694** (12) (1964).

Weiss, A., and W. Thielepape, Improvement of Technical Properties of Kaolinite Through Preparation of Intercalated Compounds, *Intern. Clay Conf., Stockholm,* (2), 427–429 (1965).

Miscellaneous

Corbett, W. J., J. H. Burson, and R. A. Young, Gamma-Irradiation of Kaolinite, Proceedings of the 10th National Clay Conference, pp. 344–354, Pergamon Press, New York (1963).

De Keyser, W., Dielectric Losses in Clay Minerals, *Silicates Ind.,* **24**: 615–617 (1959).

McEuen, R. B., Dielectrophoretic Behavior of Clay Minerals, I, Dielectrophoretic Separation of Clay Mixtures, Proceedings of the 12th National Clay Conference, pp. 549–556, Pergamon Press, New York (1964).

Osipov, Yu. B., Behavior of Clay Suspension in Magnetic Field, *Vestn. Mosk. Univ., Ser. IV, Geol.,* **20**(5): 81–87 (1965).

Worrall, W. E., and M. K. Basu, Viscosity Changes Induced in Clay Slips by Ultrasons, *Trans. Brit. Ceram. Soc.,* **64**: 61–69 (1965).

Thirteen
Origin and Occurrence of the Clay Minerals

Synthesis of the clay minerals

Introduction. The synthesis of the clay minerals at elevated temperatures and pressures from oxides and hydroxides and from various minerals, particularly the feldspars, in the presence of acids and alkalies has been studied in considerable detail. Some investigation has been made of the synthesis of the clay minerals at ordinary temperatures and pressures, but the amount of such work is inadequate for general conclusions. Consideration is given herein only to syntheses investigations of a sort which throw light on problems of the occurrence and origin of the clay minerals in nature.

The identification of the crystalline phases formed in syntheses experiments is often very difficult because of the very small particle size, the poor crystallinity, and the contamination with large amounts of unreacted or unchanged materials. Usually, X-ray diffraction provides the only trustworthy method of identifying the new phases. The identifications made in the years before the development of X-ray-diffraction techniques and before adequate diffraction data were available for the pure clay minerals are, therefore, often open to serious question. Even with the use of diffraction methods, there may be important uncertainties in the identifications reported in the literature. Thus, some of the material reported as kaolinite may actually be halloysite, since in few cases have the identifications been checked by electron micrographs.

Syntheses from Mixtures of Oxides and Hydroxides at Elevated Temperatures and Pressures. The reaction products in the system Al_2O_3-SiO_2-H_2O appear to have been studied first with modern methods of phase identification by Noll.[1] Later Roy and Osborn[2,3] investigated the phase equilibrium relationships in the system Al_2O_3-SiO_2-H_2O under conditions of independently controlled temperature and water-vapor pressure. The reaction products obtained by Noll at temperatures between 250 and 500°C with varying amounts of alumina and silica, the pressure bomb containing a constant percentage of water, are given in Table 13-1. Silica gel and alumina gel were used

Table 13-1   Reaction products in the system Al_2O_3-SiO_2-H_2O

(After Noll[4])

Temp., °C (Pressure, atm)	Molecular ratio Al_2O_3:SiO_2 (H_2O constant)					
	1:0	>1:2	1:2	1:4	<1:4	0:1
500 (530–540)	Corundum		Pyrophyllite + ?			
400 (300)	Boehmite	Kaolinite + boehmite	Pyrophyllite + boehmite + kaolinite ?	Pyrophyllite	Pyrophyllite + SiO₂ (amorphous)	SiO₂ (amorphous)
350 (168) 300 (87) 250 (41)			Kaolinite	Kaolinite	Kaolinite + SiO₂ (amorphous)	

[4] Noll, W., Ueber die Bildungsbedingungen von Kaolin, Montmorillonit, Sericit, Pyrophyllit, und Analcim, *Mineralog. Petrog. Mitt.*, **48**: 210–246 (1936).

[1] Noll, W., Mineralbildung im System Al_2O_3-SiO_2-H_2O, *Neues Jahrb. Mineral. Geol.*, *Beil. Bd.A*, **70**: 65–115 (1935).

[2] Roy, R., and E. F. Osborn, Studies in the System Alumina-Silica-Water, "Problems of Clay and Latitude Genesis," pp. 76–80, American Institute of Mining and Metallurgical Engineers (1952).

[3] Roy, R., and E. F. Osborn, The System Al_2O_3-SiO_2-H_2O, *Am. Mineralogist*, **39**: 853–885 (1954).

in the experiments; the latter crystallized rapidly to bayerite and boehmite when the gel was aged prior to the experimentation. Most of the alumina used was in the form of a mixture of boehmite and bayerite. When the two were used separately, the only observable difference in the results was a difference in the rate of reaction, boehmite reacting more readily. The rate of cooling and the degree of filling of the bomb had little, if any, effect on the nature of the product.

The data in Table 13-1 indicate that, in the absence of silica, boehmite is formed up to 400°C, and at 500°C corundum is formed. According to Roy and Osborn,[2] at very low temperatures gibbsite would form, with a transition to boehmite at about 120°C. Bayerite is a metastable phase formed during the rapid crystallization of alumina hydrate from sodium aluminate solutions. Also, according to these authors, diaspore instead of boehmite would be the monohydrate phase above 280°C under certain conditions of elevated pressure. Roy and Osborn showed that γ-alumina, a metastable phase, intervenes in the transition of boehmite to corundum (α-alumina).

Noll's data show that, with molecular $Al_2O_3:SiO_2$ ratios of greater than 1:2, kaolinite and boehmite are formed up to 400°C, and an unidentifiable phase develops at 500°C. With a ratio of 1:2, only kaolinite is formed up to 350°C; above that temperature kaolinite, pyrophyllite, and boehmite are formed up to 400°C; and at 500°C, kaolinite and boehmite disappear, with pyrophyllite and an unidentifiable phase the only components present. With a ratio of 1:4, kaolinite is the only crystalline phase up to 350°C, and at 400 and 500°C pyrophyllite is the only phase developed. With molecular alumina-to-silica ratios less than 1:4, the results of the synthesis are the same as when the ratio is exactly 1:4, except that amorphous silica is present at all temperatures. In the absence of alumina, no crystalline silica phase is formed under the conditions of Noll's experiments. Noll pointed out that the rate of reaction varies greatly with the temperature; e.g., kaolinite formed at 300°C in 1 hr, whereas 111 hr were required at 200°C.

Roy and Osborn[3] synthesized the kaolin minerals from coprecipitated Al_2O_3-SiO_2 gels at 150 to 400°C under varying vapor pressures. Brindley and DeKimpe[5] reported the synthesis of kaolinite at 150°C and about 5 atm of vapor pressure, using gibbsite as a source of alumina with various forms of silica.

It would seem from the work of Roy and Osborn[2,3] that, at very low temperatures and pressures, halloysite (2H₂O) may be the resulting phase in the system Al_2O_3-SiO_2, rather than kaolinite; however the mineral could not be identified definitely.

The data in Table 13-2, after Noll,[4] show the reactions obtained in the

[5] Brindley, G. W., and C. DeKimpe, Attempted Low-temperature Synthesis of Kaolin Minerals, *Nature,* 190(4772): 254 (1961).

Table 13-2   Reaction products in the system $(CaMg)O-(K_2Na_2)O-Al_2O_3-SiO_2-H_2O$

(After Noll[4])

		$(CaMg)O-(K_2Na_2)O-Al_2O_3-SiO_2$ (H_2O constant)			
		<0.2:1:4	0.2:1:4	0.37:1:2	1:1:4
300°C MgO	Alkali solution	Kaolinite (>) Smectite (<)	Smectite	. . . .	Smectite
300°C CaO	Alkali solution	Kaolinite (>) Smectite (<)	Smectite	. . . .	?
300°C Na₂O	Alkali solution	Kaolinite (>) Smectite (<)	Smectite (+ kaolinite)	. . . .	Analcime
300°C K₂O	Alkali solution	Kaolinite (>) Smectite (<)	Smectite (+ kaolinite)	Mica	K feldspar* + ?
	Acid solution	Kaolinite	Kaolinite	Kaolinite	Kaolinite
400°C K₂O	Alkali solution	. . . .	. . . .	Mica	
	Acid solution	. . . .	. . . .	Pyrophyllite + boehmite and kaolinite	

* In pure condition at $Al_2O_3:6SiO_2$.

system $(CaMg)O-(K_2Na_2)O-Al_2O_3-SiO_2-H_2O$ on heating to 300 and 400°C for periods of 8 to 24 hr. In each case the bomb was 0.35 percent filled with water, and the pressure developed was 87 atm at 300°C and 300 atm at 400°C. As the alkalies or alkaline earths are added to a mixture of composition $Al_2O_3:4SiO_2$, smectite forms in increasing amounts at the expense of kaolinite. A concentration of about 0.2 mole of $(CaMg)O$ or $(Na_2K_2)O$ per mole of alumina with an $Al_2O_3:SiO_2$ ratio of 1:4 appears to be the optimum for the formation of smectite. At slightly higher concentrations with MgO as the alkaline earth, smectite is still the phase formed. At higher concentrations with Na₂O, analcime is formed, and at moderately higher concentrations with K₂O, muscovite is formed. The reaction products with CaO at higher concentrations could not be determined by Noll, but they were probably calcium silicates and calcium aluminates (Buckner, Roy, and Roy[6]).

At very high concentrations of K₂O with the $Al_2O_3:SiO_2$ ratio equal to 1:4, a potash feldspar was formed. At very high concentrations of MgO with the same $Al_2O_3:SiO_2$ ratio, according to Noll and also Stresse and Hof-

[6] Buckner, D. A., D. M. Roy, and R. Roy, Studies in the System CaO-Al₂O₃-SiO₂-H₂O, II, System CaSiO₃-H₂O, *Am. J. Sci.*, **258**(2): 1?2–147 (1960).

mann,[7] talc and then serpentine formed plus kaolinite or pyrophyllite instead of smectite. Pyrophyllite rather than smectite is the phase formed if the temperature used is above 400°C.

Noll showed that, for acid systems with K_2O, kaolinite rather than the mica is formed. Other data suggest the generality that kaolinite is likely to be the phase formed at 300°C in an acid system containing any of the alkalies or alkaline earths. Gruner[8] confirmed the formation of kaolinite at 300°C with K_2O in an acid system, even in the presence of a large excess of potash. However, Gruner showed that at 400°C muscovite will form in an acid system in the presence of potash, and also that kaolinite will be converted to muscovite at this temperature in the presence of an excess of KCl.

Studies in the systems Al_2O_3-SiO_2-H_2O, MgO-Al_2O_3-SiO_2-H_2O, and Na_2O-Al_2O_3-SiO_2-H_2O by various workers, for example, Ames and Sand,[9] have resulted in the synthesis of many smectites.

Using natural minerals, mixtures of oxides, glasses or coprecipitated gels, numerous detailed studies of stability in the system MgO-Al_2O_3-SiO_2-H_2O have been made (Yoder,[10] Roy and Roy,[11] and Nelson and Roy[12]). These workers reported the synthesis of smectite and chlorite, and their publications should be consulted for the conditions necessary for their formation. Laboratory studies (Roy and Romo[13]) show that primary vermiculite is unlikely to form at a temperature above 200 to 300°C.

A considerable amount of work on the synthesis of the micas has been summarized by Yoder.[14] He emphasized that the simpler polymorphic forms develop initially at lower temperatures and pressures and that with increasing

[7] Stresse, H., and U. Hofmann, Synthesis of Magnesium Silicate Gels with Two Dimensional Regular Structures, *Z. Anorg. Allgem. Chem.,* 247: 65–95 (1941).

[8] Gruner, J. W., Formation and Stability of Muscovite in Acid Solutions at Elevated Temperatures, *Am. Mineralogist,* 24: 624–628 (1939).

[9] Ames, L. L., and L. B. Sand, Factors Effecting Maximum Hydrothermal Stability in Montmorillonites, *Am. Mineralogist,* 43: 641–648 (1958).

[10] Yoder, H. S., The MgO-Al_2O_3-SiO_2-H_2O System and the Related Metamorphic Facies, *Am. J. Sci. (Bowen Volume),* pp. 567–627 (1952).

[11] Roy, D. M., and R. Roy, Synthesis and Stability of Minerals in the System MgO-Al_2O_3-SiO_2-H_2O, *Am. Mineralogist,* 40: 147–178 (1955).

[12] Nelson, B. W., and R. Roy, Synthesis of Chlorites and Their Structural and Chemical Constitution, *Am. Mineralogist,* 43: 707–725 (1958).

[13] Roy, R., and L. A. Romo, Weathering Studies, I, New Data on Vermiculite, *J. Geol.,* 65: 603–610 (1957).

[14] Yoder, H. S., Jr., Experimental Studies on Micas: A Synthesis, Proceedings of the 6th National Clay Conference pp. 42–60, Pergamon Press, New York (1959).

temperatures and pressures there is a transition to the more complicated polymorphic forms.

Romo and Roy[15] were unsuccessful at first in attempts to synthesize mixed-layer clay minerals, but later work by Iiyama and Roy[16] with the system $Na_2O-MgO-Al_2O_3-SiO_2$ indicated that randomly stacked mixed-layer phases were formed at 1 kb in the temperature range 450 to 575°C, whereas at higher pressures regular mixed layers tended to form.

Ewell and Insley[17] reported the formation of nontronite from a mixture of silica gel and ferric oxide heated at 350°C for 6 days at a pressure of 167 atm.

Syntheses from mixtures of crystalline minerals and chemical reagents at elevated temperatures and pressures. A considerable amount of work has been done on the alteration of ground feldspar subjected to elevated temperatures and pressures in the presence of acids and various other reagents. Unfortunately, little work has been done with minerals other than the feldspars.

According to Gruner,[18] kaolinite, pyrophyllite, muscovite, and boehmite are the phases that may form when microcline and albite are subjected to high temperatures and pressures in the presence of aluminum hydroxide, silica, and potassium chloride under acid conditions. In Gruner's experiments the concentration of the potassium ions and the $Al_2O_3:SiO_2$ ratio of the system were the factors determining which minerals formed from the feldspars. Kaolinite will form from feldspars and is stable below approximately 350°C, regardless of the potassium-ion concentration, provided the ratio of Al:Si is about 1:1. This means that additional easily available aluminum has to be in the system. If the additional aluminum is not present, pyrophyllite forms instead of kaolinite below about 350°C.

In the temperature range from 350 to at least 530°C, pyrophyllite formed from the feldspars if the potassium-ion concentration was low (it must not be much greater than the amount that the acid can remove from the potash feldspars). Under these conditions boehmite formed from any alumina in excess of that required for the pyrophyllite.

If the concentration of potassium and sodium ions was high but no addi-

[15] Romo, L. A., and R. Roy, Essais de synthèse des minéraux argileux dits a "couches mixtes," *Bull. Soc. Franc. Mineral.*, **78**(7–9): 433–448 (1955).

[16] Iiyama, J. T., and R. Roy, Controlled Synthesis of Heteropolytypic (Mixed-layer) Clay Minerals, Proceedings of the 10th National Clay Conference, pp. 4–22, Pergamon Press, New York (1963).

[17] Ewell, R. H., and H. Insley, Hydrothermal Synthesis of Kaolinite, Dickite, Beidellite, and Nontronite, *J. Res. Natl. Bur. Std.*, **15**: 173–186 (1935).

[18] Gruner, J. W., Hydrothermal Alteration of Feldspars in Acid Solutions between 300° and 400°C., *Econ. Geol.*, **29**: 578–589 (1944).

tional alumina was available, the feldspars remained unaltered. No true potassium feldspar was synthesized under the conditions of the experiments. If alumina, silica, and potassium chloride are added to the feldspar, muscovite and pyrophylite are synthesized without any formation of feldspar.

Muscovite formed readily from feldspar, beginning at about 350 to 525°C, the highest temperature tried, provided that the potassium and aluminum were in sufficient concentration.

Schwarz and Trageser[19] stated that orthocase and anorthite in the presence of 0.5 N HCl gave kaolinite as an alteration product at temperatures below about 400°C, pyrophyllite and boehmite at temperatures of 400 to 550°C, and corundum at about 600°C.

Badger and Ally[20] produced kaolinite when a potash feldspar was heated at 225°C for 24 hr in the presence of 5 percent hydrofluoric acid. These same authors found no reaction when the same feldspar was heated in the presence of carbonic acid for 156 hr at 60°C and a pressure of 1,800 psi.

The formation of secondary silicate compounds from solid phases like feldspar and aluminum hydroxide in the foregoing type of experiments is little understood. Schwarz and Trageser[19] believed that the hydrolysis of the original mineral is essentially complete and that the new insoluble compounds are reaction products of the hydrolyzed components. Gruner[18] suggested that the degree of ionization, particularly with respect to SiO_4 groups, must be very small in such systems. Jander and Wuhrer[21] suggested that there is only partial hydrolysis of the original mineral, with the formation of some kind of an amorphous intermediate product. Various investigators have favored a structural transformation of the primary minerals to the secondary ones with only very slight hydrolysis. Schwarz and Trageser[19] pointed out that kaolinite can form under similar conditions from a variety of parent minerals (e.g., leucite, in addition to the feldspars), which means that no single original structure is required if the change is largely a structural transformation.

Syntheses from mixtures of oxides and hydroxides at ordinary temperatures and pressures. Sedletsky[22] mixed sodium silicate and sodium aluminate and then leached with N $MgCl_2$ until the pH of the leachate was 5.6. After

[19] Schwarz, R., and G. Trageser, Ueber die Synthese des Pyrophyllis, *Z. Anorg. Allgem. Chem.*, **225**: 142–150 (1935).

[20] Badger, A. E., and A. Ally, Note on the Formation of Kaolin Minerals from Feldspar, *J. Geol.*, **40**: 745–747 (1935).

[21] Jander, W., and J. Wuhrer, Die Bildung von Magnesium-hydrosilikaten, *Z. Anorg. Allgem. Chem.*, **235**: 282–287 (1938).

[22] Sedletsky, I. D., Genesis of Minerals from Soil Colloids of the Montmorillonite Group, *Compt. Rend. Acad. Sci. URSS*, **17**: 375–377 (1937).

further washing to remove all the free chloride, the gel was held in a closed vessel at the temperature of the laboratory for 4 years, at the end of which X-ray diffraction showed the presence of a product similar to smectite.

Caillere and Henin[23] reported the synthesis of clay minerals by the electrolysis of a solution of silica or a mixture of solutions of silica and alumina. Platinum was used as the cathode and various anodes of aluminum, iron, nickel, and magnesium were tried. The product depended on the anode used; e.g., with a magnesium anode an antigorite-like mineral was formed, and with aluminum the product seemed to be kaolinite.

Stresse and Hofmann[7] reported that magnesium silicate gels obtained by boiling together solutions of $MgCl_2$ and hydrated silica developed clay-mineral structures when they were boiled further with potassium hydroxide, calcium hydroxide, or sodium hydroxide. The structures in every case were like those of smectite, except when the concentration of potassium hydroxide became fairly high; then a mica structure developed.

Teichner and Pernoux[24] pointed out that some catalysts for the Fischer process frequently are prepared by impregnating kieselguhr with nickel and that in the impregnation of the silica the catalyst develops a smectite-like structure. Unpublished reports[25] indicate that nearly all the clay minerals have been synthesized with nickel proxying for aluminum. The conditions under which the syntheses were made, however, have not been revealed.

Various clay minerals, including kaolinite, have been synthesized at low temperatures by mixing very dilute solutions of a silicate, or silicate and aluminate, with dilute solutions of magnesium and aluminum salts (Caillere and Henin[26] and Henin and Robichet[27]). Henin[28] stated that numerous attempts have shown that it seems to be impossible to crystallize coprecipitates of silicate and hydroxides at low temperatures but that by allowing dilute solutions of the components of the clay minerals to pass slowly into a flask containing distilled water it was possible to prepare substances similar to the clay minerals.

[23] Caillere, S., and S. Henin, Essais de synthèse des minéreaux argileux, *Verre Silicates Ind.,* **13:** 63–64 (1948).

[24] Teichner, S., and E. Pernoux, Texture and Surface Area of Kieselguhrs after Certain Treatments, *Clay Minerals Bull.,* **1**(5): 145–150 (1951).

[25] DeLange, J. J., Notes, *J. Sci. Instr.,* **24**(1): 20 (1947).

[26] Caillere, S., and S. Henin, Passage expérimental de la montmorillonite à une phyllite à équidistance stable de 14 Å, *Compt. Rend.,* **224:** 842–843 (1947).

[27] Henin, S., and O. Robichet, Sur les conditions de formation des minéraux argileux par voil expérimentale, a basse température, *Compt. Rend.,* **236:** 517–519 (1953).

[28] Henin, S., Synthesis of Clay Minerals at Low Temperatures, *Natl. Acad. Sci., Publ.* 456, pp. 54–60 (1956).

Henin[29] in 1957 stated that, in syntheses of the type mentioned above, there must be formed initially a hydroxide layer upon which SiO_2 can be deposited. The mechanism of the crystallization has more recently been studied in considerable detail. Thus, Gastuche et al.[30,31] concluded that the gibbsite layer plays a predominant part in the bidimensional orientation of the SiO_4 tetrahedrons and that this layer is a kind of nucleus for phyllite crystallization, with low pH and desaturated conditions favoring hexacoordination of aluminum and decrease in the solubility of SiO_2; gibbsite and kaolinite appear subsequently. The higher pH values and higher content of ions in the electrolyte stabilize the tetracoordination of aluminum and increase the solubility of SiO_2 with the consequent formation of 2:1 minerals.

Work by Okkerse and DeBoer[32] showed that silica gels prepared at an acid pH have a very high surface area, whereas those prepared in an alkaline medium have only a feeble surface. Fripiat and Gastuche[33] concluded, on the basis of laboratory studies, that under very acid conditions (pH 2) the silica is extremely depolymerized, whereas at an alkaline pH the silica gel is strongly polymerized. These authors concluded that the mechanism of the formation of one-to-one layer silicates is profoundly different from that of the formation of two-to-one layer silicates. In the first case, it is a matter of the fixation of depolymerized silica tetrahedrons on an octahedral bed, whereas in the second case, under alkaline conditions, it is a matter of the formation of silica layers and octahedral layers in a disordered stacking arrangement.

Granquist and Pollack[34] succeeded in synthesizing hectorite at atmospheric pressure and reflux temperature. The reaction mixture was the system SiO-MgO-Li_2O in a large excess of water. Lithium was found to accelerate the crystallization. Lowering of the pH of the reaction mixture by the use of alkali metal fluorides rather than hydroxides resulted in an increase in the solubility of the magnesium hydroxide which also caused the crystallization of the hectorite to proceed more rapidly.

[29] Henin, S., Synthesis of Clay Minerals, *Atti. Simp. Intern. Agrochim., 1°, Pisa,* pp. 232–241 (1957).

[30] Gastuche, M. C., J. J. Fripiat, and C. DeKimpe, Genesis of Clay Minerals of the Kaolin Family, I, Colloidal Aspect, *Centre Natl. Rech. Sci. (Paris),* **105:** 57–65 (1962).

[31] Gastuche, M. C., and C. DeKimpe, Genesis of Clay Minerals of the Kaolin Family, II, Crystallographic Aspects, *Centre Natl. Rech. Sci. (Paris),* **105:** 67–81 (1962).

[32] Okkerse, C., and J. H. DeBoer. See Fripiat and Gastuche.[33]

[33] Fripiat, J. J., and M. C. Gastuche. Réflexions sur les problèmes de synthèses, "Genèse et synthèses des argiles," *Centre Natl. Rech. Sci. (Paris),* **105:** 207–210 (1962).

[34] Granquist, W. T., and S. S. Pollack, A Study of the Synthesis of Hectorite, Proceedings of the 8th National Clay Conference, pp. 150–156, Pergamon Press, New York, (1960).

Transformations of clay minerals at ordinary temperatures and pressures.
Many investigators, including Caillere and Henin,[35] Volk,[36] Aleshin,[37] and
Barshad,[38] have shown that a material substantially like illite is produced
from smectite when all the exchange positions are occupied by potassium
ions and the material is thoroughly dried at about 110°C. Such material
does not again expand even on treatment with a polar organic liquid, and
it has substantially the same X-ray-diffraction characteristics as illite. Similar results are obtained under the same conditions by treating smectite with
NH_4^+. Vermiculite reacts with both K^+ and NH_4^+, as does smectite, and
with similar results.

Caillere and Henin[39] and Barshad[38] showed also that the treatment of
either smectite or vermiculite under certain conditions with a solution containing Mg^{++}, so that all the exchange positions are occupied by the magnesium
ions, results in a product which does not expand and has the X-ray-diffraction
characteristics of chlorite. The exact experimental conditions of the treatment process necessary to obtain chlorite rather than montmorillonite with
exchangeable Mg^{++} are not known.

Several investigators (Sawhney[40] and Barnhisel and Rich[41]) have shown
that treatment of smectites with various aluminum salts followed by aging
and/or moderate heat treatment causes an alumina interlayer to develop,
producing a 14-Å mineral resembling chlorite.

Caillere and Henin[42] reported the formation of kaolinite from montmorillonite in the laboratory. The kaolinite was obtained by treating montmorillonite with 20 percent $CaCl_2$, 0.5 percent Na_3AlO_3, and 10 percent $Al(NO_3)_3$
for 3 to 4 days and then precipitating with HCl or NH_4OH.

Recently, there has been much work on the formation of clay minerals

[35] Caillere, S., S. Henin, and S. Meriaux, Transformation expérimentale d'une montmorillonit en une phyllit à 10 Å type illite, *Compt. Rend.*, **226**: 680–681 (1948).

[36] Volk, G., Nature of Potash Fixation in Soils, *Soil Sci.*, **45**: 263–276 (1938).

[37] Aleshin, S. N., Changes of Montmorillonite into Hydro-mica, *Dokl. Akad. Nauk SSSR,*
61: 693–695 (1948).

[38] Barshad, I., The Effect of Interlayer Cations on the Expansion of the Mica Type of
Crystal Structure, *Am. Mineralogist*, **35**: 225–238 (1950).

[39] Caillere, S., and S. Henin, Experimental Formation of Chlorites from Montmorillonite,
Mineral. Mag., **28**: 612–620 (1949).

[40] Sawhney, B. L., Aluminum Interlayers in Clay Minerals, Montmorillonite, and Vermiculite: Laboratory Synthesis, *Nature,* **187**: 261–262 (1960).

[41] Barnhisel, R. I., and C. I. Rich, Gibbsite Formation from Aluminum-Interlayers in
Montmorillonite, *Soil Sci. Soc. Am. Proc.,* **27**(6): 632–635 (1963).

[42] Caillere, S., and S. Henin, Formation of a Phyllite of the Kaolinite Type by Treatment
of Montmorillonite, *Compt. Rend.,* **224**: 53–55 (1947).

by the reaction of pure water, and of water containing dissolved components, with various silicates, particularly the feldspars. Correns[43] and Keller[44] have presented summaries of this work. Many investigators have considered the mechanism of the alteration of the primary silicates and their stability. Thus Barshad[45] listed the fundamental factors controlling the stability of the primary silicates under weathering conditions. DeVore[46] indicated how the regularity or nonregularity of Al-Si in tetrahedral coordination in the feldspars may direct the alteration toward three-or two-layer clay minerals, respectively. Nash and Marshall[47] and Garrels and Howard[48] discussed the physical chemistry of silicate mineral surfaces in relation to mineral transformations.

There has been much consideration of the actual transition of the primary silicate to the clay mineral; i.e., is the transition through an intermediate amorphous substance on the silicate surface or separated from it, or is there a direct transition with an epitaxial fit between the primary and secondary mineral, with the structure of the primary mineral exerting some control of the reaction? Correns[49] reported the results of a large number of experiments on the weathering of silicate minerals, using an ultrafiltration technique. He concluded that the components of framework silicates entered the solution as ions and that there was formed on the particles, at least under suitable experimental conditions, a very thin residue layer, the thickness of which does not increase during the course of decomposition. The composition of the residue layer is such that the $SiO_2:Al_2O_3$ ratio does not correspond to that of kaolinite or that of montmorillonite. Furthermore the residue layer gives no X-ray-diffraction lines, and no crystal formation could be observed

[43] Correns, C. W., Experiments on the Decomposition of Silicates and Discussion of Chemical Weathering, Proceedings of the 10th National Clay Conference, pp. 443–459, Pergamon Press, New York (1963).

[44] Keller, W. D., Processes of Origin and Alteration of Clay Minerals, "Soil Clay Mineralogy," pp. 3–76, The University of North Carolina Press, Chapel Hill, N.C. (1964).

[45] Barshad, I., Chemistry of Soil Development, "Chemistry of the Soil," chap. I, pp. 1–70, American Chemical Society Monograph (1964).

[46] DeVore, G. W., The Surface Chemistry of Feldspars as an Influence on Their Decomposition Products, Proceedings of the 6th National Clay Conference, pp. 26–41, Pergamon Press, New York (1959).

[47] Nash, V. E., and C. E. Marshall, The Surface Reactions of Silicate Minerals, II, Reactions of Feldspar Surfaces with Salt Solutions, *Missouri, Univ., Agr. Expt. Sta., Res. Bull.* 614 (1956).

[48] Garrels, R. M,, and P. Howard, Relations of Feldspar and Mica with Water at Low Temperature and Pressure, Proceedings of the 6th National Clay Conference, pp. 66–88, Pergamon Press, New York (1959).

[49] Correns, C. W,, The Experimental Chemical Weathering of Silicates, *Clay Minerals Bull.*, 4(26): 249–265 (1961).

with electron diffraction. Recently Tchoubar[50] concluded, on the basis of electron diffraction and microscopy, that under the conditions of some of his experiments there were an epitaxial fit and structural control in the development of kaolinite from albite. As Keller[44] pointed out, presently available data indicate that the mechanism and sequence of structural changes in the development of clay minerals from primary silicates vary, depending on materials and conditions, and that further work is necessary before broad generalities can be established.

General conclusions from synthesis data. Some general conclusions regarding the environmental conditions favorable for the formation of the clay minerals can be reached from the available data from synthesis experiments. At low temperatures and pressures, acid conditions apparently favor the formation of the kaolinite type of mineral, whereas alkaline conditions favor the formation of smectite or mica, if potassium is the alkali metal and if it is present in concentration above a certain level. From the work of Noll, the minimum concentration of K_2O necessary for the formation of the mica is between 0.2 and 0.37 mole per mole of Al_2O_3. The presence of magnesium particularly favors the formation of montmorillonite.

At temperatures somewhat above about 350°C and with moderate pressures, pyrophyllite forms in place of kaolinite, with an excess Al_2O_3 forming boehmite. At more elevated temperatures and pressures, other alumina phases develop. At more elevated temperatures, the foregoing generalities for results under acidic and alkaline conditions do not always hold. Thus, mica can form under acid conditions, and kaolinite or pyrophyllite can develop in the presence of an excess of K_2O, depending on the temperature, the amount of K_2O and the $Al_2O_3:SiO_2$ ratio.

There appears to be no reported synthesis of attapulgite or sepiolite under hydrothermal or low-temperature conditions. However, Mumpton and Roy[51] reported that these minerals readily alter to smectites by mild hydrothermal treatment.

Clay minerals of hydrothermal origin

Introduction. It has been recognized for a long time that argillaceous alteration products due to hydrothermal action are frequently to be found as an

[50] Tchoubar, C., Formation de la kaolinite à partir d'albite altérée par l'eau à 200°C. Etude en microscopie et diffraction électroniques, *Bull. Soc. Franc. Mineral Crist.*, **88**: 483–518 (1965).

[51] Mumpton, F. A., and R. Roy, New Data on Sepiolite and Attapulgite, *Natl. Acad. Sci., Publ.* 566 pp. 136–143 (1958).

aureole around metalliferous deposits. Such alteration products are also found associated with hot springs and geysers, in vesicles in igneous rocks, and in fractures.

The size of the alteration mass varies widely. At Butte, Montana,[52] for example, a width of 10 ft is common in the peripheral area of the ore body, whereas the width may become 100 ft or more in the zone of intermediate mineralization intensity. In fissure veins the alteration lies parallel to the walls of the fissure and is of relatively uniform width, varying according to the size of the vein.

Bateman[53] pointed out that with epithermal veins the alteration zone is likely to be narrow and the changes may be hardly discernible, whereas with mesothermal veins the alteration halo is likely to be wide and intense. If the veins are closely spaced, the alteration halos may merge, as in the central zone at Butte and in porphyry copper deposits.

Descriptions of hydrothermal alteration products, before the development of modern X-ray, thermal, and optical methods, are of little value because of questionable clay-mineral identifications. For example, a large mass of hydrothermal halloysite at Eureka, Utah, was first described as talc. Unfortunately, even in very recent years, some students of ore deposits have failed to realize the complexities of clay-mineral identification and have published determinations which are questionable because they are based on insufficient analytical data. The study of the alteration products is complicated by the difficulty of distinguishing between hypogene and supergene products. In many cases, later supergene alteration is superimposed on earlier hypogene alteration, and the same clay mineral may be developed in both processes. Kaolinite, for example, in some instances may be formed as a hypogene product and later by supergene alteration of some other clay minerals.

The probable economic importance of wall-rock alteration has been recognized in recent years, because of the likelihood of significant relationships between kind of alteration and mineralization. As a consequence, the problem has been studied by many investigators, but much of the data have not yet been published, and the topic has not reached the point where many broad generalizations are established. Further, some large deposits of kaolinite, halloysite, and smectite are of value in industry, and for this reason they have been studied.

Types of clay minerals in hydrothermal deposits. The alteration products associated with hypothermal deposits are generally not argillaceous, and the secondary products found with some other types of metalliferous ore bodies

---

[52] Sales, R., and C. Meyer, Wall Rock Alteration at Butte, Montana, *Am. Inst. Mining Met. Engrs., Publ.* 2400 (1948).

[53] Bateman, A. M., "Economic Mineral Deposits," Wiley, New York (1942).

may contain few or substantially no clay minerals. All the various clay minerals have been authentically reported in hydrothermal bodies. The various forms of the kaolinite minerals are recorded, including anauxite, dickite, and nacrite, as well as various smectites; including nontronite and the magnesium-rich members. Takeshi[54] reported the occurrence of diaspore as an alteration product in the central zone of the Roseki ore deposit in Japan. Dickite and nacrite are associated with the diaspore. Kaolinite minerals are present in an outer alteration zone.

Cahoon[55] described an occurrence of saponite which has apparently formed by the hydrothermal replacement of dolomitic limestone at a low temperature near Milford, Utah. Konta and Sindelar[56] described saponite as a fissure-filling amphibolite associated with zeolites from an occurrence near Caslan in Czechoslovakia. The white mica found in alteration halos is frequently described as sericite and undoubtedly has similarities to illite. The beidellite which is frequently recorded, particularly in the earlier literature, is probably often a mixture[57] of clay minerals. Halloysite and allophane are often reported, and although the identifications are sometimes well established, some are based solely on optical measurements and must be questioned. Material appearing to be isotropic and therefore apparently amorphous and called allophane may be revealed by X-ray diffraction to be largely crystalline but extremely fine-grained and an aggregate of randomly oriented particles.

Skvortsova and Kopenhenova[58] described allophane as a hydrothermal alteration product in granodiorites associated with quartz-pyrite veins. Vivaldi and Vilchez[59] described an occurrence of halloysite formed from the alteration of an acidic volcanic rock near Melilla in Morocco. Gibbsite and montmorillonite are found associated with this halloysite. Donath[60] described a large deposit of halloysite from a lead and zinc ore deposit in Novo Birdo

[54] Takeshi, H., Kaolin Minerals in "Roseki" Ore Deposits in Japan, *Kobutsugaku Zasshi*, 3: 388–405 (1958).

[55] Cahoon, H. P., Saponite Near Milford, Utah, *Am. Mineralogist*, 39: 222–230 (1954).

[56] Konta, J., and J. Sindelar, Saponite from Pressure Fillings in the Amphibolites of Caslan, *Acta Univ. Carolinae Geol.*, 1: 177–186 (1955).

[57] Grim, R. E., and R. A. Rowland, Differential Thermal Analyses of Clay Mineral and Other Hydrous Materials, *Am. Mineralogist*, 27: 746–761 (1942).

[58] Skvortsova, K. V., and E. V. Kopenhenova, Formation of Allophane Under Hydrothermal Conditions, *Zap. Vses. Mineralog. Obshchestva*, 87: 695–698 (1958).

[59] Vivaldi, J. L., and F. G. Vilchez, A Study of the Halloysite from Maazza (North Morocco), "Journees Internationales d'étude des Argiles," L'Association belge pour favoriser l'étude des Verres et des Composes siliceux (1958).

[60] Donath, M., Halloysite from a Deposit of Lead-Zinc Ore in Novo Birdo, Yugoslavia, *Tonind.-Ztg. Keram. Rundschau*, 85: 426–429 (1961).

in Yugoslavia. Loughnan and Craig[61] described an occurrence of fully hydrated halloysite at Muswellbrook, New South Wales, Australia, which is considered to have originated from the rehydration of metakaolinite formed by the thermal metamorphism of well-crystallized kaolinite.

Caillere[62] described palygorskite associated with traces of chalcopyrite and cuprite in a granite from Morocco. The palygorskite is reported to have formed by hydrothermal action after the emplacement of the granite. Stephen[63] reported an occurrence of palygorskite in vertical crushed bands in a syenite in the Shetland Islands which is believed to have formed from hydrothermal alteration. Ehlmann, Sand, and Regis[64] described four widely separated occurrences of sepiolite in Utah and Nevada. In all cases the host rock is a magnesium carbonate, and the sepiolite is associated with quartz, opal, and sulfide mineralization and nearby acidic intrusive rocks. It is believed that the sepiolites are formed by low-temperature hydrothermal solutions during the last stages of the sulfide mineral deposition in the carbonate beds.

Iwao[65] reported an interesting gypsum deposit of shallow hydrothermal replacement origin situated in marine Tertiary strata that are largely pyroclastic. An alteration zone consisting of an innerchlorite zone and an outer montmorillonite zone surrounds the gypsum deposit.

Sudo and Hayashi[66] described diaspore as a hydrothermal wall-rock alteration product associated with kaolinite, chlorite, and montmorillonite associated with sulfide ore deposits in Japan. These same authors also reported the presence of regular mixed-layer montmorillonite-chlorite components in some hydrothermal wall-rock-alteration occurrences.

Heystek[67] earlier reported formation of regular-mixed-layer illite-montmorillonite structures by hydrothermal alteration of a shale composed of illite.

[61] Loughnan, F. C., and D. C. Craig, An Occurrence of Fully-Hydrated Halloysite at Muswellbrook, NSW, *Am. Mineralogist,* **45**: 783–790 (1960).

[62] Caillere, S., A Palygorskite from Tafraout (Morocco), *Compt. Rend.,* **233**: 697–698 (1951).

[63] Stephen, I., An Occurrence of Palygorskite in the Shetland Isles, *Mineral. Mag.,* **30**: 471–481 (1954).

[64] Ehlmann, A. J., L. B. Sand, and A. J. Regis, Occurrences of Sepiolite in Utah and Nevada, *Econ. Geol.,* **57**: 1085–1094 (1962).

[65] Iwao, S., Hydrothermal Gypsum Deposits of the Wanibuchi Mine, Japan, with Special Reference to Alteration of Wall Rocks, *Japan. J. Geol. Geography,* **27**: 105–132 (1956).

[66] Sudo, T., and H. Hayashi, Diaspore Found in Association with Zincblend and Pyrite Ores of Hanoaka Mine, Akita Prefecture, *Mineral. J. (Tokyo),* **2**: 187–192 (1957).

[67] Heystek, H., An Occurrence of a Regular Mixed-layer Clay Mineral, *Mineral. Mag.,* **30**: 400–408 (1954).

Bassett[68] and Varley[69] described vermiculite formed by hydrothermal action at the contact of acid intrusives and basic or ultrabasic rocks.

In some instances, hydrothermal clay-mineral bodies are of large size (of the order of 100 ft in diameter) and are composed substantially of a single clay mineral. An example is the Cornish kaolin, whose hydrothermal origin is now well established by the finding of kaolin masses which are overlain by sills and which do not reach the surface (Exley[70]). Another example is the halloysite body at Eureka, Utah, which is composed of a single clay mineral. The so-called bentonite from the island of Ponza, Italy, may be an example of a huge mass of almost pure montmorillonite formed from the alteration of an eruptive igneous rock. In other cases, the hydrothermal origin of such large bodies of clay minerals is not agreed upon. Thus, Konta and Pouba[71] concluded that the kaolinization of the granites in the Zettlitz area of Czechoslovakia is due to weathering processes rather than hydrothermal alteration.

Mode of formation and occurrence with ore bodies. In many hydrothermal clay bodies a zonal arrangement of the clay minerals has been observed. Often there is an inner halo of sericite, an intermediate zone of kaolinite, and an outer zone of smectite and chlorite, with chlorite being most abundant on the outermost fringe. The boundaries of the zones are gradational. In some other bodies the alteration products appear to be superimposed rather than zonal, as if they had been developed in several fairly distinct stages.

Because of the seeming difference in the occurrence of the argillaceous alteration products, two concepts of their formation have been suggested. On the basis of their classic study of wall-rock alteration at Butte, Montana, Sales and Meyer[52] concluded that the alteration is essentially contemporaneous with ore deposition and that different mineralogical zones in the surrounding rock may be regarded as reaction rims, representing diminishing activity of the powerful solution that deposited the ore minerals. The different mineralogical and chemical responses within the wall rock are dependent not on a drastic change in the hydrothermal fluid itself, or on a periodicity in its operation, but on continuously varying conditions of physicochemical environment within the wall rock outward from the vein.

[68] Bassett, W. A., The Geology of Vermiculite Occurrences, Proceedings of the 10th National Clay Conference, pp. 61–69, Pergamon Press, New York (1963).

[69] Varley, E. R., Vermiculite, *Colonial Geol. Mineral Resources (Gt. Brit.) Surv.* (1952).

[70] Exley, C. S., Magmatic Differentiation and Alteration in the St. Austell Granite, *Quart. J. Geol. Soc. London,* **114:** 197–230 (1958).

[71] Konta, J., and Z. Pouba, Excursion Guide, *2nd Conf. Clay Mineral Petrog., Prague* (1961).

The second concept appears in the writings of Lovering,[72] Kerr et al.,[73] Schwartz,[74] and others[75] who have described wall-rock alteration at numerous deposits which seems to have developed in stages, by differing solutions appearing at different periods separated by time intervals of relative inactivity. In some cases the products developed during a given stage show some zoning. Often there appears to be an early stage of the development of chlorite and sericitic mica, a second stage of development of halloysite, kaolinite, and smectite, and a late stage of development of sericite. In the second stage, the halloysite and kaolinite are likely to be close to the vein, and Lovering recognized a zoning in some deposits of dickite-kaolinite-halloysite outward from the vein.

It seems likely that both concepts are applicable at particular deposits. Adequate information is not yet at hand to determine what controls the operation of one or the other process at a given deposit. Also, adequate data are not available for broad generalizations on the nature of the zoning or on the sequence of clay minerals likely to develop in the periodic process.

According to Sales and Meyer,[52] the alteration of the quartz monzonite at Butte, Montana, consists of an outermost fringe in which the hornblende and biotite are changed to chlorite. Progressing toward the vein, smectite becomes the dominant alteration product developing from plagioclase, with the nontronite variety forming from hornblende. A small amount of amorphous material is also formed with the smectite. Progressing further toward the vein, kaolinite is found in increasing abundance, with the gradual reduction in the amount of smectite. In the kaolinite subzone there is always some smectite present, and frequently there is a regeneration of the chlorite. In the argillaceous alteration zone, the orthoclase is unaltered except for a faint trace of kaolinization at the innermost edge. The argillaceous zone grades veinward into the sericite zone, which is made up solely of sericite, quartz, and pyrite. In this zone all the remaining silicate minerals are altered.

Sales and Meyer showed that the abundance of lime, soda, and silica declines considerably and magnesia moderately, from the unaltered rock toward the vein. They concluded that there is a continual removal of these components during the gradual outward movement of the alteration; i.e., these

[72] Lovering, T. S., "Rock Alteration as a Guide to Ore—East Tintic District, Utah," Economic Geology Monograph I (1949).

[73] Kerr, P. F., J. L. Kulp, C. M. Patterson, and R. J. Wright, Hydrothermal Alteration at Santa Rita, New Mexico, *Bull. Geol. Soc. Am.*, **61:** 275–348 (1950).

[74] Schwartz, G. M., Hydrothermal Alteration to the "Porphyry Copper" Deposits, *Econ. Geol.*, **42:** 319–352 (1947).

[75] Anderson, A. L., Structural Control and Wall Rock Alteration at the Webber Mine, Butte County, Idaho, *Econ. Geol.*, **42:** 368–383 (1947).

components are carried outward as they are released by the breakdown of the parent minerals. The result over a period of time is a gradual pushing outward of the front of attack as the total envelope of alteration grows wider while maintaining the identity and positions relative to each other of its component zones. The smectite subzone, for example, prevailed, until the outward migration caused a depletion of alkalies and alkaline earths, when kaolinite became the stable phase.

According to Sales and Meyer, sericitization and argillization are contemporaneous processes. The potash content is nearly constant throughout the entire alteration halo. The orthoclase feldspars break down only in the sericite zone, and in fact the beginning of the sericite zone coincides with the destruction of the potash feldspar. Orthoclase is broken down only close to the vein as a consequence of the higher temperatures and the greater potency of hydrothermal acid water. With the destruction of the orthoclase near the vein, the potash necessary for the formation of sericite is available. Sericite is not formed at a distance from the vein because of the absence of available potash as a result of the stability of the orthoclase. An alkaline environment, in addition to the presence of potash, is usually required for the formation of white mica, but Gruner[8] showed that the mica can form in an acid environment at temperatures of the order of 350°C in the presence of an excess of potassium.

Lovering[72] described wall-rock alteration in the Tintic district of Utah as a periodic development. The first stage was chloritization of the volcanic rocks and dolomitization of the limestones, both involving the addition of magnesium. The second stage, or argillitic stage, witnessed the development of the clay minerals other than chlorite. The clay minerals are better developed in the volcanics than in the limestones, and Lovering was able to detect a zoning within the argillic zone consisting of dickite-kaolinite-halloysite $(2H_2O)$-beidellite-halloysite $(4H_2O)$, reading progressively away from the ore deposit. The final stage was extensive sericitization and silicification.

Kerr and colleagues[73] described the hydrothermal alteration of intrusives at Santa Rita, New Mexico, as occurring in several stages. The initial stage of clay-mineral development is chloritization of biotite and hornblende. The next stage is one of "hydromica-argillic alteration," with the development primarily of hydromica from the biotite and chlorite, and of sericite and kaolinite from the plagioclase feldspars. The final stage is one of sericitization and silicification, in which nearly all the primary silicates are altered. The absence of smectite at Santa Rita is noteworthy.

Peterson and colleagues[76] described three stages of alteration at Castle Dome, Arizona. The earliest stage is the development of chlorite and sericite

---

[76] Peterson, N. P., C. M. Gilbert, and G. L. Quick, Hydrothermal Alteration in the Castle Dome Copper Deposit, Arizona, *Econ. Geol.*, **41**: 820–846 (1946).

with some nonclay minerals. This is followed by an argillic stage in which smectite and "hydromica" are the important constituents. Kaolinite is also developed in the argillic stage, but it is of minor abundance. The final phase is one of intense sericitization and silicification.

Schwartz[74] summarized the data for the alteration associated with a series of porphyry coppers and showed that there is great variation in the kind and amount of alteration in various ore bodies. In some deposits, notably at Ajo, Arizona, there are substantially no clay-mineral alteration products, whereas, in others, e.g., at Morenci and Miami, Arizona, hypogene clay minerals are extensively developed. He also showed that the identity and relative abundance of the clay minerals differ widely in different deposits. In most deposits, however, the development of argillic material seems to be followed by a final period of intense sericitization and silicification.

Iwao[77] studied many hydrothermal clay deposits in Japan and arrived at the following general conclusion. Such deposits situated at depths less than a kilometer show the following vertical and horizontal zoning in mineral assemblages: vertical zoning from upper to lower; silica, alunite, kaolinite, pyrophyllite, and diaspore: horizontal zoning, inner to outer; pyrophyllite, sericite, montmorillonite, chlorite. The zoning was presumably formed because of the reaction between preexisting rocks and acid mineralizing solutions.

Schwartz[78] summarized further the early literature. He presented a table showing the sequence of alteration at 12 well-known mining localities and pointed out that the general order in which the minerals tend to form is by no means constant from district to district and that this is to be expected since temperature, pressure, and composition would not be entirely similar in widely spread districts. He emphasized that in some places, for example, at Butte, Montana, the evidence is convincing for the simultaneous growth of argillic and sericitic zones, whereas in other places, for example, at Morenci in Arizona where a large mass of rock has been argillized, the evidence seems to favor an initial stage of sericitization followed by a later stage in which the principal minerals developing were kaolinite, halloysite, allophane.

Gorlich[79] concluded that the sequence of formation in the course of increasing pH and decreasing temperature is as follows: dickite, nacrite, sericite, kaolinite, illite, smectite.

Stringham[80] reported seven stages of hydrothermal activity attending

[77] Iwao, S., Geologic Problems on Some Hydrothermal Clayey Deposits in Japan, *Mining Geol. (Japan)*, **2**: 120–130 (1953).

[78] Schwartz, G. M., Argillic Alteration and Ore Deposits, *Econ. Geol.*, **51**: 407–414 (1956).

[79] Gorlich, E., Origin of Clay Minerals, *Arch. Mineral. (Poland)*, **21**: 409–426 (1957) (Pub. 1959).

[80] Stringham, B., Granitization and Hydrothermal Alteration at Bingham, Utah, *Bull. Geol. Soc. Am.*, **64**: 945–992 (1953).

ore deposition in the disseminated copper deposit at Bingham, Utah. Stage 1 saw the formation of kaolinite and illite. In stage 2 biotite and sericite were widely developed. In stage 3 chlorite and biotite were formed in restricted areas. In stage 4 quartz and sericite formed over wide areas. In stage 5 quartz was deposited in open fissures without accompanying alteration. Stage 6 was the sulfide stage, and stage 7 saw the development of sericite and allophane in open fissures.

Bonorino[81] found six patterns of hydrothermal alteration in the Front Range mineral belt of Colorado. They ranged from the simplest type with a single zone of hydromica to the most complex type with five zones characterized by a phlogopite carbonate (fringe zone), montmorillonite, kaolinite, orthoclase, and hydromica (innermost zone). A regular mixed-layer illite-smectite mineral may substitute for the hydromica either in all or in the outer part of the innermost zone. The most complex patterns were associated preferably with the tungsten-pyritic-gold and telluridic gold deposits. The origin of the complex pattern is explained in terms of chemical fronts developing as the altering fluids migrated away from the conduit.

It is clear that there is considerable variation in the character of hydrothermal alteration products associated with ore bodies and in their sequence of formation. As a consequence, only very broad generalities, as indicated above, are justified.

Nature of the hydrothermal solutions. Information concerning the composition of the hydrothermal solutions is scant, but in many, perhaps most, cases, such waters are thought to be acid to begin with, carrying chlorine, sulfur, carbon dioxide, and/or silica. The composition of the solutions must change as reactions take place with the host rock, and in the case of silicate rocks the solutions would become alkaline as a consequence of such reactions. The alkaline constituents would probably be transported outward by the solutions, and the alkalinity of the solutions would persist only so long as alkalies or alkaline earths were being released by the breakdown of the parent rock. As the action of the acid water proceeded, therefore, the acid-alkaline front would move outward as the alkaline earths and alkalies were carried outward. As Sales and Meyer[52] postulated, the later breakdown of a resistant mineral, such as orthoclase, which had persisted through the beginning of the acid encroachment until temperatures or solution concentration became sufficiently high to break it down, can cause a reversal to alkaline conditions.

It is by no means established that hydrothermal solutions are always acid initially. Andreatta,[82] for example, described a sequence of hydrother-

[81] Bonorino, F. G., Hydrothermal Alteration in the Front Range Mineral Belt, Colorado, *Bull. Geol. Soc. Am.,* **70:** 53–90 (1959).

[82] Andreatta, C., Study of the Hydrothermal Stratum of the Clay Minerals of Capalbia, Italy, *Ind. Ceram. Silicati,* **2:** 17–21 (1949).

mal clay minerals developed in the order micaceous clay mineral–montmoril-lonite–halloysite–kaolinite; this sequence he believed to have formed from warm water originally alkaline and rich in potassium which became neutral and finally acid with progressive cooling.

Iwao[77] concluded that the temperatures of the altering solutions ranged from 150 to 600°C. According to Kerr,[83] the temperatures under which the alteration takes place, interpreted from minerals formed, appear to lie in the range of 100 to 400°C, with emphasis on the middle temperature zone.

Relation to parent material. If the hydrothermal alteration is intense, as a result of relatively high temperatures and concentrations, and long contin-ued, so that all the original minerals have been affected, and if there has been considerable transportation of components, such as alkalies and alkaline earths released by the breakdown of the parent minerals, the alteration prod-ucts will tend to be the same regardless of the parent rocks, except for car-bonate and quartzitic rocks. In the case of limestones, the tendency is toward silicification. In the case of dolomitic rocks, silicification may also be the result, but the reaction between the magnesia and the silica sometimes forms a magnesium-rich clay mineral. The deposit of the smectite mineral hectorite, at Hector, California, probably had such an origin.

If the hydrothermal alteration is relatively slight, the character of the primary mineral largely determines the alteration product. With very slight alteration, the magnesium-rich minerals such as the hornblendes and biotite tend to change to chlorites. With slight alteration and in the presence of alkalies and alkaline earths, except potassium, the micas, the ferromagnesium minerals, and the plagioclase feldspars are likely to yield smectite. The pres-ence of magnesium particularly favors the formation of smectite. The pres-ence of potash, either from feldspars or primary micas, favors the development of secondary micas. For example, Zsolny[84] described alteration in adjacent rhyolite and andesite; the former shows illitic secondary products and the latter, alteration to smectite. The literature has many references to smectite types of alteration products in basic igneous rocks, indicating the frequency of smectitic alteration in this type of rock.

Kaolinite may form from any constituents, if the alkalies and alkaline earths are removed approximately as fast as they are liberated from the parent rock or if the environment is acid and the temperature is moderate. When present in the environment, calcium particularly appears to retard the formation of kaolinite.

Kerr[83] pointed out that such factors as permeability, porosity, and con-

[83] Kerr, P. F., Hydrothermal Alteration and Weathering, *Geol. Soc. Am., Spec. Paper* 62, pp. 525–544 (1955).

[84] Zsolny, L. M., Illite, Montmorillonite, Halloysite and Volcanic Ash as Whiteware Body Ingredients, *J. Am. Ceram. Soc.*, 29: 254–260 (1946).

duits permitting fluid penetration are important in influencing alteration. Similar alteration may result from rocks as dissimilar as diorite and alaskite, while tremendous differences in the magnitude of the alteration halo may be noted between porous tuff and compact quartz monzonite.

Tooker[85] emphasized that the character of the host rock is very important in localizing the zonal arrangement of the alteration. An example of the influence of parent material was given by Knizek[86] who described numerous occurrences of kaolinite of hydrothermal origin in Mexico. In some cases the kaolinite is associated with cristobalite. Knizek found that the presence of glassy material in the parent rock favors the development of cristobalite.

Sales and Meyer,[52] Ross and Hendricks,[87] and others have shown that the clay minerals themselves alter to other clay minerals if the hydrothermal environmental conditions change. Such alterations take place in accord with the generalities mentioned above.

Relation to mineralization. The impetus for much of the recent study of wall-rock alteration associated with ore bodies has been the hope that correlations of type of alteration with mineralization would provide keys for prospecting. In recent years, many publications have disclosed a correlation between type of alteration and mineralization for particular ore deposits. It is obvious from the following that there is no universal correlation, at least on presently available data.

Lovering[72] showed that at Tintic, Utah, ore values are not found when the late stage of sericitization and silicification has not been operative. Kerr and colleagues[73] showed that at Santa Rita, New Mexico, at least part of the mineralization is contemporaneous with the argillic alteration. These authors also pointed out that the feldspars in the area of copper mineralization show prominent kaolinite alteration, whereas in the zinc area the dominant alteration product is montmorillonite. Sales and Meyer[52] observed a general close association of sulfides and the sericite alteration at Butte, Montana.

Leroy[88] reported four stages of hydrothermal alteration at the Santa Rita porphyry copper deposit in New Mexico, starting with nearly fresh rock followed by a second stage of smectite-illite alteration, a third stage of ka-olinitic alteration, and a fourth stage of sericitization. Stage 3 appears to

[85] Tooker, E. W., Altered Wall Rocks Along Vein Deposits in the Central City–Idaho Springs Region, Colorado, *Natl. Acad. Sci., Publ.* 456, pp. 348–361 (1956).

[86] Knizek, I., Cristobalita en arcillas, *Bol. Soc. Espan. Ceram.,* 11: 5–24 (1963).

[87] Ross, C. S., and S. B. Hendricks, Minerals of the Montmorillonite Group, *U.S. Geol. Surv., Profess. Paper* 205B, pp. 23–79 (1945).

[88] Leroy, P. G., Correlation of Copper Mineralization with Hydrothermal Alteration in the Santa Rita Porphyry Copper Deposit, New Mexico, *Bull. Geol. Soc. Am.,* 65: 739–768 (1954).

be most favorable for ore since, besides having a porosity and pyrite content high enough to facilitate secondary enrichment, this stage also furnished the hypogene copper minerals that constitute the protore of the deposit.

Nagasawa[89] found two stages of mineralization accompanied by wall-rock alteration at the Mikawa mine in northeastern Japan. The earlier stage containing galena, sphalerite, and quartz is associated with sericite and kaolinite alteration. The later stage at a lower level containing chalcopyrite and pyrite is associated with kaolinitic alteration.

In the Midnite mine, near Spokane, Washington, Barrington and Kerr[90] found that argillic alteration in the form of kaolinite and illite preceded uranium deposition, with illite forming in the higher-temperature regions. Smectite formed during and following the precipitation of uraninite when the temperatures were lower and the environment possibly became slightly alkaline.

Abdel-Gawad and Kerr[91] found that the argillic alteration accompanying the emplacement of uranium in the Chinle siltstone of the Colorado Plateau consisted of the removal of iron and of the transformation of the smectitic component of mixed-layer clay into the mica clay mineral, with the corresponding increase of the 2M over the 1M mica polymorph.

Shcherban[92] reported that mercury mineralization occurred after the formation of clay-mineral alteration products in the Aktash deposit in Siberia. Numerous other authors (e.g., Tooker[85] and Kerr[83]) have pointed out that frequently most of the alteration is accomplished before the ore minerals are deposited.

Clay minerals associated with hot springs, fumaroles, etc. The action of hot springs, fumaroles, etc., provides an excellent opportunity to study hydrothermal alteration. Fenner,[93] studying samples from boreholes in the Yellowstone area, found that beidellite was the alteration product at depths greater than about 95 ft when the solutions were alkaline. At shallower depths kaolinite was found, and it was believed to have resulted from descending meteoric acid waters. The acid resulted from the oxidation of sulfides.

[89] Nagasawa, K., Mineralization at the Mikawa Mine, Northeastern Japan, *J. Earth Sci., Nagoya Univ.,* 9(1): 129–172 (1961).

[90] Barrington, J., and P. F. Kerr, Uranium Mineralization at the Midnite Mine, Spokane, Washington, *Econ. Geol.,* 56: 241–258 (1961).

[91] Abdel-Gawad, A. M., and P. F. Kerr, Alteration of Chinle Siltstone and Uranium Emplacement, Arizona and Utah, *Bull. Geol. Soc. Am.,* 74: 23–46 (1963).

[92] Shcherban, I. P., Hydrothermal Argillitization of Country Rocks in the Aktash Deposit, *Geol. i. Geofiz., Akad. Nauk. SSSR, Sibirsk. Otd.,* 9: 48–59 (1962).

[93] Fenner, C. N., Bore Hole Investigations in Yellowstone Park, *J. Geol.,* 44: 225–315 (1916).

Steiner[94] studied in detail the hydrothermal rock alteration at Wairakei in New Zealand. In this area numerous boreholes have been drilled to provide steam for the generation of electric power. The cores from the holes, together with information concerning the temperature and composition of the hydro-thermal emanations, have provided a unique opportunity for the study of the development of clay minerals by hydrothermal process. Steiner found that there is a superficial zone containing kaolinite and alunite extending down to depths of about 100 ft. At greater depths up to about 250 ft, with temperatures increasing to about 150°C, there is a zone in which smectite is the dominant alteration product. Below about 250 ft and with tempera-tures reaching 600 to 700°C, zeolite and adularia are the alteration products. At greater depths sericite is formed. Steiner concluded that the initial hydro-thermal agents are acidic and that they become neutralized and then alkaline as they proceed to the surface. The surficial kaolinite is perhaps, in part at least, produced by supergene alteration.

Chen[95] described the alteration of andesitic lavas and pyroclastic rocks in Taiwan by postvolcanic emanations of gases, steam, and hydrothermal waters. He reported that kaolinite is the dominant clay mineral and that associated with it are hydrated halloysite, allophane, smectite, and mixed-layer koalinite-smectite. Minto[96] described the occurrence of kaolinite, hal-loysite, and smectite associated with fine-grained opal, and in some cases vermiculite, as alteration products of hot spring waters at Beppu, Kyushu, in Japan. These waters were determined to have a pH of 3 to 6 and a temperature of 60 to 90°C.

Hayashi and others[97] identified smectite, vermiculite, and randomly inter-stratified combinations of chlorite-vermiculite and regular interstratified com-binations of smectite-vermiculite in amygdules of some basalts from Japan.

Miyamoto[98] found iron-rich saponite as a thin film covering the inner walls of amygdules in some basalts from Japan.

Mitchell[99] found saponite in vesicular cavities in andesitic lavas of the

[94] Steiner, A., Hydrothermal Rock Alteration at Wairakei, New Zealand, *Econ. Geol.* **48:** 1–13 (1953).

[95] Chen, Pei-Yuan, Post-volcanic Alteration of Andesite Lavas and Pyroclasts in Tatun Volcanic Group, Paipei-Hsien, Taiwan, *Sci. Repts. Natl. Taiwan Univ. Acta Geol. Taiwan.,* **9:** 19–38 (1961).

[96] Minato, H., An Occurrence of Clay by the Reaction with Hot Springs, *Nendo Kagaku No Shimpo,* **1:** 85–93 (1959).

[97] Hayashi, H., A. Inaba, and T. Sudo, Complex Clay Mineral Mixtures Occurring in Amygdules of Basalt, *Clay Sci.,* **1:** 12–18 (1961).

[98] Miyamoto, N., Iron-rich Saponite from Maze, Niigata Prefecture, Japan, *Mineral. J. (Sapporo),* **2:** 193–195 (1957).

[99] Mitchell, W. A., An Occurrence of Saponite in Vesicular Lava, *Congr. Intern. Sol, 3rd, Paris,* **1:** 75–79 (1954).

old red sandstone in Scotland. He also found the same mineral in small irregular patches in the ground mass of the rock.

Lyakhovich[100] identified palygorskite associated with zeolites and opal in amygdules in some gabbro diabases from Siberia.

As in the case of hydrothermal alteration associated with ore bodies, it is apparent that other alteration products can be equally varied.

Soils and weathering

Modern investigations by Kelley, Hendricks, MacEwan, Henin, Marshall, Jackson, Barshad, and many other students of soils have provided a tremendous amount of information on all phases of clay mineralogy. Each of these investigators and many others have extensive bibliographies dealing with clay mineralogy, so that it is not feasible to list even the major publications of the most active workers. Soil investigators were the first to apply extensively modern methods of clay-mineral analysis to their problems. In this section, the distribution of clay minerals in various soil types, the environmental conditions leading to the formation of particular clay minerals during weathering processes, and the mechanism of the alteration will be considered.

Factors controlling weathering processes. No arrangement of the factors controlling weathering processes in a general order of importance is possible, since the factors are closely interdependent and since their relative importance is not always the same.

Parent Rock. The composition and texture of the parent rock are important in initial stages of weathering, but their importance decreases as the duration of weathering increases. Under conditions of drastic alteration under humid conditions, as in the development of podsolic and lateritic soils, the influence of parent rock is relatively short-lived, whereas in aridic zonal soils it prevails almost indefinitely. It has been proved that soils containing kaolinite and soils containing smectite can both develop from the same parent rock under different conditions of climate, topography, and time.[101] Also it has been shown[102] that the same type of soil with a characteristic clay-mineral

[100] Lyakhovich, V. V., Minerals of the Palygorskite Group from Yakutia, *Dokl. Akad. Nauk SSSR,* **112:** 322–324 (1957).

[101] Hosking, J. S., The Soil Clay Mineralogy of Some Australian Soils Developed on Granitic and Basaltic Parent Material, *J. Council Sci. Ind. Res.,* **13:** 206–216 (1941).

[102] Humbert, R. P., and C. E. Marshall, Mineralogical and Chemical Studies of Soil Formation from Acid and Basic Igneous Rocks of Missouri, *Missouri, Univ., Agr. Expt. Bull.* 359 (1943).

composition can develop after a relatively long period of time from rocks of widely different composition and texture.

The most important factor of composition in parent rocks is the content of alkalies and alkaline earths. Rocks containing no alkalies, such as kaolin clays, can yield only kaolinite or lateritic weathering products unless ground-water movement brings alkalies and/or alkaline earths to the environment. Igneous rocks, shales, slates, schists, and argillaceous carbonates can yield a variety of weathering products, at least in the initial stages of weathering, because of their content of alkalies and alkaline earths in addition to alumina, silica, etc.

Structural aspects of the parent rock may be important. Thus, Dorfman[103] found that chemical weathering of nepheline syenites of the Khibiny mass of the U.S.S.R. has progressed to a depth of 500 m along fractures, and that there is a difference depending on whether the fractures are steeply or gradually sloping. Kaolinite, hydromica, and nontronite are reported as forming in a steeply sloping fractures, whereas montmorillonite and saponite are forming in those with a gradual slope.

Climate. The temperature and the rainfall, particularly its seasonal distribution, are the significant climatic factors. Decay of the parent minerals is most rapid in warm, humid climates. The direction of the movement of water through the weathering zone and the leaching of soluble salts from the material depend on the amount of rainfall and its seasonal distribution. Thus, in a continuously wet climate the movement of the water is downward, tending to remove downward the components liberated by the decay of the parent minerals. In a continuously dry climate or seasonal one with long dry seasons, the prevailing water movement may be upward, and the decay components would not be removed from the decay zone. In a seasonal wet-dry climate, as compared with a continuously wet one, oxide and hydroxide products (e.g., of iron) may be fixed in the dry interval so that they resist migration during the wet season. The climate determines the amount and kind of vegetation and, further, the amount and kind of organic alteration products yielded by the decay of the organic material. Thus in a humid cool climate an abundance of organic material decays slowly to produce an abundance of active organic acids and other compounds to react with the parent material, whereas in a hot humid climate with seasonal rainfall the organic material is rapidly destroyed by oxidation and alteration. As a consequence, the kind and amount of organic products available for leaching are very different in the two environments.

Topography. Topography determines whether or not there is active verti-

[103] Dorfman, M. D., Linear (Fractural) Weathering Zone in Nepheline Syenites of Khibiny, *Kora Vyvetrivaniya, Akad. Nauk SSSR, Inst. Geol. Rudnykh Mestorozhden., Petrogr., Mineralog., i Geokhim.,* no. 3, pp. 219–245 (1960).

cal movement of water through the weathering material. This is particularly significant in regions of relatively high rainfall in which low, flat areas may be saturated with water almost to the surface. The prevalence of oxidizing or reducing conditions and the depth to which active oxidizing conditions penetrate depend in part on the water saturation of the soils. Poorly drained areas saturated with water for long periods of the year generally have only reducing or feebly oxidizing conditions, or at best a thin surficial zone of oxidation.

Topography, by its control of the vertical movement of ground water, also influences the leaching processes. Thus, in low, flat areas with little movement of water through the soils, there would be relatively little leaching.

Mohr and Van Baren[104] observed in Indonesia that laterite and smectite were formed a short distance from each other from the same parent volcanic material under the same climate. The laterite site was well drained whereas the smectite site was water-logged and silicon metal ions were not removed. Mitchell[105] pointed out that, in the acid soils of Scotland, leaching causes the formation of kaolinite whereas in areas of little leaching expandable minerals are present in the soils.

Topography also influences the rate of erosion at the surface, and hence the rate of removal of the products of weathering and the rate at which fresh parent material is brought close to the surface into the most active zone of alteration.

Vegetation. The significant factors with regard to vegetation are its amount and kind and the amount and kind of resulting products of its decay. In an arid area where vegetation is absent or in an area with long hot, dry intervals where vegetation is quickly oxidized, few or no organic acids are produced from the vegetation to enter into the alteration processes. Also in countries of eucalyptus rather than deciduous trees, there would be scant accumulation of leaf litter and scant production of organic decay products, even though conditions favored the abundant growth of vegetation.

Soil investigations suggest that, under cool, humid conditions with deciduous forest growth, organic decomposition products which have very drastic effects on parent silicate minerals develop in the soil. Under temperate conditions and prairie grass cover, the resulting vegetal decomposition products are different and appear to have less drastic effects on the parent minerals.

Lovering[106] emphasized the significance in weathering processes of plants

---

[104] Mohr, E. C. J., and J. Van Baren, "Tropical Soils," Interscience, New York (1954).

[105] Mitchell, B. D., The Influence of Soil Forming Factors on Clay Genesis, "Genèse et synthèse des argiles," pp. 139–147, Centre National de la Recherche Scientifique, Paris (1962).

[106] Lovering, T. S., Significance of Accumulator Plants in Rock Weathering, *Bull. Geol. Soc. Am.,* **70**(6): 781–800 (1959).

which accumulate certain major elements such as silicon, aluminum, calcium, magnesium, and iron. For example, many kinds of plants, especially in the tropics, contain several percent of silica by dry weight. He computed that a forest of silica-accumulating plants averaging 2.5 percent silica and 16 tons dry weight of new growth per year would extract about 2,000 tons of silica per acre in 5,000 years, equivalent to the silica in 1 acre-foot of basalt. Lovering pointed out the probability that silica-accumulator plants would enhance substantially the amount of silica swept into the drainage system.

Bloomfield[107] showed that extracts of organic material from leaves of different trees have different degrees of deflocculating effects on kaolinite and smectite suspensions and that this could thereby influence weathering processes.

Rozanov[108] concluded that the difference in the migration ability of iron and aluminum in podsolization and burozem processes (dark-brown soils) was related to difference in the action of the various conifers and wide-leaved trees on soil formation.

Time. Since weathering processes are relatively slow, time is a significant factor in the development of alteration products. In the initial stages of the weathering of some rocks, alkalies and alkaline earth may be present in the weathering zone, and as a consequence a certain kind of alteration product will develop. After a long time interval all the alkalies may have been removed by leaching, and a different type of alteration product will form. As a consequence, weathering products from a given type of parent material may be quite different in an initial stage and in a later stage. Bray[109] showed, for example, the great difference in the mineral composition of weathering products of loess in Illinois as a consequence of time. Manifestly, the factor of time is most important when weathering in general is moderately severe and when the parent rock has a composition permitting the formation of a variety of alteration products.

Soil-profile development. Weathering processes, especially where there is vertical (downward or upward) movement of moisture, tend to develop a sequence of horizons within the zone of alteration. The horizons frequently grade sharply into adjacent ones and have thicknesses measured in inches or a few feet, although there are some very mature soils with horizons many feet in thickness. The horizons show differences in degree of breakdown of parent

---

[107] Bloomfield, C., A Study of Podzolization, Part I, The Mobilization of Iron and Aluminum by Scots Pine Needles, *J. Soil Sci.*, **4**(1): 5–23 (1953).

[108] Rozanov, B. G., Aluminum and Iron Migration During Soil Formation in Dense Forests, *Vestn. Mosk. Univ., Ser. VI, Biol., Pochvoved.*, **16**: 67–78 (1961).

[109] Bray, R. H., Chemical and Physical Changes in Soil Colloids with Advancing Development in Illinois Soils, *Soil Sci.*, **43**: 1–14 (1937).

material, in kind and relative abundance of secondary mineral components, in particle-size distribution in organic material, and/or in pH and character of alkali and alkaline-earth content. Vertical movement of water causes the translocation of soil constituents, thereby developing the sequence of varying horizons. A fairly typical profile of a podsolic type of soil is given in

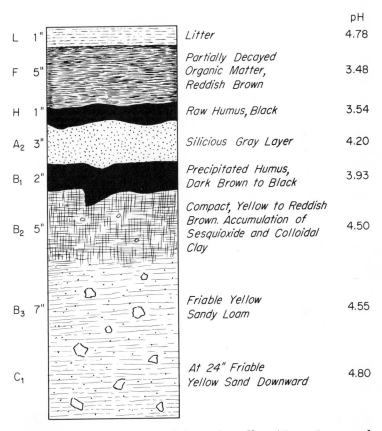

			pH
L	1"	Litter	4.78
F	5"	Partially Decayed Organic Matter, Reddish Brown	3.48
H	1"	Raw Humus, Black	3.54
A₂	3"	Silicious Gray Layer	4.20
B₁	2"	Precipitated Humus, Dark Brown to Black	3.93
B₂	5"	Compact, Yellow to Reddish Brown. Accumulation of Sesquioxide and Colloidal Clay	4.50
B₃	7"	Friable Yellow Sandy Loam	4.55
C₁		At 24" Friable Yellow Sand Downward	4.80

Fig. 13-1. Well-developed podsolic soil profile. (From Lyons and Buckman.[110])

Fig. 13-1, from Lyons and Buckman.[110] As a consequence of the difference in composition and texture, the physical properties of plasticity, tilth, etc., vary from horizon to horizon in the profile.

The soil horizons are designated from the surface downward, A, B, C,

[110] Lyons, T. L., and H. O. Buckman, "The Nature and Properties of Soils," Macmillan, New York (1943).

D. The nomenclature of soil horizons outlined by the Soil Survey Staff of the U.S. Department of Agriculture, as summarized by Winters and Simonson,[111] is as follows: The A horizon is defined as a layer of maximum accumulation of organic matter and/or as the surface layer that is lighter in color than the underlying horizon and has lost some of its clay and sesquioxides. The B horizon is defined as the soil layer that contains an accumulation of clay and sesquioxides with small amounts of organic material and/or has a blocky or prismatic structure and a stronger color than the overlying A horizon. The C horizon is defined as a layer of unconsolidated and partly weathered material relatively unchanged by soil-forming processes. The D horizon is unaltered parent material.

In descriptions of weathering products it is obviously necessary to describe the profile as a whole as well as the composition and properties of the individual horizons.

Classification and description of great soil groups. Students of soils have given much attention to the classification and nomenclature of soils (see Soil Classification—a Comprehensive System, U.S. Department of Agriculture, 1960). It will suffice herein to use a very general classification that will permit a consideration of clay-mineral composition in relation to mode of occurrence and environment of origin. The various types of soil may be considered under three main headings as follows:

1. Zonal Soils. After soil-forming processes have been at work for a considerable time, soils over wide areas, underlain by rocks of varying kinds, tend to become alike in general characteristics if the climatic influences are reasonably uniform and continuous, and if erosion is not too vigorous. Such soil groups are called zonal and constitute the Great Soil Groups of the World. In these soils, climate, with its associated kind of vegetation and organic decomposition material, is the major factor in soil genesis.

2. Intrazonal Soils. These soils are associated with the zonal groups and are those that reflect the influence of some local conditions such as poor drainage, alkali salts, or some other unique characteristic superimposed on the general characteristics resulting from the climate and vegetation factors. Such soils frequently cross zonal boundaries and thus are called intrazonal.

3. Azonal Soils. These are soils without profile development, such as alluvial and colluvial soils. There is little or no alteration of parent material because of their youth or environmental setting.

The classification of the Great Soil Groups with some of the more important associated intrazonal soils suggested by Lyons and Buckman is given in

[111] Winters, E., and R. W. Simonson, The Subsoil, *Advan. Agron.*, **3:** 2–92 (1951).

Table 13-3. A map showing the world distribution of the Great Soil Groups is presented in Fig. 13-2. A somewhat idealized diagrammatic presentation of representative profiles of five zonal soils is given in Fig. 13-3.

Table 13-3   Classification of Great Soil Groups of the World
(After Lyons and Buckman[110])

I. Zonal humid soils
    Tundra
        Dark gray peaty accumulation over gray mottled mineral horizons.   Substratum
           ever frozen and often the subsoil as well
    Podsolic
        Typical Podsol (gray forest soil)
        Brown podsolic
        Gray-brown podsolic
        Red and yellow podsolic (lateritic)
        Prairie dark brown and reddish brown
    Lateritic
        Typical and well-developed laterites
        Yellowish-brown lateritic soils
        Reddish-brown lateritic soils
        Terra rossa
    Associated intrazonal groups
        Planosols—strongly leached surface soils over clay pans
        Rendzinas—dark-colored grassland soils from soft limy material
        Brown forest soils, similar to gray-brown podsolic but neutral instead of acidic
II. Zonal aridic soils*
        Chernozem (black earth)
        Chestnut and reddish chestnut
        Brown and reddish-brown aridics
        Reddish desert soils
        Noncalcic brown soils
    Associated intrazonal groups
        Saline soils (Solonchak)—high content of soluble salts of Ca, Mg, and Na.   Mildly
           alkaline
        Alkali soils (Solonetz)—moderate content of soluble salts but with those of Na domi-
           nant.   Strongly alkaline

* Direction of arrow indicates a progressive reduction in effective rainfall.

The general relation of the great soil groups to climate conditions is shown in Fig. 13-4 after Millar and Turk.[112]   A Pedalfer soil is one in which there is a tendency for alumina and ferric iron to accumulate in the profile.   A Pedocal soil is one in which there is a tendency for calcium carbonate to accumulate in the profile.

[112] Millar, C. E., and F. M. Turk, "Fundamentals of Soil Science," Wiley, New York (1943).

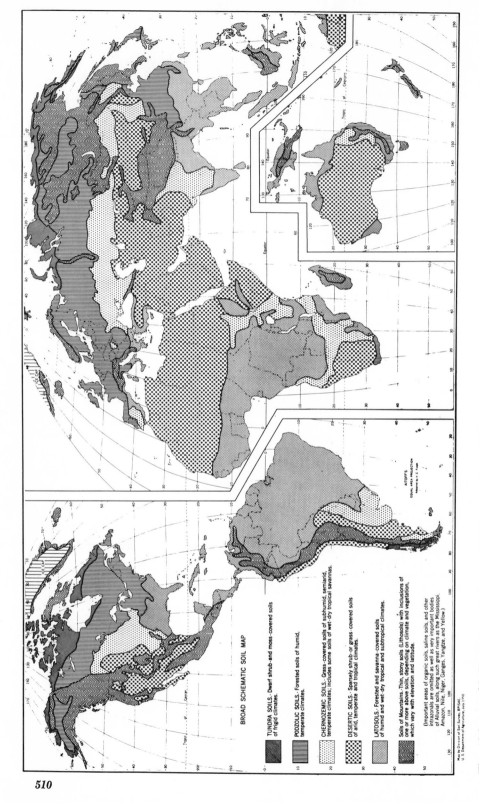

BROAD SCHEMATIC SOIL MAP

TUNDRA SOILS.- Dwarf shrub-and moss-covered soils of frigid climates.

PODZOLIC SOILS.- Forested soils of humid, temperate climates.

CHERNOZEMIC SOILS.- Grass-covered soils of subhumid, semiarid, temperate climates; includes some soils of wet-dry tropical savannas.

DESERTIC SOILS.- Sparsely shrub- or grass-covered soils of arid, temperate and tropical climates.

LATOSOLS.- Forested and savanna-covered soils of humid and wet-dry tropical and subtropical climates.

Soils of Mountains.-Thin, stony soils (Lithosols) with inclusions of one or more above soils, depending on climate and vegetation, which vary with elevation and latitude.

(Important areas of organic soils, saline soils, and other intrazonals are omitted as well as very important bodies of Alluvial soils, along such great rivers as the Mississippi, Amazon, Nile, Niger, Ganges, Yangtze, and Yellow)

Map by Division of Soil Survey, BPISAE.
U. S. Department of Agriculture, July 1940

ATOFFS
EQUAL AREA PROJECTION
Adapted by C. Fogel

Fig. 13-2. World distribution of Great Soil Groups. (From Winters and Simonson.[111])

510

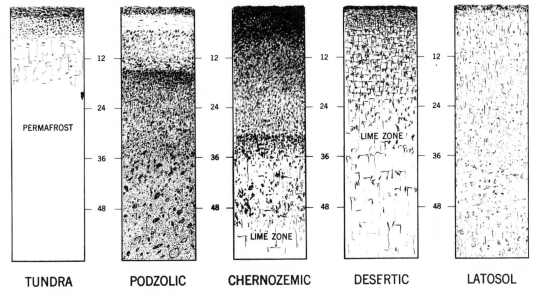

TUNDRA   PODZOLIC   CHERNOZEMIC   DESERTIC   LATOSOL

Fig. 13-3.   Idealized profiles of five zonal soils.   Depth in inches.   (From Winters and Simonson.[111])

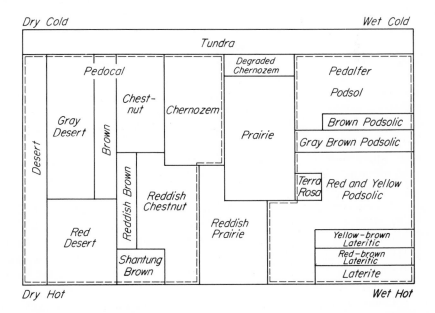

Fig. 13-4.   Classification of soils in relation to climatic conditions.   (After Millar and Turk.[112])

Zonal humid soils. These soils are found in regions wherein the climate varies from cold to tropical, with rainfall greater than about 25 in./year. The vegetation is forest and grasses in warmer regions, and mosses, lichens, and shrubs in tundra areas. The soils are characterized generally by a concentration of iron and aluminum in certain horizons. Ordinarily there is no concentration of carbonate when the soils have reached a mature stage.

Tundra soils. These soils are characterized by an accumulation of organic matter giving a peaty surface overlying a bluish-gray sticky and compact subsoil. The frozen subsoil causes poorly drained, boggy conditions for most of the year. The clay-mineral content of tundra soils is frequently rather low.

Podsolic soils. Such soils are developed under cool and damp climatic conditions, providing an abundant surface accumulation of organic material. Forests are the most common vegetative cover, although podsolization occurs with grass cover. The covering vegetation must be such as to yield on decomposition organic acids and other compounds of great potency, and there must be enough rainfall to provide active leaching conditions. The environment is acid, and pH values as low as 3.5 are common in the humus and 4.0 to 4.5 in the clay component.

Under such conditions carbonates are quickly dissolved and alkalies and alkaline earths are removed from the profile. The base saturation ranges from about 20 to 80 percent and is commonly 40 to 60 percent. The organic and inorganic colloidal components are translocated from the upper horizon downward to about the water table. In addition, leaching conditions are so intense that the inorganic colloidal complex becomes unstable in the surface horizons, and the iron and alumina are carried away to be reprecipitated at about the water table. As a consequence of the removal of aluminum and iron, a gray, highly siliceous, acid horizon develops directly beneath the surface accumulation of organic material. Because of the gray color and ashy appearance of the highly siliceous horizon, the term Podsol (Russian *pod,* meaning under, and *zola,* ash) has been applied to such soils. An example of a well-developed Podsol profile is shown in Fig. 13-1.

Brown podsolic soils are found in the Northern Hemisphere to the south of the true Podsols, and where leaching is less intense. The siliceous gray A_2 horizon (Fig. 13-1) is replaced by a friable light-brown horizon containing considerable organic material. There is little breakdown of the colloidal silicates and little translocation of either the colloidal silicates or the aluminum and iron.

Gray-brown podsolic soils lie to the south of the brown podsolic soils in the Northern Hemisphere and show very mild leaching, primarily because of a vegetative cover which yields less-potent organic decomposition products and a climate which yields a smaller quantity of such products because of more complete oxidation. The gray siliceous layer is replaced by a gray or yellow-brown horizon, and considerable exchangeable calcium remains on the clay minerals. There is some downward movement of colloids, but the organic acids are too feeble to cause appreciable disintegration of the silicates.

Red and yellow podsolic soils are transitional with laterite soils in warm areas and have formed under mild climate, abundant rainfall, and a forest cover which is either mixed or largely deciduous. Leaching is marked, but the breakdown of the silicates is small. There is some translocation of colloidal silicates. The relatively rapid oxidation of the iron checks its downward movement, so that it tends to remain in the upper part of the profile.

Prairie soils are sometimes classed separately from the Podsols. They form under climatic conditions similar to the Gray-brown Podsols but under grass instead of forest cover. Apparently the grass cover produces decomposition products of less potency than does the forest cover. The A horizon is extremely rich in organic material, and downward leaching is moderate. Long times are required to remove the exchangeable calcium and other ions, and there is little breakdown of the silicates. There is frequently considerable downward movement of colloidal silicates to form impervious horizons (pans) in the profile.

Laterite soils. These soils develop under tropical conditions of high temperature and high rainfall that is often seasonal. The primary silicates are quickly broken down, and the extensive rainfall causes the quick removal in solution of any alkalies and alkaline earths. Iron tends to be oxidized, particularly under seasonal rainfall, and its movement along with alumina is retarded. The organic material is very rapidly oxidized so that it does not accumulate, and downward-seeping waters carry little organic acid. The downward-moving waters are almost neutral, or more often slightly on the alkaline side as a result of the presence of alkalies or alkaline earths from the primary silicates. Under such conditions of slight alkalinity the silica tends to be dissolved and removed, with attendant concentration of the alumina and ferric oxide in the upper part of the weathering zone.

Zonal aridic soils. In zonal aridic-soil regions the rainfall is less than about 25 in./year, and the vegetative cover is grass or desert plants. There is little or no disintegration of the silicates or migration of the colloidal components. These soils are characterized by the movement of carbonate and by its concentration in some particular horizon. As the climate becomes drier, the zone of carbonate concentration gradually comes closer to the surface (Fig. 13-5).

Chernozem soils are found in areas of 12 to 25 in. of rainfall under grass cover and in a cool climate. Because of the cool climate, the organic material is oxidized very little, and a thick black surface horizon rich in organic material develops. There is little disintegration of the silicates, and leaching is very mild. The leaching removes the carbonates from the A_2 horizon, and they are reprecipitated and concentrated in the B horizon.

The *chestnut-brown* aridic soils are found where rainfall is 10 to 15 in./year and the climate is warm. The vegetation is scant, and as a result the surface color is gray brown, instead of black as in Chernozems. Also the low rainfall permits slight leaching, and the horizon of carbonate concentration is close to the surface.

Gray or Red Desert soils develop under conditions of high aridity and in the general absence of a cover of vegetation. Because of the scant rainfall

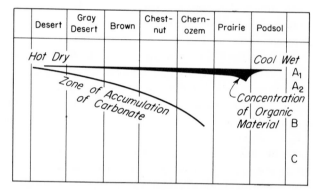

	Desert	Gray Desert	Brown	Chest-nut	Chern-ozem	Prairie	Podsol	

Fig. 13-5. Zones of concentration of organic material and carbonate in relation to soil types and climate. (After Millar and Turk.[112])

there is no downward leaching. The dominant movement of moisture through the soil is upward, owing to the dominance of evaporation, with the accumulation of carbonates and other salts at the surface.

Clay-mineral composition of soils. An early study of the clay-mineral composition of soils of various groups by Alexander, Hendricks, and Nelson[113] indicated that kaolinite is the dominant clay-mineral component of red and gray podsolic soils, with illite frequently present but in quantities usually less than about 10 percent. In Prairie soils both illite and kaolinite are usually present, with the former slightly the more abundant. Small amounts of smectite are often present in the hardpan horizon of Prairie soils. In the series

[113] Alexander, L. T., S. B. Hendricks, and R. A. Nelson, Minerals Present in Soil Colloids, II, Estimation in Some Representative Soils, *Soil Sci.*, **48**: 273–279 (1939).

of Chernozem soils studied by these investigators, illite was the dominant component, and kaolinite and smectite were also present in about equal amounts; and in a series of Desert soils from California, illite and smectite were the dominant clay minerals.

Knox[114] made a compilation of the clay-mineral determinations reported in the literature for about 100 soils. Kaolinite was the only clay mineral reported in more than half of a total of over 20 Red and Yellow Podsols. In the remainder, kaolinite was the dominant component, but small amounts of illite and smectite were also reported. Kaolinite was reported as the dominant constituent in 6 out of 12 Gray-brown Podsols that have been studied. The remaining 6 Gray-brown Podsols are reported to contain illite and smectite in varying amounts, as well as kaolinite. The analyses of six Planosols were reported in the literature, and they all contained smectite. In all but one of the Planosols, illite was also present in about equal abundance with the smectite, and in two of them small amounts of kaolinite were present also. A series of three Rendzina soils all showed illite, kaolinite, and smectite, but in varying amounts. Kaolinite was the component of all the laterites whose clay-mineral composition was recorded.

The clay-mineral analyses of 15 Prairie soils were compiled by Knox, and illite is reported in 11, smectite in 12, and kaolinite in only 4 of them. It seems clear from Knox's data that illite and smectite are the characteristic clay minerals in Prairie soils.

Knox's work shows definitely the predominance of smectite and illite in the soils of arid regions. Smectite was recorded in all seven Chernozem soils that have been reported. Illite is reported in four of the Chernozems, with small amounts of kaolinite in only two of them. Illite is reported as the only clay-mineral component in 5 out of 10 Solonetz soils that have been analyzed. Smectite is reported in five of these soils and is the dominant component in two of them. No kaolinite was found in any of the Solonetz soils. In a series of five soils classed as White Alkali and Black Alkali, illite was present in all of them and was the only clay mineral in three of them. Smectite was present in two of these alkali soils, and kaolinite was completely absent in all of them.

Winters and Simonson[111] published compilations of data on the clay-mineral composition of the B horizon of the Great Soil Groups which agree with the findings of Knox. According to them, the composition of the B horizon of various soil groups is as follows: Tundra soils contain mainly mica[115] and/or smectite types of clay minerals; podsolic soils contain dominantly

---

[114] Knox, E., personal communication.

[115] The authors refer to a 2:1 type of clay mineral, which would include chlorite, illite, vermiculite, and smectite clay minerals.

illite, with lesser amounts of smectite and occasional minor amounts of kaolinite; lateritic soils are dominantly kaolinitic with minor amounts or none of the other clay minerals; chernozemic soils are largely composed of illite and smectite and Desert soils contain largely smectite and illite clay minerals.

Jackson[116] assembled data on the frequency distribution of clay minerals in the major Great Soil Groups. He concluded that inherited minerals such as illite, quartz, feldspars, ferromagnesian minerals, carbonates, and gypsum are most abundant in clays of little-weathered parent materials and in the zonal Desert, Brown, Chestnut, and Tundra soils as well as in intrazonal mountain groups and azonal Regosols and Lithosols. Secondary layer-silicate minerals such as vermiculite, secondary chlorite, smectite, kaolinite, and halloysite are most abundant in clays of moderately weathered parent material in soils of the zonal Chernozem, Prairie, gray-brown podsolic, Podsol, red-yellow podsolic, and low humic Latosol groups as well as intrazonal Planosol, Rendzina, dark Magnesium soil, and Wesenboden groups. Secondary sesquioxide minerals such as hematite, goethite, and gibbsite, and allophane, anatase, and residual resistant primary minerals such as ilmenite and magnetite, predominate in the more highly weathered materials and the zonal ferruginous humic Latosols, hydrol humic Latosols, latosolic brown, Ando soils, and laterites as well as the intrazonal tropical savannah and ground water Podzol ortstein soils.

Investigations by Millot[117] and by the author indicated that the palygorskite-sepiolite clay minerals are likely to be significant components of Desert soils.

Heystek and Schmidt[118] reported attapulgite and montmorillonite as weathering products on a basalt in South Africa. Ravikovitch et al.[119] identified palygorskite in soils of arid areas in Israel. Elgabaly[120] identified attapulgite in soils of the western desert area of Egypt. Pendias[121] reported the presence of palygorskite as a weathering product of basalts in lower

[116] Jackson, M. L., Frequency Distribution of Clay Minerals in Major Great Soil Groups as Related to the Factors of Soil Formation, Proceedings of the 6th National Clay Conference, pp. 133–143, Pergamon Press, New York (1959).

[117] Millot, G., Relations entre la constitution et la genèse des roches sédimentaires argileuses, "Geol. appliq. et prosp. min.," vol. 2, Nancy, France (1949).

[118] Heystek, H., and E. R. Schmidt, The Mineralogy of the Attapulgite-Montmorillonite Deposit in the Springbok Flats, Transvaal, *Trans. Geol. Soc. S. Africa,* **56:** 99–115 (1953).

[119] Ravikovitch, S., F. Pines, and M. Ben-Yair, Composition of the Colloids in the Soils of Israel, *J. Soil Sci.,* **11:** 82–91 (1960).

[120] Elgabaly, M. M., The Presence of Attapulgite in Some Soils of the Western Desert of Egypt, *Soil Sci.,* **93:** 387–390 (1962).

[121] Pendias, H., Weathering of Basalts in the Region of Legnica and Luban Slaski (Lower Silesia), *Kwart. Geol. (Poland),* **5:** 429–445 (1961).

Silesia.   Taldykin[122] reported the occurrence of sepiolite as a weathering product of serpentine in the northern Caucasus.

A further generality of importance regarding soils is reported by Beavers.[123]   According to him, the organic material in many soils accumulates only up to a certain maximum amount.   Additional quantities in excess of the equilibrium amount do not tend to persist.   He has found that the attainable maximum may be relatively high for smectite soils and that it is low for kaolinitic soils.

Discussion of weathering products formed from various types of rock under varying conditions.   Adequate information on the composition of soils, the specific influence of the various factors of weathering, and the environmental conditions favorable for the formation and occurrence of particular types of clay minerals is not available for complete and positive generalizations on the information and occurrence of the clay minerals during weathering. However, the data are adequate for some generalizations which appear reasonably well established.   These may be summarized as follows:

When a basic igneous rock contains considerable magnesium and weathers under conditions which, because of poor drainage or low rainfall, permit the magnesium to remain in the weathering zone after it is released by the breakdown of the parent mineral, smectite will be the alteration product.   If, however, because of high rainfall and good drainage, the magnesium is removed as soon as it is released from the parent mineral, kaolinite will be the weathering product.   If the leaching of the magnesium is relatively rapid, smectite will be an initial stage of weathering and kaolinite a later stage.   Under long-continued weathering by processes removing magnesium, where such processes evolve in a cold, wet climate producing an abundance of organic acids (e.g., podsolic alteration) and hence a potent, acid, leaching environment, the clay minerals will ultimately be broken down, with the aluminum and iron carried downward and the silica concentrated near the surface.   Under long-continued weathering by processes removing magnesium, but evolving in a wet, hot climate producing no organic acids and hence a neutral or slightly alkaline environment (lateritic alteration), the silica is carried away, and iron and alumina are concentrated at or near the surface.

An acid igneous rock containing considerable quantities of potassium as well as magnesium, under weathering conditions permitting the potash and magnesia to remain in the weathering environment after the breakdown of the parent minerals, will yield illite and smectite as the alteration products.

[122] Taldykin, S. I., Fibrous Sepiolite from the Weathering Crust in the Malka Serpentine Massif of the Northern Caucasus, *Materialy Vses. Nauchn.-Issled. Geol. Inst.*, no. 26, pp. 31–34 (1959).

[123] Beavers, A. H., personal communication.

If the content of magnesium is low, illite will be the only product, and if the content of potash is low, smectite will be the only product. Rapid removal of the potash and magnesia leads to the formation of kaolinite. The same sequence of events noted for long-continued weathering of the basic igneous rocks applies to the acid igneous rocks. The other alkalies and alkaline earths which may be present appear to play a relatively minor role, except that there is a suggestion that calcium favors the formation of smectite and that kaolinite does not form so long as there is any appreciable calcium in the environment.

It will be shown later that the clay-mineral components, and thus perhaps the chief silicate components, present in calcareous sediments (including dolomites) are illites and smectites. In the weathering of calcareous sediments, there is substantially no alteration of the silicates until the carbonate is completely broken down and the calcium removed from the environment. After the removal of the calcium, alterations may be podsolic, with the development of a silica-rich surficial material, or lateritic, with the development of a zone of aluminum and iron concentration.

The weathering of a shale progresses about like that of the acidic igneous rocks. In the case of a calcareous shale, the calcium must be removed before there is any appreciable alteration of the primary silicates.

In the weathering of kaolin clays in which there is no appreciable content of alkalies or alkaline earths, there can be no formation of illite or smectite, since potash and magnesia are necessary for their formation, unless some alkalies or alkaline earths are brought into the weathering zone, as, for example, by the upward movement of moisture under hot, dry climatic conditions.

In summary, then, significant conditions in weathering are the presence of alkalies and alkaline earths, particularly potassium and magnesium, in the environment of alteration and the length of time they remain in the environment after their liberation from the parent minerals. The kind of alkali or alkaline earth is also important, since potash leads to the formation of illite, magnesium to the formation of smectite, and calcium probably to the formation of smectite, with an added tendency to block the formation of kaolinite. Carbonate tends to retard the disintegration of the primary silicates.

The pH of the water moving through the soils is also important, particularly under conditions of active leaching and long periods of time. Under the latter conditions, acidity favors the removal of iron and aluminum and the concentration of silica, whereas neutral or alkaline conditions favor the accumulation of iron and aluminum near the surface.

Reversion of weathering cycle. A drastic change in the environment of alteration can lead to the regeneration of clay minerals or to the formation of clay minerals more characteristic of an earlier stage of weathering. Hardy et al.[124] and Alexander et al.[125] showed that gibbsite developed in lateritic soils may be resilicated with the formation of kaolinite. Gordon, Tracey, and Ellis[126] described the formation of kaolinite from gibbsite in bauxite deposits in Arkansas.

Regeneration of the clay minerals can be a result of man-made activities. Volk[36] and Wood[127] showed independently that illite is formed in soils under certain conditions as a consequence of the addition of potash fertilizers. Further, the formation of illite may take place quickly (in a matter of a year or so), illustrating the rapidity with which clay materials can adjust themselves to environmental changes.

Occurrence of halloysite and allophane in soils. Several investigators have recorded the presence of halloysite in weathering products. Thus Bayley[128] described halloysite in some deep kaolins of North Carolina and ascribed its presence to weathering processes. Bates[129] confirmed the identification of halloysite in these same kaolins. Alexander et al.[130] and Allen[131] described weathering products which contain halloysite. Although the presence of halloysite in weathering products is well established, it is a rare component of such materials, and peculiar conditions must be required for its formation. Alexander et al. stated that plagioclase feldspar in a neutral or slightly acid environment and in the presence of water can weather to the hydrous form of halloysite.

It is now well established that some soils contain a component amorphous to X-ray diffraction which can be classified as allophane. For example, Mitchell and Farmer[132] showed that in well-leached soils in Scotland an amor-

[124] Hardy, F., and G. Rodrigues, Soil Genesis from Andesite in Grenada, British West Indies, *Soil Sci.,* **48:** 361–383 (1939).

[125] Alexander, L. T., S. B. Hendricks, and G. T. Faust, Occurrence of Gibbsite in Some Soil-forming Materials, *Soil Sci. Soc. Am. Proc.,* **6:** 52–57 (1941).

[126] Gordon, M., J. I. Tracey, and M. W. Ellis, Geology of the Arkansas Bauxite Region, *U.S. Geol. Surv., Profess.* Paper 299 (1958).

[127] Wood, L. K., The Chemical Effects of Soluble Potassium Salts on Some Illinois Soils, Ph.D. thesis, University of Illinois (1941).

[128] Bayley, W. S., Kaolin in North Carolina, with a Brief Note on Hydromica, *Econ. Geol.,* **15:** 236–246 (1920).

[129] Bates, T. F., personal communication.

[130] Alexander, L. T., G. F. Faust, S. B. Hendricks, H. Insley, and H. F. McMurdie, Relationship of the Clay Minerals Halloysite and Endellite, *Am. Mineralogist,* **88:** 1–18 (1943).

[131] Allen, V. T., Formation of Bauxite from Basaltic Rocks of Oregon, *Econ. Geol.,* **43:** 619–626 (1948).

[132] Mitchell, B. D., and V. C. Farmer, Amorphous Clay Minerals in Some Scottish Soil Profiles, *Clay Minerals Bull.* **5**(28) : 128–144 (1962).

phous component is favored by a high organic content, and Beutelspacher and Van der Marel[133] presented evidence for the presence of amorphous material in a wide variety of soils. It cannot, however, be concluded that an amorphous component is present in all soils and is the "essence" of soils generally.

Allophane appears to be particularly prevalent in soils derived from volcanic ash which have a very low content of alkalies and alkaline earths. Halloysite is frequently associated with the allophane. Thus, Aomine et al.[134] reported halloysite and allophane as weathering products from volcanic ash when the magnesium content is low; otherwise montmorillonite develops. Hay[135] identified halloysite and allophane as weathering products of a volcanic ash on St. Vincent island in the British West Indies. Beutelspacher et al.[133] concluded that halloysite, perhaps from an intermediate allophane stage, develops under tropical weathering conditions where there are not definite dry intervals.

Nature of the alteration process. Correns and Von Engelhardt[136] concluded that weathering proceeds by the breakdown of the parent mineral to ionic solutions and that secondary minerals are reaction products of such ionic solutions. Niggli[137] stated that, in the alteration of feldspar, under certain conditions, the breakdown reaches a colloidal state before the formation of the secondary products. Demolon and Batisse[138] concluded that in the weathering of a granite there is a spontaneous change to the clay minerals.

It is certain that the transition from a primary mineral or minerals to a clay-mineral weathering product does not always take place in exactly the same way. It can take place by means of the formation of ionic solutions with the later precipitation of the clay minerals, by the formation of colloidal mixtures with later crystallization of the clay minerals, or by a direct structural change from the parent mineral to the clay mineral.

The direct structural change can involve merely a relatively slight modi-

[133] Beutelspacher, H., and H. W. Van der Marel, Über die Amorphen Stoffe in den Tonen Verschiedener Boden, *Acta Univ. Carolinae Geol. Suppl.,* 1: 97–114 (1961).

[134] Aomine, S., and T. Higashi, Clay Minerals of Alluvial Soils in Kyushu, Part II, *J. Fac. Agr., Kyushu Univ.* 10: 355–364 (1956).

[135] Hay, R. L., Rate of Clay Formation and Mineral Alteration in a 4000-year-old Volcanic Ash Soil on St. Vincent, B.W.I., *Am. J. Sci.,* 258: 354–368 (1960).

[136] Correns, C. W., and W. Von Engelhardt, Neue Untersuchungen über die Verwitterungen des Kalifeldspates, *Chem. Erde,* 12: 1–12 (1939).

[137] Niggli, P., Zusammensetzung und Klassification der Locherstein, Vortag Gehalten am Erdbaukurs, *Eidgenöss. Tech. Hoch. Zürich,* 5 (1939).

[138] Demolon, A., and E. Batisse, First Stages of the Spontaneous Change of a Granite and the Origin of Clayey Colloids, *Compt. Rend.,* 233: 115–118 (1946).

fication of the structure, as illustrated by the formation of smectite from the micas, or a substantial change, as illustrated by the formation of kaolinite and micas from feldspars; smectite from ferromagnesium minerals; etc. The former change is commonly through an intermediate mixed-layer transition stage (Droste et al.[139] and Millot and Camez[140]). A given secondary mineral, e.g., kaolinite, is not always formed by the same mechanism. Variations in the environment of alteration, e.g., efficiency of leaching, pH, etc., and variations in the structure and composition of the primary minerals are factors determining the mechanism of alteration. The literature on this subject is very extensive, and the following can be considered only a brief introduction.

In an extensive consideration of the chemistry of soil development, Barshad[45] listed the following as factors determining the relative stability of aluminosilicate minerals:

1. The size of the cations in octahedral and tetrahedral positions in relation to the size of the interstices. For example, aluminum with a radius of 0.57 Å cannot fit into tetrahedral positions without some distortion. Aluminum and ferrous iron are too small and too large, respectively, to fit into octahedral positions without some distortion.

2. The degree of basicity, degree of linkage of tetrahedrons, relative number of alumina and silica tetrahedrons, and other factors that induce a lowering of the basicity of the mineral and a destruction of bonds linking the tetrahedrons.

3. The presence of ferrous iron or other cations that oxidize during weathering.

4. The closeness of packing of the oxygens about the cations other than those in tetrahedral positions.

5. The occurrence of empty positions in certain parts of the structure.

6. The relation of the size of the unit cell to that of the particle.

7. The nature of associated minerals.

8. The total cation-exchange capacity.

Devore[46] discussed the hydrous decomposition of alkali feldspars and the formation of micaceous minerals as decomposition products in terms of the surface chemistry, the Al-Si ordering of the feldspars, and a crystal growth

[139] Droste, J. B., N. Bhattacharya, and J. A. Sunderman, Clay Mineral Alteration in Some Indiana Soils, Proceedings of the 9th National Clay Conference, pp. 329–342, Pergamon Press, New York (1962).

[140] Millot, Georges, and T. Camez, Genesis of Vermiculite and Mixed-layered Vermiculite in the Evolution of the Soils of France, Proceedings of the 10th National Clay Conference, pp. 90–95, Pergamon Press, New York (1963).

model for the micas. He concluded that the kaolinite minerals would be more likely to form than muscovite, and the possibility of further decomposition to produce boehmite would be increased for disordered as compared with ordered feldspars. The three-layer clay minerals are more likely to be the decomposition products from the ordered feldspars.

Garrels and Howard[48] concluded, on the basis of a laboratory study of the reaction of feldspar and mica with water at low temperature and pressure, that the first result of the reaction of mica and adularia with water is a surface layer that grades from an outer portion that is structurally disrupted to an inner portion that retains the original silicate structure but with hydrogen substituted for potassium.

In the present state of knowledge, it is clear that potassium feldspar may alter to kaolinite by way of intermediary mica under certain conditions, but directly to kaolinite under other conditions. Also, this feldspar sometimes alters to halloysite by way of a gel phase, possibly including transitory allophane, or directly to gibbsite.

Bates[141] pointed out that the weathering of basaltic rocks in the Hawaiian Islands is largely a matter of desilication and the removal of bases, with the production of large amounts of halloysite and gibbsite together with variable quantities of allophane, nontronite, iron and titanium oxides, and amorphous gel materials. The nature and relative abundance of intermediate and end products of weathering depend primarily upon the amount of rainfall, the angle of slope, and the texture of the rocks. No kaolinite was observed, its absence being interpreted as due to the absence of fine-grained mica as an intermediate weathering product.

Delvigne[142] concluded that, in the development of laterites under intense and continuous leaching conditions, feldspars alter directly to gibbsite and that any kaolinite results from the silication of the gibbsite.

Altschuler[143] and others have traced in detail the transition of kaolinite developing from smectite in some Florida soils. They have shown a progressive transition from the edge to the center of the smectite flakes.

Jackson[144] stated that the accumulation of aluminum possibly with some iron, magnesium, and allophane as interlayer components in three-layer minerals of soils is a genetic stage in the transition of these minerals to kaolinite and halloysite.

[141] Bates, T. F., Halloysite and Gibbsite Formation in Hawaii, Proceedings of the 9th National Clay Conference, pp. 315–328, Pergamon Press, New York (1962).

[142] Delvigne, J., "Pedogenèse en zone tropicale," Dunod, Paris (1965).

[143] Altschuler, Z. S., Transformation of Montmorillonite to Kaolinite During Weathering, *Science*, **141**(3576): 148–152 (1963).

[144] Jackson, M. L., Chemical Composition of Soils, "Chemistry of the Soil," chap. II, pp. 71–141, American Chemical Society Monograph (1964).

Keller[44] emphasized that the clay minerals may enter into the mechanism of decomposition. After some clay has been formed so that it becomes a significant part of the soil and is present as a coating of particles of undecomposed silicate minerals, that clay acts as an agent of weathering to produce more clay from silicate parent minerals and rocks. The clay coating of silicate mineral particles may serve as a single-stage bridge between rootlet and mineral particle. The clay coating contains cations from the parent minerals. As these are removed by plant roots, the clay goes back for more cations from the silicate minerals.

The character of organisms present may influence the alteration processes as illustrated by the following; Fripiat and others[145] pointed out that in tropical soils, where there is a concentration of termites and consequently of calcareous material, the soils tend to be relatively rich in smectite. Anderson, Jonas, and Odum[146] investigated the effect on clay minerals by their passage through the digestive tracts of marine organisms. They found substantial degrading effects for the three-layer clay minerals, whereas kaolinite was not affected. Antipov-Karataev and Tsyurupa[147] showed that the destruction of albite and muscovite was twice as fast in a soil when bacteria are present as in sterile clay, irrespective of the salts present.

Rate of alteration. The rate of alteration under weathering conditions obviously varies widely, depending on factors of climate, parent rock, etc. Weathering processes are relatively slow but may not be as slow as commonly considered. Thus, Mortland, Lawton, and Uehara[148] reported that, in a synthetic soil in which biotite was the only source of potassium, four crops of wheat caused a change of biotite to vermiculite. De Keyser et al.[149] found that in polders in Belgium, developed from the eleventh to the nineteenth century, muscovite has altered to illite which in turn has altered to smectite

[145] Fripiat, J. J., M. C. Gastuche, L. Vielvoye, and Ch. Sys, Les Argiles des sols de la région d'Elisabethville, *Pedologie,* **7:** 12–18 (1957).

[146] Anderson, A. E., E. C. Jonas, and H. T. Odum, Alteration of Clay Minerals by Digestive Processes of Marine Organisms, *Science,* **127:** 190–191 (1958).

[147] Antipov-Karataev, I. N., and I. G. Tsyurupa, The Weathering of Primary Minerals (Albite, Muscovite) Under the Influence of Silicate Bacteria and the Character of the Secondary Compounds, *Tr. Shestogo Soveshch. po Eksperim. i Tekhn. Mineralog. Petrogr., Akad. Nauk SSSR, Inst. Geol. Rudn. Mestorozhd., Petrogr. Mineralog. i Geokhim., Inst. Khim. Silikatov, Leningrad,* pp. 75–80 (1961). (Publ. 1962).

[148] Mortland, M. M., K. Lawton, and G. Uehara, Alteration of Biotite to Vermiculite by Plant Growth, *Soil Sci.,* **82:** 477–481 (1956).

[149] De Keyser, W., J. van Keymullen, and F. Hoebsch, Les minéraux argileux des polders et de quelques autres régions naturelles, Comptes rendus de reserches verslagen over navorsingen travaux du comite pour l'et ablissement de la carte des sols et de la végétation de la Belgique (Centre Roentgenographique) **14:** 47–128 (1955).

plus a little kaolinite.   Hamdi[150] pointed out that, in soils of the Nile River Delta, illite altered to smectite in the course of 12,000 years.

Under tropical conditions alteration processes may be extremely rapid, particularly on volcanic ash.   For example, in New Zealand, there has been substantial development of secondary minerals in volcanic ash that has fallen in historic times.[151]   Hay[135] showed that a 4,000-year-old pyroclastic deposit on St. Vincent island of the British West Indies has been weathered to a depth of 6 ft.   Halloysite and allophane have formed from vitric andesitic ash and lapilli.

The possible rapid rate of clay-mineral formation is illustrated also by the regeneration of illite as a consequence of the application of potash fertilizers (see page 220).

Clay-mineral composition in relation to the agricultural properties of soils. The soil literature (e.g., Bear and others[152]) reviews in detail the relation of clay-mineral composition to the tilth and fertilization characteristics of soils.   Fieldes[153,154] discussed in detail the relation between the characteristics of the clay minerals and the fertility and other properties of soils.

Only very little research has been reported on the influence of specific clay minerals on the growth of specific crops.   For example, Franc de Ferriere et al.[155] studied the relation of clay-mineral composition to the nutrient requirements of wheat and potatoes and to variations in rainfall.   Morita[156,157] showed that the nutrients needed by rice are more available from kaolinite than illite soils.

Another interesting and little studied area of clay-mineral soil research concerns the possible influence of specific clay minerals on plant diseases.

[150] Hamdi, H., Alterations in the Clay Fraction of Egyptian Soils, *Z. Pflanzenernähr. Düng.*, **84**: 204–211 (1959).

[151] Fieldes, M., personal communication.

[152] Bear, F. E., and others, "Chemistry of the Soil," American Chemical Society Monograph (1964).

[153] Fieldes, M., The Role of Clays in Soil Fertility, *New Zealand Inst. Agr. Sci. Proc., Soil Bur. Publ.* 167: pp. 84–102 (1958).

[154] Fieldes, M., The Nature of the Active Fraction of Soils, *Intern. Soil Conf., New Zealand, Soil Bur. Publ.* 277, pp. 3–19 (1962).

[155] Franc de Ferriere, J. J., Th. Camez, and G. Millot, Influence des types d'argile des sols et de la pluviosite sur la nutrition potassique de la pomme de terre, *Annal. Agron.*, **1**: 51–72 (1958).

[156] Morita, S., Nutrient Absorption by Rice from Media Containing Different Types of Clay Minerals, *Soil Sci.*, **86**: 336–342 (1958).

[157] Morita, S., Clay Mineral and Rice Growth, *Nendo Kagaku No Shimpo*, **1**: 295–311 (1959).

Stotzley and others[158] found a correlation between clay-mineral composition and susceptibility of banana plants to fusarium wilt. Smectite clay minerals are present in all moderate- to high-resistant soils and were detected in only two highly susceptible soils. Galvez[159] suggested that the spread of cadang disease of coconut trees is slower in plantations where both smectite and kaolinite types of clay exist in the soil than in plantations where kaolinite types are not found.

Additional references

Synthesis of clay minerals

Barrer, R. M., Contributions to Synthetic Mineral Chemistry, *Proc. Intern. Symp. Reactivity Solids, Gothenburg,* 1952, I, p. 373 (1954).

Bowen, N. L., and O. F. Tuttle, The System $MgO \cdot SiO_2 \cdot H_2O$, *Bull. Geol. Soc. Am.,* **60:** 439–460 (1949).

Crowley, M. S., Effect of Starting Materials on Phase Relations in the System $CaO-Al_2O_3-H_2O$, *J. Am. Ceram. Soc.,* **47:** 144–148 (1964).

De Keyser, W. L., Silica-Alumina System, *Bull. Soc. Franc. Ceram.,* **62:** 19–33 (1964).

De Kimpe, C., Low-temperature Syntheses of Kaolin Minerals, *Am. Mineralogist,* **49:** 1–16 (1964).

DeKimpe, C., M. C. Gastuche, and G. W. Brindley, Ionic Coordination in Alumino-Silica Gels in Relation to Clay Mineral Formation, *Am. Mineralogist,* **46:** 1370–1381 (1961).

Gastuche, M. C., The Octahedral Layer, Proceedings of the 12th National Clay Conference, pp. 471–493, Pergamon Press, New York (1964).

Gastuche, M. C., and C. DeKimpe, Synthesis of Kaolin Clays at Normal Temperature and Pressure Conditions, *Bull. Classe Sci., Acad. Roy. Belg.,* **45:** 1087–1104 (1959).

Gruner, J. W., Hydrothermal Alteration of Montmorillonite to Feldspar at Temperatures from 245°C to 300°C, *Am. Mineralogist,* **21:** 511–515 (1936).

Morey, G. W., and E. Ingerson, The Pneumatolytic and Hydrothermal Alteration and Synthesis of Silicates, *Econ. Geol.,* **32:** 607–760 (1937).

Schwarz, R., and G. Trageser, Ueber die kunstliche Umwandlung von Felspat in Kaolin, *Z. Anorg. Allgem. Chem.,* **215:** 190–200 (1933).

Siffert, B., Quelques réactions de la silica en solution: la formation des argiles, *Mem. Serv. Carte Geol. Alsace Lorraine,* **21:** 1–86 (1962).

Tchoubar, C., Investigation by Microscopy and by Electron Diffraction of the Formation of Kaolinite from Albite Altered by Water at 200°C, *Bull. Soc. Franc. Mineral. Crist.,* **88**(3): 483–518 (1965).

Yusopova, S. M., The Characteristics of Montmorillonite, *Compt. Rend. Acad. Sci. URSS,* **51:** 631–634 (1946).

Hydrothermal alteration

Alderman, A. I., Clay Derived from Sillimanite by Hydrothermal Alteration, *Mineral. Mag.,* **29:** 271–279 (1950).

[158] Stotzley, G., J. E. Dawson, R. T. Martin, and C. H. H. Ter Kulei, Soil Mineralogy as a Factor in the Spread of Fusarium Wilt of Banana, *Science,* **133:** 1483–1485 (1961).

[159] Galvez, N. L., Clay Mineral Contents of Some Soils from the Bicol Region, *Philippine Agriculturalist,* **46:** 398–401 (1962).

Burbank, W. S., Problems of Wall Rock Alteration in Shallow Volcanic Environments, *Quart. Colo. School Mines,* **45:** 286–326 (1950).

Butler, B. S., Influence of Replaced Rocks on Replacement Minerals, *Econ. Geol.,* **27:** 1–27 (1932).

Chukhrov, F. V., and N. A. Kozlova, Halloysites from the Deposits Berkara and Karagaily, *Compt. Rend. Acad. Sci. URSS,* **27:** 1011–1012 (1940).

Da Silva, D. J., Montmorillonite dans des pegmatites granitiques et le problème de la montmorillonitisation, *Port. Serv. Fom. Mineral.,* **4:** 48–52 (1948).

Exley, C. S., Some Factors Bearing on the Natural Synthesis of Clays in the Granites of South-West England, *Clay Minerals Bull.,* **5**(32): 411–426 (1964).

Keller, W. D., Hydrothermal Kaolinization (Endellitization) of Volcanic Glassrock, Proceedings of the 10th National Clay Conference, pp. 333–343, Pergamon Press, New York (1963).

Kerr, P. F., Alteration Studies, *Am. Mineralogist,* **32:** 158–162 (1947).

Lindgren, W., Copper Deposits of the Clifton-Morenci District, Arizona, *U.S. Geol. Surv., Profess. Paper* 43 (1905).

Lovering, T. S., The Origin of the Tungsten Ores of Boulder, Colorado, *Econ. Geol.,* **36:** 229–279 (1941).

Lovering, T. S., Geochemistry of Argillic and Related Types of Alteration, *Quart. Colo. School Mines,* **45:** 231–260 (January, 1950).

Morey, G. W., and R. O. Fournier, The Decomposition of Microcline, Albite, and Nepheline in Hot Water, *Am. Mineralogist,* **46:** 688–699 (1961).

Nakamura, T., Mineralization and Wall-rock Alteration at the Ashio Copper Mine, Japan, *J. Inst. Polytech., Osaka City Univ.,* **5**(1): 53–127 (1961).

Prider, R. T., and W. F. Cole, The Alteration Products of Olivine and Leucite in Leucite Lamprophyres from the West Kimberly Area, Western Australia, *Am. Mineralogist,* **27:** 373–384 (1942).

Sudo, T., et al., Application of Clay Mineralogy to Studies of Wall-Rock Alteration, *Kozan Chishitsu,* **11:** 356–363 (1963).

Tomlinson, W. H., and A. J. Meier, On the Origin of Montmorillonite, *Am. Mineralogist,* **22:** 1124–1126 (1937).

Tooker, E. W., Altered Wall-Rocks in the Central Part of the Front Range Mineral Belt, Gilpin and Clear Creek Counties, Colorado, *U.S. Geol. Surv., Profess. Paper* 439 (1964).

Soils and weathering

Barshad, I., A Pedologic Study of California Prairie Soils, *Soil Sci.,* **61:** 423–442 (1946).

Brown, A. L., and A. C. Caldwell, Clay Mineral Content of Some Colloidal Material Extracted from a Solodi Soil Profile, *Soil Sci. Soc. Am. Proc.,* **11:** 213–215 (1916).

Camez, T., The Evolution of Clay Minerals in Soils of Temperate Regions, *Mem. Serv. Carte Geol., Alsace Lorraine,* **20:** 1–90 (1962).

Claridge, G. G. C., Clay Mineralogy and Chemistry of Some Soils from the Ross Dependency, Antarctica, *New Zealand J. Geol. Geophys.,* **8**(2): 186–220 (1965).

Craig, D. C., and F. C. Loughnan, Chemical and Mineralogical Transformations Accompanying the Weathering of Basic Volcanic Rocks from New South Wales, *Australian J. Soil Res.,* **2**(2): 218–234 (1964).

Day, J. H., and H. M. Rice, The Characteristics of Some Permafrost Soils in the Mackenzie Valley, Northwest Territories, *Arctic,* **17**(4): 223–236 (1964).

De Leenheer, L., Les Propriétés sorptives des sols et leur interprétation minéralogique, avec application aux sols argileux des polders marin, *Ann. Soc. Geol. Belg., Bull.,* **57:** 299–320 (1948).

Feustel, I. C., A. Dutilly, and M. S. Anderson, Properties of Soils from North American Arctic Regions, *Soil Sci.*, **48**: 183–199 (1939).

Gieseking, J. E., The Clay Minerals in Soils, *Advan. Agron.*, **1**: 159–204 (1949).

Hay, R. L., Zeolitic Weathering in Olduvai Gorge, Tanganyika, *Bull. Geol. Soc. Am.*, **74**: 1281–1286 (1963).

Jacob, A., The Composition of the Clay Fraction of Soils, *Bodenkunde Pflanzenernähr*, **29**: 219–231 (1943).

Jenny, H., "Factors of Soil Formation," McGraw-Hill, New York (1941).

Kelley, W. P., W. H. Dore, and J. B. Page, The Colloidal Constituents of American Alkali Soils, *Soil Sci.*, **51**: 101–124 (1941).

Kuzvart, M., Problems of Rock Weathering, *Vestn. Ustredniho Ustavu Geol.*, **40**: 55–69 (1965).

Mattson, S., The Laws of Soil Colloidal Behavior, IX, Amphoteric Reactions and Isoelectric Weathering, *Soil Sci.*, **34**: 209–239 (1932).

Nagelschmidt, G., The Mineralogy of Soil Colloids, *Imp. Bur. Soil Sci., Tech. Commun.* **42** (1944).

Nikitina, A. P., Formation and Types of Crust of Weathering on Rocks of the Crystalline Basement of the KMA (Kursk Magnetic Anomaly), *Kora Vyvetrivaniya, Akad. Nauk SSSR, Inst. Geol. Rudnykh. Mestorozhd., Petrogr., Mineralog. i Geokhim.*, **6**: 102–124 (1963).

Nikitina, A. P., Yu M. Korolev, and V. G. Vorontsov, Palygorskite and Saponite in the Crust of Weathering of the KMA (Kursk Magnetic Anomaly), *Kora Vyvetrivaniya, Akad. Nauk SSSR, Inst. Geol. Rudnykh Mestorozhd., Petrogr. Mineralog. i Geokhim.*, **6**: 48–54 (1963).

Patterson, S. H., Halloysitic Underclay and Amorphous Inorganic Matter in Hawaii, Proceedings of the 12th National Clay Conference, pp. 153–172, Pergamon Press, New York (1964).

Pearson, R. W., and L. E. Ensminger, Types of Clay Minerals in Alabama Soils, *Soil Sci. Soc. Am. Proc.*, **13**: 153–156 (1948).

Sedletsky, I. D., Mineralogic Composition of Clays and Their Genesis, *Soviet Geol.*, **8**: 82–89 (1940).

Seeger, M., Fossil Weathering Products on the Swabian Alb. A Contribution to the Knowledge of Red Bolus Clay, *Jahresh. Geol. Landesamtes Baden-Wuerttemberg*, **6**: 421–459 (1963).

Van der Merwe, C. R., and H. W. Weber, The Clay Minerals of South African Soils Developed from Granite under Different Climatic Conditions, *S. African J. Agr. Sci.*, **6**: 411–454 (1963).

Van der Merwe, C. R., and H. W. Weber, The Clay Minerals of South African Soils Developed from Dolerite under Different Climatic Conditions, *S. African J. Agr. Sci.*, **8**: 111–141 (1965).

White, J. L., X-ray Diffraction Studies on Weathering of Muscovite, *Soil Sci.*, **93**: 16–21 (1962).

Fourteen
Origin and Occurrence of the Clay Minerals
(cont'd)

Recent sediments

Until recently only a very limited study had been made of the clay-mineral composition of sediments accumulating in various environments at the present time. The data are still inadequate for broad generalities, but many studies in recent years have greatly enhanced our knowledge in this area. Recent sediments frequently are composed of complex mixtures of relatively poorly crystallized minerals, and their analysis is particularly difficult. Unfortunately many of the determinations, particularly those reported in the earlier literature, are incomplete or are based on inadequate analytical data.

The differences in the conclusions arrived at by various investigators derived from the study of their analytical data, e.g., the importance of diagenesis in the marine environment, undoubtedly are a consequence, in part at least, of the lack of precision in the analytical data and the fact that the analytical data of the investigators are not comparable. Unless the clay-mineral analyses are made by a combination of methods—X-ray diffraction, differential thermal, chemical, etc.—and unless the procedures are detailed, the completeness and precision of the results are questionable.

Marine environment. On the basis of their *Challenger* studies, Murray and Renard[1] suggested that the fine-clay fraction of the sediments accumulating on the present ocean floors was noncrystalline. At that time the idea that many clays were composed of amorphous components was widely held (see Chap. 2), largely because there were no adequate techniques available to study them.

Using modern analytical methods, Revelle[2] in 1936 and Correns[3] in 1937 first showed that the fine-clay fraction of Recent marine sediments is substantially all crystalline. Correns[3] and his colleagues found that mica (probably illite) was present in all the bottom samples collected by the *Meteor* in the South Atlantic. In addition to the mica, the clay fraction in some of the samples was composed predominantly of kaolinite, in others of halloysite, and in a few of smectite. A map was presented showing the distribution of the predominant clay minerals, omitting from consideration the mica. Smectite was found to be limited to relatively small regions where there were neighboring areas of basic volcanic activity. Correns himself pointed out that the differentiation of halloysite and kaolinite was not certain, and the presence of halloysite in Recent marine sediments has not yet been definitely established. Some of the *Meteor* samples were cores, and Correns reported no variation in the clay-mineral composition of any of his cores from the top downward. He found no evidence of diagenetic changes in the clay-mineral components of his samples.

Leinz[4] in a further study of the *Meteor* samples reported that all samples of Blue Clay showed mica and that most of them showed also a kaolinite type of mineral. Smectite was found in some of the Blue Clay samples. All samples of globigerina ooze also contained mica, but only a few showed a kaolinite type of mineral reported as halloysite. Smectite was also found in some of them.

Dietz[5] determined the clay-mineral composition of 39 samples collected from various oceans of the world. Most of his samples were collected by the *Challenger*. Dietz found illite in all his samples, and it was the predomi-

[1] Murray, J., and A. F. Renard, Reports on Deep-sea Deposits Based on Specimens Collected during the Voyage of H.M.S. *Challenger* in the Years 1872–1876, Challenger Reports (1891).

[2] Revelle, R., Marine Bottom Samples Collected in the Pacific Ocean by the *Carnegie* on its Seventh Cruise, Ph. D. thesis, University of California (1936).

[3] Correns, C. W., et al., Die Sedimente des äquatorialen atlantischen Ozeans, Deutsche atlantische Expedition *Meteor* 1925–1927, *Wiss. Ergeb.*, 3: Teil 3 (1937).

[4] Leinz, V., Die Mineralfazies der Sedimente des guinea Beckens, Deutsche atlantische Expedition *Meteor* 1925–1927, *Wiss. Ergeb.*, 3: Teil 3, 245–262 (1937).

[5] Dietz, R. S., Clay Minerals in Recent Marine Sediments, Ph.D. thesis, University of Illinois (1941).

nant clay mineral in 23 of them. Illite was particularly abundant in the deep-sea clays. Kaolinite, although generally less abundant than illite, was present in all the samples and predominant in nine of them, most of which were relatively close to the shore. Small amounts of smectite were found in the near-shore samples, but this mineral was not detected in most deep-sea sediments. Again it should be stated that adequate methods for detecting small amounts of smectite in mixture of poorly crystallized material had not been developed when Dietz did his work. Dietz presented strong evidence for the development of illite on the sea floor in sediments accumulating at the present time. It was his concept that the illite develops largely by the alteration of smectite from the source material being carried to the sea.

Grim, Dietz, and Bradley[6] studied a series of samples collected by the *E. W. Scripps* from the Gulf of California and the Pacific Ocean off the California coast. They found that all the samples contained illite, smectite, and kaolinite and that generally illite was the most abundant and kaolinite the least abundant clay-mineral type. A chloritic clay mineral could be identified definitely in some samples. The clay minerals were very complex mixtures, including mixed crystallizations as well as mechanical mixtures of discrete phases. In general the crystallinity was lower, the individual size smaller, and the intergrowth more intimate in these Recent sediments than in ancient sediments which these authors had studied.

Grim and his colleagues concluded from their data that kaolinite is very slowly being lost in these marine sediments, probably by alteration to illite or chloritic mica. They postulated that potassium is being taken up by the accumulating sediments mostly by "degraded illite," which is illite that reaches the environment of deposition with a relatively low potassium content as a result of weathering processes, and that magnesium is also being taken up by the accumulating sediments, probably in the formation of illite or chloritic micas. The widespread occurrence of smectite indicates that this clay mineral is not lost completely or quickly in diagenetic processes.

Suzuki and Kitazaki[7] studied the Red Clays in the Western Pacific and concluded that they were dominantly volcanic in origin and consisted of amorphous glass in various stages of devitrification.

Zen[8] studied the composition of marine-bottom sediments off the coast

[6] Grim, R. E., R. S. Dietz, and W. F. Bradley, Clay Mineral Composition of Some Sediments from the Pacific Ocean off the California Coast and the Gulf of California, *Bull. Geol. Soc. Am.*, **60**: 1785–1808 (1949).

[7] Suzuki, I. K., and U. Kitazaki, Mineralogical Studies on the Red Clays from the Western Pacific Ocean, East of the Bonin Islands—Studies on the Clay Mineral Fractions of the Recent Marine Sediments, *Japan. J. Geogr.*, **24**: 171–180 (1954).

[8] Zen, E-An, Mineralogy and Petrography of Marine Bottom Sediment Samples off the Coast of Peru and Chile, *J. Sediment. Petrol.*, **29**(4): 513–539 (1959).

of Peru and Chile. He found volcanic glass, in various stages of devitrification, a very common component in nearly all his cores. The clay minerals identified were kaolinite, chlorite, illite, and a mixed-layer complex predominantly composed of smectite.

Oinuma and others[9] identified illite, kaolinite, and smectite in Recent sediments in the Western Pacific. Illite was the principal clay-mineral component in all specimens except the globigerina oozes and was relatively more abundant in the Red Clays.

Cloud and others[10] identified kaolinite, illite, smectite, and chlorite in lime muds accumulating west of Andros Island in the Bahamas.

El Wakeel and Riley,[11] in mineralogical analyses of a series of samples from the Atlantic, Pacific, and Indian Oceans and the Mediterranean Sea, found a mainly terrigenous origin for deep-sea argillaceous sediments and that illite or chlorite with lesser smectite were the principal clay minerals in deep-sea sediments; the near-shore clays were relatively rich in kaolinite. These same authors presented in another publication[12] a comparison of the composition of a fossil Red Clay with Recent Red Clay. They found that the only difference was a lower content of iron and a higher content of potassium in the fossil clays.

Eroshchev-Shak[13] reported that in deep-sea deposits of the Atlantic Ocean illite predominates at latitudes corresponding to subarctic, temperate, and subtropical climates; kaolinite prevails in the tropical, subequatorial, and equatorial climatic zones. Smectite slightly disturbs the regular zonation of the argillaceous mineral distribution in the sea. Smectite was found in the Azores volcanic area and was shown to be unrelated to continental runoff. Based on its crystal chemical features and paragenesis with products of volcanic activity, smectite was considered as originating by the alteration of volcanic material.

In another paper this same author[14] reported the presence of halloysite

[9] Oinuma, K., K. Kobayashi, and T. Sudo, Clay Mineral Composition of Some Recent Marine Sediments, *J. Sediment. Petrol.*, **29**(1): 56–63 (1959).

[10] Cloud, P. E., Jr., P. D. Blackmon, F. D. Sisler, H. Kramer, J. H. Carpenter, E. C. Robertson, L. R. Sykes, and M. Newell, Environment of Calcium Carbonate Deposition West of Andros Island, Bahamas, *U.S. Geol. Surv., Profess. Paper* 350, 138 pp. (1962).

[11] El Wakeel, S. K., and J. P. Riley, Chemical and Mineralogical Studies of Deep-sea Sediments, *Geochim. Cosmochim. Acta,* 110–146 (1961).

[12] El Wakeel, S. K., and J. P. Riley, Chemical and Mineralogical Studies of Fossil Red Clays from Timor, *Geochim. Cosmochim. Acta,* **24**: 260–265 (1961).

[13] Eroshchev-Shak, V. A., Zoning of the Argillaceous Mineral Distribution in Deposits of the Atlantic Ocean, *Tr. Inst. Okeanol., Akad. Nauk SSSR,* **56**: 59–69 (1962).

[14] Eroshchev-Shak, V. A., Clay Minerals in Bottom Sediments of the Atlantic Ocean, *Okeanologiya,* **2**(1): 98–105 (1962).

and chlorite and complex mixed-layer minerals in bottom sediments of the Atlantic Ocean in small amounts, and he stated that they cannot be separated into independent zones as can the other clay minerals.

Biscaye[15] made clay-mineral analyses of approximately 500 deep-sea core samples from the Atlantic, Antarctic, and Western Indian Oceans and adjacent seas. He determined the relative abundances of smectite, illite, kaolinite, and chlorite and indicated the presence of mixed-layer clay minerals. He concluded that the bulk of Recent Atlantic Ocean deep-sea clays are of continental origin or derived from the continents and that the formation of minerals *in situ* on the ocean bottoms is of relatively minor importance in the Atlantic but may be significant in parts of the Southwestern Indian Ocean. He pointed out that mineralogical analysis of the fine fraction of the Atlantic Ocean deep-sea sediments is a useful indicator of sediment provenance. Kaolinite, gibbsite, mixed-layer minerals, and chlorite contribute the most unequivocal provenance information because they have relatively restricted sources of continental origin.

There has been an intensive study of sediments accumulating in the Gulf of Mexico, particularly those in the vicinity of the Mississippi River Delta (Grim and Johns,[16] Milne and Shott,[17] Milne and Earley,[18] Johns and Grim,[19] McAllister,[20] Welby,[21] Griffin,[22] and Taggart and Kaiser[23]). All investigators agree that the sediments are composed of variable amounts of illite, chlorite, kaolinite, smectite, and mixed-layer minerals, and that close to shore the clay-mineral composition reflects, in general, variations in the source material.

[15] Biscaye, P. E., Mineralogy and Sedimentation of Recent Deep-sea Clay in the Atlantic Ocean and Adjacent Seas and Oceans, *Bull. Geol. Soc. Am.,* **76**(7): 803–832 (1965).

[16] Grim, R. E., and W. D. Johns, Jr., Clay Mineral Investigation of the Sediments from the Northern Gulf of Mexico, *Natl. Acad. Sci., Publ.* **327**, pp. 81–103 (1954).

[17] Milne, I. H., and W. L. Shott, Clay Mineralogy of Recent Sediments from the Mississippi Sound Area, *Natl. Acad. Sci., Publ.* **566**, pp. 253–265 (1958).

[18] Milne, I. H., and J. W. Earley, Effect of Source and Environment on Clay Minerals, *Bull. Am. Assoc. Petrol. Geol.,* **42**: 328–338 (1958).

[19] Johns, W. D., Jr., and R. E. Grim, Clay Mineral Composition of Recent Sediments from the Mississippi River Delta, *J. Sediment. Petrol.,* **28**: 186–199 (1958).

[20] McAllister, R. F., Jr., Clay Minerals of Recent Marine Sediments to the West of the Mississippi Delta, Ph. D. thesis, A. and M. College of Texas (1958).

[21] Welby, C. W., Occurrence of Alkali Metals in Some Gulf of Mexico Sediments, *J. Sediment. Petrol.,* **28**: 431–452 (1958).

[22] Griffin, G. M., Clay-Mineral Facies—Products of Weathering Intensity and Current Distribution in the Northeastern Gulf of Mexico, *Bull. Geol. Soc. Am.,* **73**: 737–768 (1962).

[23] Taggart, M. S., Jr., and A. D. Kaiser, Jr., Clay Mineralogy of the Mississippi River Deltaic Sediments, *Bull Geol. Soc. Am.,* **71**: 521–530 (1960).

Murray and Harrison[24] showed that abyssal deeps contain relatively larger amounts of smectite. McAllister[20] claimed to have identified halloysite in a few samples.

There is considerable difference of opinion regarding changes in composition as the fresh-water muds accumulate in the marine environment. Thus Grim and Johns[16] reported the gradual loss of smectite material with the ultimate formation of poorly crystalline illitic and chloritic phases in sediments from the vicinity of Rockport, Texas. These same authors (Johns and Grim[19]) stated that thoroughly degraded micaceous material is being regraded to illite and chlorite in Mississippi River Delta sediments.

Milne and Shott[17] found that chemical analyses of sediments from Mississippi Sound and the Mobile Bay area along a section from fresh water to the Gulf showed a progressive increase in magnesium, potassium, and sodium and a decrease in calcium. Milne and Earley[18] reported the apparent formation of illite from smectite in area of very slow sedimentation in the Mississippi River Delta. They also stated that Pleistocene clays in cores from depths of more than 500 ft off the Louisiana coast contain more illite than the present bottom sediments in the same area, and the increase in illite is attributed to diagenesis.

Welby[21] determined the abundance of the alkali metals in 130 samples from the Gulf of Mexico. The results showed only a slight enrichment of rubidium with respect to potassium in the sediments. While there is present a significant enrichment of cesium with respect to both potassium and rubidium, the potassium-sodium ratios indicate that these two elements are present in the same proportion as in igneous rocks. According to Welby, the alkali-metals content of the sediments apparently does not vary systematically with depth below the present sedimentary interface or with distance from shore.

McAllister[20] studied sediments in the Gulf of Mexico west of the Mississippi River Delta and concluded that there was only minor change in composition on contact with saline waters. Griffin[22] reported on an investigation of clay minerals in sediments of the northeastern Gulf of Mexico and concluded that alteration in the clays produced by marine waters was minor and took the form of (1) slight reduction of the smectite-to-kaolinite ratio, (2) reduction of interlayer water in smectite, and (3) reduction in swelling capacity of degraded vermiculite.

Taggart and Kaiser[23] found no appreciable alteration as the result of diagenesis in late Quaternary Mississippi River Delta sediments off the coast of Louisiana. They further concluded that clay-mineral composition of

[24] Murray, H. H., and J. L. Harrison, Mineral Composition on and Adjacent to a Seamount in Sigsbee Deep, *Bull. Geol. Soc. Am.*, **66**: 1601 (abstr.) (1955).

source sediments determines the clay-mineral composition of these marine sedimentary rocks.

Studies of sediments have been made in other areas where there is a transition from fresh-water to marine conditions, also with variable results. Thus Griffin and Ingram[25] found kaolinite, chlorite, and illite in samples from the Neuse River estuary along the Atlantic Coast of the United States. They found that kaolinite was dominant in the source material and decreased in abundance relative to illite and chlorite in the saline waters, and that chlorite and illite were developed at the lower end of the estuary, with illite apparently increasing faster than chlorite.

Powers[26] determined that the predominant clay minerals in the waters and rivers in the area of Chesapeake Bay consist of a well-formed to weathered illite, with minor amounts of kaolinite and a trace of weathered chlorite. Mixed-layer clay minerals were also observed in the estuaries. Chlorite-like and vermiculite clay minerals were reported as forming in estuaries along the Atlantic Coast. The thermal stability of the diagenetic chlorite increased with increasing salinity of the environment and to a lesser extent with depth in the sediment. The chloritic material is believed to arise principally from the diagenesis of weathered illite in the Atlantic Coastal environment and probably passes through a vermiculite stage. Powers[26] reported on the study of Recent sediments from many areas. According to him, smectite and to a lesser extent illite are altered to a chloritic mineral in the Gulf of Mexico and along the Pacific Coast. Chemical data lend support to the hypothesis that chloritic-like and vermiculite clay minerals are being constructed by the alteration of weathered illite and smectite. On the other hand, Ingram and others[27] investigated the clay-mineral composition of Carolina Bay sediments and found no evidence of alteration since deposition.

Grim and Vernet[28] studied some bottom sediments from the Mediterranean Sea and found that the smectite was of volcanic origin and was being altered to mixed-layer structures which in turn were altered to illite. Kaolinite was also identified but no variation in its abundance was observed.

Grim and Loughnan[29] investigated the clay-mineral composition of sedi-

[25] Griffin, G. M., and R. L. Ingram, Clay Minerals of the Neuse River Estuary, *J. Sediment. Petrol.*, **25**: 194–200 (1955).

[26] Powers, M. C., Adjustment of Land Derived Clays to the Marine Environment, *J. Sediment. Petrol.*, **27**: 355–372 (1957).

[27] Ingram, R. L., M. Robinson, and H. T. Odum, Clay Mineralogy of Some Carolina Bay Sediments, *Southeastern Geol.*, **1**: 1–10 (1959).

[28] Grim, R. E., and J. P. Vernet, Etude par diffraction des minéraux argileux de vases Méditerranéenes, *Bull. Suisse Mineral. Petrog.*, **41**: 65–70 (1961).

[29] Grim, R. E., and F. C. Loughnan, Clay Minerals in Sediments from Sydney Harbour, Australia, *J. Sediment. Petrol.*, **32**: 240–248 (1962).

ments from Sydney Harbour, Australia, and found evidence that illite develops by the replacement of ferrous iron with potassium in vermiculitic material that is carried into the harbor from the adjacent land area.

Gorbunova[30] found a correlation of the clay-mineral composition of sediments from the Bering Sea with that of source areas. She found that in a vertical sedimentary column there was a decrease in smectite and an increase in chlorite with increasing depth, which she concluded was probably caused by diagenetic alterations of the minerals.

In summary, all types of clay minerals have been identified in present-day marine sediments. The occurrence of halloysite is not unequivocal, and references to attapulgite and sepiolite are extremely scant (Heezen et al.,[31] Hathaway and Sachs[32]). Illite and chlorite are abundant components, with kaolinite commonly abundant near shore in localized areas. The distribution of smectite is irregular; it is frequently very abundant where volcanic-ash debris has been an important source material. In general, the clay-mineral composition is likely to reflect climatic conditions in the source. It is not certain, on the basis of available data, if there is a correlation between type of sediment, e.g., Red Clay, ooze, etc., and clay-mineral composition.

Diagenetic processes. There has been much discussion in recent years regarding diagenesis in marine sediments. It is generally agreed that diagenesis should include changes taking place in a sediment when it moves from one environment to another, e.g., fresh-water to marine. It is not generally agreed when diagenetic processes should be considered to end and metamorphic processes to take over. Also it is not generally agreed as to the degree of structural change that must take place before it is considered diagenesis. For example, is the regeneration of degraded chlorite a diagenetic process; is the irreversible collapse of smectite following the adsorption of potassium diagenesis? Or must there be a more fundamental structural change (see Weaver,[33] Grim,[34] Millot,[35] Keller[36])?

[30] Gorbunova, Z. N., Clayey and Other Highly Dispersed Minerals in Sediments of the Bering Sea, *Okeanologiya,* **2:** 1024–1034 (1962).

[31] Heezen, B. C., W. D. Nesteroff, A. Oberlin, and G. Sabatier, Attapulgite in Deep-lying Sediments of the Gulf of Aden and of the Red Sea, *Compt. Rend.,* **260**(22): 5819–5821 (1965).

[32] Hathaway, J. C., and P. L. Sachs, Sepiolite and Clinoptilolite from the Mid-Atlantic Ridge, *Am. Minerologist,* **50:** 852–867 (1965).

[33] Weaver, C. E., A Discussion of the Origin of Clay Minerals in Sedimentary Rocks, *Natl. Acad. Sci., Publ.* 566, pp. 159–173 (1958).

[34] Grim, R. A. Concept of Diagenesis in Argillaceous Sediments, *Bull. Am. Assoc. Petrol. Geologists,* **42:** 246–254 (1958).

[35] Millot, G., "Geologie des argiles," Masson et Cie, Paris (1963).

On the basis of chemical considerations, it would seem that conditions in the marine environment should be reflected in sediments accumulating therein. Thus, the marine environment is alkaline; there is no leaching; and the water contains a good deal of dissolved calcium. These environmental conditions favor the formation of smectite, illite, or chlorite clay mineral rather than kaolinite. Millot[37] presented evidence from a study of the composition of many sediments to show that the presence of Ca^{++} tends to block the formation of kaolinite. A study of weathering processes indicates that kaolinite does not form from calcareous parent material until all the carbonate has been removed.

Correns[38] pointed out that, in environments in which the pH is less than 5, silica is insoluble and R_2O_3 is relatively soluble. At pH values from 5 to 9, the R_2O_3 is very slightly soluble and the silica becomes more and more soluble; and at pH values above 9, silica and alumina are both soluble. Therefore, at low pH values, there would be little silica in solution, and kaolinite is likely to be the stable form, whereas at higher pH values, with much silica in solution, the clay minerals with higher proportions of silica to R_2O_3 are likely to be the stable form.

Millot[37] showed that silica sols are relatively stable and that a considerable cation content is necessary for flocculation, whereas alumina sols are not very stable, being flocculated by a relatively low cation content. Therefore, according to Millot, in an environment of low cation content, there would be a relatively large amount of available alumina with a consequent clay mineral of low silica-to-alumina ratio, e.g., kaolinite. In an environment of high cation concentration, the available silica would be relatively high, because of the flocculation of the alumina, with a consequent clay mineral of high silica-to-alumina ratio, e.g., smectite.

The land-derived muds carried into the sea from many areas probably consist to a considerable extent of degraded illite, degraded chloritic mica and kaolinite. The degraded minerals represent material that has been partially leached of its constituent alkalines and alkaline earths, but not sufficiently to transform it into new minerals. Such degraded minerals would have deficiencies of alkalies and alkaline earths and would quickly adsorb available potassium and magnesium, with an attendant increase in the perfec-

[36] Keller, W. D., Diagenesis in Clay Minerals—A Review, Proceedings of the 11th National Clay Conference, pp. 136–157, Pergamon Press, New York (1963).

[37] Millot, G., Rélations entre la constitution et la genèse des roches sédimentaires argileuses, "Géologie Appliquée et Prospection Minière," vol. II, Nancy, France (1942).

[38] Correns, C. W., T. F. Barth, and P. Eskola, "Die Enstehung der Gesteine," Julius Springer, Berlin (1939).

tion of the mica structures. It is generally agreed that such regeneration of micas takes place but it is not generally agreed that this should be considered diagenesis.

It seems certain (e.g., Milne and Earley[18] and Johns and Grim[19]) that illite and chlorite form from smectite in the marine environment following adsorption of potassium and magnesium, respectively, and collapse of the structure. Whitehouse and McCarter[39] provided supporting evidence for this conclusion, based on extensive laboratory investigations of the reaction of artificial sea water and various clay minerals. According to these authors, the initiation of the observed diagenetic modification of smectite material occurs during the flocculation and settling stages in sea water. Progressive development toward discrete chloritic-illitic clay modification continues within the deposited material. The extent of modification of the clays is more dependent upon the magnesium-potassium ratio in sea water than upon the total salt concentration of such water, but the presence of other inorganic ions probably promotes or accelerates the alteration of the clay by partial control of the ion-exchange mechanism. Organic material in sea water may retard or inhibit the initiation of diagenesis during the flocculation and settling stages of the clay material.

Examples of the formation of new clay minerals during marine diagenesis representing significant structural changes have been described in the literature. Thus, Grim, Dietz, and Bradley[6] concluded that kaolinite was lost and altered probably to illite or a chlorite mineral in the Gulf of California and off the coast of California. Griffin and Ingram[25] arrived at a similar conclusion for sediments of the Neuse River estuary along the Atlantic Coast. Powers[26] concluded that a chlorite-like mineral is forming in Chesapeake Bay estuaries from degraded illite.

In a general survey of the clay-mineral composition of sediments, Rateev[40] concluded that kaolinite predominates in deposits of continental origin while the littoral facies of fresh-water origin show kaolinite mixed with hydromica. Hydromica mixed with smectite and chlorite predominate in marine sediments. The author concluded that kaolinite disappears partially or completely in marine sediments owing to diagenetic transformation. Millot[35] presented extensive evidence for significant changes in marine diagenesis.

Other authors (e.g., Weaver[33]) have concluded that significant mineral

---

[39] Whitehouse, U. G., and R. S. McCarter, Diagenetic Modification of Clay Mineral Types in Artificial Sea Water, *Natl. Acad. Sci., Publ.* 566, pp. 81–119 (1958).

[40] Rateev, M. A., Clay Minerals and Their Facial Occurrence in Water Basins of the Humid Zones, *Issled. i Ispol 'z. Glin, L'vovsk. Gos. Univ., Materialy Soveshch., Lvov,* pp. 117–132 (1957).

changes during diagenesis are unimportant. In this connection the experiments of Quaide[41] are interesting. He kept muds composed of chlorite, smectite, and micas from San Francisco Bay and from nearby salt concentration ponds at nearly constant conditions of pH, Eh, and salinity for periods of 7 to 27 years and found no substantial alteration in the structure that could be detected by X-ray diffraction.

Riviere and Visse[42] concluded that variations in the settling rates of the various clay minerals were adequate generally to explain the variations in clay-mineral compositions in ancient argillaceous sediments that have been attributed to diagenesis.

Riviere[43] found that the presence of small amounts of humic material makes kaolinitic clays relatively more resistant to flocculation by sea water. In another paper, Riviere and Vernhet[44] found that, contrary to what occurs with illitic and smectite clays, traces of humic matter greatly retard the flocculation of kaolinite clays by divalent ions, notably by those in sea water, and that this may explain the diminution in the amount of kaolinite with increasing distance from fresh water.

Whitehouse and Jeffrey[45] determined the settling velocity of kaolinite, smectite, and illite in synthetic sea water with increasing chlorinity. They found that the settling velocity of illite and kaolinite increased very rapidly with slight increases in the salinity, with illite settling slightly faster than kaolinite. Smectite was relatively slightly influenced by increased salinity and had, in general, a much slower rate than illite or kaolinite. In mixtures of illite and kaolinite, the illite settled out appreciably faster than the kaolinite; however, if smectite was also present, it tended to reduce the settling velocity of the illite.

Van Andel and Postma[46] observed illite to be the dominant clay mineral in the Gulf of Paria off South America. They also noted that kaolinite was distributed fairly uniformly, but smectite was more abundant in the open

[41] Quaide, W., Clay Minerals from Salt Concentration Ponds, *Am. J. Sci.,* **256:** 431–437 (1958).

[42] Riviere, A., and L. Visse, L'origine des minéraux (argileux) des sédiments marins, *Bull. Soc. Geol. France,* **4:** 467–473 (1955).

[43] Riviere, A., Influence des matières organiques sur la sèdimentation des argiles, *Intern. Geol. Congr. 19th, Resumé,* p. 27 (1952).

[44] Riviere, A., and S. Vernhet, Sur la sédimentation des minéraux argileux en milieu marin en présence de matières humiques; consequences géologiques, *Compt. Rend.,* **233:** 807–808 (1951).

[45] Whitehouse, U. G., and L. M. Jeffrey, Chemistry of Marine Sedimentation, A. and M. College of Texas, Department of Oceanography (1953).

[46] Van Andel, Tj., and H. Postma, Recent Sediments of the Gulf of Paria, *Verhandel. Koninkl. Ned. Akad. Wetenschap., Afdel. Natuurk. Eerste Reeks, Bull.* **20** (1954).

gulf than in the delta portion. They suggested that illite and kaolinite were flocculated earlier than montmorillonite when river water carrying them was mixed with saline water and that this accounted for the observed clay-mineral distribution.

Loughnan and Craig,[47] studying the clay-mineral composition of Recent sediments off the east coast of Australia, found evidence for differential settling as a contributing factor in the distribution of clay minerals.

Recently potassium-argon age determinations of the mica component of sediments have thrown much light on diagenesis. Hurley and associates[48] concluded that the illite in the Mississippi River Delta sediments was detrital, showing ages in excess of 200 million years. They found that the illite and potassium-bearing materials in the Rappahannock River sediments had an age in excess of 300 million years. Pelagic illitic clays from deep ocean cores and bottom samples showed various ages that suggested an airborne derivation from continental source areas. In another paper, Hurley and associates[49] reported that the age of illitic material of dominantly silt size from the Mississippi River Delta averaged 280 million years with little horizontal or vertical variation. The clay-size fraction of the same sample, however, averaged only 166 million years. They concluded that this age is lower than would be expected of detrital micas from Paleozoic shales in the central and eastern part of the source area, but that it might be due to mixtures of younger mixed-layer illite-smectite released from younger sediments in a western source area. Another interpretation is the adsorption of younger potassium by the clay-size fraction. These same authors pointed out that clay soils forming over ancient shales can contain illite of the same age as the underlying shales.

Hower et al.[50] determined the age of various size fractions of an Ordovician shale. They found that the whole shale had an age of 460 million years, which is in good agreement with the accepted age of the Upper Ordovician. The age, however, varied over approximately 100 million years with particle size. In the coarser sizes where significant amounts of 2M mica, and thus presumably detrital, are present the potassium-argon age is too old. With

[47] Loughnan, F. C., and D. C. Craig, A Preliminary Investigation of the Recent Sediments off the East Coast of Australia, *Australian J. Marine Freshwater Res.*, 13: 48–56 (1962).

[48] Hurley, P. M., W. H. Hart, D. G. Pinson, and H. W. Fairbairn, Authigenic Versus Detrital Illite in Sediments, *Geol. Soc. Am., Program Annual Meeting* (1958).

[49] Hurley, P. M., D. G. Brookins, W. H. Pinson, S. R. Hart, and H. W. Fairbairn, K-Ar Age Studies of Mississippi and Other River Sediments, *Bull. Geol. Soc. Am.*, 72: 1807–1816 (1961).

[50] Hower, J., and staff, The Dependence of K/Ar Age on the Mineralogy of Various Particle Size Ranges in a Shale, NYO-3942 Ninth Annual Progress Report for 1961, U.S. Atomic Energy Commission Contract AT (30-1)-1381, 137–141 (1961).

a decrease in the relative amount of detrital mica the age goes down; however, it does not level off at the correct age with decreasing particle size. The finer sizes give much too young an age. Either the finer particle sizes have lost argon, or some of the illite formed much later than the time of sedimentation.

Hurley and others[51] in a still later paper reported the age of a considerable variety of argillaceous materials. They concluded that the potassium-argon ages of Cretaceous and Miocene shales are greater than the true age, indicating the presence of older detrital components. All Recent sediments from river mouth to pelagic areas show clay fractions with the potassium-argon ages greater than 100 million years, which are much greater than the true geologic ages. Results indicate that the clay fractions consist of mixtures of (1) more ancient detrital grains with older potassium-argon ages and (2) authigenetic or epigenetic illite with younger potassium-argon values than the time of sedimentation. In the younger samples, the effect of the first situation predominates, and in the older, the second is the dominant factor.

It is clear that major structural changes sometimes take place in the accumulation of marine sediments. At other times, the evidence for such changes is not unequivocal. It is of interest to consider the possible factors that may control the alterations. Oppenheimer[52] studied the bacterial activity in sediments and concluded that such activity may be primarily responsible for much of the diagenesis that takes place at or close to the surface of accumulation. This author pointed out that bacterial activity is greater in sands with large interstices than in clays with small pores and that diverse bacterial activity influences the pH, Eh, gas production, concentration of trace elements, precipitation of carbonates, diagenesis of silicates, and the formation of organic complexes.

Marine conditions vary from oxidizing to reducing, depending on the topography of the ocean floor and hence on water movement, on the rate of sedimentation, and perhaps on other factors. It is not known exactly how the variation from oxidizing to reducing conditions would influence diagenetic processes. Perhaps oxidizing conditions, by tending to fix the R_2O_3 in a relatively inert form, would retard diagenetic changes.

Variations in the concentration of potassium and magnesium would also affect diagenesis. Deficiencies of the alkalies and alkaline earths would retard the formation of the micas. Diagenetic changes would be expected to be most intense near shore, where the sediments first encounter the marine environment.

[51] Hurley, P. M., J. M. Hunt, W. H. Pinson, Jr., and H. W. Fairbairn, K-Ar Age Values on the Clay Fractions in Dated Shales, *Geochim. Cosmochim. Acta,* **27:** 279–284 (1963).

[52] Oppenheimer, C. H., Bacterial Activity in Sediments of Shallow Marine Bays, *Geochim. Cosmochim. Acta,* **19:** 244–260 (1960).

The nature of the marine environment and diagenesis are such that variations in the character of the source material *might be* reflected in the sedimentary accumulation.[53] For example, a kaolinitic marine sediment means a kaolinitic source area, since kaolinite is not formed in the sea. Also it means relatively rapid accumulation with the persistence of kaolinite through an environment fundamentally unfavorable for it.

A completely micaceous marine sediment means a source area producing little kaolinite, or possibly extremely slow accumulation. It seems likely that source material containing a considerable amount of kaolinite would give rise to a marine sediment in which at least some kaolinite persisted.

Glauconite. Glauconite is a rather unique (see page 97) clay mineral that is formed during marine diagenesis. Many investigators have recorded the frequent association of glauconite with organic residues, and it has been generally concluded that the presence of organic material is necessary for the formation of the mineral.

Galliher[54] showed that some glauconite has formed as an alteration product of biotite, with the alteration taking place under reducing conditions and slow deposition. Unless deposition is slow so that alteration takes place before burial, the biotite mica persists. Hendricks and Ross[55] reviewed the literature which shows that minerals other than biotite may alter to glauconite in a marine environment.

Hadding,[56] who made an extremely complete study of glauconite, showed that glauconite occurs in flakes and as pigmentary material, in addition to the well-known large rounded pellets. According to Hadding, glauconite forms in the shallow sea in agitated waters that are not highly oxygenated. It forms during times of decreased deposition of detritus and often during negative sedimentation. Revelle[2] also emphasized the glauconite is currently forming off coasts lacking important rivers, where deposition is slow.

Hendricks and Ross,[55] after a careful study of the variation in the composition of glauconites concluded that the mineral is formed in a reducing environment maintained by bacterial action. The environment probably remains unchanged for long intervals of time. The very uniform content of magnesium and the rather constant ratio of Fe^{3+} to Fe^{++} indicate that a critical content of magnesium and a particular oxidation-reduction potential might be required for the mineral to form. The ratio of Na^+ to K^+ in interlayer

[53] Grim, R. E., The Depositional Environment of Red and Green Shales, *J. Sediment. Petrol.*, **21**: 226–232 (1951).

[54] Galliher, E. W., Glauconite Genesis, *Bull. Geol. Soc. Am.*, **46**: 1351–1366 (1936).

[55] Hendricks, S. B., and C. S. Ross, Chemical Composition and Genesis of Glauconite and Celadonite, *Am. Mineralogist*, **26**: 683–708 (1941).

[56] Hadding, A., The Pre-Quaternary Rocks of Sweden, IV, Glauconite and Glauconite Rocks, *Medd. Lunds. Geol.-Mineral. Inst.*, **51** (1932).

positions in the structure of glauconite is rather distinctive, suggesting that a certain concentration of these ions may be required for its formation.

Cloud[57] listed the following limits for glauconite formation: stratigraphic range, Cambrian to present; present areal distribution from 65° south latitude to 80° north latitude; occurring off most ocean coasts and mainly on the continental shelves away from large streams. It is known to originate only in marine waters of normal salinity. Its formation requires at least slightly reducing conditions and seems to be facilitated by the presence of decaying organic material which results in reducing conditions. Its present occurrence is rare in sediments at depths greater than 1,000 fathoms or less than 5 fathoms, and it is uncommon above 10 and below 400 fathoms. Its formation is probably favored by markedly warm waters, and micaceous minerals or bottom muds of high iron content seem most favorable source materials. It seems to form best where sedimentation is not rapid. Formation is favored on continental shelves and swells or banks off the coast of crystalline land areas that lack important rivers. It is found mainly in calcareous detrital sediments such as calcareous sandstones and impure granular limestones. It is rare in pure clay rock, pure quartz sandstones, or chemically precipitated carbonates. It is commonly associated with the remains of fecal pellets of sediment-ingesting organisms or as internal fillings of foraminifera. It is rare or absent from beds that are rich in algae, coral, or bryozoans, and was probably reworked or transported where so found. Ehlmann and others[58] offered the following interpretation of the genesis of glauconite. Soon after the death of marine organisms the shell sinks to the bottom of the ocean. Within the shell an internal molded clay material forms by flocculation of the colloidal material or by impregnation of detrital clay. In the sample investigated, the earliest-stage material had an Fe_2O_3 content of about 20 percent and was regarded as a potassium-deficient prototype of glauconite with iron already incorporated into the structure. There begins a process of potassium fixation and an expulsion of water with the resulting formation of the glauconite. In the early stages kaolinite may be present but this disappears as potassium fixation progresses.

Pratt[59] investigated the origin and distribution of glauconites on the sea floor off the coast of southern California. He found clay aggregates, which probably would be designated as glauconite, grading from relatively soft pale yellow-green highly expandable smectite types to dark-green illitic types.

[57] Cloud, P. E., Jr., Physical Limits of Glauconite Formation, *Bull. Am. Assoc. Petrol. Geologists*, 39(4): 484–492 (1955).

[58] Ehlmann, A. J., N. C. Hulings, and E. D. Glover, Stages of Glauconite Formation in Modern Foraminiferal Sedimentation, *J. Sediment. Petrol.*, 33(1): 87–96 (1963).

[59] Pratt, W. L., Origin and Distribution of Glauconites and Related Clay Aggregates on Sea Floor off Southern California, *Bull. Am. Assoc. Petrol. Geologists*, 46(2): 275 (1962).

Maximum concentrations occurred in sediments in the outer shelf and upper slope areas, but the distribution was patchy both areally and vertically within these sediments. Glauconite was rare in water depths of less than 100 ft.

Burst[60,61] pointed out, in two very significant papers, that material described as glauconite varied in composition from illite-smectite mixed-layer structures with relatively large amounts of smectite to substantially pure illites with 1M to 1Md morphology. In general, the amount of mixed layering decreased and the ordering of the illite increased from Recent to ancient glauconites. Further, the potassium and iron content tended to increase as the mixed layering decreased and as the ordering increased. Chlorite and kaolinite were identified occasionally as admixed constituents. Glauconite was considered to form as the result of reactions between degraded layer-silicate lattices and the environment, the reaction being triggered by local reducing conditions caused by organic matter. The characteristics of the environment and the completeness or incompleteness of the reaction control the nature of the material.

Using Burst's model of glauconization, Hower[62] found that pellets in clean sandstones, limestones, and dolomites are almost exclusively monomineralic, consisting only of mixed-layer illite-smectite with a low percentage of expandable layers. Pellets in argillaceous sandstones and marls frequently contain extraneous clay minerals, and the illite-smectite contains a relatively high percentage of expandable layers. Because of the well-known observation that glauconite forms only at relatively low sedimentation rates, it is concluded that the pellets occurring in argillaceous sandstone and marls represent less glauconized material than those pellets occurring in clean sandstones, limestones, and dolomites. It is pointed out that these observations fit Burst's model for glauconization, which asssmes a degraded 2:1 layer lattice structure as a parent material, and that the glauconization process is a gradual substitution of iron for alumina in the octahedral positions, a concomitant gradual increase in lattice charge arising from the octahedral layer, and a consequent increase in interlayer potassium. It is also concluded that potassium increases with octahedral charge and is independent of tetrahedral charge, a phenomenon predictable from Burst's model.

Gorbundova[63] studied glauconite from the Lower Cretaceous sediments of Dagestan and concluded that glauconite is formed mainly during the regres-

[60] Burst, J. F., "Glauconite" Pellets: Their Mineral Nature and Application to Stratigraphic Interpretations, *Bull. Am. Assoc. Petrol. Geologists,* **42:** 310–327 (1958).

[61] Burst, J. F., Mineral Heterogeneity in "Glauconite" Pellets, *Am. Mineralogist,* **43:** 481–497 (1958).

[62] Hower, J., Some Factors Concerning the Nature and Origin of Glauconite, *Am. Mineralogist,* **46:** 313–334 (1961).

[63] Gorbundova, L. I., Glauconite from Lower Cretaceous Sediments of Dagestan, *Dokl. Akad. Nauk SSSR,* **130:** 846–849 (1960).

sion stages of sedimentation and fades out in the transgressive phases. The glauconite cements the sandstone grains and also forms granular aggregates in veinlets filled with quartz and feldspar fragments.

Triplehorn[64] described the morphology of a large number of glauconites. The great majority of his samples were simple ovoidal or irregular pellets and consisted of cryptocrystalline aggregates in random orientation. He also observed pellets which were clearly casts of foraminifera and others that were alterations of books of micas. In other cases there was a replacement of organic materials and sometimes concentric coatings on sand grains. He concluded that it did not seem likely that all these types had formed by the same process or from the same parent material and, therefore, that glauconite has a multiple origin and the exact process of formation varies with the type of parent material.

Hurley and others[65] investigated the age of glauconites by the potassium-argon and rubidium-strontium methods and found that the age showed consistent variation with geological age with moderately small scattering. In general, the results appear to fall 10 to 20 percent short of ages measured on micas associated with dated igneous rocks. In no samples measured did the glauconites have ages that were definitely higher than expected limits in the time scale.

Nonmarine conditions. Lacustrine. The paucity of clay-mineral data for Recent nonmarine sediments is so great that few conclusions can be reached. Cuthbert[66] showed that illite is the dominant clay mineral in sediments accumulating in Lake Erie.

Moore[67] found that the average clay-mineral composition of samples from the bottom of northeastern Lake Michigan was illite, 50 percent; mixed-layer minerals, 30 percent; and chlorite, 20 percent. A comparison of the average clay-mineral composition of lake sediments and glacial till from the adjacent shore showed little variation for the type of minerals present or their relative abundance. A definite change in the clay-mineral distribution was noted in a few samples; e.g., some dredge and core samples taken from shallow-water shelf areas had a higher proportion of mixed-layer materials than illite. In addition, many of these same samples showed a much higher content of

[64] Triplehorn, D. M., The Petrology of Glauconite, Ph.D. thesis, University of Illinois, (1961).

[65] Hurley, P. M., R. F. Cormier, J. Hower, H. W. Fairbairn, and W. H. Pinson, Jr., Reliability of Glauconite for Age Measurement by K-Ar and Rb-Sr Methods, *Bull. Am. Assoc. Petrol. Geologists,* **44**: 1793–1808 (1960).

[66] Cuthbert, F. L., Clay Minerals in Lake Sediments, *Am. Mineralogist,* **29**: 378–388 (1944).

[67] Moore, J. E., Petrography of Northeastern Lake Michigan Bottom Sediments, *J. Sediment. Petrol.,* **31**: 420–436 (1961).

poorly crystallized chlorite. The observed differences in composition were considered to be related to the texture of the sediment, weathering during transport, and weathering in place. In many cases the clay fraction of the sandy clay, silty and clayey sand, and sand zones in the shelf environment contain a higher proportion of mixed-layer minerals and also poorly crystallized kaolinite. These coarser-grained zones probably would be more susceptible to subaqueous weathering action by waves and currents. This weathering could result in the degradation of chlorite and illite with a partial leaching of potassium and magnesium from the clay minerals. On the other hand, sediments found in deeper water in the valley and ridge areas would probably not be subject to such weathering action.

Rateev[68] showed that the sediments from southern Lake Baikal in Siberia were composed of a variety of clay minerals. The author found no evidence of diagenetic changes.

Sarmiento and Kirby[69] reported the presence of kaolinite, illite, montmorillonite, and chlorite in sediments of Lake Maracaibo.

Brydon and Patry[70] studied the composition of the sediments in the Champlain Sea and identified micas, chlorites, and small amounts of interstratified illite-smectite clay minerals. They reported no evidence of any diagenetic changes.

In a study of sediments of Lake Mead, Rolfe[71] found that smectite-type clays dominate in the dispersed clay fractions upstream, whereas illite is predominant downstream. The author concluded that this variation in composition is due to conditions of deposition rather than any diagenetic changes.

Brousse and Maurel[72] reported the presence of hydrated halloysite in a deposit of lacustrine origin in France.

It would seem that there would be few diagenetic changes in fresh-water lake sediments and that the sediments would reflect the source material supplied to the lake. Lacustrine environments in which there is any appreciable movement of water through the sediments, with some consequent removal of alkalies and alkaline earths, might be expected to show the development of kaolinite as a diagenetic mineral. In general, lakes in which there is no accumulation of lime would probably show a tendency toward the develop-

[68] Rateev, M. A., Clay Minerals in Soil Sediments of the Southern Lake Baikal, *Dokl. Akad. Nauk SSSR,* **82:** 981–1031 (1952).

[69] Sarmiento, R., and R. A. Kirby, Recent Sediments of Lake Maracaibo, *J. Sediment. Petrol.,* **32:** 698–724 (1962).

[70] Brydon, J. E., and L. M. Patry, Mineralogy of the Champlain Sea Sediments and a Rideau Clay Soil Profile, *Can. J. Soil Sci.,* **41:** 169–181 (1961).

[71] Rolfe, B. N., Surficial Sediments in Lake Mead, *J. Sediment. Petrol.,* **27:** 378–386 (1957).

[72] Brousse, R., and P. Maurel, A New Deposit of Hydrated Halloysite, *Bull. Soc. Franc. Mineral. Crist.,* **85**(2): 128 (1962).

ment of kaolinite. In lakes where lime is accumulating, the lime would tend to block the formation of kaolinite, and any diagenetic changes would probably be very slight.

Investigations by De Lapparent,[73] Millot,[37] and unpublished work by the author have shown that the sepiolite-attapulgite minerals are frequently found in Recent sediments accumulating in dry desert basins. Bonython[74] reported the presence of palygorskite in sediments of Lake Eyre in South Australia.

Grim, Droste, and Bradley[75] reported the occurrence of a range of mixed-layer clay-mineral sequences of the corrensite type in evaporites of Permian age in New Mexico. Droste[76] investigated the clay-mineral composition of sediments from numerous desert lakes in Nevada, California, and Oregon and found that the clay-mineral components were predominantly illite and smectite, with minor amounts of chlorite and kaolinite, and that the clay-mineral components reflected the source area. No evidence for diagenetic changes was found. In another report, Droste[77] concluded that the clay-mineral composition of desert saline sediments is controlled almost entirely by the composition of the source area. The high sodium and calcium environment does not seem to produce significant diagenetic changes in the clay minerals deposited in these desert areas. Magnesium was a relatively minor component in the environment studied.

Grim et al.[78] found smectite, illite, and kaolinite in sediments from the Great Salt Lake, Utah, and concluded that the minerals were detrital.

Hay and Moiola[79] reported the formation of authigenic zeolite minerals in Pleistocene sediments of Searles Lake, California.

[73] De Lapparent, J., Structure et origine des terres naturelles susceptibles d'etre utilisées pour la décoloration des huiles minéral, *Compt. Rend. Congr. Mondial Petrole Congr., Paris* (1937).

[74] Bonython, C. W., The Salt of Lake Eyre, Its Occurrence in Madigan Gulf and Its Possible Origin, *Trans. Roy. Soc. S. Australia*, **79**: 66–92 (1956).

[75] Grim, R. E., J. B. Droste, and W. F. Bradley, A Mixed-layer Clay Mineral Associated with an Evaporite, Proceedings of the 8th National Clay Conference, pp. 228–236, Pergamon Press, New York (1960).

[76] Droste, J. B., Clay Mineral Composition of Sediments in Some Desert Lakes in Nevada, California, and Oregon, *Science*, **133**: 1928 (1961).

[77] Droste, J. B., Clay Minerals in Sediments of Owens, China, Searles, Panamint, Bristol, Cadiz, and Danby Lake Basins, California, *Bull. Geol. Soc. Am.*, **72**: 1713–1722 (1961).

[78] Grim, R. E., G. Kulbicki, and A. V. Carozzi, Clay Mineralogy of Sediments of the Great Salt Lake, Utah, *Bull. Geol. Soc. Am.*, **71**: 515–520 (1960).

[79] Hay, R. L., and R. J. Moiola, Authigenic Silicate Minerals in Pleistocene Sediments of Searles Lake, California, *Geol. Soc. Am., Program Annual Meeting (Houston), Abstr.*, 68A (1962).

In the case of sepiolite-attapulgite minerals, diagenesis appears to be significant in sediments accumulating in saline lakes. Probably a particular composition, chiefly in regard to the abundance of magnesium, leads to the formation of these minerals. The desert areas surrounding saline lakes would be expected to have soils containing smectite, illite, and/or chlorite; these clay minerals would be "at home" in saline lakes, and substantial changes would not be expected. Substantial amounts of kaolinite would not be likely in such sediments.

Rivers. Holmes and Hearn[80] investigated the clay-mineral composition of Recent alluvium deposited by the Mississippi River and its tributaries, with the results shown in Table 14-1.

Table 14-1 *Mineralogical composition of alluvial material deposited by the Mississippi River and its tributaries*

(From Holmes and Hearn[80])

	Kaolinite	Smectite	Illite
Eastern tributaries: Ohio, Cumberland, Tennessee, Duck, and Clinch Rivers	10–20	10–15	65–75
Mississippi River	5–15	25–45	40–60
Western tributaries: Milk, Yellowstone, Missouri, Platte, and Arkansas Rivers	10–20	20–45	40–60

Values given are percentages.

The clay minerals in fluviatile sediments should reflect the composition of the soils from which they are derived. The data of Holmes and Hearn indicate that there is a relatively larger amount of smectite in the western tributaries of the Mississippi River, which derive their material from dry areas which favor the formation of smectite soils. Lafond[81] investigated the clay-mineral composition of sediments in the Vilaine River in France and found that they reflected the character of the parent rock. Packham et al.[82] arrived at the same conclusion for the sediments in nine rivers in England.

In rivers themselves, one might expect a certain amount of leaching action; and if this is the case, degraded micaceous minerals and kaolinite would tend to develop, otherwise any changes would probably be insignificant.

---

[80] Holmes, R. S., and W. E. Hearn, Chemical and Physical Properties of Some of the Important Alluvial Soils of the Mississippi Drainage Basin, *U.S. Dept. Agr., Tech. Bull.* 833 (1942).

[81] Lafond, R. L., Mineralogical Study of Actual Clays from the Vilaine Basin, *Compt. Rend.,* **252:** 14–16 (1961).

[82] Packham, R. F., D. Rosaman, and H. G. Midgely, A Mineralogical Examination of Suspended Solids from Nine English Rivers, *Clay Minerals Bull.,* **4**(25): 239–242 (1961).

Glacial Deposits. A considerable study of the composition of the glacial deposits in North America has been made by students of soils. In unweathered till and loess, the dominant clay mineral is illite, frequently with considerable amounts of chloritic mica. Kaolinite and smectite are occasionally present in small amounts.

Lafond and others[83] investigated the mineral composition of some glacial clays from Switzerland and found that they reflected the mineral composition of the parent rocks. This would be expected to be a general conclusion for sediments of glacial origin.

The composition of varved glacial lake sediments is not well known, but Cooling[84] has shown that, in at least some varved sediments in Canada, illite is the dominant clay mineral, with smectite also present in the dark layers but not in the light layers.

Ancient sediments

In recent years a tremendous literature has accumulated on the clay-mineral composition of a vast variety of ancient sediments. Within the space available it is possible to present only a brief summary of important generalizations.

Clay-mineral composition in relation to mode of origin. Millot[37] presented thorough clay-mineral analyses of a large series of sediments from eastern France and the adjacent sections of Germany and Switzerland whose mode of origin is well established on paleontological and stratigraphic evidence. In sediments of *marine origin,* Millot found illite invariably present, making up from 50 to 100 percent of the total clay-mineral content. The illite was particularly abundant in calcareous marine sediments, in which it made up from 70 to 100 percent of the total clay-mineral content. Kaolinite was often present and made up from 0 to 50 percent of the clay-mineral content. Millot emphasized the point that calcareous sediments are likely to have little or no kaolinite and that calcium tends to inhibit the formation of this clay mineral. Pyritiferous marine sediments were particularly likely to contain appreciable kaolinite. Chloritic mica and vermiculite were frequently present as minor constituents.

In 26 sediments of *lagoonal origin,* illite was found to be the dominant

---

[83] Lafond, L. R., A. Riviere, and S. Vernhet, The Mineralogic Composition of Some Glacial Clays, *Compt. Rend.,* **252:** 3310–3312 (1961).

[84] Cooling, L. F., personal communication.

clay mineral in 19, smectite in 4, and attapulgite-sepiolite in 3. Chloritic mica and vermiculite were frequent minor clay-mineral components. Kaolinite was completely absent in 19 samples and in no case totaled more than 30 percent of the clay-mineral content. It is probable that the sediments containing attapulgite-sepiolite represent lacustrine rather than lagoonal conditions.

In sediments of *lacustrine origin* where the lake is believed to have contained "aggressive" water, i.e., where there was actual leaching of alkalies and alkaline earths in the accumulating sediments because of active movement of water through them, or because of relatively low pH of the water, the dominant clay mineral is kaolinite. In such sediments the total content of illite, chloritic mica, and smectite ranged from 0 to 30 percent.

In sediments of *lacustrine origin* of the "nonaggressive" type, where salts and carbonate are likely to accumulate because of slight movement of water through the lake and/or relatively alkaline water, the dominant clay-mineral components are illite, smectite, and sepiolite-attapulgite. The latter type of clay mineral is particularly prevalent in sediments that have accumulated in dry desert lakes.

Numerous other investigators have reported a correlation between clay-mineral composition and mode of origin. Thus, Strahl[85] analyzed a large number of shales from the United States and Europe and found that the amount of illite relative to kaolinite is greater in marine than in continental shales. Oberlin and Freulon,[86] studying sediments from the central Sahara area, found that large well-formed particles of kaolinite are characteristic of continental deposits and small poorly formed plates are characteristic of clays of marine origin. Weaver[87] found in the study of the Late Cretaceous rocks of the Washakie Basin in Wyoming that the kaolinite content was usually much greater in sandstones than in adjacent shales and that it increased from marine to continental environments.

Pryor and Glass[88] studied the clay-mineral composition of Cretaceous and Tertiary sediments from the Upper Mississippi embayment. They found that clays deposited in a fluviatile environment were dominantly kaolinitic. Those in the outer neritic environment were dominantly smectite, whereas those in the inner neritic environment were composed of nearly equal amounts

[85] Strahl, E. O., The Relations Between Selected Minerals, Trace Elements, and Organic Constituents of Several Black Shales, *U.S. At. Energy Comm.*, NYO-7908, 155 pp. (1958).

[86] Oberlin, A., and J. M. Freulon, Mineralogical Study of Some Paleozoic Clays of Tassili N'Ajjer and Fezzan (Central Sahara), *Bull. Soc. Franc. Mineral. Crist.*, **81**: 186–189 (1958).

[87] Weaver, C. E., Clay Mineralogy of the Late Cretaceous Rocks of the Washakie Basin, *Wyoming Geol. Assoc., Guidebook 16th Ann. Field Conf.*, pp. 148–154 (1961).

[88] Pryor, W. A., and H. D. Glass, Cretaceous-Tertiary Clay Mineralogy of the Upper Mississippi Embayment, *J. Sediment. Petrol.*, **31**: 38–51 (1961).

of kaolinite, illite, and smectite. They concluded that segregation of the clay minerals in the depositional environments was responsible for these variations in composition.

Parham[89] reported that, in Pennsylvanian underclays of the Illinois Basin, kaolinite was more abundant in near-shore phases than illite, and Hayes,[90] in a study of the clay-mineral composition of Mississippian strata from southeastern Iowa, found that, in general, kaolinite was confined to near-shore phases, chlorite in an intermediate position, and illite concentrated farther from shore.

Degens and others,[91] studying a series of carboniferous sediments, found that boron and rubidium were more abundant in marine sediments than in those deposited in fresh water and that gallium was more abundant in freshwater sediments. The abundance of these elements is related to clay-mineral compositions.

Frederickson and Reynolds[92] found that the boron content of sea water increases linearly with salinity within the range tested. The boron content of the illite minerals in the clay-mineral size fractions of many types of sedimentary rocks has been shown to increase with the geologically known salinity prevailing during deposition of the rocks. It is believed that boron is abstracted by the illitic clay minerals from sea water in proportion to the salinity and that the boron content is not substantially changed by other geological processes normally operative in sedimentary rocks. These authors concluded that the boron content of illite minerals in sedimentary rocks can be used to evaluate the salinity of the sea prevailing at the time the sediment was deposited. Harder[93] arrived at similar conclusions.

Tourtelot and others,[94] investigating the Pierre shale, which is entirely marine, found a wide range of boron contents, indicating a relation between boron content and total abundance of clay minerals, source materials, boron

[89] Parham, W. E., Clay Mineral Facies of Certain Pennsylvanian Underclays, Ph.D. thesis, University of Illinois (1962).

[90] Hayes, J. B., Clay Mineralogy of Mississippian Strata of Southeast Iowa, Proceedings of the 9th National Clay Conference, pp. 413–425, Pergamon Press, New York (1962).

[91] Degens, E. T., E. G. Williams, and W. F. Keith, Environmental Studies of Carboniferous Sediments, Part I, Geochemical Criteria for Differentiating Marine from Fresh Water Shales, *Bull. Am. Assoc. Petrol. Geologist,* **41:** 2427–2455 (1957).

[92] Frederickson, A. F., and R. D. Reynolds, Jr., Geochemical Method for Determining Paleosalinity, Proceedings of the 8th National Clay Conference, pp. 203–213, Pergamon Press, New York (1960).

[93] Harder, H., Einbau von Bor in detritische Tonminerale, *Geochim. Cosmochim. Acta,* **21:** 284–294 (1961).

[94] Tourtelot, H. A., L. G. Schultz, and C. Huffman, Jr., Boron in Bentonite and Shale from the Pierre Shale, South Dakota, Wyoming, and Montana, *U.S. Geol. Surv., Profess. Paper* 424-C, pp. 288–292 (1961).

content of sea water, and rate of sedimentation of clay minerals from sea water.

Walker and Price[95] reported that natural illites contain a small amount of boron which probably substitutes for silicon in the structure. The exact amount of boron taken into the structure is determined by the potassium content and grain size of the illite and the salinity at the time of deposition. In illites of constant grain size from the same salinity environment, there is an antipathetic relationship between boron and potassium. Departure curves were constructed from the experimental data which were used to correct for variations in boron concentration due to changes in composition and grain size of illites. The residual variations in boron concentration are believed to be directly related to paleosalinity.

As Weaver[96] pointed out on the basis of clay-mineral analyses of many sediments, any of the major clay minerals can occur in any of the major depositional environments. No particular clay mineral is restricted to a particular environment. Illite, smectite, and mixed-layer illite-smectite can occur in abundance, frequently as the only clay mineral present in any of the major depositional environments. Kaolinite is dominant mainly in fluviatile and near-shore environments although it can occur in abundance in all environments. Chlorite is seldom if ever a dominant clay mineral. Nonmarine shales are seldom if ever monomineralic.

Clay minerals in relation to geologic age. Analyses in the author's laboratory of many ancient sediments from the United States and elsewhere have shown that smectite is generally absent in sediments older than the Mesozoic. Except for a few beds of probable bentonitic origin, smectite is rare in the older sediments. This point is of particular importance, for smectite is abundant in many Mesozoic and Cenozoic sediments, in Recent marine sediments, and in present-day weathering products.

It seems likely that metamorphic processes would cause the alteration of smectite to a mica type of mineral, and, therefore, the mineral would tend to disappear in more ancient sediments. When the structure of smectite is completely collapsed so that *all* the water is removed from between the unit layers, rehydration of the mineral is difficult, and the mineral has many of the attributes of mica. Further in the course of geologic time, smectite would probably slowly pick up potassium and magnesium from ground waters, which might in turn obtain such constituents from the disintegration of ferromagnesian minerals and feldspars. The adsorption of potassium and magne-

[95] Walker, C. T., and N. B. Price, Departure Curves for Computing Paleosalinity from Boron in Illites and Shales, *Bull. Am. Assoc. Petrol. Geologists,* **47**: 833–841 (1963).

[96] Weaver, C. E., A Discussion on the Origin of Clay Minerals in Sedimentary Rocks, *Natl. Acad. Sci., Publ.* **566**, pp. 159–173 (1958).

sium would favor a shift to a mica type of mineral. Thus in the course of time, because of dehydration due to compaction, smectite would tend to be altered in the direction of the micas.

Keller[97] pointed out that the content of sodium and potassium is nearly the same in the average igneous rock (3.8 per cent Na_2O and 3.13 percent K_2O), but after weathering, transportation, and sedimentation the resulting average shales contain more than twice as much K_2O (3.24) as Na_2O (1.30). The sodium goes in solution to the ocean where about 48 times more sodium than potassium is dissolved in the water. The potassium is held by the clay on the land and in marine clay sediments. In this connection, the dominance of illite in older Paleozoic shales has been reported by many investigators.

Burst[98] studied the clay-mineral relationships in the subsurface Wilcox of Eocene age in Louisiana to depths of 13,000 ft and found that smectite becomes less evident below 3,000 ft and is not normally found in an unmixed state below about 10,000 ft of overburden. At depths between 3,000 and 14,000 ft smectite lattices are commonly interspersed with illite components. The frequency of illite increases to the virtual elimination of smectite below 14,000 ft.

Powers[99] observed from several deep wells in the Tertiary sediments of the coast of the Gulf of Mexico that clay minerals at depths of less than about 5,000 ft consist predominantly of smectite, with only small amounts of illite and chlorite and mixed layers. With increasing depth there is a slight increase in the amount of illite and a pronounced increase in illite-smectite mixed layers. There is no discrete smectite below 9,000 to 12,000 ft.

Bailey and others[100] separated the illite from several shales and clays of Pennsylvanian age into $2M_1$ and 1Md polytypes on the basis of particle size. The potassium-argon date showed the $2M_1$ component to be considerably older than Pennsylvanian age, indicating it to be detrital. The potassium-argon age of the 1Md component is less than half the age of Pennsylvanian sediments. The low age may be due to preferential argon loss because

[97] Keller, W. D., Processes of Origin and Alteration of Clay Minerals, "Soil Clay Mineralogy," The University of North Carolina Press, Chapel Hill, N.C. (1964).

[98] Burst, J. F., Jr., Postdiagenetic Clay Mineral Environmental Relationships in the Gulf Coast Eocene, Proceedings of the 6th National Clay Conference, pp. 327–341, Pergamon Press, New York (1959).

[99] Powers, M. C., Adjustment of Clays to Chemical Change and the Concept of the Equivalence Level, Proceedings of the 6th National Clay Conference, pp. 309–326, Pergamon Press, New York (1959).

[100] Bailey, S. W., P. M. Hurley, H. W. Fairbairn, and W. H. Pinson, Jr., K-Ar Dating of Sedimentary Illites Polytypes, *Bull. Geol. Soc. Am.,* **73**: 1167–1170 (1962).

of the small particle size involved, or to the reorganization and potassium fixation in smectite and degraded micas in post-Pennsylvanian time. Increasing temperature and pressure conditions accompanying deep burial apparently cause a transformation of the 2Md to the 2M polymorph in dioctahedral micas (Maxwell[101]).

There is, however, another aspect to this problem of the absence of smectite in ancient sediments. It appears that certain times and certain places in the geologic past have been either particularly favorable or particularly unfavorable for the formation of smectite. Thus, the Cretaceous formations of the Great Plains and the Gulf Coast of the United States contain large amounts of bentonite and smectite clays and shales. Volcanic activity producing ash which altered to smectite is largely responsible for the flood of smectite in these sediments. Millot[37] studied many Cretaceous sediments from France and found little smectite in them. Thus the abundance of smectite in some ancient sediments varies geographically because of peculiar, relatively local conditions.

Smectite is developing in many soils widely scattered over the land today; hence the argillaceous material carried to many areas of accumulation at the present time has a fairly large content of smectite. Recent sediments, therefore, often have a fairly high smectite content. This probably was not always the case, in the geologic past. An interval of geologic time without much volcanic activity, without large areas of relatively low rainfall, and/or with large areas of lateritic weathering would produce little smectite. Thus, because of variations of weathering conditions and/or volcanic activity, either on a world-wide scale or in restricted areas, smectite would be expected to be absent or abundant in sediments of a particular geologic interval or of a particular geographic area.

Fuechtbauer and Goldschmidt[102] found that depth of burial may cause an increase in the crystallinity of kaolinite. Also, there is a suggestion from available data that kaolinite is less abundant in very ancient sediments than in those deposited after, say, the Devonian. Kaolinite may therefore be lost in ancient sediments, because of metamorphic processes complemented by adsorption of alkalies and alkaline earths, with the formation of micas. This would be expected to be a much slower process than the alteration of smectite to the micas.

Since kaolinite is formed during weathering under certain conditions and

[101] Maxwell, D. T., Late Diagenetic and Early Metamorphic Polymorphic Changes in Dioctahedral Mica Found in the Precambrian Belt Series, Western Montana and Northern Idaho, *Geol. Soc. Am., Program Annual Meeting (Miami)*, p. 128 (1964).

[102] Fuechtbauer, H., and H. Goldschmidt, Clay Mineral Diagenesis, *Proc. Intern. Clay Conf., Stockholm*, **1**: 99–111 (1963).

not under others, it would be expected that ancient sediments of a particular age and/or from a particular area might have either a flood of kaolinite or none of it.

Since smectite and kaolinite tend to disappear in sediments of increasing age, it follows that very ancient sediments must be composed largely of the illite and chlorite types of clay minerals. Eckhardt[103] studied the distribution of chlorite in sediments and emphasized the relative increase in abundance with increasing age. Attapulgite-sepiolite also appears restricted to relatively young sediments. No information is available on the distribution of vermiculite in sediments. Halloysite is generally absent in sedimentary rocks. It has been suggested that the halloysite minerals would change under metamorphic processes to kaolinite.

Clay minerals in relation to lithology. *Shales.*

The dominant clay mineral in many, probably most, shales that have been studied is illite. Chlorite mica is frequently present, and smectite is a common component in many shales of Mesozoic or younger age. Kaolinite is a common component of some shales, but usually in minor amounts. In shales of relatively high kaolinite content, the lamination is frequently not well developed.

Yaalon[104] assembled the analyses of 10,000 shales and found that their average composition was clay minerals, 59 percent, predominantly illite; quartz and chert, 20 percent; feldspar, 8 percent; carbonates, 7 percent; iron oxides 3 percent; organic material, 1 percent; others, 2 percent.

Slates. The mica clay minerals are the dominant components in the slates that have been studied. Fairbairn[105] reported kaolinite in some slates and found that kaolinitic slates have relatively low schistosity.

It appears that the illite and chloritic clay minerals in the slates have a higher crystallinity than these same clay minerals in clays. This means that the clay minerals are composed of layer units more regularly oriented and with fewer structural defects. The increased crystal perfection developed in the metamorphism of illite yields a material approaching muscovite or biotite, depending on the original composition and structure of the illite.

No montmorillonite, attapulgite-sepiolite, or halloysite clay minerals have been reported in slates. Vermiculite may occur in some slaty materials, but it is usually difficult to establish that it is not an alteration product.

[103] Eckhardt, F. J., Chlorites in Sediments, *Geol. Jahrb.,* **75**: 437–474 (1959).

[104] Yaalon, D. H., Mineral Composition of the Average Shale, *Clay Minerals Bull.,* **5**(27): 31–36 (1962).

[105] Fairbairn, H. W., X-ray Petrology of Some Fine-grained Foliated Rocks, *Am. Mineralogist,* **28**: 246–256 (1943).

Clays. Material which can be classed as clay and containing each of the clay minerals in substantially pure form has been described. Except for mixtures of kaolinite with sepiolite-palygorskite, any combination of the various clay minerals in such sediments has been described.

Most clays contain mixtures of clay minerals, but there are many that are substantially monomineral, and the origin of some such sedimentary clays is difficult to understand. The kaolins in the Upper Cretaceous of Georgia and South Carolina are examples of such sediments. Kesler[106] proposed the following mode of origin for these quite pure kaolinite clays. The sediments were derived from the crystalline rocks of the Piedmont area by vigorous erosion of a youthful surface, and they accumulated in coalescing deltas as coarse feldspathic sands. Most of the iron in the source rock was taken into solution at the source in mildly acid ground and surface waters and was removed to the sea. Kaolinite formed by decomposition of the detrital feldspar in the exposed parts of the delta and through low-grade erosion of the sands collected in ponds and formed as cut-off segments of the distributaries. So-called "soft kaolin" was deposited slowly in fresh acidic waters. At times wave erosion breached some of the ponds, admitting sea water in which "hard kaolin" was coagulated rapidly.

Underclay. Many beds of coal, particularly of Carboniferous age, the world over are immediately underlain by a nonbedded carbonaceous clay characterized by many small slickensided fracture surfaces. Such material, frequently called underclay, contains either kaolinite or illite, or a mixture of these clay minerals. Occasionally the underclay is calcareous, and then illite is usually the only clay-mineral component.[107]

Brindley and Robinson[108] showed that the English underclays contain kaolinite of the poorly crystalline variety. The kaolinite in the plastic underclays in Illinois is also the poorly crystalline variety. In several instances very small amounts of dickite[109] have been identified in underclays. This dickite could not have had the usual hydrothermal origin for the mineral, and its genesis is obscure.

In recent years, these underclays have been discussed in great detail, and various concepts of origin have been suggested to explain their unique

[106] Kesler, T. L., Environment and Origin of the Cretaceous Kaolin Deposits of Georgia and South Carolina, *Econ. Geol.*, **51**: 541–554 (1956).

[107] Grim, R. E., and V. T. Allen, Petrology of the Pennsylvanian Underclay of Illinois, *Bull. Geol. Soc. Am.*, **49**: 1485–1513 (1938).

[108] Brindley, G. W., and K. Robinson, X-ray Study of Some Kaolinitic Fireclays, *Trans. Brit. Ceram. Soc.*, **46**: 49–62 (1947).

[109] Honess, A. P., and F. J. Williams, Dickite from Pennsylvania, *Am. Mineralogist*, **20**: 462–466 (1935).

character. They were thought to be fossil soils, but this mode of origin does not seem to be completely satisfactory. McMillan[110] concluded that the underclays he studied in Kansas are fossil gleys and that the characteristics of the underclay are due to the action of organic compounds that have moved downward from the overlying peat by diffusion.

Schultz[111] studied a large number of underclays of the United States and found that they are composed principally of kaolinite of a poorly crystallized type, illite, mixed-layer illite-smectite, and in many places a 14-Å component that ranges from vermiculite to chlorite. Underclays from the central parts of the various basins contain less kaolinite and more 14-Å clay mineral than those of shelf areas. Stratigraphically, the Lower Pennsylvanian underclays are more kaolinitic than those of the Upper Pennsylvania. Field observations led Schultz to the conclusion that the underclays were formed before the deposition of coal-forming material began and, therefore, cannot be considered residual soils on which the coal-forming flora grew.

Scheere[112] described the clays associated with the coal seams of Belgium. These clays vary from illite-kaolinite mixtures to substantially pure kaolinite. In some cases, the kaolinite beds are extremely pure and contain vermicular worms, which the author concluded are not of sedimentary origin but of neoformation from postsedimentary solutions.

Nicholls and Loring[113] found evidence that the potassium content of the illite in British coal-measure underclays has increased by the postdepositional adsorption of potassium.

Loughnan[114] found that dickite and gibbsite, in addition to kaolinite, are present in the clays associated with the coal measures in New South Wales, Australia. These contain vermicular forms of kaolinite which the author considered to have been formed by the resilicification of bauxite minerals in the environment of accumulation.

Roeschmann[115] found that the underclays of the Ruhr Valley in Germany were composed of illite, kaolinite, and chlorite and that the kaolinite varied

[110] McMillan, N. J., Petrology of the Nodaway Underclay (Pennsylvanian) Kansas, *State Geol. Surv. Kansas, Bull.* 19, pt. 6, pp. 191–247 (1956).

[111] Schultz, L. G., Petrology of Underclays, *Bull. Geol. Soc. Am.*, **69:** 363–405 (1958).

[112] Scheere, J., Kaolinite of the Hard-coal Seams of Belgium, *Silicates Ind.*, **24:** 475–483 (1959).

[113] Nicholls, G. D., and D. H. Loring, Some Chemical Data on British Carboniferous Sediments and Their Relationship to the Clay Mineralogy of These Rocks, *Clay Minerals Bull.*, 4(24): 196–207 (1960).

[114] Loughnan, F. C., Some Tonstein-like Rocks from New South Wales, Australia, *Neues Jahrb. Mineral. Abhandl.*, 99(1): 29–44 (1962); *Univ. New South Wales, Publ.* 61 (1962).

[115] Roeschmann, G., Underclays of the Ruhr Carboniferous, *Fortschr. Geol. Rheinland Westfalen*, 3: 497–524 (1962).

from well crystallized to poorly crystallized, with the degree of crystallinity increasing with age and degree of coalification. It was determined that the content of B_2O_3 was distinctly higher in illite-rich samples than in kaolinite-rich samples.

In a study of the clay-mineral composition of Pennsylvanian underclays in the Illinois Basin, Parham[89] found that kaolinite was more abundant in near-shore phases than illite. Earlier Malkovsky[116] found the same sort of clay-mineral distribution in the claystones of Carboniferous age in the Ostrava Basin of Czechoslovakia.

Patterson and Hosterman[117] studied the underclays of parts of Kentucky and found that they varied from substantially pure kaolinite with the characteristics of flint clay to plastic clays containing mixtures of kaolinite, illite, and mixed-layer clay minerals. The structure of the kaolinite ranged from highly crystalline to very poorly crystalline, and the degree of crystallinity and the hardness of the clay varied inversely with the amount of illite and mixed-layer clay minerals present. The nearly pure kaolinite is believed to have formed by the removal of alkalies and some silica from mixtures of kaolinite, illite, and mixed-layer clays, by leaching in swamps. In another paper Huddle and Patterson[118] concluded that these Pennsylvanian underclays are "seat earths," that the formation took place in a water-logged environment, and that the development was generally followed by the immediate accumulation of peat.

Flint Clay. Flint clay is a dense, hard, nonslakable, massive, nonplastic clay with flint-like characteristics. Flint clay is frequently associated with underclays, but other occurrences, as, for example, in association with the diaspore clays of Missouri (see page 570), are also known. The flint clays that have been studied in detail are composed of well-crystallized kaolinite in extremely small particle sizes. Sometimes excess silica or alumina is present and may serve to cement the kaolinite particles together. In many cases, however, the precise factors that give flint clays their distinctive physical properties have not been established.

A similar clay called tonstein is found in association with Carboniferous coals of Europe. Von Eckhardt and Von Gaertner[119] believe postdepositional alteration and leaching due to humic acids are responsible for the characteris-

[116] Malkovsky, M., Jilove mineraly ostravskych vrstev (Clay Minerals of "Ostrava Beds"), *Acta Univ. Carolinae Geol.,* pp. 197–210 (1959).

[117] Patterson, S. H., and J. W. Hosterman, Geology and Refractory Clay Deposits of the Haldeman and Wrigley Quadrangles, Kentucky, *U.S. Geol. Surv., Bull.* 1122-F (1962).

[118] Huddle, J. W., and S. H. Patterson, Origin of Pennsylvanian Underclay and Related Seat Rocks, *Bull. Geol. Soc. Am.,* 72:. 1643–1660 (1961).

[119] Von Eckhardt, F. J., and H. R. Von Gaertner, Zur Entstehung und Umbildung der Kaolin-Kohlentonsteine, *Fortschr. Geol. Rheinland Westfalen,* 3(2S): 623–640 (1962).

tics of the clay. Smith and O'Brien[120] reached a similar conclusion for some American flint clays, and Moore[121] concluded that products of microbiological degradation of organic matter influenced the *in situ* development of well-crystallized kaolinite.

Sand and Sandstones. Any of the clay minerals can be present in sands and sandstones. Kaolinite is frequently most abundant, and dickite is fairly common (Ferrero and Kubler,[122] Bayliss et al.[123]). Further postdepositional alteration may lead to relatively degraded forms and to the neoformation of new mineral species (see pages 563–565).

Limestones and Dolomites. An analysis of 35 Paleozoic limestones from Illinois[124] showed illite in all of them. Kaolinite in small amounts was present in about half of them. No smectite was determined in these limestones, but the analyses were made before the development of the glycol method for the determination of smectite and small amounts of this clay mineral would probably have been missed.

The determination of the clay minerals in limestones and dolomites is particularly difficult since it is frequently necessary to dissolve away the carbonate in order to concentrate the clay minerals to get adequate analytical data, and some of the clay minerals themselves are fairly soluble in acid (see pages 434–446).

Rateev and Kotel'nikov,[125] studying a series of calcareous sediments of Carboniferous age from the Soviet Union, identified sepiolite and palygorskites and found that the sepiolite was mostly associated with dolomites, whereas the palygorskites occurred more abundantly in marine marls in formations that had a considerable content of silica.

Ostrom[126] investigated the clay-mineral composition of a large number of carbonate rocks from the Paleozoic section of Illinois and found that the dominant clay mineral was illite. Smectite was identified only in mixed-layer

[120] Smith, W. H., and N. R. O'Brien, Middle and Late Pennsylvanian Flint Clays, *J. Sediment. Petrol.,* **35:** 610–618 (1965).

[121] Moore, L. R., The Microbiology, Mineralogy and Genesis of a Tonstein, *Proc. Yorkshire Geol. Soc.,* **34**(3): 235–292 (1964).

[122] Ferrero, J., and B. Kubler, Présence de dickite et de kaolinite dans les grès cambriens d'Hassi Messaoud, *Bull. Serv. Carte Geol. Alsace Lorraine,* **17**(4): 247–261 (1964).

[123] Bayliss, P., F. C. Loughnan, and J. C. Standard, Dickite in the Hawkesbury Sandstone of the Sydney Basin, Australia, *Am. Mineralogist,* **50:** 418–426 (1965).

[124] Grim, R. E., J. E. Lamar, and W. F. Bradley, The Clay Minerals in Illinois Limestones and Dolomites, *J. Geol.,* **45:** 829–843 (1937).

[125] Rateev, M. A., and D. D. Kotel'nikov, New Occurrences of Sepiolite in the Carboniferous of the Russian Platform, *Dokl. Akad. Nauk SSSR,* **109:** 191–194 (1956).

[126] Ostrom, M. E., Clay Mineralogy of Some Carbonate Rocks of Illinois, Ph.D. thesis, University of Illinois (1959).

assemblages with illite. The clay minerals associated with deposits believed to have been more significantly influenced by terrestrially derived materials were less well crystallized than those found in distinctly marine deposits. In addition, kaolinite was commonly a major component of the former environment and of rare occurrence in the latter.

In an extremely interesting study of the mineralogy and petrology of the Upper Mississippian carbonate rocks of the Cumberland Plateau in Tennessee, Peterson[127] found that, with only minor exceptions, the mineral assemblages obeyed the Gibb's phase rule. Compositional variation of the expandable chloritic-mineral component was between the composition of vermiculite and that of corrensite, the ordered end member, and is consistent with the topologic requirements of the phase diagram of the system $MgO\text{-}CaO\text{-}Al_2O_3\text{-}SiO_2\text{-}CO_2\text{-}H_2O$. He concluded that the minerals have formed an equilibrium assemblage at some time in the diagenetic history of the rock.

Tooker[128] found an interesting association between the clay-mineral composition and the lithology of a series of sediments from the Oquirrh formation of Pennsylvanian age in Utah. Illite and chlorite occur in limestone; illite, chlorite, and mixed-layer clay materials in cherty bioclastic and sandy limestones and in calcareous quartzites; chlorite and illite in dolostones and in calcareous quartzites; chlorite and illite in dolomitic quartzite; chlorite, illite, and mixed-layer corrensite type of minerals in dolomitic limestones; and kaolinite, illite, and chlorite in quartzites.

Coal. Hicks and Nagelschmidt[129] showed that a two-layer mineral was the dominant clay mineral in English coals and that this mineral appeared to be halloysite in the anthracites and kaolinite in the bituminous coals. This relationship is difficult to understand, since it is contrary to the usual concept of the transition of halloysite to kaolinite by metamorphic action. Some illite is also present in the English coals. The clod and roof shale of the coals is largely composed of illite containing some sodium in interlayer positions. Nagelschmidt found that the sodium content of the illite decreased from about 1.8 to 0.6 percent with a decrease in the rank of the coal.

Endell and Endell[130] showed that illite is the dominant clay mineral

[127] Peterson, M. A., Mineralogy and Petrology of Upper Mississippian Carbonate Rocks of the Cumberland Plateau in Tennessee, *Geol. Soc. Am., Program Annual Meeting,* p. 175 (1960).

[128] Tooker, E. W., Clay Minerals in Rocks of the Lower Part of the Oquirrh Formation, Utah, Proceedings of the 9th National Clay Conference, pp. 355–364, Pergamon Press, New York (1962).

[129] Hicks, D., and G. Nagelschmidt, The Chemical and X-ray Diffraction Analysis of the Roof and Clod of Some South Wales Coal Seams and of the Mineral Matter in Coal, *Med. Res. Council, Spec. Rept.* 244 (1943).

[130] Endell, K., and J. Endell, Ueber die Bestimmung der Röntgenfeinstruktur mineralischer Bestandteile von Kohlen und Aschen sowie ihre technische Bedeutung, *Feuerungstech.,* 31: 137–143 (1943).

in the German coals. However, later work (Burger[131]) showed that kaolinite is prevalent in many German coals, and Eckhardt[132] found an increase in the crystallinity of the kaolinite with an increase in degree of coalification. Van Tassel[133] reported nacrite in some Belgian coals. Leskevich et al.[134] reported kaolinite in coals of the western Donbass of the U.S.S.R. Mukaiyama[135] found a variety of clay minerals in some Japanese coals.

Ball[136] found that in Illinois coals kaolinite was the dominant clay mineral, both in fractures and cavities and in the coal itself. A relatively small amount of illite is usually present also in these coals. It may be concluded that kaolinite is commonly the dominant clay mineral in coals, but, particularly in younger coals, other clay minerals may also be relatively abundant.

Salines. Unpublished work by the author and studies by Millot[37] and De Lapparent[73] show that the sepiolite-palygorskite-attapulgite minerals are frequent in the argillaceous sediments associated with salines. Smectite and illite clay minerals are also common. Frequently the clay minerals in such sediments are poorly crystallized, and their precise identification is difficult.

Kuhn[137] studied the composition of the Zechstein salt in Germany and found that illite is the dominant clay-mineral component. He believed the mineral to be of diagenetic origin. He reported also that the crystallinity of the illite is quite low in these sediments.

Braitsch[138] found muscovite, chlorite, and corrensite as clay-mineral components of the Stassfurt salt deposits in Germany.

Rateev and Osipova[139] described sepiolite and palygorskite in a series

[131] Burger, K., F. J. Eckhardt, and G. Staaler, Zur Nomenklatur und Verbreitung der Kaolin-Kohlentonsteine im Ruhrkarbon, *Fortschr. Geol. Rheinland Westfalen,* **3**:525-540 (1962).

[132] Eckhardt, F. J., Effect of Temperature on the Crystallographic Classification of Kaolinite, *Proc. Intern. Clay Conf., Stockholm, 2nd,* pp. 137–145 (1963).

[133] Van Tassel, R. Occurrence de nacrite dans les fissures de veinules de charbon, *Ann (Bull.) Soc. Geol. Belg.,* **85**: 319–321 (1961–62).

[134] Leskevich, I. Ye, and Savchuk, S. V., Kaolinite in the Lower Carboniferous Coals of the Western Donbass, *Dokl. Acad. Sci. USSR, Earth Sci. Sect.,* **137**: 424–426 (1962).

[135] Mukaiyama, H., Clay Minerals in Sedimentary Rocks in the Karatsu Coal Field, *Kyushu Kozan Gakkai-shi,* **33**(2) : 57–64 (1965).

[136] Ball, C., Kaolinite in Illinois Coal, *Econ. Geol.,* **29**: 767–776 (1934).

[137] Kuhn, R., Ueber die Mineralgehalt der Salztone, *Schriften Mineral. Petrog. Inst. Kiel,* p. 6 (1938).

[138] Braitsch, O., Mineralfaragenesis und Petrologie der Stassfurtsalze in Reyerhausen, *Kali Steinsalz.,* Heft 1, pp. 1–14 (1960).

[139] Rateev, M. A., and A. I. Osipova, Clay Minerals in Sediments of the Arid Paleogene Zone of Fergana, *Dokl. Akad. Nauk SSSR,* **123**: 166–169 (1958).

of sediments from the arid paleogene zone of Fergana in the Soviet Union. They pointed out that palygorskite is completely absent in noncarbonate marine rocks rich in smectite and that sepiolite occurs in sediments with a particularly high content of dolomite. They also found that such sediments are particularly free of kaolinite.

Loughnan[140] described an occurrence of palygorskite in the Redbank Plains formation near Ipswich in Australia. He considered that the clay originated in a playa-lake environment in which smectite, a weathered derivative of the associated basalts, formed the parent material for the magnesium-rich assemblage. It was considered that the controlling influence was the magnesia concentration in the playa lake. As the magnesia content increased, there was a transition from smectite to palygorskite to sepiolite.

Von Muller[141] described an occurrence of palygorskite and sepiolite in a Tertiary marl from Saudi Arabia. It was concluded that these minerals formed under marine conditions from a magnesium-rich sedimentary debris derived from weathered basic igneous material in the source area.

Glacial Material. Any of the clay minerals may be found in glacial till, depending on the character of the source material, since there is likely to be substantially no alteration of the clay minerals during transportation and deposition. Older tills might be expected to have a relatively higher mica content, because later metamorphic changes probably would be in the direction of the formation of well-crystallized micas from any degraded mica or smectite. It seems likely that kaolinite would tend to persist unless the tills were very old or had been metamorphosed substantially.

The Pleistocene till of North America has been extensively studied, and illite is the dominant clay mineral in it. Chlorite, smectite, and kaolinite are frequently present, but of subordinate importance. Willman, Glass, and Frye[142] investigated the mineral composition of the clay fraction of tills in Illinois. They found that tills from the northwest were distinguished by relatively high percentages of smectite; from the Lake Michigan Basin by relatively high percentages of illite; from the Green Bay area by the presence of abundant vermiculite; and from the Saginaw–Lake Erie lobes by relatively high percentages of illite.

Gumbotil, found at the surface of some pre-Wisconsin tills in North America, contains illite, with considerable amounts of smectite and occasion-

[140] Loughnan, F. C., Further Remarks on the Occurrence of Palygorskite at Redbank Plains, Queensland, *Proc. Roy. Soc. Queensland,* **71**(3): 43–50 (1960).

[141] Von Muller, G., Palygorskit und Sepiolith in tertiären und quartären Sedimenten von Hadramaut (S-Arabien), *Neues Jahrb. Mineral. Abhandl.,* **97**(2): 275–288 (1961).

[142] Willman, H. B., H. D. Glass, and J. C. Frye, Mineralogy of Glacial Tills and Their Weathering Profiles in Illinois, Part I, Glacial Tills, *Illinois State Geol. Surv., Circ.* 347 (1963).

ally more kaolinite than the normal till. It also has a higher clay-mineral content than the unweathered till as a consequence of the concentration of colloidal material by downward-moving waters.

Cresseneuil[143] studied a series of varied clays in France and found that the clay-mineral component was illite. He did not detect any variation between the light- and dark-colored layers.

Lysenko[144] found that smectite and hydromicas prevailed as the clay-mineral components of loess silts from the Donetz Basin in the Soviet Union. Yuskpova and Skavletskii[145] obtained similar results for loess from the Dangarinsk depression.

Earlier, Beavers, Johns, Odell, and Grim[146] pointed out that smectite was the dominant clay mineral in the loess of Illinois and that it was relatively much more abundant than in associated tills.

Fuller's Earth. This name is used for any natural earthy material which will decolorize mineral or vegetable oils to a sufficient extent to be of economic importance. The name as generally used at the present time has no genetic or mineralogical significance. An exception is the use of the term in England as a stratigraphic term for certain beds long used as commercial bleaching clays. Some fuller's earths are composed of attapulgite,[147] smectite,[148] and kaolinite.[149] Some fine silts of relatively low clay content have high decolorizing ability. On the other hand there are many silts and clays composed of smectite and kaolinite and possibly also of attapulgite which are not fuller's earths.

Neoformation of clay minerals. It is now generally recognized that clay minerals may be precipitated directly from solutions or crystallize from colloidal

[143] Cresseneuil, J. Sarrot-Reynauld de, Observations sur la nature physico-chimique des argiles d'Eybens (Isere), *Grenoble, Univ. Lab. Geol.*, Tr. t. 31, pp. 243–246 (1954).

[144] Lysenko, M. P., Loess Soils of the Donetz Basin, *Vestn. Leningr. Univ.* 16, p. 12; *Ser. Geol. i Geograf.*, **2**: 84–94 (1961).

[145] Yuskpova, S. M., and E. N. Skavletskii, Mineralogical Characteristics of the Loess Rocks of the Dangarinsk Depression, *Dokl. Akad. Nauk. Tadzh. SSR*, **7**(8): 20–23 (1964).

[146] Beavers, A. H., W. D. Johns, R. T. Odell, and R. E. Grim, Clay Minerals in Some Illinois Soils Developed from Loess and Till Under Grass Vegetation, *Natl. Acad. Sci., Publ.* 395, pp. 356–372 (1955).

[147] De Lapparent, J., Les Argiles des terres à foulon, *Congr. Intern. Mines Met. Geol. Appl.*, **1**: 381–387 (1935).

[148] Grim, R. E., Petrography of the Fuller's Earth Deposit, Olmsted, Illinois, with a Brief Study of Some Non Illinois Earths, *Econ. Geol.*, **28**: 344–363 (1933).

[149] Piersol, R. J., J. E. Lamar, and W. H. Voskuil, Anna "Kaolin" as a New Decolorizing Agent for Edible Oils, *Illinois State Geol. Surv., Rept. Invest.* 27 (1933).

sols of poorly organized material. Keller[150] emphasized this mode of origin in his study of the flint clays of Missouri, and Tarr and Keller[151] years ago described kaolinite filling geodes in some shales of Mississippian age which must have had this mode of origin. Botinelly and Fischer[152] provided an example of the crystallization of a clay mineral from solution involving replacement wherein a vanadiferous clay mica replaces otherwise fairly pure and clean quartz in the Navajo sandstone of Colorado.

Scheere[112] described vermicular worms of kaolinite in clay associated with Belgian coals which he considered to have resulted from the reaction of colloidal silica and alumina hydrogels.

Millot and his associates, in a series of papers which Millot[35] later summarized, considered in great detail the neoformation of the clay minerals. As the result of detailed studies of many sediments from various areas in West Africa, this group concluded that neoformation takes place to a considerable extent in basin areas where the neighboring land area has undergone intense weathering. The material supplied to the basin is largely in solution with little or no detrital material. Smectite, palygorskite, and sepiolite are believed to be precipitated under such conditions, with the relative magnesium, silicon, and aluminum content of the waters controlling which of these clay minerals is formed. Except for this work of Millot and his colleagues, little effort has yet been made to investigate the possible nondetrital clay minerals in sediments.

Postdepositional alteration of clay minerals. There has been a growing appreciation in recent years of the widespread occurrence and importance of the postdepositional formation and alteration of the clay minerals. Some of the alterations take place in the environment of accumulation and are properly considered under diagenesis (see pages 535–544); others are definitely the result of metamorphic processes; still others seem to be due mainly to circulating fluids, and apparently the alteration can begin with accumulation or commence much later under conditions little related to the environment of accumulation. It is this latter clay-mineral formation that we shall consider now, and the following will serve as examples.

[150] Keller, W. D., Observations on the Origin of Missouri High-Alumina Clays, "Problems of Clay and Laterite Genesis," pp. 115–134, American Institute of Mining Engineers (1952).

[151] Tarr, W. A., and W. D. Keller, Occurrences of Kaolinite Deposited from Solution, *Am. Mineralogist,* **22:** 933–935 (1937).

[152] Botinelly, T., and R. P. Fischer, Mineralogy and Geology of the Rifle and Garfield Mines, Garfield County, Colorado, *U.S. Geol. Surv., Profess. Paper* 320, pp. 213–230 (1959).

Kulbicki,[153] concluded that the following changes have taken place in sediments in the Aquitaine Basin in France. Solution and reprecipitation have produced vermicular kaolinite and halloysite from nuclei of other clay minerals. Montmorillonite has formed from halloysite or formed together with halloysite from kaolinite, by the action of basic siliceous waters.

Keller[154] described the occurrence of a glauconitic mica in the Morrison formation in Colorado and concluded that the mineral was formed by downward-moving solutions percolating through partly montmorillonized and still permeable ash, converting the montmorillonite in part to hydrous mica and at shallower depths locally to glauconitic mica.

Salger[155] described kaolinite in pseudomorphs after muscovite and feldspar that had been produced by the postdepositional alteration of these minerals in arkoses by circulating ground waters.

Maurel and Brousse[156] concluded that postdepositional alteration in some arkoses in France caused smectite to form where the alteration was of short duration and kaolinite to form over long periods of time. Borovko[157] described small veinlets of halloysite in a reddish-brown Oligocene sandy clay from the Ufa River in Siberia, and he concluded that the halloysite was deposited from solutions.

Mellon[158] described authigenic minerals from Lower Cretaceous sedimentary rocks of Alberta in Canada and found quartz-kaolinite assemblages in rocks low in volcanic detritus, whereas chlorite, quartz, kaolinite, and smectite were present in rocks containing abundant volcanic detritus.

Kulbicki and Millot[159] described the postdepositional alteration in a series of Cambro-Ordovician argillaceous sands in the central Sahara. These authors found an important correlation between the clay-mineral composition,

[153] Kulbicki, G., Phénomènes de diagenèse dans les sediments argileux, *Clay Minerals Bull.,* **2**(12): 183–187 (1954).

[154] Keller, W. D., Glauconitic Mica in the Morrison Formation in Colorado, *Natl. Acad. Sci., Publ.* 566, 120–128 (1958).

[155] Salger, M., Mineralogical and Sedimentary Petrographic Investigations on the Kaolin Profile of Boring Kick No. 9 near Schnaittenbach/Oberpfalz, *Geol. Bavarica,* no. 37, pp. 1–84 (1958).

[156] Maurel, P., and R. Brousse, Sur les phyllites contenues dans quelques arkoses, *Bull. Soc. Franc. Mineral.,* **82**: 1–3, 87–90 (1959).

[157] Borovko, N. G., Halloysite from the Tertiary of Bashkiria, *Mem. All-Union Mineral. Soc. J.,* **89**: 361–364 (1960).

[158] Mellon, G. B., Authigenic Processes in the Lower Cretaceous Blairmore Group, Alberta, *Geol. Soc. Am., Program Annual Meeting (Abstr.)* (1960).

[159] Kulbicki, G., and G. Millot, L'évolution de la fraction argileuse grès pétroliers Cambro-Ordoviciens du Sahara central, *Bull. Serv. Carte Geol. Alsace Lorraine,* **13**(4): 147–156 (1960).

the presence of oil, and saline waters. Where petroleum was present, the original kaolinitic cement was preserved, but where saline water was present, illite has been formed from the kaolinite. They pointed out that the presence of petroleum in the sand blocked the secondary development of illite from the kaolinite. Nelson[160] presented evidence from a study of the Maccrady formation of Mississippian age that well-crystallized muscovite and chlorite formed after vitrification where an excess of silica-deficient clay phase existed and in the presence of saline pore fluids.

Smoot[161] studied a large number of samples of pre-Pennsylvanian sandstones from the Illinois Basin and found that the clay minerals from the permeable sandstones were characterized by a heterogeneous mixture of degraded illites, chlorites, kaolinite, and occasionally minor amounts of smectite. Those from the shales were characterized by well-crystallized illite and chlorite, minor amounts of degraded illite and chlorite, and rarely traces of kaolinite. These differences were apparent in intimately associated sediments that came from the same general area and had approximately the same environment of deposition and the same originally introduced clay-mineral suites. Therefore, it was concluded that heterogeneity of the sandstone clay-mineral suites was brought about by degradation after lithification by circulating formation fluids. The low permeability of the shales prohibited such circulation from taking place so that their clay minerals have been protected from degradation. Numerous authors (e.g., Weaver,[33] Kopeliovich and Menyailenko[162]) have emphasized that postdepositional alterations are best observed in clastic rocks.

Towe[163] pointed out that postdepositional transformation of smectite and/or mixed-layer illite-smectite to illite was a possible source of silica cement in some sedimentary rocks. He believed the depth of burial was an important factor in influencing this change.

It is clear that postdepositional alteration can yield widely different results, depending on conditions of alteration and starting material. The original minerals may be degraded, regraded, or new species formed. Undoubtedly, the composition of the fluids (saline versus fresh), permeability, presence

[160] Nelson, B. W., Authigenic Muscovite and Chlorite in the Maccrady Formation, *Geol. Soc. Am. Program Annual Meeting (Abstr.)* 1960.

[161] Smoot, T. W., Clay Mineralogy of Pre-Pennsylvanian Sandstones and Shales of the Illinois Basin, Part I, Relation of Permeability to Clay Mineral Suites, *Illinois State Geol. Surv., Circ.* 286 (1960).

[162] Kopeliovich, A. V., and P. A. Menyailenko, Secondary Alterations in New Mineral Formations in Rocks of the Semiluki Horizon of the Archedy Region, *Izv. Vysshikh Uchebn. Zavedenii, Geol. i Razvedka,* **5:** 72–84 (1962).

[163] Towe, K. M., Clay Mineral Diagenesis as a Possible Source of Silica Cement in Sedimentary Rocks, *J. Sediment. Petrol.,* **32:** 26–28 (1962).

of organic material, and time, as well as parent minerals, are significant controlling factors.

Miscellaneous clay materials. Under this heading will be considered the origin and occurrence of clay minerals in argillaceous materials of unique origin or whose origin is not well established.

Bentonite. The term bentonite was first applied by Knight[164] to a particular, highly colloidal, plastic clay found near Fort Benton in the Cretaceous beds of Wyoming; it has the unique characteristic of swelling to several times its original volume when placed in water, and it forms thixotropic gels with water even when the amount of bentonite in such gels is relatively small. It was subsequently shown by Hewett[165] that this clay was formed by the alteration of volcanic ash *in situ*. Later, Ross and Shannon[166] studied a number of clays which had been formed by the alteration of ash and redefined the term bentonite to limit it to clays produced by the alteration of volcanic ash *in situ*. They pointed out that such clays are largely composed of smectite clay minerals and that they are generally highly colloidal and plastic. The term as now used by many mineralogists and geologists has no reference to the physical properties of the clay.

Until about 1930 bentonite was known in few places outside of the original Wyoming area, and until about 1940 the term bentonite, especially in industrial usage, was synonymous with the Wyoming material. Even today (1967) the term bentonite to most people, other than some mineralogists and geologists, carries with it the connotation of a highly plastic, colloidal, swelling clay. Also, the term is used frequently for clays with these properties without reference to a particular mode of origin. Thus, smectite clay from the island of Ponza, Italy, is generally called bentonite and sold commercially as such, even though it is apparently of hydrothermal origin.

Because of the high commercial value of bentonite, a search for it has been made in all quarters of the globe since about 1930, and now clays composed largely of smectite and formed by the alteration of volcanic ash *in situ* have been found in almost all countries and in rocks of a wide variety of ages. Bentonites are particularly widely distributed in the Tertiary, as, for example, in California and the Gulf Coast states of the United States, in Italy, in New Zealand, in North Africa, in Argentina, and the U.S.S.R. Bentonites are also very widely distributed in the Upper Cretaceous, as, for example, in the Gulf Coast and High Plain states of the United States, in

[164] Knight, W. C., Bentonite, *Eng. Mining J.,* **66**: 491 (1898).

[165] Hewett, D. F., The Origin of Bentonite, *J. Wash. Acad. Sci.,* **7**: 196–198 (1917).

[166] Ross, C. S., and E. V. Shannon, Minerals of Bentonite and Related Clays and Their Physical Properties, *J. Am. Ceram. Soc.,* **9**: 77–96 (1926).

England, in Germany, and in Canada. Beds older than the Cretaceous appear to contain little bentonite. Some so-called metabentonite has been found in the Paleozoic rocks (e.g., the Ordovician of Illinois and of the Appalachian states and rocks of similar age in Canada and Norway). In such metabentonite the smectite usually has been partially collapsed by compaction and metamorphism. Further, the potassium content has increased relatively and mixed-layer illite-smectite structures have developed. As a consequence, the mineral swells very little, and the clay does not have the usual highly colloidal properties of bentonite.

Usually the volcanic-ash parent material of bentonites can be detected by the persistence of shard structures of the ash as pseudomorphs in the clay. Also the presence of a suite of nonclay minerals, such as feldspars, biotite, hornblende, etc., characteristic of igneous material, may provide evidence for the origin from ash.

Beds of bentonite vary in thickness from a fraction of an inch up to many feet. Rarely are beds thicker than about 10 ft, but a few are known which reach a thickness of 50 ft. The clay called bentonite which has a hydrothermal origin is found in irregular masses without any bedding. Bentonite varies in color from white to gray, yellow, green, blue, and black but is perhaps most often yellow or yellowish green. Frequently a yellow color develops on the weathered outcrop, and the weathered material may have physical properties quite different from these of the fresh material. Also the weathering outcrop frequently develops a characteristic cellular structure, because of the swelling and shrinking on wetting and drying, which permits it to be recognized easily in the field. The bentonite frequently has a characteristic waxy appearance when freshly dug. On exposure to the air, it often develops a characteristic "jigsaw puzzle" set of fractures accompanying the shrinking on drying.

Bentonites interbedded with shales, limestones, and sands are known. Perhaps most frequently it is found in a section of marine beds. Often the bentonite is immediately underlain by a hard siliceous bed in an association that suggests that there has been an enrichment of silica in the underlying bed accompanying the formation of the bentonite.

As stated above, the dominant clay-mineral component of bentonites is smectite. Other clay minerals, particularly illite and kaolinite, are present in many bentonites, sometimes in amounts up to about 50 percent of the total clay-mineral content. Nonclay minerals characteristic of igneous material as well as of detrital origin are also present in varying amounts. Some bentonites are substantially pure smectite, but rarely is there less than about 10 percent of nonclay minerals. Cristobalite[167] is frequently present in varying amounts up to about 30 percent of the total rock. This mineral is often

---

[167] Gruner, J. W., Abundance and Significance of Cristobalite in Bentonite and Fuller's Earth, *Econ. Geol.*, **35:** 867–875 (1940).

intimately mixed with smectite and is very difficult to identify, so that it has certainly been missed in many clay-mineral analyses of bentonites. It undoubtedly has formed from the ash at the same time as the smectite.

In a few cases zeolites[168] are found in bentonites intimately mixed with the smectite. Recently rocks composed almost wholly of zeolites resulting from the alteration of volcanic ash have been described (Deffeyes[169]).

The composition of the smectite itself varies greatly in different bentonites, as has been shown in detail by Ross and Hendricks.[170] The variation may be within the smectite lattice itself or in the nature of the exchangeable cations. There is considerable variation in the relative abundance of magnesium and iron and also apparently in the R_2O_3-to-silica ratio. That is to say, there is considerable variation in the population of the octahedral and tetrahedral layers. With regard to exchangeable cations, most bentonites that have been described carry Ca^{++} as the most abundant ion. Only a few are known which carry Na^+ as the dominant ion, and of these the Wyoming bentonite is the main example. Also only a few bentonites that carry H^+ or K^+ as the dominant ions have been reported. Mg^{++} is frequently present as an exchangeable ion in relatively small amounts, and this is particularly the case when Ca^{++} is the dominant ion. McAtee[171] pointed out that in some cases there is a regular distribution of exchangeable cations between the smectite unit layers rather than a random arrangement; i.e., there is a kind of mixed-layer structure in the smectite particles.

The geologic column contains many beds of volcanic ash which have not altered to bentonite. In order for bentonite to form, it is probably necessary for the ash to fall in water. The kind of water, i.e., fresh or saline, undoubtedly is important in determining whether bentonite forms at all and, if it does, the precise character of the resulting smectite. Since much bentonite is associated with marine formations, it seems certain that the alteration can take place in sea water. Whether or not the alteration can also take place in even more saline waters or in fresh water is not known definitely. The composition of the ash is also an important factor. Apparently the ash must have a moderate content of MgO, since ash devoid of magnesia does not seem to alter to smectite.

The evidence strongly indicates that the alteration of the ash to smectite

[168] Kerr, P. F., and E. N. Cameron, Fuller's Earth of Bentonitic Origin from Tehachapi, California, *Am. Mineralogist,* **21:** 230–237 (1936).

[169] Deffeyes, K. S., Zeolites in Sedimentary Rocks, *J. Sediment. Petrol.,* **29**(4): 602–609 (1959).

[170] Ross, C. S., and S. B. Hendricks, Minerals of the Montmorillonite Group, *U.S. Geol. Surv., Profess. Paper* 205-B, pp. 23–27 (1945).

[171] McAtee, J. L., Jr., Heterogeneity in Montmorillonite, *Natl. Acad. Sci., Publ.* 566, 279–288 (1958).

takes place soon after accumulation or possibly almost contemporaneously with accumulation. It does not seem, at least in most cases, to be a later process and certainly is not a weathering process. The Cretaceous bentonites in Mississippi frequently grade upward without a break into glauconitic sands. The basal parts of these sands contain rounded masses of bentonite, indicating that the bentonite was already formed when the sand began to accumulate.

The process of formation of the smectite is essentially a devitrification of the natural glass of the ash and the crystallization of the smectite. The ash probably in most cases contains an excess of silica and alkalies. The silica in some cases undoubtedly remains in the clay and gives rise to the cristobalite. Also the excess silica is probably the source of the free silica in the hard siliceous beds immediately underlying many bentonites. The excess alkalies and alkaline earths are probably carried away in solution in the water in which the ash fell. The population of the exchangeable-cation positions would be a consequence of the composition of the original water and the alkalies and alkaline earths from the ash which dissolved in it.

Bentonites possess important and unique properties which give them great commercial value in decolorizing oils, in the manufacture of catalysts, in bonding molding sands, in the preparation of oil-well drilling muds, and in many other relatively minor uses. The properties of individual bentonites vary widely, and not all of them are suitable for commercial use. Also many bentonites have properties fitting them for one particular use, whereas they cannot be used at all for other purposes. Thus, the Wyoming type of bentonite is excellent for drilling muds and foundry use but is of no value for oil decolorizing or the making of catalysts. Similarly, some bentonites from Mississippi are good for decolorizing clays and for foundry use but are of no value for drilling muds. The cause of the variation in properties fundamentally resides in the composition of the smectite itself and in the character of the exchangeable ions. Much remains to be learned of the factors controlling the properties of bentonites, and the subject cannot be pursued at any length here. However, some examples of pertinent correlations between properties and composition can be given. The very high swelling and highly colloidal bentonites carry Na^+ as the principal exchangeable cation. The Wyoming bentonite is an example of such a bentonite. That the presence of Na^+ is not the sole cause of the properties of the Wyoming clay is shown by the fact that other clays treated so that they will carry Na^+ as the exchangeable cation do not fully develop the properties of the Wyoming material. Another cause for the relatively unique properties of the Wyoming bentonite must reside in some feature within the smectite lattice composing it.

Bentonites with smectite low in iron are required for catalyst manufacture. Decolorizing clays are usually those carrying Ca^{++} and/or Mg^{++} as exchangeable cations. Again, the smectite itself must be important, for good decolorizing clays cannot always be prepared by a simple ion-exchange reaction. Smectite clays carrying either Na^+ or Ca^{++} are suitable for foundry molding-sand use, but they have quite different properties in such sands. Thus a sodium smectite provides relatively high dry strength and lower green strength, whereas calcium smectite gives high green strength and lower dry strength.

Highly Aluminous and Ferruginous Clays. It is not proposed in this work to discuss in detail the highly aluminous or ferruginous material, such as bauxites and laterites, since adequate volumes[172-176] on these subjects are already available. The usual clay mineral associated with deposits of the oxides or hydroxides of aluminum and iron is kaolinite. Halloysite has been reported[177,178] in some high-alumina clays. There appears to be kaolinite of two modes of origin in such materials: one which has formed prior to the development of the hydrates and is the parent material from which the aluminum hydrates has developed, and another[179] which has formed after the development of the aluminum hydrates by the reaction of silica with such hydrates. Depending on weathering conditions, the parent silicate minerals may alter directly to a hydrate or to an intermediate silicate (Keller[36]).

A particularly distinctive type of high-alumina clay is the diaspore and burley clay of Missouri, in which kaolinite is associated with diaspore. The origin of these clays is not entirely clear. McQueen[180] and Allen[177] suggested that they have formed by some kind of leaching action associated with the development of the sinkholes in which they are found. The genetic relation of the kaolinite to the diaspore is not clear.

Keller[159] concluded that postdepositional leaching played a minor role

[172] Fox, C. S., "Bauxite," Crosby Lockwood & Son, Ltd., London (1927).

[173] Harder, E. C., Stratigraphy and Origin of Bauxite, *Bull. Geol. Soc. Am.,* **60:** 887–908 (1949).

[174] Harrassowitz, H., Laterite, *Fortschr. Geol. Paleontol. Mineral.,* **14:** 253–566 (1926).

[175] Sabot, J., Les Laterites, *Congr. Geol. Intern. Algiers* (*ASGA*), pt. 2e, pp. 181–192 (1954).

[176] Valeton, I., Petrographie und Genese von Bauxitlagerstatten (Ein Uberlick über die Bauxitliterature), *Geol. Rundschau,* **52:** 448–474 (1963).

[177] Allen, V. T., Mineral Composition and Origin of Missouri Flint and Diaspore Clays, *Missouri Geol. Surv., Ann. Rept.* 58 (1935).

[178] Bardossy, G., The Relation of the Kaolinitic Clay Deposits of the Bakony Mountains to the Deposits of Bauxite, *Acta Geol. Acad. Sci. Hung.,* **6:** 333–340 (1960).

[179] Goldman, M. I., and J. I. Tracey, Relations of Bauxite and Kaolinite in the Arkansas Bauxite Deposits, *Econ. Geol.,* **41:** 567–576 (1946).

[180] McQueen, H. S., Geologic Relations of the Diaspore and Flint Fire Clays of Missouri, *J. Am. Ceram. Soc.,* **12:** 687–697 (1929).

and that direct precipitation from colloidal solutions was a major factor in the formation of these clays. Slatkine and Heller[181] described a similar high-alumina clay from the Jurassic section in the Negev of Israel. They suggested that the clay was formed by selective leaching of a weathered basalt flow in the presence of organic matter.

Indiana-type Halloysite Deposits. The deposits of halloysite near Bedford in Indiana, first studied by Logan[182] and later in great detail by Callaghan,[183] typify an occurrence of halloysite known in several other places. The halloysite occurs as lenses and cavity fillings at the contact of the Mansfield sandstone of Pennsylvanian age with the underlying Chester clastics. According to Callaghan, the halloysite was precipitated from solution from ground water which derived its alumina and silica from the residual soils on the Chester beds. The formation of the halloysite was post-Mansfield. The waters were believed to be acid in nature because of a sulfate content which was derived from the oxidation of pyrite in the Mansfield sandstone.

The halloysite is relatively pure in some places but is often mixed with a good deal of alunite. Also the clay sometimes contains P_2O_5 $(10\%\pm)$ in a mineral structurally similar to alunite. Some of the halloysite contains allophane as an additional component.

Supergene Kaolinite Deposits. Occurrences of supergene kaolinite are frequently found associated with metalliferous sulfide ore bodies. Such clays are developed during the downward movement of acid water produced by the oxidation of the sulfides. The conditions of origin of such clays would seem to be quite similar to those of the Indiana-type halloysite noted above. It is not known why halloysite is produced in one case and kaolinite in the other.

Igneous Clay Minerals. A few occurrences of clay minerals in igneous rocks have been described which suggest that some clay minerals can occur as primary igneous components. Thus Rosetti[184] described an association of anauxite and cristobalite in the spherulites of a rhyolite from Sardinia which seems to have this origin. Prider and Cole[185] described some clay minerals in a lamproite from West Kimberly, Australia, which they believe are of late magmatic origin.

[181] Slatkine, A., and L. Heller, A Petrological Study of the Flint Clay from Makhtesh Ramon, Israel, *Intern. Geol. Congr., 21st, Rept. Session, Norden,* pt. 24, pp. 88–107 (1961).

[182] Logan, W. N., "Kaolin of Indiana," Division of Geology, Indiana (1919).

[183] Callaghan, E., Endellite Deposits in Gardner Mine Ridge, Lawrence County, Indiana, *Indiana Dept. Conserv., Geol. Surv., Bull.* 1 (1948).

[184] Rosetti, V., Anauxite Rhyolite of Salamura, Sardinia, *Periodico Mineral. (Rome),* **20:** 61–92 (1951).

[185] Prider, R. T., and W. F. Cole, The Alteration Products of Olivine and Leucite in the Leucite Lamproites from the West Kimberly Area, Australia, *Am. Mineralogist,* **27:** 373–384 (1942).

It is manifestly difficult to distinguish with certainty between clay minerals of hydrothermal origin and those of possible igneous origin. There seems to be no reason why the clay minerals could not be formed directly in some igneous rocks.

A related occurrence of clay minerals is found in the filling of vesicles in some basalts. Thus, smectite and a zeolite fill the vesicles of some basalts in the Columbia River area of Oregon. These minerals are not weathering products but reaction products of the gases and vapors entrapped in the cooling basalt and forming the vesicles with the basalt itself.

Additional references

Recent sediments

Andre, K., Die Diagenese der Sedimente, *Geol. Rundschau,* **2:** 61–74, 117–130 (1911).

Eroshchev-Shak, V. A., Distribution of Clay Minerals in Deep-water Sediments of the Atlantic Ocean, *Okeanol. Issled., Sb. Statei,* no. 8, pp. 125–136 (1963).

Griffin, J. J., and E. D. Goldberg, Clay Mineral Distributions in the Pacific Ocean: The Sea, vol. 3, pp. 728–741, Interscience, New York (1963).

Hirst, D. M., The Geochemistry of Modern Sediments from the Gulf of Paria, II, The Location and Distribution of Trace Elements, *Geochim. Cosmochim. Acta,* **26:** 1147–1187 (1962).

Hummel, K., Green Earths of South Tyrol and Halmyrolytic Iron Silicates, *Chem. Erde,* **6:** 468–551 (1931).

Kennedy, V. C., Mineralogy and Cation-exchange Capacity of Sediments from Selected Streams, *U.S. Geol. Surv., Profess. Paper* 433-D (1965).

Lundquist, G., Depositional Environment of Lake Sediments, *Sveriges Geol. Undersokn. Arsbok,* 36 (1942).

Norris, R. M., Sediments of Chatham Rise, *New Zealand Inst. Oceanog., Mem.,* **26:** 9–39 (1964).

Potter, P. E., N. F. Shimp, and J. Witters, Trace Elements in Marine and Fresh-water Argillaceous Sediments, *Geochim. Cosmochim. Acta,* 27: 669–694 (1963).

Rateev, M. A., Distribution and Genesis of Clay Minerals in Recent and Ancient Marine Basins, *Tr. Geol. Inst., Akad. Nauk SSSR,* **12:** 5–288 (1964).

"Recent Marine Sediments," P. D. Trask, ed., American Association of Petroleum Geologists, Tulsa, Okla. (1939). (Has a series of papers by many authors on all aspects of Recent sediments.)

Sato, T., Y. Iwabuchi, A. Iijima, K. Oinuma, and K. Kobayashi, Geological Research on the Bottom Sediments Sampled by the Fifth Japanese Antarctic Research Expedition, *Sci. Rept. Japan. Antarctic Res. Expedition* 1956–1962, Ser. C, 4: 1–42 (1965).

Strakhov, N. M., Diagenesis of the Black Sea Deposits, *Litologiya i Polezn. Iskop.* **1:** 7–27 (1963).

Ancient sediments

Albareda-Herrera, J. M., and R. Munoz, Fenomenos de Ordenacion y reoanisotropia de Arcillas, *Anales Inst. Espan. Edafol. Ecol. Fisiol. Vegetal.,* **6:** 475–541 (1947).

Bates, T. F., Investigation of the Micaceous Mineral in Slate, *Am. Mineralogist,* **32:** 625–636 (1947).

Bentor, Y. K., W. Bodenheimer, and L. Heller, A Reconnaissance Survey of the Relationship between Clay Mineralogy and Geological Environment in the Negev (Southern Israel), *J. Sediment. Petrol.,* **33**(4): 874–903 (1963).

Correns, C. W., Die Tone, *Geol. Rundschau,* **29:** 201–219 (1938).

Davidson, C. F., The Archean Rocks of the Rodil District, South Harris, Outer Hebrides, *Trans. Roy. Soc. Edinburgh,* **41:** 71–112 (1942).

De Lapparent, J., Les Argiles d'El Golea (Sahara), *Bull. Serv. Carte Geol. Algerie,* ser. 5, no. 3 (1937).

Endell, K., U. Hofmann, and E. Maegdefrau, Ueber die Natur des Tonanteils in Rohstoffen der deutschen Zementindustrie, *Zement,* **24:** 625–650 (1935).

Grim, R. E., The Constitution of Various Ceramic Clays, *J. Am. Ceram. Soc.,* **19:** 307–315 (1936).

Kerr, P. F., and J. Barrington, Clays of Deep Shale Zone, Caillou Island, Louisiana, *Bull. Am. Assoc. Petrol. Geologists,* **45:**(10): 1697–1712 (1961).

Knechtel, M. M., and S. H. Patterson, Bentonite Deposits in Marine Cretaceous Formations, Hardin District, Montana and Wyoming, *US. Geol. Surv., Bull.* 1023 (1956).

Krumm, H., Mineral Composition and Genesis of the Keuper and Liassic Clays of Franconia (Southern Germany), *Beitr. Mineral. Petrog.,* **11**(2): 91–137 (1965).

Long, G., S. Neglia, and L. Favretto, Geochemical Contribution to Research for the Reconstruction of the Paleogeography of a Sedimentary Basin, *Intern. Ser. Monographs Earth Sci.,* **15:** 239–259 (1964).

Lucas, J., The Transformation of Clay Minerals During Sedimentation. Studies of Triassic Clays, *Mem. Serv. Carte Geol. Alsace Lorraine,* **23:** 1–202 (1962).

Rateev, M. A., The Relation of Allauthigenic and Authigenic Clay Formations in the Lithogenesis of Various Types, *Lithologiya i Polezn. Iskop.,* **2:** 39–62 (1964).

Riviere, A., Contribution a l'étude des sédiments argileux, *Bull. Soc. Geol. France,* **16:** 43–55 (1946).

Ross, C. S., Minerals and Mineral Relationships of the Clay Minerals, *J. Am. Ceram. Soc.,* **28:** 173–183 (1945).

Sedletsky, I. D., Mineralogical Composition of Clays and Their Genesis, *Soviet Geol.,* **8:** 82–89 (1940).

Slaughter, M., and J. W. Earley, Mineralogical and Geological Significance of the Mowry Bentonites, Wyoming, *Geol. Soc. Am., Spec. Paper* 83 (1965).

Ventriglia, U., Mineralogical Constitution and Minerogenic Classification of Clays, *Periodico Mineral. (Rome),* **17:** 189–217 (1948).

Bentonites

Bechtner, P., Bentonite, "Industrial Minerals and Rocks," chap. V, American Institute of Mining and Metallurgical Engineers, New York (1949).

Bhola, K. L., Bentonite in India, *Quart. J. Geol. Mining Met. Soc. India,* **19:** 55–77 (1947).

Bordas, A. F., Argentinian Bentonites, *Rev. Minera, Geol. Mineral., Soc. Arg. Mineria Geol.,* **14:** 3–60 (1943).

Conley, J. E., Bentonite, Its Properties, Mining, Preparation, Use, *US. Bur. Mines, Tech. Paper* 609 (1940).

Deribere, M., "Le Bentonite," Dunod, Paris (1943).

Endell, K., Bentonite: Properties and Uses, *Tonind-Ztg.,* **77:** 122–126 (1950).

Hagner, A. F., Adsorptive Clays of the Texas Gulf Coast, *Am. Mineralogist,* **24:** 67–108 (1939).

Kerr, P., Bentonite from Ventura, California, *Econ. Geol.,* **26:** 153–168 (1931).

Lee, D. E., Bentonite and Bleaching Clays in Japan, *Supreme Command Allied Powers, Natl. Resources Sect., Rept.* 139 (1951).

Rios, E., "Bentonitas Espanolas," Cousejo Superior de Investigaciones Cientificas, Madrid (1949).

Savelli, F., "La Bentonite," Ulrico Hoepti, Milan (1943).

Sazonova, N. K., Bentonite Clays in U.S.S.R., *Tr. Lab. Keram.,* **7:** 217–244 (1939).

Schroter, G. A., and I. Campbell, Geologic Features of Some Bleaching Clays, *Am. Inst. Mining Met. Engrs., Tech. Publ.* 1139, pp. 1–31 (1940).

Schultz, L. G., Nonmontmorillonitic Composition of Some Bentonite Beds, Proceedings of the 11th National Clay Conference, pp. 169–177, Pergamon Press, New York (1963).

Spence, H. S., Bentonite, *Can. Dept. Mines, Resources Mines Branch, Bull.* 626 (1924).

Appendix
Chemical Analyses

A vast number of chemical analyses of clay materials have been published. In many cases no attempt was made to determine the mineral composition of the material analyzed, or the material is clearly a mixture of minerals. There are, however, many analyses of materials whose clay-mineral composition is well established and which seem to be monomineral. In the tables to follow an attempt is made to present a sampling of these analyses. The objective has been to show the spread of the composition of the various clay minerals so that their range in composition is illustrated.

In the case of the vermiculite and chlorite minerals, there are substantially no data available for clay-mineral specimens, and the data given are for larger specimens.

Table A   Chemical analyses of kaolinite minerals

	1	2	3	4	5	6	7
SiO_2	46.90	44.81	45.20	46.77	44.59	54.32	48.80
Al_2O_3	37.40	37.82	37.02	37.79	36.83	29.96	35.18
Fe_2O_3	0.65	0.92	0.27	0.45	1.14	2.00	1.24
FeO	.....	......	0.06	0.11			
MgO	0.27	0.35	0.47	0.24	0.39	0.14	
CaO	0.29	0.43	0.52	0.13	1.02	0.32	0.22
K_2O	0.84	......	0.49	1.49	0.32	.....	0.40
Na_2O	0.44	......	0.36	0.05	0.13	0.37	0.25
TiO_2	0.18	0.37	1.26	.....	2.17	.....	0.61
H_2O-	.....	1.10	1.55	0.61	......	0.84	1.16
H_2O+	12.95	14.27	13.27	12.18	13.63	11.80	12.81
Total	99.92	100.07	100.47	99.82	100.22	99.75	100.67

Kaolinite

1. Zettlitz, Czechoslovakia
2. Mexia, Texas
3. Macon, Georgia
4. St. Austell, England
5. Anna, Illinois

Anauxite

6. Bilin, Czechoslovakia
7. Ione, California

Analyses 1, 2, 6, and 7 from C. S. Ross and P. F. Kerr, *U.S. Geol. Surv., Profess. Paper* 165E (1931); 3 and 4 from P. F. Kerr et al., Rept. 7, American Petroleum Institute Project 49, Columbia University, New York (1950); 5 from R. E. Grim, *Econ. Geol.*, **29:** 659–670 (1934).

Table B   Chemical analyses of nacrite, dickite, and halloysite

	1	2	3	4	5	6	7
SiO_2	44.75	46.86	40.80	44.75	43.79	44.46	43.98
Al_2O_3	39.48	37.12	35.70	36.94	36.10	36.58	38.46
Fe_2O_3	0.53	1.43	. . . . .	0.31	0.11	0.36	
FeO	. . . . .	0.06	0.04	. . . . .	0.01	0.07	0.03
MgO	0.19	0.09	Tr	. . . . .	0.18	0.18	Tr
CaO	0.13	0.22	3.96	0.11	0.36	0.19	0.32
K_2O	. . . . .	0.60	0.20	}0.60	. . . . .	0.51	0.48
Na_2O	. . . . .	0.07	0.20		. . . . .	0.01	0.14
TiO_2	. . . . .	0.51	0.02	. . . . .	0.01	0.15	0.01
H_2O-	0.61	0.29	1.95	2.53	5.92	4.05	2.58
H_2O+	14.40	12.77	17.68	14.89	13.53	13.38	14.59
Total	100.09	100.02	100.55	100.13	100.01	99.94	100.59

Nacrite

1. Brand, Saxony

Dickite

2. Anglesey, Wales
3. Ouray, Colorado

Halloysite

4. Liége, Belgium
5. Cadouin, France
6. Bedford, Indiana
7. Eureka, Utah

Analyses 1 from C. S. Ross and P. F. Kerr, *U.S. Geol. Surv., Profess. Paper* 165E (1931); 2, 3, 5, 6, and 7 from P. F. Kerr et al., Rept. 7, American Petroleum Institute Project 49, Columbia University, New York (1950); 4 from C. S. Ross and P. F. Kerr, *U.S. Geol. Surv., Profess. Paper* 185G (1934).

Table C   *Chemical analyses of smectite minerals*

	1	2	3	4	5	6	7	8
SiO_2	52.09	50.30	50.20	51.14	55.44	57.55	49.91	45.12
Al_2O_3	18.98	15.96	16.19	19.76	20.14	19.93	17.20	28.24
Fe_2O_3	0.06	0.86	4.13	0.83	3.67	6.35	2.17	4.12
FeO	.....	.....	.....	.....	0.30	0.95	0.26	
MgO	3.80	6.53	4.12	3.22	2.49	3.92	3.45	2.32
CaO	3.28	1.24	2.18	1.62	0.50	1.94	2.31	0.88
K_2O	.....	0.45	0.16	0.11	0.60	0.59	0.28	
Na_2O	.....	1.19	0.17	0.04	2.75	0.33	0.14	
TiO_2	.....	.....	0.20	....	0.10	0.32	0.24	
H_2O-	14.75	}23.61	15.58	14.81	}14.70	.....	15.77	}18.72
H_2O+	7.46		7.57	7.99		8.53	7.70	
Total	100.42	100.14	100.50	99.52	100.69	100.41	99.43	99.40

Montmorillonite

1. Tatatilla, Mexico
2. Otay, California
3. Polkville, Mississippi
4. Montmorillon, France

5. Upton, Wyoming
6. Pontotoc, Mississippi
7. Chambers, Arizona
8. Nashville, Arkansas

Analyses 1 to 5 from C. S. Ross and S. B. Hendricks, *U.S. Geol. Surv., Profess. Paper* 205B (1945); 6 from R. E. Grim and R. A. Rowland, *Am. Mineralogist*, **27**: 746–761 (1941); 7 from P. F. Kerr et al., Rept. 7, American Petroleum Institute Project 49, Columbia University, New York (1950); 8 from C. S. Ross and S. B. Hendricks, *U.S. Geol. Surv., Profess. Paper* 205B (1945).